Vegetables and Spices

for

JRF, SRF, ARS, NET, PhD, IARI & SAUs Entrance

About the Authors

Dr Dheeraj Singh is working as Principal scientist and Head, Krishi Vigyan Kendra, Central arid zone research institute, Pali. He had earlier worked as Assistant Professor, Pomology and as Assistant Professor Post Harvest Technology in different Agricultural universities. He had completed his M.Sc. from SKNRAU, Jobner and doctorate from Indian Agriculture Research Institute, Pusa New Delhi. He had also written three books and contributed 99 research papers, 21 book chapters, 71 popular articles and 62 folders and pamphlets on different aspects of horticultural crops. He was awarded best extension scientist award 2015 by ICAR, New Delhi, conferred Young scientist award in 2016 and recognized by PPVFRA, New Delhi in 2016 for facilitating the Kharchi community for the prestigious Plant Genome Souvenir Community Award. He was selected as our common future (OCF) fellow by Volkswagen foundation, Germany and has represented India for scientific deliberation in different international forums and conferences at Philippines, Kenya, Ethiopia, China, UK, Germany, Italy, Switzerland and Sweden.

Dr. Gograj Singh is presently working as a Scientist (Senior Scale) in the area of pre-breeding, genetic improvement and breeding for protected environment in bitter gourd and tomato at Division of Vegetable Science, ICAR, New Delhi. He did his M.Sc. and Ph.D. (Horticulture-Vegetable Science) from ICAR-IARI, New Delhi. He has qualified JRF, SRF and NET in Horticulture. He has received the prestigious Jawaharlal Nehru Award for outstanding doctoral thesis research in Horticulture by Indian Council of Agricultural Research, New Delhi during 2018. He is also the recipient of Dr B. R. Barwale Young Researcher Award for best Ph.D. thesis in horticultural biotechnology by Indian Academy of Horticulture and Dr G. S. Randhawa Award from Society for Promotion of Horticulture, ICAR-IIHR, Bengaluru. He has published several research papers in journals of national and international repute, popular articles, technical articles, book chapters and one edited book.

Dr. Chandan Kumar is working as Subject Matter Specialist (SMS) Horticulture, ICAR-Central Arid Zone Research Institute (CAZRI), Krishi Vigyan Kendra (KVK), Pali-Marwar (Rajasthan) India. He has done his graduation from Marathawada Agriculture University (MAU), Parbhani (Maharasthra) in 2008, Post-graduation from G.B. Pant University of Agriculture & Technology, Pantnagar (Uttarakhand) in 2010 and Doctorate from Banaras Hindu University (BHU), Varanasi (Uttar Pradesh) India in 2015. He has contributed in 12 book chapters, 24 research paper and 37 popular articles. He has a qualified ICAR-NET and SRF in vegetable science in Horticulture. He has awarded young scientist award and best extension worker award. He specializes in vegetable breeding and its production technology.

Vegetables and Spices

for

JRF, SRF, ARS, NET, PhD, IARI & SAUs Entrance

Dr. Dheeraj Singh
Principal Scientist and Head,
ICAR-Central Arid Zone Research Institute (CAZRI),
Krishi Vigyan Kendra (KVK),
Pali-Marwar (Rajasthan) India.

Dr. Gograj Singh
Scientist
Division of Vegetable Science,
ICAR, New Delhi.

Dr. Chandan Kumar
Subject Matter Specialist (SMS) Horticulture,
ICAR-Central Arid Zone Research Institute (CAZRI),
Krishi Vigyan Kendra (KVK),
Pali-Marwar (Rajasthan) India.

Published by
SCIENTIFIC PUBLISHERS (INDIA)
5 A, New Pali Road, P.O. Box 91
Jodhpur 342 001 (INDIA)
E-mail: info@scientificpub.com
Website: http://www.scientificpubonline.com

ISBN: 978-93-92590-48-1 (PB)
ISBN: 978-93-92590-56-6 (EBook)

Printed in India

डॉ. नरेन्द्र सिंह राठौड़
कुलपति
Dr. Narendra Singh Rathore
Vice-Chancellor

Phone : 91-294-2471101 (O), 2470682 (Fax), 2463839 (R) and 9414166961 (M)
E-mail : vc_mpuat@yahoo.co.in and vc@mpuat.ac.in

Maharana Pratap University of Agriculture & Technology
University Campus, Udaipur-313001 (Rajasthan), INDIA

महाराणा प्रताप कृषि एवं प्रौद्योगिकी विश्वविद्यालय
विश्वविद्यालय परिसर, उदयपुर–313001 (राजस्थान), भारत

No. PS/VC/MPUAT/2021/227.
Date- 7th June, 2021

FOREWORD

Agriculture and allied sciences are becoming increasingly popular for career building. Therefore, the number of incumbents appearing in competitive examinations like ICAR AIEEA, JRF, SRF and ARS and for employment in state agriculture and horticulture departments and universities is increasing year after year. However, such examinations do not guarantee success to even the best of students securing copiously in curriculum examinations because the model of competitive examination is different than the institutional examinations. Even the syllabus of each subject is moulded at national level in contrast to regional course outlines in different universities. Nowadays, competition is getting tougher and the doorway into good academic institutions for higher studies as well as class jobs is just not cushy. There are a lot of challenges in cracking these examinations out of which one is selection of quality and smart study material. No matter how hard you work while preparing for competitive exams, availability of quality and well-groomed study material in tune with syllabus is crucial and choice of study material with astuteness is essential for smart learning.

Horticulture is an important subject of agriculture discipline at school and college levels. Within horticulture, production of vegetables and spices has a special significance. This is substantiated with the fact that out of total estimated horticultural production of 326.58 million tonnes during 2020-21, alone vegetable production has been estimated to be 193.61 million tonnes and that of spices to an extent of 10.24 million tonnes. The spotlight on these commodities is due to their importance in human health as they contain many of the dietary factors like vitamins, minerals and amino acids they are considered as protective supplementary food. Despite availability of volumes of reference material on these aspects, a need is felt to have a all-inclusive book containing in-depth information on individual crops. In this context this book is a timely attempt to provide updated and witty subject matter to the students desirous of attempting various examination including AIEEA, JRF, SRF and ARS, universities entrance examination at various levels, employment examinations by central and state public commissions and universities.

I congratulate the authors Dr. Dheeraj Singh, Dr. Gograj and Dr. Chandan Kumar for bringing out the revised edition of the book entitled 'Vegetable Science' containing chapters on individual vegetable and spice crops along with model MCQs for the benefit of the users.

(Dr. Narendra Singh Rathore)
Vice-Chancellor

PREFACE

Vegetables and spices form the backbone of Horticulture. As the spectrum covered by these crops is too wide, the students had to study different book, assimilate their contents and prepare suitable notes. Often this task is not easy and in case of less time and shortage of books it is really impossible. Keeping these facts in mind we had made an effort to make available a book suitable for preparation and revision for competitive examinations, at the same time maintaining academic rigour and faculty accuracy. All the chapters in the book had been properly covered according to their families and relative importance. Questions given in the books are very relevant to the competitive examinations and most of the previous year questions asked in different exams had been incorporated. The appendix given at the end carries vast information on different aspects of vegetables and spices in short tabular form. This book fulfill the need of a good reference-cum text book for the student community.

I express my sincere thanks to scientists from ICAR institutes/SAUs for their inputs in terms of material collection, compilation and correction. Every effort has been made to ensure that the information collected from various sources such as individuals, institutions, organizations and publications are accurate. We hope that this book will be a useful reference material for the teachers, practitioners and students searching their careers in ICAR Institutes, State Agricultural Universities, State Agriculture/Horticulture Departments and for competitions like State Public Services/UPSE/ARS/JRF/SRF/NET along with entrance examinations for B.Sc ,M.Sc and Ph.D Degree in Horticulture .

We hope that this effort will help the readers in their endeavours. All the suggestions in making the book further informative are always welcome.

Jan 2022

Author

CONTENTS

VEGETABLE SCIENCE

1. INTRODUCTION

- The term Horticulture first appeared in written language in 17th century and word derived from the Latin words *Hortus* and *colere,* meaning garden and to cultivation.
- The term Olericulture derived from Latin words *Oleris* (Pot herb) and *Culture* (Raising of plants).
- Olericulture is the science of vegetable crops.
- By definition a vegetable is "a herbaceous plant cultivated for food, as the cabbage, potato, bean, etc; also, the edible part or parts of such plants, as prepared for market or table." Vegetables are generally classified according to the source of the edible plant part(s):

  Root (e.g. beet, carrot, turnip)
  Bulb (e.g. leek, onion, garlic)
  Stem (e.g. asparagus, kohlrabi)
  Flower (e.g. cauliflower, broccoli)
  Tuber (e.g. potato)
  Immature fruit (e.g. cucumbers, green peas, summer squash, sweet corn)
  Mature fruit (e.g. muskmelon, watermelon, tomato, winter squash)
  Leaf (e.g. cabbage, lettuce, spinach)
  Petiole (e.g. celery)
  Seed (e.g. bean)
- Vegetable area in India – 10.26 M. Ha (2017-18)
- Vegetable production In India – 184.4 M.T. (2017-18)

Vegetables	productivity	% area	**% production**
Beans	9.99	2.22%	1.23%
Bitter gourd	11.72	0.95%	0.62%
Bottle gourd	17.09	1.53%	1.46%
Brinjal	17.54	7.12%	6.94%
Cabbage	22.65	3.89%	4.90%
Capsicum	13.58	0.23%	0.18%
Carrot	16.99	0.95%	0.89%

Cauliflower	19.13	4.42%	4.70%
Cucumber	15.37	0.80%	0.68%
Chillies	11.62	3.01%	1.95%
Elephant foot yam	25.80	0.29%	0.42%
Okra	11.97	4.96%	3.31%
Onion	18.10	12.53%	12.62%
Pointed gourd	15.50	0.19%	0.17%
Peas	10.04	5.26%	2.94%
Potato	23.95	20.88%	27.83%
Radish	14.65	2.04%	1.66%
Pumpkin	21.97	0.76%	0.93%
Sweet Potato	11.45	1.28%	0.81%
Tapioca	28.61	1.69%	2.68%
Tomato	25.04	7.69%	10.72%
Others	14.13	15.40%	12.10%

- Major nutrients (5) present in vegetables and their uses –
  1. Carbohydrates, protein and fat – All three are required in large quality for supplying energy.
  2. Minerals and vitamin – These are required in small quality for physiological process and metabolic activity of body.
- Major mineral elements obtained from vegetables are Ca, P and Iron.
- Most of leafy vegetables are rich in Carotene, riboflavin and minerals.
- Vegetable known as God sent vegetable - Winged bean
- The man-made vegetable is- Rutbaga
- The oldest known spice crop is - Cassia
- Most expensive spice crop in world – Saffron
- Viviparous nature is present in vegetable - Chowchow
- Flavour compound in vegetables–
  1. Volatiles - Aldehydes, ketones, alcohols, sulphur compound and aromatic hydrocarbons.
  2. Non-Volatiles – Organic acids, sugar, fatty acids and amino acids.
- Characteristic flavour of Cole crop is due to volatile compound – Dimethyl trisulphide.
- Pungency and flavours of Onion – Allyl- Propyl –Disulphide.
- Pungency and flavours of Garlic – Diallyl –Disulphide.
- Mango ginger is botanically - *Curcuma amada*
- Bitter component present in bitter guard - Momardicine

- Toxic Substance – Trypsin inhibitors present in beans particularly in raw soybean which inhibit proteolytic enzyme activity.
- Phytic acid (Inositol Hexa-phosphoric acid) – Peas and beans.
- Colocasia – Oxalic acid (also present in elephant foot which is responsible for bitter taste).
- Most ancient type of vegetable garden – Kitchen Garden.
- Very intensive method of vegetable cultivation – Market Garden.
- Very extensive method of vegetable cultivation – Truck Garden.
- The Research and development programme in Horticulture received an impressive support from the Eight Five-year plan onwards.
- Horticulture sector covers only 8% of the total crop area in the country, contributing 24.5% to the GDP and 54.55% to export earnings in the agriculture sectors.
- In 1945 Horticulture section was established in division of Botany IARI, New Delhi. A full-fledged division of Horticulture was created at the ICAR Headquarter, headed by DDG during 1987.
- The percentage share of horticulture output in Agriculture has become 33%.
- Under the purview of Agriculture & allied activities, the share of plan outlay for Horticulture which was 3.9% during IX Plan, has increased to 4.6% during the XII Plan.
- Over the last decade, the area under horticulture grew by 2.6% per annum and annual production increased by 4.8%.
- The production of vegetables has increased from 101.2 Million Tonnes to 184.40 Million Tonnes since 2004-05 to 2017-18.
- During 2017-18 the area under vegetables was 10.26 Million Hectares with a production of 184.40 Million Tonnes and productivity of 17.97MT/Ha in India.
- For this period the total vegetable production was highest in case of Uttar Pradesh (283.16 Million Tonnes) followed by West Bengal (276.95 Million Tonnes).
- Percentage Share of vegetable Production in Total Horticulture production for last five Years is 59.2.
- Agricultural Research Institute established at Pusa, Bihar in 1905 and renamed as Imperial Agricultural Research Institute in 1911. Later shifted to New Delhi in 1936.
- ICAR was established on July 16, 1929 under the Societies Registration Act 1860 with its headquarter at New Delhi.
- Horticulture research in India received very little attention till the 3rd Five Year Plan.
- In 1945 Horticulture section was established in division of Botany IARI, New Delhi.
- The establishment of the Indian Institute of Horticultural Research at Bangalore and starting of eight All India Coordinated Crop Improvement Projects to cover different horticultural crops was a landmark in the history of horticulture in 4th Five Year Plan (1969-74).
- Establishment of first Agricultural University at Pantnagar in 1960.

- Placement of different agricultural research institutes under the purview of ICAR in 1966.
- Creation of Department of Agricultural Research and Education (DARE) in the Ministry of Agriculture in 1973.
- Opening of first Krishi Vigyan Kendra in Puducherry in 1974.
- Establishment of Agricultural Research Services (ARS) and Agricultural Scientist Recruitment Board (ASRB) in 1975.
- Establishment of National Gene Bank in New Delhi in 1996.
- The Protection of Farmers Variety and Farmers Rights (PPV&FRA) authority was established in November, 2005 in New Delhi.
- National Food Security Mission (NFSM) – 2007-08.
- Kisan Credit Card (KCC) – 1998.
- Rastriya Krishi Vikas Yojna (RKVY) – 2007.
- Kisan Call Centre – 2004.
- AICRP for improvement of Vegetable crops was started in 1970-71 with headquarters at IARI, New Delhi
- Rapid expansion of infrastructure took place in 7th and 8th Plans. Division of Horticulture was carved out of crop division in 1981 and in 1985; a position of Horticulture Commissioner was also created.
- National Horticulture Board was established in 1984 at Gurgaon (Haryana).
- Indian Institute of Vegetable Research, Varanasi (U.P.) and Project Directorate of Vegetable Research was started in 1987.
- A full-fledged division of Horticulture was created at the ICAR Headquarter, headed by DDG during 1987.
- All India Vegetable Improvement Project was shifted from IARI, New Delhi to Varanasi in 1992.
- 17 Centres work on Horticulture.
- Horticulture Division of ICAR spearheads the horticulture research in India, which is carried out through 10 Central Institutes, 6 Directorates, 7 NRCs, 13 AICRPs and 6 Network Projects / Outreach programmes.
- Spice board works under Ministry of Commerce.
- India occupies IInd position in Cauliflower, onion and cabbage production.
- Trypsin inhibitor is also present in sweet potato.
- Rutabaga is a man-made vegetable (*Brassica napus* var - *napobrassica)*
- The Salem belt of Tamil Nadu and Smalkot (Andhra Pradesh) are known for Cassava production as industrial crop.
- ISTA (International Seed Testing Association) was established in 1924 at Norway (Switzerland).

Indian Vegetable productivity compared to other country.

Crop	India's Productivity (t/ha.)	Major producer country	Highest productivity Country (t/ha.)
Potato	20.55	China	49.02 (USA)
Brinjal	18.90	China	61.00 (Iraq)
Tomato	24.21	China	90.29 (USA)
Cabbage	22.56	China	73.01 (Korea Republic)
Onion	16.18	India	56.40 (USA)
Cauliflower	19.25	China	30.30 (Egypt)
Okra	11.35	India	14.37 (Saudi Arabia)
Peas	10.04	China	18.2 (Belgium)

Production share of leading vegetable producing states in India

S. No.	State	Production (in 000 MT)	% share	Position
1.	UP	28316	15.4	I
2.	West Bengal	27695	15.03	II
3.	MP	17545	9.52	III
4.	Bihar	15863	8.61	IV
5.	Gujarat	12254	6.65	V
6.	Maharashtra	12306	6.63	VI

- Of the ten top vegetable crops of India, Okra is only crop where India figures among top countries in order of their productivity.

Major Vegetable producing country

Country	% production
China	50.52%
India	11.15%
USA	3.08%
Turkey	2.27%
Egypt	1.56%
Russia	1.52%
Other Countries	23.22%

In vegetable productivity of the world, Spain occupies 1st position with 44.28 t/ha. (world productivity 18.86 t/ha.).

China has 1st position in production of all vegetables mentioned above except Tapioca (Nigeria 1st), Yam (Nigeria 1st), Okra (India 1st) and Sprout Broccoli (USA 1st).

Vegetables	**Major Producing States**	**National Productivity (MT/ha)**
Tomato	Andhra (13.89%) Gujarat (6.87%)	25
Brinjal	West Bengal (23.65%); Orissa (15.73%)	17.52
Okra	Gujarat (15.12%); West Bengal (15%)	11.97
Onion	Maharashtra (38%); Madhya Pradesh (15.9%)	18
Cabbage	West Bengal (25.23%); Orissa (11.72%)	22.68
Cauliflower	West Bengal (21.87%); Madhya Pradesh (11.37%)	19.6
Potato	Uttar Pradesh (30.32%); West Bengal (24.91%)	23.96
Sweet Potato	Orissa (25.39%); Kerala (22.73%)	11.49
Tapioca	Tamil Nadu (57.83%); Kerala (34.87%)	28.64
Water melon	Uttar Pradesh (24.59%); Andhra Pradesh (14.29%)	24.93
Radish	West Bengal (17.61%); Haryana (17.36%)	14.68
Peas	Uttar Pradesh (46.32%); Madhya Pradesh (17.73%)	10.03
Musk Melon	Uttar Pradesh (45%); Andhra Pradesh (25.6%)	22.75
Chillies (Green)	Karnataka (19%); Madhya Pradesh (18.6%)	11.64
Cucumber	Haryana (21.78%); Madhya Pradesh (12.3%)	15.36
Carrot	Haryana (27%); Punjab (12%)	17.00
Capsicum	Karnataka (20%): Himachal Pradesh (18%)	13.41

Among the Vegetables tapioca has the highest productivity i.e. 28.61 Mt/Ha.

1. Vegetable Consumption Per Capita

1. Recommendation – 300 gm/day
   125 gm. Green leafy vegetables.
   100 gm. Root vegetables.
   075 gm. Other vegetables.
2. Present availability – 393.76 gm. (In India)

- India is the IInd largest producer of vegetable after China (51%) producing 11% of total world production.
- Vegetable crops in India occupy only 2.8% of the total cropped area.
- Productivity of vegetables in India – 17.90 tonnes/ha.
- State having largest area under vegetables – U P.
- State having maximum produce under vegetables – U P.
- State having maximum productivity of vegetables – A P.
- Green leafy vegetables are rich source of "Folic acid".
- Vegetables are called "protective food".
- Daily Requirement –

Cereals	475g.	Vitamin D	5 mg.
Calories	2800	Niacin	19 mg.
Pulses	80g.	Vita B_{12}	1.0 mg.
Protein	55g.	Fruits	120 g.
Calcium	450mg.	Milk	200 ml.
Iron	20mg.	Sugar	40 mg.
Vitamin C	50mg.	Oil	40 ml.
Thiamine	1.4mg.	B-Carotene	3000 mg.
Riboflavin	1.5mg	Folic Acid	100 mg.

3. According to FAO, in most developing countries per capita vegetable consumption is 25 kg/annum, whereas dieticians recommended that a person should consume 100-120 kg vegetable/ annum.

- Highest consumer of vegetable (per capita)
  1. China – 377.17 Kg/Year.
  2. Armenia – 336.18 Kg. /Year.
  3. India – 80 Kg. /Year.

India Scenario in vegetable: India is the IInd largest producer of vegetable after China (51%) producing 11% of total world production. During 1950-1951 under vegetable area was 2.84 million ha with production of 16.5 million tonnes and productivity of 5.8 t/ha.

Area ('000Ha)	Production ('000 MT)	Productivity (tonnes/ha)	In area growth trends (2017-18)	Inv production growth trends (2017-18)
10.26	**184.40**	**17.90**	**0.20%**	**3.50%**

Largest area	Highest production	Major Vegetables	Productivity (t/ha)
U.P. (14.2%) W. B. (13.65%) M. P. (8.6%)	1. U.P . (15.40%) 2. W. B. (15.04%) 3. M P (9.52%)	1. Potato (28%) 2. Onion (13%) 3. Tomato (11%)	A. P. (28.42) H.P. (26.49) Kerala (22.71)

Production statistics of different states in India

S.No.	Crop	Leading state	Followed by
1	Cucumber	Haryana	Madhya Pradesh
2	Cauliflower	West Bengal	Madhya Pradesh
3	Brinjal	West Bengal	Orissa
4	Cabbage	West Bengal	Orissa
5	Capsicum	Karnataka	Himachal Pradesh
6	Carrot	Haryana	Punjab

Productivities of different vegetables

Vegetable	Highest productivity in world	Productivity in India
Brinjal	Iraq (61.00)	17.5 t/ha
Cabbage	Japan (66 t/ha)	21.5 t/ha
Cauliflower & broccoli	Pakistan (24.8 t/ha)	18.3 t/ha
Tomato	Spain (74 t/ha)	19.1 t/ha
Potato	USA (49.02)	22.7 t/ha
Onion	USA (56.40)	t/ha

In Tamil Nadu productivity of tapioca is 41.9% which is highest amongst all vegetables (28.1 Mt/Ha).

Productivity values (Rs./ha) of different crops

Cereals	Pulses	Sugarcane	Cotton	Fruit & Vegetable	All Crops
19,470	12,897	78,038	33,868	1,35,000	34,247

India ranks first in production of okra in the world (73% of world production) and second in other vegetables such as brinjal (27.55%), cabbage (13%), cauliflower & broccoli (36%), onion (19.90%), potato (13%) and tomato (11%).

Largest Producer in World –

S.No.	Vegetable
1.	Potato
2.	Brinjal
3.	Tomato
4.	Cabbage
5.	Onion
6.	Cauliflower
7.	Peas
8.	Sweet Potato
9.	Okra

Name of the vegetable	Leading producer country	Followed by	Highest productivity (t/ha)
Onion	India	China	USA (56.40)
Potato	China	India	USA (49.02)
Okra	India	Nigeria	Saudi Arabia (14.37)
Brinjal	China	India	Iraq (61.00)
Cabbage	China	India	Korea (73.01)
Cauliflower	China	India	Egypt (30.30)

China has 1st position in production of all vegetables mentioned above except Tapioca (Nigeria Ist), Yam (Africa Ist), Okra (India 1st) and Sprout Broccoli (USA 1st).

- By the end of 2030, according to an estimate we will need 151-193 M.T. of vegetables to meet our requirement.
- Horticulture sector covers only 8% of the total crop area in the country, contributing 24.5% to the GDP and 54.55% to export earnings in the agriculture sectors.

Agroclimatic Zone for vegetable crops

Zone	Details
Zone-I	Western Himalayas region state like J&K, H.P, U. K
Zone-II	Humid region Bengal Assam basin, West Bengal, Assam
Zone-III	Humid eastern Himalayan and Bay Island. Sikkim, Meghalaya, Manipur, Nagaland, Arunachal Pradesh
Zone-IV	Sub humid Sutlej Ganga alluvial plains Punjab, U.P. Bihar
Zone-V	Sub humid to humid Eastern and south east upland Chhattisgarh, Orissa, A.P.
Zone-VI	Arid western plains Rajasthan, Gujrat, Haryana, Delhi
Zone-VII	Semi-arid and central Island M.P, Maharashtra
Zone-VIII	Humid and semi-arid western ghats Karnataka, Tamil Nadu, Kerala

Nutritive Value of vegetables-

1.	Vitamin A	**Bathua leaves**	11300 IU/100g.
	Calocassia leaves	10278 IU/100g.	
	Beet leaves	9770 IU/100g.	
2.	Vitamin B_1 (Thiamine)	**Chillies**	0.55 Mg./100gm.
		Calocasia leaves	0.22 Mg./100gm.
		Tomato (Red)	0.12 Mg./100gm.
3.	Vitamin B_2 (Riboflavin)	**Fenugreek leaves**	0.31 Mg./100gm.
		Amaranthus	0.30 Mg./100gm.
4.	Vitamin C (Ascorbic Acid)	**Drumstick leaves**	220 Mg./100gm.
		Coriander leaves	135 Mg./100gm.
		Chillies	111 Mg./100gm.
		Brocoli	109 Mg./100gm.
		Tomato	31 Mg./100gm.
5.	Carbohydrate	**Tapioca**	38.1%
		Sweet Potato	28.2%
		Potato	22.6%
6.	Protein	**Lima Bean**	7.9g./100gm.
		Peas	7.2g./100gm.
		Cowpea	4.3g./100gm.

7.	Fat	**Potato**	118mg./100gm.
8.	Fiber	**Potato**	752mg./100gm.
		Chilli	6.8g./100gm.
		Amaranthus	1.0g./100gm.
		Beet leaf	0.7g./100gm.
		Spinach	0.6g/100gm.
9.	Calcium	**Agathi**	1130mg/100g
		Curry leaves	813 mg/100g
		Amaranthus and Fenugreek leaves	395 mg/100g
10.	Phosphorus	**Amaranthus**	800 mg/100g
		Garlic	187 mg/100g
11.	Iron	**Agathi**	83.9 mg/100g
		Amaranthus	22.5 mg/100g
12.	Potassium	**Spinach**	605 mg/100g
		Amaranthus	230 mg/100g

Sl. No.	Type of mineral/vitamin & their role	Name of the vegetables	Deficiency symptoms
1.	Carbohydrates: Provide	Tuber vegetables *ie.,* potato, sweet potato, tapioca and yams.	Retarded growth
2.	Proteins: Made up of amino acids, growth and repair of the body	Immature seeds of lima bean, broad bean, peas, garlic, onion etc.	Retarded growth, retarded mental development, discoloration of skin,swelling of leg and feet, fatty liver (kwashiorkor)
3.	Calcium: Important for bones, teeth, blood clotting, resistance against infection	Amaranthus, cauliflower, drumstick leaves, lettuce, methi, carrot, onion, turnip, green peas, tomato, coriander, spinach, cabbage.	Irritability, retarded growth and bone weakness.
4.	Iron: Essential part of red blood corpuscles.	Moringa-leaves, amaranthus, methi, mint, coriander, drumstick, spinach.	Anaemia, pale smooth tongue, pale lips.
5.	Phosphorus: Cell multiplication, proper maintenance of liquid content in the tissue, role in oxidation of carbohydrates	Potato,carrot,tomato,cucumber, spinach,cauliflower,lettuce, onion.	Retarded growth

6.	Vitamin A: Provides general health	Carrot, spinach, palak, leafy vegetables, sweet potato (yellow), pumpkin (yellow).	Night blindness, respiratory infections, formation of stones in kidney, rough skin,growth in children retarded
7.	Vitamin B complex: a) Thiamin (B1) b) Riboflavin(B2) c) Niacin (B5) d) Pyridoxin(B6)	Peas, broad bean, lima bean, garlic, asparagus, corns, tomatoes	a) Beriberi : loss of appetite b) Red coloured mouth cracks in the mouth c) Sore tongue, pellagra d) Ulcer
8.	Vitamin C: Essential for growth and resistance against diseases.	Turnip, green chilli, Brussels sprout, mustard, green leafy vegetables, cole crops, bitter gourd, radish	Scurvy, bleeding gums and mucous membrane, cold, loss of energy, delay in wound healing
9.	Vitamin D: Essential for bone and teeth	Green leafy vegetables	Bone and teeth weakness.
10.	Vitamin E: Antisterility and essential for reproduction	Cabbage, lettuce, methi, Spinach and vegetable oils	Fertility is affected
11.	Roughage: (Cellulose and fiber) Add digestion and prevent constipation	Leafy vegetables (Cabbage, Spinach, Lettuce), most root crops	Indigestion and constipation

Properties and content in vegetable colours

Vegetable colours	Properties	Contents
White	To strengthen the immune system	Allicin, Potassium
Green	To purify with detoxification	Folic acid, Lutein, Vitamin C, Potassium.
Red	To improve heart and blood health	Lycopene, Anthocyanins and Phytochemicals
Yellow	To fortify skin elasticity	Beta-carotene or Pro-vitamin A
Orange	To prevent cancers	Vitamin C, Potassium
Purple	To increase longevity	Antioxidants and Phytochemicals

- Major mineral present in vegetables – K.
- Fat percent in vegetable – <0.1%.

The deficiency of vitamin 'A' causes night blindness.

- The deficiency of vitamin 'B1 causes beriberi disease.
- Scurvy disease is due to deficiency of vitamin 'C'.
- Vitamin E is also known as antisterility vitamin.

- Agathi (seobania) is the richest source of protein and calcium.
- Rajira leaves are the richest source of Vitamin A followed by Colocasia
- Radish leaves are the richest source of riboflavin followed by fenugreek leaves.
- Vitamin 'D' is essential for prevention of rickets, osteomalacia and dental diseases.
- The synthesis of prothrombin and normal blood clotting regulate by the vitamin C
- The deficiency of calcium in body causes rickets and osteomalacia.
- Phosphorus is essential in human diet for cell multiplication of bones and soft tissue.
- The goitre disease in human is due to deficiency of iodine.
- For good health, the requirement of vitamin 'A' per day is more than 2000 IU.
- The requirement of vitamin '13' per capita/day for good health is above 0.17 mg.
- For good health per capita/day requirement of vitamin 'C' is above 20 mg.
- Iron requirement per capita/day for good health is above 3.0 mg.
- Calcium requirement per capita/day is above 20 mg.
- Magnesium is implicated to have role in cardiovascular diseases.
- Zinc deficiency in human body leads to growth failure and poor development of body growth.
- Chromium deficiency in body leads to impaired glucose tolerance.
- Manganese deficiency in body leads to abnormality in skeletal bone mineralization.
- Excess molybdenum intakes in human body may increase the risk of gout.
- Selenium deficiency in human body is also implicated as a risk factor in cancer.
- Megaloblastic anaemia in living organism is due to deficiency of vitamin B12.
- Inadequate intake of vitamin like riboflavin results in soreness of the tongue (glossaries), cracking at the angles of mouth (angular stomatitis), redness of the eye and burning sensation in eyes, scaliness of the skin in the region between the nose and the angles of the lips (seborroic dermatitis).
- Psychomotor development in children may be impaired in riboflavin deficiency.
- Vitamin A is incorporated in rhodopsin (eye pigment)
- Tapioca is the richest source of carbohydrate (38.1g) 100 g edible part) and calories followed by sweet potato.
- Chilli is the richest source of fiber (6.8 g / 100 g edible part).
- Giant chillies are the richest source of thiamine followed by peas.
- Parsley is the richest source of vitamin C (281mg/ 100 g edible part).
- Sweet potato is the richest source of vitamin A (14190 11g/100g edible part).
- China is the largest producer of vegetables in world followed by India. Vegetables are called low cost **"protective food"** (Role in alleviating the malady).
- Vegetables contain all five nutrient components these are Carbohydrate, Protein, Minerals, Vitamin and Fat.
- Vegetables are alkaline in nature and supply enough alkali to neutralize the harmful action of proteaceous foods, i.e. provide alkaline reactions for normal metabolism.

- Vegetables also contain significant levels of biologically active component, i.e. antioxidant, mainly vitamin 'K' prevent oxidation of β- Carotene, vitamin 'A' and fatty acid.
- Kale contain highest antioxidant, then others crucifers. The antioxidant scavenging the free radicles and reactive oxygen species (ROS).

Major Acid present in Vegetables.

1. Citric acid – Tomato, Potato, Legumes, Beetroot and Leafy Vegetables.
2. Malic acid – Carrot, Celery, Lettuce, Onion and Broccoli.
3. Oxalic acid – Amaranthus, Portulaca, Celosia and Basella.

Aroma containing compounds

1. Non adienal – Cucumber.
2. Allyl-iso-thiocynate – Raw Cabbage.
3. Dimethyl disulphide – Cooked Cabbage.
4. Dimethyl pyarzin – Potato.
5. Iso-thiocynate – Radish.
6. Allyl- Propyl –Disulphide – Onion
7. Diallyl –Disulphide – Garlic

Toxic Substance in Vegetables

1. Haemaglutine French bean.
2. Saponine Spinach, Tomato, Asparagus.
3. Serotonine Water Melon.
4. Salosodine Brinjal.
5. Apiin Celery.
6. Cholic ertrass Pumpkin.
7. Zea-Xanthin Corn.
8. Sinigrin Cole crops.
9. HCN (Dhurin) Cassava.
10. Dioscorine Yams.
11. Trypsin inhibitor (Anti vit 'E' Factor) Pea, Beans and Sweet Potato
12. Ca Oxalates Colocasia and some leafy vegetable
13. Cucurbitacins Cucurbits
14. Solanin Green Potato

Vegetable in disease prevention/ medicinal properties –

Bioactive components	Source	Potential benefits
Carotenoids (Alpha-carotene/ beta-)	Carrots, Fruits, Vegetables	Neutralize free radicals which may cause damage to cells.

Carotene		
1. Lutein	Green vegetables	Reduce the risk of muscular degeneration.
2. Lycopene	Tomato products (ketchup, sauces)	Reduce the risk of prostate cancer.
Prebiotics and Probiotics (Fructo-oligosaccharides (FOS))	Jerusalem artichokes, shallots, onion powder	Improve quality of intestinal microflora; gastrointestinal health.
Phenolics (Anthocyanidins, Flavones, Lignans)	Fruits/vegetables	Neutralize free radicals; reduce risk of cancer.
Cheratin	Bitter gourd	Treating diabetes
Two triterpenes (Alunsenol and Mutiflorenol)	Ash guard	For weak nervous systems
Bioflavonoids (sulphar compounds)	Onion and Garlic	Reducing harmful blood cholesterol
3-n-Butyl Pthalide	Celery	Effective against Hypertension
Diosgenin	Dioscorea	Manufacture of cortisone and contraceptive drugs

1. Bitter gourd – Treating diabetes.
2. Ash guard – For weak nervous systems.
3. Bioflavonoids –Compound closely associated with Vitamin-C and found in several Vegetables eg. Onion and Garlic (quercetin –protection against cancer and heart disease).
4. Other compounds

   Onion and Garlic –
   1. Allicin
   2. Allistalin
   3. Garlicin
   4. Diallyl Disulphide
   5. Diallyl Trisulphide
   6. Allyl Propyl-disulphide

   (All are sulphur compounds effective in reducing harmful blood cholesterol)
5. Cheratin – Isolated from bitter gourd (effective against diabetes disease).
6. Diphenyl amino – Onion (used in curing Diabetes).
7. 3-n-Butyl Pthalide – Celery (effective against Hypertension).
8. Diosgenin – Dioscorea (Manufacture of cortisone and contraceptive drugs).

- By the end of 2030, according to an estimate we will need 151-193 M.T. of vegetables to meet our requirement.

Classification of vegetables

- Monocot Vegetables –

  Family

(a) Alliaceae — Onion, Garlic, Leek.
Multiplier Onion - *Allium cepa* var. *aggregatum*.
Top Onion -*Allium cepa* var. *viviparum*.
Welsh Onion - *Allium cepa* var. *fistulosum*.
Shallat - *Allium cepa* var. *ascaloricum*.
Chive - *Allium cepa* var. *schoenoprasum*.

(b) Liliaceae — Asparagus (*Asparagus officinalis*).

(c) Araceae
1. Elephant foot yam (*Amorphosphorus companulatum*).
2. Taro (*Colocasia esculenta*).
3. Tannia (*Xanthosoma sagittifolium*).

(d) Dioscoreaceae
1. Greater yam – *Dioscorea alata*.
2. Lesser yam – *Dioscorea esculenta*
3. White yam – *Dioscorea rotundata*

- **Dicot Vegetables:-**

(a) Aizoaceae
1. New Zealand spinach (*Tetragouia expansa*).

(b) Chenopodiaceae
1. Beetroot or Sugarbeet Palak (*Beta vulgaris*)
2. Swiss Chard (*Beta vulgaris* var. *cicla*).
3. Spinach (*Spinacea oleracea*).
4. Palak or Spinch beet (*Beta vulgaris* var. *bangalensis*).

(c) Compositae
1. Lettuce – *Lactuca sativa*
2. Chicory - *Chichorium intybus*.
3. Endive – *Chichorium endive*.
4. Globe artichoke - *Cynara scolymus*.
5. Jerusalem artichoke – *Helianthus tuberosus*

(d) Convonlvulaceae
1. Sweet potato (*Ipomea botatas*).

(e) Cruciferae
1. Cole crops 2. Radish.

(f) Cucurbitaceae
1. All cucurbitaceous vegetables.

(g) Euphorbiaceae
1. Tapioca (*Manihot esculenta*).
2. Chekurmanis (*Sauropus androgynous*).

(h) Leguminaceae
1. Pea 2. Beans. 3. Cow pea (*Vigna unguiculata*).
4. Winged Beans (*Pisophocarpus tetragonolobus*).
5. Agathi (*Sesbania grandiflora*).
6. Fenugreek (*Trigonella foenugraceum*)

(i) Malvaceae
1. Okra (*Abelmoschus esculentum*).

(j) Polygonaceae
1. Rhubarb – (*Rheum rhaponticum*).
2. Sorrel – (*Rumex vesricarius*).

(k) Solanaceae — Potato, Tomato, Brinjal, Chilli.

(l) Umbeliferae
1. Carrot 2. Parsely 3. Corinder
4. Celery (*Apium graveolens*).

(m) Labitae Chinese potato (*Coleus peruviflorus*).
(n) Moringaceae Drum Stick (*Moringa Oleifera*)
(o) Lauraceae Bay leaf (*Paurus nobilis*)
(p) Aizoaceae New Zealand Spinach (*Tetragonia tetragonioides*)
(q) Basillaceae Indian spinach (*Basella rubra*)

Photosynthesis based classification

- C4 plant – Amaranthus, Globe Amaranthus (Artichoke).
- C3 plant – Lettuce, Sugar beet, Carrot, Tomato, Potato, Sweet potato.

Photoperiod based classification

1. **Long Day Plant:** Flowering is induced by period longer than critical units. eg; Cabbage, Cauliflower, Radish, Carrot, Beet, Turnip, Palak, Spinach, Lettuce, Onion, Potato.
2. **Short Day Plant:** Flowering is induced by period shorter than critical units. eg; Sweet Potato, Indian Spinach, Dolichus bean, Cluster bean, Winged bean.
3. **Day Neutral Plant:** Flowerings induced by a period of 10-18 hours of even continuous illumination. eg: Tomato, Brinjal, Chilli, Okra, Cucurbits, Amaranthus, French bean, Cow pea, Sweet Pepper, Squash, Pumpkin

Temperature requirement-based classification

On the basis of temperature, vegetable crops may be grouped into cool and warm season crops. In cool season crops, the edible parts are mainly root, stem, leaf and immature flower parts. Whereas, in warm season crops edible part is mainly fruit with exception of pea and broad bean which are cool season crops. The cool season crops grow well when the monthly mean temperature does not exceed 210 C. They thrive best if the monthly mean temperature is 15-170 C. The warm season crops on the other hand grow best when the monthly mean and average maximum temperature are 50 to 60C higher than for the cool season crops. The average monthly minimum temperature should not be below 90C to 100C for the warm season crops.

1. ***Cool season vegetable crops*** Cole crops, Root crops, Bulb crops, Lettuce, Methi, Pea, Potato
2. ***Warm season vegetable crops*** Beans(mostly) Solanaceous vegetables, Gourds, Okra, Cassava, Summer squash, Sweet potato

1. 18-25^0 C - Cabbage, Cauliflower, Radish, Carrot, Lettuce, Potato.
2. 20-25^0 C - Tomato, Brinjal, Pepper, Onion, Garlic, Cucumber, Pumpkin.
3. 25-27^0 C - Okra, Chilli, Sweet Potato, Water Melon, Musk Melon.

Respiration rate based classification

1. **Very low** – Potato, Onion.
2. **Low** – Sweet Potato, Cabbage

3. **Moderate** – Tomato, Chilli, Carrot.
4. **High** – Peas, Beans.
5. **Very High** – Leafy greens, Green onion, Musk Melon, Water Melon, Cauliflower, Asparagus.

Respiration based classification

(A) **Climacteric** - Tomato, Muskmelon, Watermelon,

(B) **Non climacteric** - Cucumber.

Classification based on parts used as vegetable

(a) **Leaves or stems:** Cabbage, asparagus, amaranthus, knolkhol

(b) **Fruits :** Melon, tomato, egg-plant, beans, peas, all cucurbits, okra, chilli etc.

(c) **Flower parts:** Cauliflower, broccoli, Ageti, globe artichoke

(d) **Underground:** Sweet potato, beet, radish, carrot, turnip, onion (roots, tubers, bulbs and corms) and garlic

Growing season.

1. **Kharif:** Cucurbitaceous vegetables which complete their life cycle during rainy season.
2. **Rabi:** Root crops, Cole crops, Potato, Lettuce etc., which grow from October to February
3. **Summer:** Melons etc., which grow from February to May.

Methods of raising:

1. **Direct sown crops:** Okra, Carrot, Radish, Beans, Peas, Garlic.
2. **Transplanted crops***:* Tomato, Brinjal, Chillies, Cabbage, Cauliflower.
3. **Vines and cuttings***:* Sweet potato, Cassava, Pointed gourd, Coccinea.
4. **Bits of tubers and corms:** Potato, Yams.

Hardiness:

Vegetables which can tolerate frost are known as hardy and vice versa are known as tender. The degree of frost tolerance makes the crop hardy, semi hardy and tender.

Hardy vegetable crops	**Semi hardy vegetable crops**	**Tender vegetable crops**
Asparagus	Beet root	Amaranthus
Broccoli	Carrot	Okra
Brussels sprout	Cauliflower	Chilli
Cabbage	Celery	Tomato
Garlic	Globe artichoke	Cluster bean
Knol khol	Lettuce	Cowpea

Leek	Palak	Cucurbits
Onion	Potato	Snap bean
Peas		Sweet potato
Radish		Tapioca
Spinach Turnip		Yams

Rooting depth:

Shallow rooted (30-40 cm)	Moderately deep rooted (50-60 cm)	Deep rooted (80-100 cm and above)
Broccoli	Beet	Artichoke
Brussels sprout	Carrot	Asparagus
Cabbage	Cucumber	Cluster bean
Cauliflower	Brinjal	Cow pea
Garlic, Leek, Lettuce	Muskmelon	French bean
Onion,Potato,Radish	Pea, Pepper	Lima bean
Spinach	Summer squash, Turnip	Pumpkin, Sweet potato, Tomato, Water melon

Types of vegetable gardening

Vegetable gardening can be classified into the following types according to the purpose for which they have been developed.

1. Home / Kitchen Gardening
2. Commercial vegetable gardening
   a. Market gardening
   b. Truck gardening
   c. Vegetable forcing
   d. Vegetable processing
   e. Floating gardening

Soil Tolerance

Soil Acidity (pH – 5-6.8)			Soil Salinity (7.1-8.5)		
Less	**Moderate**	**Tolerant**	**Less**	**Moderate**	**Tolerant**
Cole crops, Okra, Musk Melon, onion, leafy vegetables	Brinjal, Chilli, Tomato, Peas Carrot, Pumpkin, Garlic, Cucumber	Potato, Rhubarb, Sweet Potato, Watermelon, Fennel	Potato, Sweet Potato, Brinjal, Radish, Peas, Beans	Tomato, Chilli, Cucumber, Cauliflower, Cabbage, Onion	Beet, Palak, Bitter gourd, Asparagus, Lettuce, Ash gourd

Important vegetables cultivation for one-hectare area

S. No.	Vegetables	Number of nursery bed (3 × 1 m)	Seed rate per bed (g)
1.	Tomato	10	40-50
2.	Brinjal	15	35-40
3.	Capsicum	12	115-120
4.	Chilli	8	125-150
5.	Cauliflower	10	70-75
6.	Cabbage	10	70-75
7.	Onion	60	175-200

Water Requirement

High	Moderate	Low	Very Low
Leafy vegetable, Cole crops, Radish, Beet, Asparagus	Tomato, Chilli, Brinjal, Potato, Onion, Cucumber	Beans, Ridge Gourd, Peas	Melons, Pumpkin

Storage Temperature

S. No.	Vegetables	Temp (°C)	RH (%)	Storage life (Weeks)
1.	Potato	3-4.4	85	34
2.	Tomato ripe	7.2	90	1
3.	Tomato Unripe	8.9-10	85-90	4-5
4.	Brinjal	10-11	92	3-4
5.	Okra	8.9	90	2
6.	Onion	0	70-75	20-24
7.	Cole crops	0-1.7	92-95	4-6
8.	Garlic	0	65	28-36
9.	Carrot, radish	0.4-4	93-99	20-24
10.	Cucumber	10-11	92	2
11.	Pumpkin	1.7-11.6	70-75	24-36
12.	Peas, Leafy Vegetables	0	95	-
13.	Chilli green	7.2	-	-
14.	Chilli ripe	5-7	-	-
15.	Bitter gourd	0.6-1.7	90-95	8-14
16.	Coriander leaves	0-1.7	90	5
17.	Water melon	7.2-15.6	80-90	2

Estimated of important vegetables post harvest losses

S. No.	Vegetables	% Loss
1.	Potato	5-40
2.	Onion	16-35
3.	Tomato	5-50
4.	Cauliflower	49
5.	Cabbage	37

Critical stages of irrigation in some vegetable crops

S. No.	Vegetables	Stages of growth
1.	Tomato, Brinjal	Flower development, fruit set and after each harvest
2.	Chilli	Tenth leaf to flower and after each harvest
3.	Potato	Stolon formation, tuberization
4.	Cucurbits	Flower bud development and early fruit development
5.	Cabbage	Head formation and enlargement
6.	Cauliflower	Whole vegetation period
7.	Onion	Bulb formation and enlargement
8.	Garlic	Bulbing
9.	Peas, Beans	Flowering and pod development
10.	Okra	After fruit set

Suitable crops for various methods of preservation

(i) **Canning:** Tomatoes, Peas, Beans, Okra, Pointed gourd, Sweet corn, Asparagus

(ii) **Freezing:** Peas, Sweet corn, Lima beans, Asparagus, Cauliflower, Spinach.

(iii) **Dehydration:** Onion, Potato, Cauliflower, Peas, Cauliflower, etc.,

(iv) **Pickling and fermentation:** Turnip, Cucumbers, Cabbage, Cauliflower, Carrot, Chilli, and Radish.

Plant reproduction

- For genetic improvement of any plant species knowledge of mode of reproduction is essential.
- Self-fertilizing plants are genetically homozygous while cross-fertilizing plants are characterized with high degree of heterozygosity.
- Each flower has four basic organs viz., sepals, petals, stamens and pistil.
- A flower having all the basic organs is called as complete flower (pea, tomato, cabbage) while incomplete flower lacks one or more basic organs.
- Stamens and pistils are the indispensable organs as these organs produce male and female gametes and flowers having both these organs are called perfect flowers e.g.

pea, tomato etc.

- Imperfect flower lacks one of the sex organs (either staminate or pistillate).
- Plants in which staminate and pistillate flowers are borne on the same individual are called "monoecious"; while plants in which the staminate and pistillate flowers are borne on different individuals are called "dioecious".
- Apparently, the perfect flowers are hermaphrodite as these include both male and female organs in the same flower and practically all complete flowers are perfect flowers contrary the imperfect flowers are unisexual as these are either staminate (male) or pistillate (female).
- **Pollination** is the process of transfer of pollen grains from the anther to the stigma of the female reproductive organ.
- Transfer of pollen from an anther to a stigma within the same flower or to a stigma of another flower on the same plant is referred to as ***self pollination*** or ***autogamy***.
- Hermaphrodite or bisexual flowers and homogamy (simultaneous maturation of anthers and stigma of a hermaphrodite flower) are pre-requisites for self-pollination in self-pollinated plant species. But practically, self-pollination is not complete and cross-pollination may take place up to certain extent.
- The degree of cross-pollination in self-pollinated species is affected by several factors, *e.g.*, variety, environmental conditions like temperature and humidity, and location.

Pollination mechanism in Vegetables –

1. Self pollination – Tomato, Pea, Cowpea, Fenugreek, Lettuce, French bean, Lablab bean, Broad bean
2. Often Cross pollination – Okra, Brinjal, Chilli, Lima bean, Sweet pepper
3. Cross Pollinated – Cole Crops
   Root Crops → All these Entomophillus (Insect Pollination)
   Cucurbits
   Onion
   Spinach
   Amaranthus
   Beet leaf (Palak) → All these Anemophillus (Wind Pollination)
   Garden Beet
   Beet root

Isolation Distance

Type of Pollination	Foundation Seed (m)	Certified Seed (m)
Self Pollinated	50	25
Often cross Pollinated	200	100
Cross Pollinated	800-1000	400

Sex Forms in vegetables –

1. Monoecious – Cucurbits, Amaranthus, Sweet corn, Cassava.
2. Dioecious – Pointed gourd (JRF, 2008), Scarlet gourd, Spine gourd, Spinach, Yams, Beet root, Asparagus.
3. Andromonoecious – Musk Melon
4. Hermaphrodite – Satputia (Ridge Gourd), Cucumber – Original Form
5. Gynoecious – Cucumber
6. Trimonoecious – Cucumber, Musk Melon, Ridge Gourd

Brief description of plant reproduction system

S.No.	Terminology	Characteristics
1a.	Asexual reproduction	Multiplication of plants through vegetative means bypassing sexual process
(i)	Natural	Multiplication by rhizome, tuber, corm, bulb, runner, suckers and stolon Example- Potato, Cassava, Yam, Taro, Elephant-foot-yam, Garlic
(ii)	Artificial	Multiplication by stem and root cuttings, grafting, layering and budding Example- Pointed gourd, Sweet Potato
1b.	Apomixis	Development of embryo without fertilization (sexual fusion)
(i)	Parthenogenesis	Development of embryo from egg cell without fertilization
(ii)	Apogamy	Development of embryo either from synergids or from antipodal cells
(iii)	Apospory	Origin of embryo from diploid egg cell of another embryosac developed from diploid tissues
(iv)	Adventive embryony	Origin of embryo sac directly from diploid cells of either nucellus or integument
2	Sexual reproduction	Multiplication of plants through the union of male and female gametes (fertilized embryos)
a	Autogamy (Self pollination)	Fertilization of ovules by the pollens of same flower
b	Allogamy (Cross pollination)	Fertilization of ovules by the pollen of another plants through wind (anemophily), water (hydrophily) or insects (entomophily)

Self pollination leads to very rapid increase in homozygosity and these species do not show inbreeding depression but may exhibit considerable heterosis. There are certain mechanisms in the flowering plants that encourage self-pollination which are as under:

1. ***Cleistogamy***: When pollination and fertilization take place in an unopened flower bud or flower do not open at all this condition ensures self-pollination. Ex.-Lettuce
2. ***Chasmogamy:*** In some crop plants flower opens but only after completion of pollination and this condition favours the self-pollination. Ex.- Tomato, Capsicum
3. ***Position of anthers:*** In some vegetable crops (tomato, brinjal), stigma is enclosed by anthers in such a manner that self-pollination is ensured.

Cross-pollination is very common in almost all vegetable crops which preserve and promote heterozygosity in any population. These plant species are highly heterozygous and exhibits considerable heterosis in certain combinations and severe inbreeding depression on selfing. Therefore, breeding methods should be devised for exploiting heterosis and development of hybrids and synthetics.

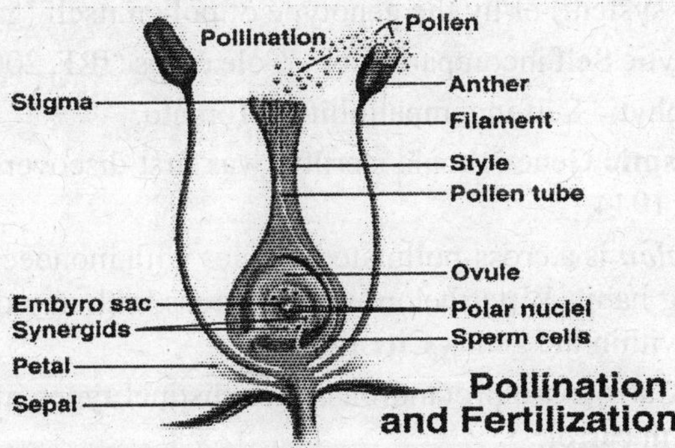

In natural, Cross-pollinated population various morphological, genetical, physiological and biochemical mechanisms are operative individually or jointly that promotes cross-pollination and restricts self-pollination in different manner which are as follows:

1. ***Dicliny or unisexuality***: In such condition flowers are either staminate (male) or pistillate (female). It can be further divided in to two classes first, when male or female flowers borne separately on the same inflorescence or on the same plant this condition is termed as monoecy. In some vegetable crops like gourds and cucumber, staminate (male) and pistillate (female) flowers are borne separately on the same plant. Second, when staminate and pistillate flowers are present on separate individuals, it is referred as dioecy.
2. ***Dichogamy***: The male and female reproductive organs mature at different times even in a perfect flower. Dichogamy is of two types:
   a. ***Protogyny:*** Stigma matures earlier than anthers of the same flower so that allogamy is facilitated. Ex.- Cole crops and Elephant Foot Yam
   b. ***Protandry:*** Anthers mature earlier than the stigma of the same flower thus promoting allogamy as reported in carrot and onion. Ex.- Onion, Carrot, Beet, Parsley, Leek
3. **Male sterility:** A condition in which either pollen is absent or nonfunctional in flowering plants while female gametes functions normally.
   (i) **Genetic Male Sterility:** Pollen sterility which is caused by nuclear genes. Ex.- Tomato, Musk Melon, Water Melon, Cole crops, Beet, Onion

(ii) **Cytoplasmic Male Sterility:** Pollen sterility which is caused by cytoplasmic genes. Ex.- Carrot, Sweet Pepper, Cucumber, Turnip, Chinese cabbage

(iii) **Cytoplasmic Genetic Male Sterility:** Pollen sterility that is controlled by both cytoplasmic and nuclear genes. Ex. Onion, Carrot, Radish, Beet, Tomato

4. **Self-incompatibility:** The inability of a plant with functional pollen to set seeds when self pollinated. It is mainly of two types: (i) Heteromorphic and (ii) Homomorphic.

(i) **Heteromorphic system:** Self incompatibility due to differences in floral morphology. The incompatibility reaction of pollen is determined by the plant producing them.

(ii) **Homomorphic system.** Presence of self incompatibility due to physiological or genetic causes rather than differences in floral morphology. The incompatibility reaction of pollen may be controlled by the genotype of plant on which it is produced (sporophytic system) or by the genotype of pollen itself (gametophytic system).

- Sporophytic Self incompatibility – Cole crops (JRF, 2008), Radish, Turnip, Beet. Gametophytic Self incompatibility – Tomato.
- **Cytoplasmic** Genetic male sterility was first discovered in onion by Jones and Davis in 1944.
- ***Watermelon*** is a cross-pollinated species with monoecious or andromonoecious flowering habit. Watermelon will not cross with any other cucurbits except for species within the genus **Citrullus.**
- ***Carrot***: Carrots are protandrous. Two distinct types of male sterile flowers are reported in carrot.

Asexual Reproduction:

- **Potato:** Potato often flowers only in cooler regions in the hills where it is grown in summer under long-day condition. In normal growing season in the plains under short-day condition it generally fails to flower. Even if flowers emerge, fruit set does not occur. In this short-day condition in plains, it can be forced to flower and set fruits with the arrangement of artificial light to lengthen the photoperiod for at least 4 hours. At present, potato is being cultivated commercially, through to a very lesser extent, from true seeds.
- **Cassava:** It is highly heterozygous and cross-pollinated crop and monoecy i.e., production of both male and female flowers on the same inflorescence render to cross-pollination. Male sterility, which is commonly found in cassava, also favours cross-pollination. Sexually propagated seedlings are highly variable in character due to heterozygosity and at the same time it takes long time to give potential yield. However, hybridization following clonal selection is an important method to develop a variety.
- **Sweet Potato:** Flowering itself is a major constraint in this crop and a number of elite clones do not flower. Flowers are bisexual but sterility of different intensities is found because of hexaploidy and accumulation of recessive sublethal mutations through long clonal propagation. It is highly heterozygous due to cross-pollination. Hence, the sexual seedlings are highly variable in nature. Sexual reproduction is only encouraged for breeding purpose.

- **Yam:** All the members of yams are dioecious hence; the male and female flowers are borne on separate plant. Flowering and seed production are greatly limited due to shy flowering and poor seed set.
- **Taro:** Generally, flowering is rare in most the taro types. Inflorescence, the spadix, is cylindrical with female flowers at the base and a longer protruding male section of staminate flowers with aborted flowers in-between.
- **Elephant-foot-yam:** The crop cycle of one to three years causes the emergence of inflorescence, although very rarely. The spadix is extended bearing male flowers at the top and female flowers at the bottom and the neuters are absent. There is complete absence of seed under natural condition due to extreme protogyny coupled with the delayed opening of the spadix.
- **Pointed gourd:** Seed are recalcitrant in nature, which lose viability very soon. The seedlings normally do not flower in the first year. The seedlings from the open pollinated seeds are highly variable due to heterozygosity and generally low yielder than the mother clone. The seedling population gives 50 per cent each of male (unproductive) and female (productive) plants due to involvement of a distinct pair of chromosomes for determination of dioecious sex mechanism. Hence, seed propagation is normally avoided.
- **Garlic:** Most of the garlic types are non-bolting where inflorescence does not emerge. Very few cultivars bear fertile flowers and viable seeds that germinate. In most of the cases, the inflorescence, if emerge, contain sterile flowers and bulbils (aerial bulbs).

Importance of vegetable and spices crops

- Vegetables constitute an important component of a balanced diet for man.
- Vegetables like carrot, pumpkin, peas, turnip, beetroot, tomato, sweet potato, cabbage, greens like spinach, fenugreek, green onion etc., are important source of vitamin A.
- Ascorbic acid or vitamin C is present in appreciable quantities in chilli, Palak, methi (fenugreek), cabbage, tomato, cauliflower, bitter gourd etc.
- Vitamin D is available in greens which are good source of Vitamin E also. Green leafy vegetables are good source of the other essential Vitamins *viz.*, niacin and vitamin K.
- Vegetables are also rich sources of minerals. Of the ten minerals needed, the most important are calcium, phosphorus and iron. Calcium is obtained from vegetables like beans, carrot, cabbage, cauliflower, onion, peas, tomatoes, amaranthus, fenugreek etc., Vegetables like potatoes, carrot, tomatoes, cucumber, cauliflower and spinach are good sources of phosphorus. Iron is supplied to the body by vegetables like cabbage, peas, beans, amaranthus, tomato, carrot, bitter gourd, onion. Iodine which is essential for the human body is provided by onion, okra, summer squash etc.
- Many of the vegetables are of the nature of roughage aiding in digestion. Most of the vegetables with the exception of tomato are alkaline in reaction which is very beneficial to neutralize the acid reaction of non-vegetarian foods.

- Vegetables like potato, sweet potato, cassava, yams etc., are good sources of carbohydrates and therefore are important subsidiary foods while others like peas and beans of different types are rich in protein.
- Spicy vegetables like onion, garlic, coriander, fenugreek, mint etc., add taste and flavour to our food. The added advantage of these spicy vegetables is their protective nature against diseases. A recent report in Spice India Journal goes an "An armament of onions, cloves, mustard and fenugreek with a splash of celery and coriander is what the body needs to ward off cancer and bacteria invading our body say nutrition scientists. The national institute of nutrition in Hyderabad reports that most spices prevent cancer and attach of harmful bacteria and fungi while some reduce blood sugar levels, help digestion and reduce cholesterol levels".
- Besides their importance in human nutrition, vegetable crops are commercial crops supporting market gardening, truck gardening or commercial gardening. Raising of vegetable crops also supports many other industries like the processing industry, seed industry, fertilize and plant protection chemical industries, the weedicide industry, farm machinery and implements and packing and marketing industries.
- Vegetable crops in comparison to field cops are high output crops. Being short term crops, their cultivation is intensive and in year, two or three crops with yields varying from 40 to 60 tonnes per ha per annum can be raised. The ease of cultivation, their versatility for growing in plains and hills of different altitudes and the high returns they fetch make them choice crops for commercial growing.
- They are also crops which can be fitted into many remunerative crop rotation and cropping patterns like inter cropping, multiple cropping and companion cropping.
- Vegetable growing enables maximum utilization of land since it can be taken up in small vacant spaces or strips around our homes to large scale commercial growing. By virtue of the foregoing characteristics, vegetable growing also provides employment for labour round the year unlike growing of cereal and other crops which are seasonal in nature.
- Although India is mainly a vegetarian country, we are no self-sufficient in vegetables production. The current production of vegetables is adequate to provide hardly 120g. as against 300g. of vegetables required per day per adult. This situation of low production is mainly on account of the dearth of technical known-how on scientific methods of cultivation of these crops.
- With the tremendous rate of population growth in India and at the global level in general, there is an urgent and imperative need to improve the vegetable crops productivity per unit area so that self sufficiently can be achieved in vegetable production.

Recent Achievements in vegetable crops -

- One hybrid Pusa Kartik Shankar and 4 varieties Pusa Meghna, Pusa Sharad, Pusa Shukti and Pusa Paushja have been released and are being commercially cultivated.
- Two varieties of tomato Pusa Rohini and Pusa Sadabahar were released and are under commercial cultivation.

- Red carrot variety Pusa Rudhira, Black carrot variety Pusa Asita and heat tolerant variety Pusa Vrishti have been developed and these seed are provided to farmers in bulk quantity.
- Cytoplasmic male sterility system has been established for the first time in tropical carrot and tropical carrot hybrid Pusa Vasudha has been developed which is the first from any public sector institution.
- One hybrid DBHL-20 and two varieties Pusa Shymala and DBL-02 of brinjal have been released/ identified by AICRP (Vegetable crops) and are being grown commercially.
- Nine varieties/hybrid of cucurbits namely Pusa Samridhi and Pusa Santushti of bottle gourd, Pusa Hybrid-2 of bitter gourd, Pusa Uday and Pusa Barkha of cucumber, Pusa Nutan of ridge gourd, Pusa Sneha of sponge gourd, Pusa Ujwal of ash gourd and Pusa Shandar of snap melon have been released and seeds are being provided to farmers.
- Seven varieties of different vegetables namely Pusa Mridula, Pusa Jamuni and Pusa Gulabi of radish; Pusa Sukomal of cowpea, Pusa Udit of broad bean, Pusa Soumya of bunching onion and Pusa sag-1 of leafy mustard have been released/ identified for commercial cultivation.
- Gynoecious lines have been developed in bitter gourd and are being used in hybrid development.
- Two molecular markers RAPD and ISSR were developed as close flanking to black rot resistant gene at 1.6 cM distance in cauliflower.
- Onion variety Sel-126 and Pusa Riddhi (Sel-383) has been identified for commercial cultivation.
- *Ogura* CMS system has been established in all the maturity group of cauliflower and is being used in hybrid development.
- Detailed investigation in the germplasm of cauliflower led to the development of the theory that Indian cauliflower was a separate group and it was distinctly different from the European types. Different maturity groups originated by simple selection, and later perhaps of re-combination, as a result of natural crossing between different European types.
- Self-incompatible lines in Indian cauliflower have been developed and are being used in hybrid seed production.
- Detailed karyotype analysis of cucumber and muskmelon led to the establishment of the theory that cucumber and muskmelon followed two different courses of evolution in Asia and Africa, respectively. Electrophoretic analysis of enzymes revealed that there was little similarity between two species of the genus *Cucumis*.
- Male sterile and maintainer lines have been isolated in short day onion Pusa Red and this will help in development of F1 hybrids in near future.
- Recessive gene marker has been isolated in pumpkin, which can be used for hybrid seed production on commercial scale.

New Released Varieties –

1. Tomato – Pusa Rohini, Pusa Sadabahar
2. Brinjal – Pusa Shyamla, DBL-02, DBHL-20,
3. Cauliflower – Pusa Meghna, Pusa Sharad, Pusa Kartik Sankar (F1 hybrid), Pusa Paushja, Pusa Shukti
4. Bottle gourd – Pusa Samridhi, Pusa Santushti
5. Bitter gourd - Pusa Hybrid-2
6. Sponge gourd - Pusa Sneha
7. Ridge gourd - Pusa Nutan
8. Cucumber - Pusa Uday, Pusa Barkha
9. Snap melon - Pusa Shandar
10. Radish - Pusa Mridula, Pusa Jamuni, Pusa Gulabi,
11. Carrot - Pusa Rudhira, Pusa Asita, Pusa Vrishti, Pusa Vasuda
12. Onion - Sel- 126, Pusa Riddhi
13. Cowpea - Pusa Sukomal
14. Vegetable mustard - Pusa Sag-1
15. Bunching Onion - Pusa Soumya
16. Broad Bean - Pusa Udit

Variety development and testing:

- **Variety:** A variety is a group of plants that has common qualities and it should be distinct from other groups of plants of the same species. To be a variety it should have three criteria. They are: -
- **Distinctness:** The variety should be different from other varieties.
- **Uniformity:** The variety should be uniform.
- **Stability:** The characteristics of the varieties should be stable from one generation to other generation.

Types of varieties:

- **Traditional varieties:** Traditional varieties have been usually started by the farmer themselves. Such varieties are being used by many generations of farmers.
- **Landraces:** Landraces are the groups of traditionally used plants that do not satisfy the rigid criteria of a variety.
- **Improved varieties:** These varieties have been developed from systematic plant breeding work and trueness of the variety is maintained by the trained personnel.
- **Open pollinated varieties:** Seeds of these varieties are produced by natural pollination through successive generations.
- **Hybrids:** Hybrids are varieties that result from controlled pollination involving 2, 3 or 4 parental lines.

- **Pure line:** Pure line is the progeny of a single homozygous self – fertilized individual plant. Its progenies remain true to the type and uniform. Selection within pure line is ineffective.
- **Clone:** Clone is a progeny of one vegetatively propagated plant.
- **Strain:** A variety may have several strains with minor differences but having all the basic characteristics of the original varieties.

Characteristics of Improved Varieties

- **Productivity:** One of the most remarkable achievements of the modern plant breeding is the improvement of crop productivity. It is measured in terms of yield/ area per time per fertilizer. It has been increased with the development of varieties having high yield potential, earliness, improved plant architecture and high harvest index.
- **Adaptability:** Varieties have wide adaptability have been developed by the plant breeders.
- **Reliability:** The crop performance should be reliable year after year.
- **Quality:** It is the most important factor that determines the acceptance of a variety. The new varieties should meet the market standard. E.g. Colour and firmness in tomato, dry matter content in onion 46% dry matter content is preferable in industries, non – bitter cucumber production.

Variety development:

Varieties are developed through a process called plant breeding. It is the genetic adjustment of plants to the service of man. Selection is the major part of plant breeding. Selection is processing of selecting for the desired type in a variable population of plants. The method of selection varies according to the nature of pollination of the crops.

Breeding methods for self – pollinated crops:

- **Mass selection:** Mass selection is a form of selection in which individual plant is selected and the next generation continued from the aggregate of seeds of the selected plants. This is done by rouging out the undesired plants from the field and harvesting the seed crop of the remaining plants in bulk or mass. Another approach is to tag or mark the desirable plants and harvest them in bulk or mass.
- **Pure line selection:** Pure line breeding generally includes 3 steps, sufficient number of selections are made from the originally available variable source population or material. The second step includes growing of progeny rows from the individual plant's selection obtained in the first step. The final and the third step include the evaluation of remaining lines in the replicated trials to compare the selections among themselves and with established commercial variety. The outstanding line is finally selected and the seeds of that line are collected for further multiplication.
- **Pedigree method:** This method of breeding consists of selection among individual plants and their progeny during inbreeding following crosses between selected donor parents. The selected plants are crossed in a single cross system. The F_2 populations

are subjected to desired character. The F_3 progenies are the selected F_2 plants are grown in rows. The most desirable F_3 progeny rows are selected and harvested seed from superior individual plants from each of the selected rows. This practice of selection continuous till F_5. F_6 generation in which desirable rows are bulked within themselves and the bulks are evaluated in an implicated trial. A proper time of record is maintained of each generation of the plants.

- **Single seed decent method:** In this method all the plants of the F_2 generation are advanced to the next generation the harvesting and growing one seed from each plant. This process is continued till F_5. In F_6 generation superior individual plants are selected finally. Adequate level of variability is maintained in all the generation. No potentially superior plant is estimated in early generation. It is cost effective as resources are required to follow the scheme of selection.
- **Bulk method:** The growing of genetically diverse population of self-pollinated crops in a bulk plot or without mass selection followed by single plant selection in F_5 or F_6 generation is known as bulk breeding. In this method plants of segregating population are harvested together and the sample seed is used to the next generation. In F_5 or F_6 generation, selected individual plants are harvested and evaluated in individual plant progeny from which superior rows are selected and seeds are used to evaluate in the replicated trial. The major advantage of this method is that there is no problem in selecting the seeds in every generation.
- **Back-cross method:** In this method, recurrent back crosses are made to one of the parents of hybrid for a specific characteristic known as back cross breeding. It is an efficient procedure to reconstitute most of the genotype of the recurrent parents. This method allows the introduction of new character from non-recurrent parents.

Advantages:

- It requires a small number of plants to be handled.
- There is no role of environment on this breeding.
- It is a stepwise breeding program.

The performance of the derived line is predictable but this method of breeding excludes the combination of genes from the parental cultivar which is a common feature.

Multiline breeding: Extension of back cross breeding.

Mutation breeding.

Breeding method for cross-pollinated crops:

- **Mass selection:** It is also known a positive mass selection. In this method, the best individual plants are uprooted and transplanted in the next field for seed production. Then the seeds are composite or bulked for raising next generation. The selection is based on the phenotype performance of the mother plant, without any progeny test. This process is repeated for several cycles till substantial achievement are obtained.
- **Mass pedigree method:** In this method the best individual plant with desired characters are selected on the basis of phenotypic performance from variable sources.

Open pollinated seed of selected individual plants are divided into 2 halves. In the second year replicated progeny row trial is conducted using one set of half seeds from each plant. Based on the progeny performance, the best parental individuals are identified. The remaining or remnant half seeds from the superior parental lines are mixed and grown in isolation in the third year.

- **Inbreeding:** Self-fertilization is the most intense form of inbreeding. In plant breeding, nearly homozygous lines are produced by continued self-fertilization for 5 – 6 generations. This method can be used in those crops which do not show inbreeding depression. e.g. cucurbits.
- **Recurrent selection:** A number of plants are self-pollinated in a source of population in the first year. Superior plants base in phenotypic performance are selected. In the second year, seeds produce by self-fertilization of the selected plants are sown & crossed in all possible combination. The seeds are collected & bulked. Such cycle of selection & crossing is repeated.
- **Synthetic variety:** Synthetic variety is maintained from open pollinated seed following its synthesis by hybridization in all combination among a number of selected genotypes which have been tested for combining ability. The component of synthetic variety could be inbred, clones, mass selected population or various other materials. The component units are maintained so that the synthetic may be reconstituted at regular intervals.
- **Development of composite varieties:** Composite varieties are derived from the varietal crosses in advanced generations. These are developed from open pollinated varieties or other heterozygous populations or germplasm which have originally not been subjected to inbreeding or have not been elaborately tested for their combining ability. Usually, they involved open pollinated varieties, synthetic, double crosses, etc. selected for yield performance, maturity, resistant to disease & pests. These composites often show a high order heterosis in F1 when widely diverse populations are crossed.

Development of hybrid varieties:

There are 2 steps for the development of hybrid varieties.

Step 1: Development of inbreed lines: All the selected plants are self-pollinated & evaluated as a pure line selection until they reach a highly developed homozygosity. The final product of this process is a pure line.

Step 2: Combining ability test: The inbreed lines are tested or for their combining ability to give an acceptable progeny in a hybrid combination with other inbreed lines.

Use of Hybrids

- In Japan, commercial Fl hybrids became available in 1935 in cucumber and in 1940 in tomato.
- In USA, the hybrid seeds of vegetables came in existence in the year 1945.
- In India, first report of hybrid vigour appeared in 1933 in chilli at Indian Agricultural Research Institute (IARI), New Delhi.

- First hybrid "Pusa Meghdoot" of bottle gourd was developed at IARI, Regional Station, Katrain and released in 1971.
- Fl hybrids of summer squash "Pusa Alankar" and cucumber "Pusa Sanyog" were also developed and released for cultivation in 1975.
- Hybrid technology has been most successful in tomato resulting in large area coverage in Karnataka, Maharashtra, Gujarat and some parts of Andhra Pradesh.
- Hybrid cabbage in next in area coverage to tomato.

Table: List of some public and private sector bred hybrids identified/released for cultivation

Crop/Hybrids	Developing centres	Year of identification
Brinjal		
Pusa Hybrid-6	IARI, New Delhi	1990
Pusa Hybrid-5	IARI, New Delhi	1992
NDBH-1	NDUA & T, Faizabad	1993
Pusa Hybrid-9	IARI, New Delhi	1997
VRBHR-1	IIVR, Varanasi	2003
Tomato		
Pusa Hybrid-2	IARI, New Delhi	1993
KTH 2	CSUAT, Kanpur	2003
Bell Pepper		
KT-1	IARI (Katrain)	1990
DARL-202	DARI, Almora	2003
Cabbage		
Pusa Synthetic	IARI (Katrain)	1992
Cauliflower		
Pusa Hybrid-1	IARI, New Delhi	1992
DCH-541	IARI, New Delhi	2003
Muskmelon		
M-3 Hybrid	IARI, New Delhi	1990
MHY-5	Durgapura	2002
Watermolon		
Arka Jyoti	IIHR, Bangalore	1981
Bottle Gourd		
PBOG-2	GBPUA & T, Pantnagar	1998
PBOG-1	GBPUA & T, Pantnagar	1999
NDBH-4	NDUA & T, Faizabad	2001
Okra		
DVR-1	IIVR, Varanasi	1998

DVR-2	IIVR, Varanasi	1998
DVR-3	IIVR, Varanasi	2001
DVR-4	IIVR, Varanasi	2002
Bitter Gourd		
Pusa Hybrid-2	IARI, New Delhi	2003

Realizing potentials of vegetable hybrid technology in India, Indian Council of Agricultural Research (ICAR) initiated a network project entitled "Promotion of Hybrid Research in Vegetable Crops" during 1995-96. This project was initiated in nine vegetables, viz., tomato, brinjal, chilli, capsicum (bell pepper), okra, onion, cabbage, cucumber and bitter gourd.

Major Constraints: High cost of F_1 hybrid seed is one of the major handicaps of successful hybrid vegetable technology. This is because underutilization of available genetic tools for economic F_1 seed production. In the entire country, mostly hand emasculation and hand pollination technique are being followed (with very few exception) to develop F_1 hybrid seeds in most of the vegetable crops. Besides, there are several other constraints prevailing, which act as barriers in the rapid spreading of hybrid vegetable technology in this country. Some of these constraints are:

- Lack of awareness among the growers about hybrid crop production techniques.
- Unorganised marketing system for fresh vegetables.
- Lack of post harvest management techniques.
- Non-availability of other inputs at proper time.
- Non-availability of biotic stress resistant hybrids.

The achievements in hybrid seed production were made possible by utilization of genetic male sterile lines.

- Punjab Agricultural University (PAU), Ludhiana, has taken lead in developing muskmelon hybrid "Punjab Hybrid" from Cross MS-1 × Hara Madhu.
- In chilli hybrid "CH-1" has been developed from cross MS-12 × ELS. "CH-1" hybrid exhibited 160 per cent heterosis for yield besides having tolerance to some diseases.
- MS-12 is a genic male sterile line in the back ground of "Punjab Lal", developed by incorporating male sterile gene introduced from France.
- Utilizing self incompatibility and gene cytoplasmic male sterility system, two hybrids viz., "BRH-5" and "H-64", respectively have been developed in cabbage at Katrain.
- In cucumber "Pusa Sanyog" (Japanese Gynoecious line × Green Long) has been developed at IARI.
- Gene cytoplasmic male sterility system has been used to develop "Arka Pitambar" and "Arka Kirtiman" hybrids in onion at Indian Institute Horticultural Research, Bangalore.
- In carrot cytoplasmic male sterility and in muskmelon genic male sterility are being used to develop low cost hybrid seeds at many places.
- **Plant introduction:** Vegetable breeding programs are global on nature & every country has her own program. The best variety developed anywhere are introduce,

acclimatized and cultivated in commercial scale. Plant introduction plays an active role as a method of breeding.

- **Seed viability test:** Viability test is popularly recognized as an accurate means of estimating alive seed.
- **Tetrazolium test:** Tetrazolium test distinguishes between viable & dead tissues of the embryo on the basis of their relative respiration rate. About 50 – 100 seeds are taken from submitted samples. The seeds are soaked for 3 – 4 hrs. in water. Cut the seeds with the sharp blade exposing the main structures of the embryo. The one halves of the seeds are placed in petri-dish & covered with 1% tetrazolium solution. Keep the petri-dish in dark at the temperature of 35 – 400C for 2 hrs. Then the tetrazolium solution is removed & filled with water. The live cells of the embryo become dark red stained & are evaluated under the magnifying glass.
- **Vigour test:** Vigour is defined as the potential for rapid uniform germination & fast seedlings growth. The indirect method of measuring vigor is:
- **Seedling growth rate:** In this parameter, we can measure the plant height & dry matter content.
- **Speed of germination:** Mathematical expression using germination speed as a measure of seedling Vigour is:

  Vigour = No. of normal seedlings (first count) / days of first count + No. of seedlings at last count / days of last count.

The direct method of vigor test includes:

- **Brick gravel test:** In this method, the germinating seeds are covered with 13 mm layer of brick and gravel of 3 mm size. Only the vigorous seedlings emerge out from the gravel
- **Paper piercing test:** A paper sheet of 0.4 mm thick & 90 gm/m^2 weight is placed above the germinating seeds. Only the vigorous seedlings come out piercing the paper sheet.
- **Moisture test:** The moisture content of the seed sample is the loss in weight when it is dried. It is expressed as the percentage of the weight of the original sample. The hot air oven method is the routine method. Moisture content is determined by removing the moisture from the seeds in a hot air oven. There are 2 methods:

1. **Low constant temperature oven method:** In this method the oven is maintained at 103^{0}C and the sample is dried for 17 hrs. This method is used for the seeds having higher percentage of oil content. After drying, the sample is taken out & put in the desiccators for 35 mins. After cooling the sample is weighed.
2. **High constant temperature oven method:** The oven is maintained at the temperature of 130 – 133^{0}C for 1 – 2 hrs. This method is used for such crop seed like cucurbits, carrot, lettuce, peas & beans. Seed moisture should be below 12% in starch & below 10% in oily seeds for ordinary stages. For sealed container, the moisture content should be lower. The storage life of the seed becomes double, for every 1% decrease moisture content. This rate is useful between 0 – 16% moisture content. Below 6% there is no viability of embryo of seed.

Planting value of seed = purity % × viability% × 100.

Nursery raising

Nursery is place where young seedlings are raised and nurtured before planting in the main field.

Requirements for raising of nursery

- **Selection of Soil:** The pH of soil should be in the range of 6 to 7.
- **Soil treatment:** Soil mulching is done with transparent polyethylene of 25-100 cm thickness during the period of high temperature and solar radiation. Formalin can also be used for soil treatment. For this, formalin solution is prepared with the ratio of 1 part formaldehyde in 100 parts of water. The five litres of this solution is sprayed uniformly on soil surface for one square meter area.

Types of nursery beds

1. **Raised nursery bed:** Raised bed during rainy season avoids losses through water stagnation because it facilitates proper drainage for excess water.
2. **Flat Nursery Bed:** During spring-summer, seedlings are raised in flat bed. Further, in areas where rainfall is not so heavy and field is well levelled and drained, flat bed is preferred.

Table: Seed Rate and Area Required for Nursery Raising in Vegetable Crops for Open pollinated varieties

Crops	Seed rate (g)	Area required (m^2)
Tomato	400-500	100-125
Brinjal	400-500	100-125
Chillies	500-600	75-100
Capsicum	400-600	100-150
Early cauliflower	700	100-150
Mid late cauliflower	400-500	75-100
Cabbage	450-500	75-100
Knol-khol	700-750	100-750
Onion	8000-10000	500

Seed treatment: In order to enhance germination, seeds may be treated with hot water, cold water and with chemicals. Seed treatment is an aid in controlling "damping off" and the rotting of the seed if germination is delayed.

- **Hot Water Treatment:** Some of the seeds need hot water treatment to kill organism commonly found inside the seed coat like black rot (Xanthomonas sp.) in cole crops and phomopsis blight of brinjal. Usually, the temperature of hot water is raised to 50-55°C for 15-30 minutes.
- **Soaking of the seeds in cold water:** Seeds with hard seed coats may be soaked in water for 18-24 hours at room temperature depending upon the kind of seeds. It is advantageous of seeds are soaked in warm water.

- **Seed treatment with chemicals:** The common seed dressers are ceresan, thiram, captan, agrosan GN and bavistin. However, 2-3 g of chemical powder per kilogram of seed is usually recommended.

Table: Vegetables and Approximate Days Taken in Seed Germination

Crops	Seed germination (days)
Tomato	5-7
Brinjal	5-6
Chilli	7-8
Cauliflower	3-4
Cabbage	3-4
Knol –khol	3-4
Savoy cabbage	4-5
Kale	3-4
Brussels sprouts	3-4
Celery	10-15
Chinese cabbage	3-4
Sorrel	8-10
Sweet marjoram	10-12
Onion	7-10
Endive	3-4
Chicory	3-4
Broccoli	3-5
Lettuce	4-5

- **Plant protection measures:** To control cut worms, timely application of sevin and spraying of malathion 50 EC (0.05 per cent) or rogor 30 EC (0.03 per cent) should be done if required. In nursery, "damping off "disease is common due to excess moisture, to control this, drenching with 0.1 per cent solution of brassicol or 0.2 percent captan or difolatan during and after germination. Bio-agents can be used by two methods. It can be applied in well prepared and pulverised nursery bed @ 10-25 g/m^2 two days before sowing. Bio-agents containing Trichoderma should be well inoculant, are mixed@6-10 g/kg seed.

This insect is controlled by spraying of 4 per cent neem kernal extract on the plants. Spraying of monocrotophos or metasystox 1.5 ml/l of water has also been recommended.

Raising of nursery for specific purposes

1. **Raising of virus free seedlings:** Treating the soil and control it by carbofuran@3-5 g/m^2 or seed treatment with imidachloprid@2.5 g/kg seed. For controlling whitefly, spray the nursery beds at 7 days interval with metasystox or monocrotophos @ 1.5 ml/litre of water.
2. **Raising of nursery for vegetatively propagated vegetables:** The vegetables like pointed gourd and ivy gourd, one year old stem cutting are used.

Categories of vegetable crops according to their ability in transplanting: On the basis of ability of transplanting, vegetable crops may be classified into three categories.

- **Easy to Transplant:** Cabbage, cauliflower, tomato, lettuce, brinjal, Brussels sprouts, and sprouting broccoli.
- **Transplanted with Care:** Onion, leek, chillies and celery.
- **Not Transplanted:** Okra, beans, peas, cucurbits, turnip, amaranths, beet leaf and fenugreek.

Bolting in onion and premature seeding in cabbage are the best examples for which planting of over-aged seedlings in the field.

Table: Suitable Age for Transplanting of Seedlings of Some Vegetable Crops

Crops	**Age in weeks**
Brinjal	3-4
Brussels sprouts	6-8
Cabbage	4-6
Cauliflower	3-4
Celery	6-8
Chillies	3-4
Leek/Lettuce/Parsley	4-5
Onion	6-8
Sprouting Broccoli	3-4
Tomato	3-4

- **Double transplanting:** In double transplanting system, young seedlings are retained for time being, then they are finally transplanted in the main field where they grow and give economic yield. Double transplanting is most successful in tomato, brinjal, cabbage and cauliflower.

Protected cultivation

- The history of vegetable production under protected cultivation started from Roman emperor Tibenus Caeser during first century AD.
- The first glasshouse in the 1700's used glass on one side only, as a sloping roof.
- The first use of polyethylene as a greenhouse cover was in 1948 when Professor Emery Myers Emmet at the University of Kentucky used the less expensive material in place of more expensive glass.
- China has 2.9 million hectares under plastic mulching for all types of vegetable crops.
- China has more than 80,000 ha under low tunnel and 48,000 ha under green houses.
- Plastics are used primarily to extend the growing season, both in spring and fall.
- **Tunnel/Row covers:** Clean plastic films or nets are stretched over low (up to 1.0 m high) hoops to provide protection to the plants against frost, winds and insect/pest. Polyethylene films of 50 micron thickness with ventilating holes (4 % surface area), are used to cover the hoops. The low tunnels permit early yields for spring crops with significantly higher total yield. The crops which are generally grown commercially

under low tunnel conditions are melons, cucumber, tomato, strawberry, pepper, beans, squash and sweet corn.

- **Floating Row cover:** A floating row cover is a plastic film fabric used without any mechanical support in the form of hoops to protect crops from insect vectors, practice winter production of tender warm season crops to enhance early maturity, improve quality and extend the growing season. Floating covers are made of spun bonded or non-woven fabrics, 10-50 g/m^2.

Table: Names and structure of some plant growth regulators used in vegetable production

Abbreviation or common name	Chemical name or structure	Major uses
ABA	Abscisic acid	Defoliant
Chlormequat (CCC)	2-chloroethyl-trimethyl ammonium chloride	LP, SR
3-CPA	2-(3-chlorophenoxy) propionic acid	FT
4-CPA	4-chlorophenoxy-acetic acid	FT
2,4-D	(2,4-dichlorophenoxy) acetic acid	Herb.
Daminozide (B 9, Alar)	Succinic acid-2, 2-demethyl hydrazine	GR
Ethephon	2-choloethyl phosphonic acid	SR
GA_3	Gibberellic acid	GE
Kinetin	6-furfuryl aminopurine	DB
IAA	Indoel-3-acetic acid	Enlarg.
IBA	Indole-3-butyric acid	Enlarg.
Maleic hydrazide	1,2-dihydro-3, 6-pyridazinedione	Herb., GR
Mepiquat chloride	N,N-dimethyl piperidinium chloride	GR
NAA	1-naphthaleneacetic acid	FT, GR
NOXA BNOA	2-naphthalenyloxyacetic acid	GS
Paclobutrazol	(2RS,\3RS)-1-(4-chlorophenyl)-4,4-dimethyl-2-(1,2,4-triazol-1-y1)-pentan-3-ol	GR
2,4,5-T	(2,4,5-trichlorophenoxy) acetic acid	Herb.
TIBA	2,3,5-triodo-benzoic acid	GR
Uniconazol	(E)-(4-chlorophenyl)-4, 4-dimethyl-2-(1,2,4-triazol-1-yl)-1-pentan-3-ol	FR, Fung.

Abbreviations:

DB = Dormancy breaker; Enlarg. = Plant cell enlarger; FT = Fruit thinner; Fung. = Fungicide; GR = Growth retardant' GS = Growth stimulant; Herb. = Herbicide; LP = Lodging preventer; GE = Grape enlarger; SR = Sugarcane ripener.

Plant growth regulators in vegetable production:

- Application of auxins, gibberellins, retardants and ethylene are known to regulate the sex in majority of cucurbits.

- In cucumber, gynoecious lines are maintained by spraying GA, at 1000 ppm at two-leaf stage, which induces staminate flowers enabling selfing and maintenance.
- Ethrel (50 to 100 ppm) spray at the stages of 2-4 true leaf stages increase pistillate and hermaphrodite flowers in muskmelon, cucumber and summer squash, subsequently yield.
- The sprays of PCPA (50-100 ppm) and CCC (500 ppm) one week before transplanting and four weeks after transplanting cause least infestation of TLVC (tomato leaf curl virus).
- Sprays of Maleic hydrazide (2500 ppm) 2-3 weeks before harvesting helps in sprout inhibition in onion.

Approaches for organic farming

Organic production refers to organically grown crops which are not exposed to any chemicals right from the stage of seed treatment to the final post harvest handling and processing.

- **Nature farming:** It was developed in Japan in 1930 by Mokichi Okada.
- **Rishi Krishi:** This system is based on Mr. Desh Pandey of Maharashtra who is promoting for 10 years.
- **Biodynamic farming:** Biodynamic faming was spawned by the late anthroposophist, Rudolf Steiner, and has grown and developed in popularity in 1922.
- **Application of vermi-compost:** The application should be repeated after one month and water should be added to keep it moist. For general use in agriculture, vermicompost should be applied at 5 t/ha.

Nutrients in vegetables

Essential nutrients

Nutrients which are required by plants for their normal growth and development and which are not replaceable in their function by any other nutrients are referred to as essential.

Criteria of essentiality of nutrients was given by Arnon and Stout in 1939.

Classification

- **Primary nutrients:** Nutrients which are required in proper ratio for a crop, ex. Nitrogen, Phosphorus and Potassium.
- **Secondary nutrients:** Next to primary nutrients, there are three elements which are known as secondary nutrients, ex. Calcium, Magnesium and Sulphur.
- **Micro nutrients:** These nutrients (seven in number) are required in plants in very small quantity, ex. Fe, Mn, Zn, Cu, B, Mo, Cl, N.
- Nitrogen is necessary for formation of Amino Acids and directly involved in photosynthesis.
- Phosphorus is involved in photosynthesis, cell division, root formation and growth.
- Potassium is involved in carbohydrate metabolism and improves quality of fruits and vegetables.

- Calcium and Boron deficiency symptoms first appear on terminal bud.
- Fe, Mn and Cu deficiency symptoms first appear on new leaves.
- N, P, K and Mo symptoms first appear on old leaves.
- There are total seventeen essential nutrients.
- Nitrogen fixing enzyme is Nitrogenase and Gene is Nif Gene.
- Fertilizer application with irrigation water by drip system in vegetables in known as Fertigation.

Importance and deficiency symptoms

Manganese:

- Manganese deficiency is reported to occur in soils having pH above 6.8. Manganese deficiency causes chlorosis followed by necrosis.

Zinc:

- In general, zinc deficiency results in shortened internodes and chlorotic areas in older leaves or many appear in younger plants also.
- Tomato leaves become smaller, chlorosis appears and the whole plant is dwarfed with small leaves curling inward.
- Onion leaves turn yellow.
- In Bakers garlic whole plants turn yellow and develop a rosette from with poor growth.

Boron:

- Boron deficiency occurs on vegetable crops with high boron requirements when grown on acidic soils.
- Hollow stems of cruciferous crops like cabbage, cauliflower and broccoli are found to result due to inadequate boron.
- Brown heart of turnip appears first in irregularly shaped water-soaked areas.
- Black heart or canker of beet occurs in internal and external necrotic areas in the root.
- In celery, cracked stem appears as cross splitting of vascular bundles in the fleshy petiole.
- In tomato, the disorder appears as open locules in the fruit and internal browning.
- "Brown heart", "heart rot" or "hollow center" occurs in roots of radish, turnip and beet due to deficiency of boron.

Table: Deficiency symptoms of Molybdenum in certain vegetable crops

Crop	Symptoms
Cauliflower	'Whiptail' develops; leaves are twisted and elongated with laminae showing various degrees of narrowness and irregularity. Laminae are of then corrugated irregularly, cupped, relatively thick, turgid and abnormally dark green or flue green. Flower curds are irregular with 'ricey' and leafy formation in slightly affected plants.
Tomato	Leaves become pale with diffused marginal and interval yellow mottling. Margins curl upward and leaflets appear rolled. A pale brown scorching being at the tip of the apical leaflet of the oldest leaf. In severe cases, the plants die.
Beetroot	Leaves die in crown which may be covered with small, deformed leaves; older leaves wilt and become necrotic, Roots show hear rot and dry rot.

Table: Important insect pest of vegetable and economic loss caused

Crops	Pests	Loss caused
Tomato	Fruit borer	22-38 per cent
Okra	Jassid	54-66 per cent
	Shoot and fruit borer	45 per cent
Brinjal	Shoot and fruit borer	54-66 per cent
Cole crops	Diamond back moth	52 per cent
	Aphid	44-54 per cent
	Cabbage borer	30-58 per cent
Cucurbits	Fruit fly	70-80 per cent
Onion	Thrips	40-50 per cent

Table: Inter crop combination effective in vegetable pest management

Crop combinations	Target pest
Cabbage + Carrot	Diamond back moth
Broccoli + Faba bean	Flea beetle
Okra + Cow pea	Yellow vein mosaic
Cabbage + French bean	Fruit fly
Cabbage + Tomato	Diamond back moth

- **Trap crop:** African marigold in tight bud stage functions as good source to trap the adults of *Helicoverpa armigera*. Maize plants applied with bait-spray trap and kill fruit fly adults when sown in combination with bitter gourd.

PROBLEMS OF VEGETABLE PRODUCTION:

- Diverse problems face the vegetable production enterprise.
- **Problems can be:**
  - o Biological
  - o Agronomic
  - o Climatic
  - o Economic
  - o Sociological
  - o Environmental
- **Biological Problems:**
  - o Perishable nature of vegetables.
  - o Pests and diseases problems.
  - o The fertility and acidity of soil.
- **Agronomic Problems:**
  - o Need for seed treatment
  - o Seed viability maintenance
  - o Lack of technology for large scale production.
  - o Use of herbicides and pesticides.
- **Climatic Problems:**
  - o Inappropriate temperature.
  - o High relative humidity.
  - o Inadequate/Excessive rainfall and distribution.
- **Economic Problems:**
  - o Marketing and distribution.
  - o Unavailability of finance.
  - o Unfavourable Government policies.
- **Sociological Problems:**
  - o Lack of Extension services.
  - o Taste and preference of consumers.
- **Environmental Problems:**
  - o Land Tenure/ unavailability of land.
  - o Lack of high- technology production packages.
  - o Lack of Research and Development.

2. SALONECEOUS VEGETABLES

"TOMATO" (No. 1 Processing vegetable)

- B. N. *Lycopersicon esculentum* (Wolf apple, Vilayti Baingan).
- Family – Solananceae.
- 2n = 24.
- Origin - Peru-Ecuador-Bolvia- (South America) Mexico.
- Type of Fruit – Berry with fleshy placenta with many small seed.
- Edible part – Fruit.
- Orissa (91010 ha) is having maximum area under tomato cultivation followed by Madhya Pradesh (84530 ha).
- In tomato production, Andhra Pradesh (274400 MT) is the leader followed by Gujarat (1357000 MT).
- Tomato is universally treated as "Protective Food".
- Nutritive value – Rich in vitamin 'C', Minerals and Lycopene (which is a powerful antioxidant that acts as an anticarcinogen). One medium ripe tomato (~145 g) can provide up to 40 percent of the recommended daily allowance of Vitamin C and 20 percent of Vitamin A.
- Medicinal Value – Pulp and juice are digestible and it is promoter of gastric secretion and blood purifier.
- Total sugar content is 2.5 per cent in ripe fruit.

  Ascorbic acid – 16-65 Mg/100g.

  Amino acid – 100-300 Mg/100g.
- Processed items – Paste, puree, syrup, sauce, ketchup, chutney and soup.
- Export quality of tomato producing area is Nasik and Pune in Maharashtra and Banglore in Karnataka.
- Processed product of tomato especially "puree" and paste have great demand in export.
- Tomatoes are members of the nightshade family and, because of this, were considered for many years to be poisonous. Indeed, many crops in this family contain highly toxic alkaloids.
- Tomatine (sometimes called tomatin or lycopersicin) is a glycoalkaloid, found in the stems and leaves of tomato plants and in the fruits at much lower concentrations (because it converted enzymatically to a non-toxic form in the fruit).
- It has fungicidal, antimicrobial and insecticidal properties.
- First known record of tomato is mentioned in the year 1544 in Italy.
- It was introduced in Europe in the early 16th century from Peru. Since year 1800 tomato are being used as food all over the world.
- Earlier the botanical name of tomato was *Solanum lycopersicon*; later miller (1978) gave the name *Lycopersicon esculentum* to cultivated tomato and *L. pimpinellifolium* to wild form of tomato.

1. *Lycopersicon esculentum*	2. *Lycopersicon pimpinellifolium*
1. Fruit are larger and solitary. 2. Self pollinated. 3. Hairy plant and strong odour.	1. Fruit are small and bunch form. 2. Self and cross pollinated. 3. Smooth plant and little odour.

- Tomato has tap root system.
- For early crop – Sandy to loam soil is best.
- Tomato is moderately tolerant to acid soil (pH 5.5).
- *L. Esculentum* has five forms –
  1. *Cerasiforme* – Probable ancestor of cultivated tomato, also known as small fruited or cherry tomato.
  2. *Pyriformae* – Pear tomato.
  3. *Commune* – Common tomato.
  4. *Grandifolium* – Potato leaf type.
  5. *Validium* – Upright tomato.
- Muller (1940) divided the genus *lycopersicon* into sub genus
  Eulycopersicon – Red Fruited type (Self Compatible).
  Eriopersicon – Green Fruited type (Self Incompatible).
- Tomato has broadly two types of varieties.

1. Determinate (restricted with appearance of terminal flower)	2. Indeterminate
1. Inflorescence occurs at every internode.	1. Inflorescence occurs at every 3rd internodes.
2. Suitable for North Indian plains.	2. Suitable for Hills and South India.
3. Suitable for mechanized harvesting.	3. Suitable for polyhouses.
4. Training and pruning not required (3-4 feet tall and Bushy type).	4. Training and pruning required (7-15 feet tall and climber type).
5. Plant ends with flower bud *i.e.* self topping.	5. Plant never ends, remains vegetative growth.
6. Require spacing 45 × 45 cm.	6. Require spacing 90-75 × 45-30 cm.

- Red colour of tomato is due to lycopene (20-50 Mg/100g).
- Lycopene is highest at 21-24° C temperature.
- Production of this pigment drops off above 27° C temperature.
- Solarization is done to reduce incidence of damping off disease in nursery.
- Seed treatment with 2, 4-D @ 2-5 ppm increase early fruit set and parthenocarpy.
- Training and Pruning – Required in Indeterminate varieties (Single stem system).
- Staking – It is done for indeterminate types.

- About 33% of total area (Tomato) is covered by F_1 hybrid which is highest among vegetables.
- In Punjab only spring summer crop is taken due to heavy incidence of tomato leaf curl virus (TLCV) in autumn.

Varieties –

S.No.	Variety	Features
1	Pusa Ruby	Indeterminate, Maturity 60-85 days. Yield 300 q/ha.
2	Pusa Early Dwarf	Determinate, Maturity 70-80 days. Yield 300 q/ha.
3	Pusa 120	Semi determinate, highly suitable for long distance transportation and processing, Maturity 110 days, Yield 300-400 q/ha.
4	Pusa Gaurav	Determinate, firm fruits highly suitable for long distance transportation and processing. Maturity 110 days, Yield 300-400 q/ha.
5	Pusa Sheetal	Semi determinate, capable of fruit setting at 8° C night temperature, Maturity 90 days ,Yield 360 q/ha.
6	Pusa Uphar	Indeterminate, resistant to fruit borer, fruits smooth round to flattish, Maturity 120 days. Yield 370 q/ha.
7	Pusa Sadabahar	Determinate, more plant per unit area, suitable for growing under wide range (8-30 °c) of night temperature. Maturity 60 days, Yield 300-400 q/ha.
8	Pusa Rohini	Determinate type, suitable for long distance transportation. Maturity 120 days, Yield 410 q/ha.
9	Roma	Determinate, suitable for transportation and hilly areas. Maturity 100-120 days. Yield 300-350 q/ha.
10	Italian Red Pear	Determinate, fruits suitable for transportation. Maturity 90-110 days. Yield 200-300 q/ha.
11	Punjab Chhuhara	Determinate, elongated fruits, low juice content, suitable for processing and transportation. Maturity 100-110 days. Yield 325 q/ha.
12	La Bonita	Determinate, Suitable for long distance transportation, maturity 100-110 days. Yield 350 q/ha.
13	S-12	Determinate, fruits round, medium size, uniform red colour at maturity, juicy highly acidic. Maturity 90-110 days. Yield 175-280 q/ha.
14	Pant T-3	Semi determinate, dark green foliage, fruits round smooth, medium size, suitable for processing, maturity 70 days. Yield 190 q/ha.
15	Pusa Hybrid -1	Determinate, fruit set up to 28°C night temperature, fruits from June to mid July during the lean period for tomato in North India. Maturity 95 days. Yield 320 q/ha.

16	Pusa Hybrid -2	Determinate tolerant to root knot nematode, Maturity 110-120 days. Yield 591 q/ha.
17	Pusa Hybrid -4	Determinate highly tolerant to root knot nematode. Maturity 120 days. Yield 600 q/ha.
18	Pusa Hybrid -8	Determinate, suitable for long distance transportation. Maturity 105 days. Yield 557 q/ha (main season).
19	Pusa Divya (F_1 hybrid)	**Only hybrid developed so far by utilizing male sterile line (Anther mutant) in India,** Indeterminate and maturity 90-100 days. Yield 350-450 q/ha.
20	Kashi Vishesh	Determinate and fruits are spherical. It is resistant to TLCV, Duration 70-75 days after transplanting and yield 450-550 q/ha.
21	Kashi Amrit	Determinate, Duration 75-80 days and yield 500-550 q/ha. Suitable for Rajasthan, Gujarat and Haryana.
22	Kashi Hemant	For MP, Chhattisgarh, AP. Orissa, Maharashtra. Yield 400-420 q/ha.
23	Kashi Sharad	Indeterminate, longer shelf life. Average Yield 400-500 q/ha. Suitable for J&K, HP and Uttaranchal.
24	Arka Saurabh	Semi determinate, High yielding, (40 tons/ha.) and suitable for transport. Kept for 8-10 days at room temperature (TSS 3.9%) and acidity (0.39%).
25	Arka Vikas	High yielding variety (35-40 tons/ha) developed by IIHR, suitable for fresh market.
26	Arka Abhijit	Semi determinate, medium fruit (65-70g) and resistant to bacterial wilt. Duration 140 days and yield 65 t/ha.
27	Arka Abha	Semi-determinate, resistant to bacterial wilt and suitable for kharif and rabi. Duration 140 days.
28	Arka Alok	Plants determinate, resistant to bacterial wilt, suitable for kharif and rabi, duration 130 days and yield 40 t/ha.
29	Arka Ashish	Plants determinate, excellent fruit colour (lycopene 10 mg/100g) with TSS 4.8%, Suitable for processing and kharif and rabi season. Tolerant to powdery mildew and fruit cracking. Duration 130 days and yield 38 t/ha.
30	Arka Ahuti	Semi determinate, oblong fruits and suitable for processing. Maturity 140 days and yield 40 t/ha.
31	Arka Meghali	Plants semi determinate, Suitable for rainfed cultivation and kharif season. Duration 125 days, yield 18 t/ha.
32	Arka Vishal	Plants indeterminate, fruits large (140g) and tolerant to cracking. Duration 165 days and yield 75 t/ha. Recommended for Karnataka, Bihar and UP.
33	Arka Vardan	Plants indeterminate, fruits large (140g), tolerant to fruit cracking. Resistant to nematodes. Duration 160 days and yield 75 t/ha.

34	Arka Shresta	Plants semi determinate. Resistant to bacterial wilt and suitable for rabi season. Yield 70 t/ha. Duration 140 days. Recommended for Karnataka.
35	Co-1	Dwarf, semi spreading habit with dark green foliage. The fruits are borne in clusters of 6 to 8, each weighting about 125g. Yield 80 q/ha.
36	Sweet 72	Cross between Pusa Red Palm and Sioux. Yield 200 q/ha. Developed at the RARI, Gwalior.
37	Arka Vishesh (H-391)	Semi-determinate with joint-less pedicle, triple disease resistance to TLCV, bacterial wilt and early blight. Recommended for summer, kharif & rabi cultivation. It has a yield potential of 43.3-90 t/ha in 140-150 days.
38	Arka Apeksha (H-385)	It is a hybrid developed by crossing ITHR 2834 and ITHR 2918. It has triple disease resistance to TLCV, BW and early blight.
39	Arka Rakshak	High yielding F_1 hybrid developed by crossing IIHR-2834 × IIHR-2833. It is also triple disease resistance to TLCV, BW and early blight. Fruits square round, large (90-100g). Suitable for fresh market and processing.
40	Arka Samrat	High yielding F_1 hybrid developed by crossing IIHR-2835 × IIHR-2832. It is also triple disease resistance to TLCV, BW and early blight. Fruits oblate, large (90-110g) and deep red. Suitable for fresh market.

Crop Improvement:

(1) Introduction

(a) Roma (Pear shaped suitable for long distant transport).
(b) Sioux.
(c) Best of all.
(d) Tip Top.
(e) Labonita.
(f) Marvel.
(g) Monkey Marker.
(h) Agethe.
(i) Marglobe.
(j) Keckruth.

All these are introductions from USA.

(2) Selection –

(a) Improved Meeruti - Suitable for autumn, winter and spring summer.
(b) HS 110 – Determinate and potato leaf type.

(c) Co-1
(d) Pant Bahar – Resistant to verticillium and fusarium wilt.
(e) Arka Saurabh
(f) Arka Vikash – Seletion from American variety Tip Top.
(g) Sonali

(3) Hybridization –

(a) Pusa Ruby – Sioux × Improved Meeruti, best combiner variety.
(b) Pusa Early Dwarf – Developed at NBPGR, New Delhi (Improved Meeruti × Red cloud), Suitable for rainy season.
(c) Pusa Gaurav – Glamour × Watch.
(d) Marglobe – Marvel × Globe.
(e) Pusa Red Plum – *L. esculentum* × *L. pimpinellifolium.*
(f) Punjab Chhuhara – EC 55005 × Punjab Tropic (Pear Shaped).
(g) Sweet – Pusa Red Plum × Sioux.
(h) Pusa Sheetal – Known as "Cold set" variety.
(i) Hisar Anmol – Hisar Arun × *L. hirsutum.*
(j) Hisar Arun – Pusa Early Dwarf × K1.
(k) Hisar Lalit – HS 101 × Resistance Banglore.
(l) Hisar Lalima – Pusa Early Dwarf × H.S. 101.
(m) Pusa Uphar – Suitable for processing.

(4) Mutation –

(a) S12 – By X-rays.
(b) Co3
(c) PKM 1
(d) Pusa Lal Meeruti – By gamma rays.
(e) Maruthan

- **Bacterial wilt resistant varieties –**
  (a) Shakti.
  (b) Arka Alok.
  (c) Arka Abha.
  (d) Sonali.
  (e) BT – 1.
  (f) BT – 10.
  (g) Arka Shreshta.
  (h) Arka Abhijit.
  (i) VC 48-1.
- Sel 120 – 1st Root knot nematode resistant variety developed in India.
- Hisar Gaurav – Leaf curl resistant variety.

- Hisar Anmol – Leaf curl resistant variety.
- **Indeterminate variety –**
  (a) Pusa Ruby – Suitable variety for processing and fresh market.
  (b) Pusa Divya.
  (c) Arka Saurabh.
  (d) Pant Bahar.
  (e) Best of All – Suitable variety for distant transport.
  (f) Sioux.
  (g) Marglobe.
  (h) DT-10
- Pusa Sheetal – For low temperature (8°C or low) region (Cold set variety).
- Pusa Hybrid 1– For high temperature region (Hot set variety).
- Pusa Sadabahar – For high and low temperature region (8-28°C temperature, fruit sets at night).
- Hisar Arun – Extreme early variety.
- Pusa Rohini – Recently released variety.
- Angur Lata – Tolerant to fruit borer.
- Mangla and Karnataka hybrid varieties are resistant to root knot nematode.
- PWR – 7 – Resistant to both root knot nematode and fusarium wilt.
- **Jawahar Tamatar – All Fruits produced at a time.**
- Pusa Red Plum (Interspecific hybrid).
- Hisar Anmol (Interspecific Hybrid).
- Arka Vikash – Suitable for drought conditions.
- Arka Meghali - Suitable for rainfed conditions.
- Variety developed by use of biotechnology- **Flaursaur** by Antisense technology **(Increased shelf life by Inhibition of ACC synthase).**
- To increase fruit set and yield foliar spray of PCPA (20 ppm) is very effective.
- "B" and "Zn" - Important micronutrient required for tomato cultivation.
- Breaker stage (10% lycopene) – Suitable for long distance transport.
- **Private sector Hybrid –**
  (1) Naveen (Resistant to fusarium and Verticilium wilt).
  (2) Avinash (Tolerant to TLCV).
  (3) Meenakshi.
  (4) Manisha.
  (5) Krishna.
- In private sector first F1 hybrid of tomato Karnataka Hybrid was released for commercial cultivation in 1973 by private seed company ''Indo-American Hybrid Seed Company" – Bangalore (Karnataka).
- Sel-32 – Determinate, Tolerant to leaf curl virus.
- Punjab Tropic – Late maturing with luxuriant growth.

- S-12-Dwarf type variety.
- Punjab Kesari - Dwarf, foliage heavy, protects fruit from sun burn.
- NDT – 108- Fruit chhuhara sized.
- **Semi-determinate varieties –**
  1. Pusa 120 – Resistant to root knot nematode.
  2. Pusa Hybrid – 2- Highly tolerant to root knot nematode.
  3. Arka Saurabh – Suitable variety for fresh market and processing.
  4. Arka Vikash - Suitable for rainfed condition.
  5. Arka Shreshta – Resistant to bacterial wilt.
  6. Arka Abhijit
  7. Arka Abha
  8. Arka Vishal
  9. Pant T-3

Other varieties – Ratna, Rupali, Rashmi gives better yield.

- Varieties suitable for fresh market- Pusa Early Dwarf (P.E.D.), Pusa Ruby, Arka Vikash, Arka Saurabh, Punjab Kesari, Pant Bahar .
- Varieties suitable for processing -Pusa Gaurav, Roma, Punjab Chuhara, Pusa Uphar, Arka Saurabh.
- In 1992 Greshof and Doy produced haploid plant of tomato by anther culture.
- Tomato is preferred for export to European countries.
- Recently three varieties of Tomato ("Swarna Anmol" for protected cultivation; "Swarna Kanchan" for processing; "Swarna Ratan" for table purposes) developed by ICAR-Research Complex for Eastern Region, Patna and two varieties (Punjab Varkha Bahar-1, Punjab Varkha Bahar-2) were developed by PAU, Ludhiana.
- Tomato is susceptible to cold winds and frost. It also affected by high temperature and low/high humidity conditions which cause incidence of foliar disease.
- Temperature for flowering and fruit setting – 25^0 C.
- Temperature for flowering and fruit setting at night – 15-20^0 C (18^0 C is best).
- Temperature for pollination – 21^0 C.
- Pigmentation in fruit (Flesh colour) –
  1. Red (Lycopene)
  2. Yellow (Carotenoids)
  3. Tangerines (Prolycopene).
- Two pair of allele give four colour as Red, Yellow, Dark tangerines and light orange.
- Below 10^0 C fruits are unable to develop red colour.
- Above 40^0 C lycopene is completely destroyed (Optimum temperature for lycopene formation 18-26^0 C).
- Seed rate – 400-500g – Open pollinated varieties.

  125-175g – Hybrid varieties.

- **Sowing Time –**

A. In Northern plain-

1. Autumn – July – August.
2. Spring summer – November.

B. In Southern part -

1. June-July
2. October-November
3. January-February.

C. In Hills – March- April.

- Close spacing is kept– For mechanical harvesting and processing.
- When excessive growth occurs in indeterminate varieties, pinching or pruning of side shoots improves size and uniformity of fruit.
- Methyl ester of long chain acid and alcohol suppress the side shoot formation.
- "B" Borax application @ 20-25Kg/ha or spray of 0.2-0.4% borax at fruit formation stage improves shape, size and colour.
- "Zn" – Zinc sulphate @ 25Kg/ha is applied for better growth and development.
- To enhance ripening ethrel @ 1000-2000ml/lit is sprayed.
- High temperature and humidity and oxygen pressure increases spoilage.
- Harvesting Stage
    1. Immature stage- Seed fully developed.
    2. Mature stage – Fully green fruit.
    3. Breaker/Turning stage – $1/4^{th}$ surface shows pink colour.
    4. Pink stage – $3/4^{th}$ surfaces shows pink colour.
    5. Hard ripe stage – Nearly red colour.
    6. Over ripe stage – Soft and fully coloured.
- Yield –
    1. 250 q/ha – Open pollinated varieties.
    2. 400 q/ha – Hybrid varieties.

Storage: Tomato is a climacteric fruit, having respiratory peak during their ripening process. Being a climacteric and perishable vegetable, tomatoes have a very short life span usually 2 to 3 weeks.

1. For longer-term storage, ripe tomatoes can be stored at temperatures of about 10–15^{0}C and 85–95% relative humidity.
2. The use of 1-methylcyclopropene (1-MCP) has been shown to suppress the action of ethylene in many fruits and vegetables.
3. High CO_2 (5%) and low O_2 (2.5%) increases shelf life and are best atmospheric storage condition.
4. After harvesting tomatoes are cooled rapidly at 13^{0}C which increases shelf life.
5. Zero energy cool chamber at breaker stage extend shelf life by 4-6 days.

Physiological disorders –

1. **BER (Blossom end rot)**

- **Symptoms** – Brown water soaked discoloration at blossom end of fruit at green stage.

  Causes

  (a) Due to "Ca" deficiency and high level of "N" in greenhouse.

  (b) Reduced/sudden change in soil moisture.

Control: Spray of $Cacl_2$ (0.5%) at fruit development and increase of irrigation frequency.

2. **Fruit cracking – Four types**

   (a) Radial – (Very severe) which is splitting of the epidermis from the stem scar towards the blossom end.

   (b) Concentric (very severe) which is splitting of the epidermis in circular patterns around the stem scar/shoulder.

   (c) Cuticular (fruit cuticle cracking) - Poor appearance (roughened skin and corky tissue) and reduces shelf life.

Causes–

(a) Effect of soil moisture, rain, dew when rain follows a long dry spell.

(b) High day temperature (expose of fruit to sun) followed by low day temperature with high relative humidity.

(c) Deficiency of "Boron" (main cause) and genetic factor.

Control: Regular irrigation, avoid pruning and staking during summer, picking before fully ripe and spray of 0.3-0.4 per cent borax.

3. **Blotchy ripening (Green Back)**

   Symptoms – Greenish-yellow to whitish patch on tomato, mostly on stem end portion.

   Causes -

   (a) Mainly by "K" deficiency.

   (b) Imbalance of "N" and "K".

   (c) High level of N, Ca and Na are associated with blotchy ripening.

4. **Puffiness** –

   Symptoms: The outer wall of the fruit is normal, but the tomato is hollow inside. One of the seed cavities is usually empty.

   1. Fruit surface flattened.
   2. Locules unfilled with pulp and seed.
   3. Affected fruit becomes light in weight or less hollow.
   4. Parenchymatous tissue retorted in growth.

   Causes –

   (a) Poor pollination

   (b) Abortion of ovules due to adverse environmental conditions (Low and High temperature)

Control: No effective controls. Puffiness declines later in the summer.

5. **Sun Scald –**

   Symptoms: Initial symptoms are shiny white or yellow areas on the sides of the fruit exposed to the sun. Later, the affected tissue dries out and collapses, forming slightly sunken, wrinkled areas. Secondary organisms invade the affected areas causing the fruit to rot.

   Causes: By Intensive sunlight (40°C or above).

   Control: Grow tomatoes in wire cages. Plants grown in wire cages provide good foliage protection. Also, control foliar diseases of tomatoes which defoliate the plants and expose the fruit to direct sunlight.

6. **Cat face–**

   Symptoms: Distortion of the blossom end rot and development of ridges, furrow, identation and blotches.

   Causes: Unfavorable climatic conditions during flowering. Cloudy and cool weather at blooming time may cause the blossom to stick to the developing fruit, resulting in catfacing. Catfacing occurs most commonly on the large-fruited varieties. Damage from the herbicide 2,4- D may produce similar symptoms.

   Control: No effective controls. Catfacing should decline with the arrival of warmer weather.

7. **Unfruitfulness –**

   Causes: Day Temperature above 32° C. Night Temperature above 20° C and <13° C adversely affecting the fruit setting of tomato.

3. SOURCE OF RESISTANCE IN TOMATO –

1. *L. pimpinellifolium* - Fruits rich in vitamin C and resistant to
   (a) Fusarium wilt.
   (b) Early blight.
   (c) LCV (Leaf curl virus).
   (d) Low temperature.
2. *L. pissisi* – Perennial, fruits green and free from bitterness.
3. *L. peruvianum* – Perennial, resistant to
   (1) TMV (Tomato Mosaic Virus).
   (2) TLCV.
   (3) Septoria blight.
   (4) Cladosporium leaf mould.
   (5) Late blight.
4. *L. hirsutum* – Perennial, fruit green, bitter and resistant to fruit borer.
5. *L. gladulosum* – Resistance to curly top disease and leaf spot.
6. *L. cheesmani* – Perennial, tolerant to salinity and crossable with *L. esculentum*.
7. *L. chilense* - Resistant to curl top, TMV, Verticillium wilt, powdery mildew, leaf curl, Gametophytic self incompatibility is present in *L. chilense*.

8. *L. pennellii* – Drought tolerant, gametophytic self incompatibility is present.
9. *L. parviforum* – Fruit small (1.0-1.4 cm in diameter).
10. *L. chmielewskii* – Plants similar to *L. parviforum*, flowers large and yellow.

Disease	Mechanism of disease resistance
Fusarium wilt	Tomatin content of root tissue increased activity of peraoxidase and polyphenoloxidase.
Root knot nematode	High content of carbohydrate in root and shoot.
Late blight	Tomatin content of leaves.
Verticilium	Antifungal compound rishitin and tomatin content.

Nature of resistance of disease in tomato –

Disease	Nature of resistance
1. Fusarium wilt 2. Root knot nematode 3. TMV 4. Rhizoctonia root rot 5. Spotted wilt virus	Monogenic dominant
1. Bacterial Canker 2. Early blight 3. Rhizoctonia root rot 4. Spotted wilt virus	Oligogenic factor
1. Bacterial Wilt 2. Bacterial canker 3. Leaf curl viruses 4. Late blight	Polygenic factor
1. Early Blight 2. Fruit rot	Additive Factor
1. Bacterial wilt	Polygenically partially dominant
2. Early Blight	Two or more recessive gene
3. Buckeye rot	Br gene
Virus	**Resistant gene**
1. Tomato mosaic	TM-2, Single dominant.
2. Tomato spotted wilt	SW-2, SW-3, SW-4, Recessive gene.
Insect	**Resistant source**
Fruit borer	1. Angurlata 2. HT 64 3. Hybrid No-37- tolerant. 4. Sabour Prabha 5. Punjab Chuhara.

4. "BRINJAL" (EGG PLANT, AUBERGINE)

B.N. – *Solanum melongena.*

Family – Solanaceae.

2n = 24.

Origin- India (Indo – Burma), China (Secondary Origin).

Fruit type – Berry.

Edible part – Fruit.

- Brinjal originated from a wild species *Solanum sysimbrifolium.*
- There are three main botanical varieties on the basis of the shape of the fruit.
  1. *S. melongena* var. *esculentum* Dunal (Nees) – Round, oval or egg-shaped.
  2. *S. melongena* var. *serpentinum* L. – Long slender fruits.
  3. S. *melongena* var. *depressum* L. – Small, miniature fruits, dwarf and early.
- In "Origin of cultivated plants" published in 1886 De Candolle, stated that the species *S. melongena* has been known in India from ancient times and regarded it as a native of Asia. Vavilov (1928) was of the opinion that its centre of origin was in the Indo-Burma region.
- The name brinjal is popular in Indian subcontinents and is derived from Arabic and Sanskrit whereas the name eggplant has been derived from the shape of the fruit of some varieties, which are white and resemble in shape to chicken eggs. It is also called *aubergine* (French word) in Europe.
- Brinjal has long bearing period when grown under mild climate of southern states but its bearing is shortened under hot summer and cold winter season of North India.
- In hills – Only summer season crop is taken.
- India stands IInd in brinjal production after China.
- Brinjal covers 8.14% of total vegetable area and 9 % of total vegetable production.
- West Bengal (163150 ha) is having maximum area under brinjal cultivation followed by Orrisa (117920 ha).
- In brinjal production, West Bengal (3027000 MT) is the leader followed by Orissa (2013000 MT).
- Brinjal fruits are good source of Vitamin-B (particularly) and also good source of Ca, P and Iron.
- Pigmented, dark-purple brinjal has more Vitamin-C than those with white skin.
- Roasting of fruit raises the nicotinic acid at expense of trigonellin.
- The de-cholestrolizing action is due to presence of poly-unsaturated fatty acid (linoleic and lemolenic) which are present in flesh and seed (65.1%).
- Bitter taste of fruit is due to presence of glycoalkaloids (such as solasodine). Generally, the high content of glycoalkaloids (20mg/100 g fresh weight) produce a bitter taste and off flavour.
- The discoloration in eggplant fruit is attributed to high polyphenol oxidase activity.
- Dry fruit is reported to contain goitrogenic principal.

- Anthocyanin pigment present in fruit is responsible for purple colour of fruit skin.
- Heterostyly is common feature in multiflower cymes except in bunchy top cultivar.
- Brinjal is a self-pollinated crop but cross-pollination (48%) also occurs. This is because it has **pronounced hetersostyly** which favours cross-pollination.
- 60-70% fruit setting happens through pollination by insect while 30-40% is by selfing.
- Insects (1) Bumble bees (*Bambus sp.*) and (2) Honey bees (*Apis sp.)* are most suitable for pollination in Brinjal.
- Lack of suitable pollinators favours selfing in crop.
- Another bee sp. *Exomalopsis biliottii* (Wild bee) has been noted in solanaceous crops especially in brinjal for pollination.
- *Mimosa pudica* plant should be planted in vicinity of brinjal to encourage pollination and visit of these bees.
- Brinjal is a day-neutral plant. It can be successfully grown as a rainy season and summer season crop.
- Optimum temperature –
  - (i) 21-30° C for growth and development
  - (ii) 20-25° C with 50% relative humidity is required for seed germination.
  - (iii) Flower abortion is favoured by natural day light reduction and high (30°C) night temperature.
  - (iv) The growth of the crop is severely affected when temperature falls below 17°C.
- The effect of maximum temperature and precipitation on fruit setting is separate from each other and governed by additive factor.
- Flowering or flower formation are affected by nutritional factor.
- *S. melongena* is more closely related to *S. incanum* than that to any other species of solanum.
- KKM – Milky white coloured cultivar of Brinjal.
- Flower opening time is 6 - 7:30 hr. in summer.
- Flower opening time is 11:15 hr. in winter.
- High temperature and high humidity in morning hasten opening and dehiscence of flower.
- Brinjal flower – Divided into four types (pentamerous, hermaphrodite and solitary.)
  1. Long styled with big ovary flower - Maximum fruit setting (70-86.7%).
  2. Medium style with medium sized ovary flower– Fruit setting (12.5 – 55.6%).
  3. Pseudo short styled with rudimentary ovary flower- No fruit setting.
  4. Short styled very rudimentary ovary flower - No fruit setting.
- For flowering at pseudo short and short styled flower spray 2 ppm 2, 4-D when few flower appear and then 60 ppm NAA and 50 ppm PCPA is used.
- All the present day cultivar belongs to three species –
  1. *Esculentum* - Round and egg shaped cultivar.

2. *Serpentinum* - Long, slender type.
3. *Depressum* - All early and dwarf brinjal cultivars.

- Filov (1940) classified both cultivated and wild forms of brinjal on agro-ecological basis.
- Five sub species are:
  1. *Agrestis* – Extremely bitter and inedible fruit.
  2. *Occidentale* – Grows in Asia Minor, adequate humidity and high temperature.
  3. *Orientale* – Origin in Japan.
  4. *Palestinicum*– Hydrophytic type can withstand high temperature.
  5. *Arabico – italicum* – Bitter fruit and only of ornamental value.
- White brinjal is best suited for the patient suffering from diabetes.
- 14% area under brinjal is covered by hybrid varieties.
- Orobanche (root parasite) is serious weed in solanaceous vegetables.
- 2, 4-D is used to control this weed Orobanche (root parasite).
- As per Indian minimum seed certification standards, an isolation distance of 300 meters is required for production of foundation seeds of varieties/hybrids in case of brinjal.
- **Sowing time –**
  1. In North India – Autumn-Winter crop (Main season crop) - May-July.
     Spring summer crop – October-November.
     Summer-Rainy season crop – February-March.
  2. In South India – The crop can be grown round the year in mild climate of Karnataka (July-August and December- January).
  3. In Hills – March- April.
- **Spacing–**
  1. For Bushy type cultivar – 50-60 cm
  2. For spreading type cultivar – 75-90 × 60-70 cm.
- In flower cluster ratio of long style and medium style ovary is 3-4: 1.

Recommended Varieties of Brinjal

S.No.	Variety	Features
1	Pusa Purple Long	Early fruiting with purple colour, 20 to 25 cm long fruits, first picking 75-80 days after sowing. Yield 300 q/ha.
2	Pusa Kranti	Fruits stocky, uniformly long, attractive dark purple. First picking 80-85 days after sowing, Yield 350 q/ha.
3	Pusa Bhairav	Resistant to phomopsis blight, fruits dark purple, glossy, attractive and 12-15 cm long. First picking 80-85 days after sowing. Yield 300 to 350 q/ha.

4	Pusa Purple Cluster	Plants erect tall, compact sturdy with purple pigmentation all over. Fruits borne in clusters of 4-6 fruits, moderately resistant to bacterial wilt. First picking 85-90 days after sowing. Yield 250 q/ha.
5	Pusa Anupam	Plants medium tall, fruits in cluster of 3 to 5, cylindrical, tender, purple colour and resistant to wilt. First picking 95-100 days after sowing. Yield 370 q/ha.
6	Pusa Uttam	Fruits oval round, glossy, dark purple and medium sized, 4-5 fruits in one kg. First picking 85 days after sowing. Yield 410 q/ha.
7	Pusa Upkar	Fruits round, glossy, dark purple and medium sized, 4-5 fruits in one kg. Fruits picking 90 days after sowing. Yield 400 q/ha.
8	Pusa Bindu	Plants pigmented all over, fruits small, oval round, glossy, dark violet purple, about 14 fruits in one kg. First picking 85-90 days after sowing. Yield 450 to 650 q/ha.
9	Pusa Ankur	Plants semi erect, fruits small, oval round, glossy, purple, about 15 fruits per cluster, each fruit weight 100g. First picking 80-85 days after sowing. Yield 450 to 650 q/ha.
10	Pusa hybrid -6	Plants semi-erect, vigorous, fruit round, glossy, attractive purple colour, each weighting 200-250g. First picking 85-90 days after sowing. Yield 400-600 q./ha.
11	Pusa Hybrid -9	Plants strong, upright branches, fruit oval round, glossy, purple, each fruit weights about 250g. First picking 85-90 days after sowing. Yield 500-600 q/ha.
12	Azad Kranti	Local cultivar used for selection, oblong thick and dark purple.
13	Shyamala	Local Selection, round and dark purple.
14	Arka Neelkanth	Tall compact plants bearing small fruits on clusters, purple fruits, resistant to bacterial wilt. 150 days duration, yield 43 t/ha.
15	Arka Keshav	Long fruits in clusters, green leaves with purple glossy fruits, resistant to bacterial wilt. 150 days duration, yield 45 t/ha. Recommended for Karnataka.
16	Kashi Sandesh	Round and purple fruits medium sized, maturity 76 DAT, yield 780 q/ha.
17	Kashi Komal	Long fruited, prolific bearing, fruits light purple, maturity 65-70 DAT, Yield 700-720 q/ha. Suitable for all brinjal growing areas.
18	Kashi Prakash	Open pollinated, plants semi upright, maturity 80-85 DAT and yield 680-700 q/ha.

19	Kashi Tarun	Open pollinated, tall and erect plants, Maturity 75-80 DAT, Yield 700-750 q/ha.
20	Arka Anand	Hybrid variety suitable for summer and rainy seasons. Resistant to bacterial wilt. Fruits are green, medium long, borne in clusters. 140-145 days duration and 60-65 t/ha.
21	Pusa Anmol	It is a new early hybrid variety with attractive dark purple oblong fruits. Fruits are ready to harvest after 70-75 DAT. The average yield is about 300-350 q/ha.
22	Krishnanagar Green Long	The fruits are long (25-30 cm). Green fleshy with scanty seeds. The fruits are ready to harvest 75-80 days after transplanting. The average yield varies from 250-300 q/ha. It is a very popular variety cultivated in South India.
23	H-4	Fruits are long (15-20 cm) and thick, average fruit weight is 125-150 g, ready to harvest in 75-80 DAT. Yield is about 250q/ha. Released from HAU, Hissar.
24	Arka Sheel	The fruits are medium long, thick with deep shining purple colour. The yield varies from 350-400 q/ha. Released from IIHR Bangalore.
25	Punjab Barsati	Fruits are 18-20 cm long and 4-6 cm in diameter. The fruits are dark purple in colour soft and shining. Released from PAU, Ludhiana.
26	Arka Nidhi	Fruits are 20-25 cm long purple in colour. Released from IIHR, Bangalore.
27	Hisar Jamuni	Fruits are light purple in colour, 10-15 cm long with 3-4 cm diameter. Released from HAU, Hisar.
28	NDB-25	Fruits are 15-20 cm long and dark red in colour. Released from ND University, Faizabad.
29	Arka Shirish	The fruits are green, extra long with very few seeds. It is an early variety and is ready to harvest after 60-65 days. The average yield varies from 350-400 q/ha.
30	Arka Kusumakar	The finger shaped fruits are borne in clusters, light green. Commonly cultivated variety in South India. Yield varies from 350-400 q/ha, fruits are harvested in 70-80 DAT.
31	Vijay Hybrid	The fruits are long, thick and deep purple in colour. They are ready to harvest in 90-100 DAT. Yield varies from 270-280 q/ha. Released by VRS, Kalyanpur, Kanpur.
32	Pant Samrat	Fruits are ready to harvest in about 70 DAT. Fruits are borne in clusters. It has resistance to bacterial wilt and phomopsis blight diseases and is also resistant to Jassids, shoot and fruit borer. The average yield varies from 300-50 q/ha. Released by GBPUAT, Pantnagar.

33	Pant Rituraj	Plants are semi-dwarf and semi-spreading. Fruits are round, medium in size with dark purple shining colour. Harvest after 60-80 DAT, yield 300 q/ha.
34	Pusa Purple Round	Fruits are round and purple coloured. Highly resistant to the 'little leaf' virus disease. Ready to harvest 80-90 DAT, yield 250-300 q/ha. Does not fruit in summer.
35	Manjri	Fruits are medium sized round, Resistant to wilt, yield is 300-350 q/ha. Recommended for Maharashtra State.
36	Surti Gola	Fruits are medium sized, round, yield 245-300 q/ha.
37	Krishnanagar Pruple Round	Fruits are large, round, slightly oval, dark purple, flesh with very less seeds, yield is 250-300 q/ha. Suitable for west Bengal.
38	Type-3	Fruits are round purple colour 300g-400g. Can be grown in spring summer also, Yield 250-300 q/ha. Developed at Kanpur.
39	Punjab-8	Fruits are round and medium is size with light purple colour. Yield is 250-300 q/ha. Developed by PAU, Ludhiana.
40	Punjab Bahar	Fruits are round with deep shining purple colour, weight 200g-300g, yield 350-400 q/ha.
41	Arka Navneet	Fruits are round, shining and dark purple in colour. Recommended from IIHR, Bangalore.
42	Hisar Shayamal	Fruits are round, short with dark colour. It is released from HAU, Hisar.
43	Arka Neelanchal Shyama	Fruits are round and green with light purple shade, weighing upto 190-200 g per fruit. It is moderately tolerant to phomopsis blight and yields 340 q/ha.
44	Arka Unnathi	High yielding bacterial wilt resistant variety derived from cross between Arka Kusumakar X IIHR-3. Fruits green long with fleshy green calyx. Yield: 340-360 q/ha in 90-110 days.
45	Arka Harshitha	High yielding bacterial wilt resistant. Fruits green long with fleshy green calyx. Yields 360-380 q/ha in 120 days.
46	Arka Avinash	High yielding bacterial wilt resistant variety. Plants tall and spreading. Fruits green long with fleshy green calyx. Yields 400-420 q/ha in 95-115 days.
47	Arka Shirish	It is a pure line selection from IIHR-194, a local collection from Karnataka. Fruits are green, extra-long with solitary bearing habit. Yield: 390 q/ha in 140-150 days.

Regional choice of varieties:-

1. North India - Dark purple, Long or round with green calyx fruit.

2. South India – Green/Purple long fruit.
3. Maharashtra - Stripped and small round with thorny calyx fruit.
4. West Bengal – Purple/stripped green, thick long or oblong fruit.
5. Eastern Uttar Pradesh – White egg shaped fruit.
6. Bihar and Karnataka – Green, long slender fruit.
7. Odisha – Round green fruit.

Crop Improvement –

Nishio et al (1984) classified 11 *Solanum* species into three groups on the basis of their inter specific compatability:

(1) *S. melongena*, *S. incanum*, and *S. macrocarpon*;

(2) *S. integrifolium*, *S. gilo* and *S. nodiflorum*;

(3) *S. indicum*, *S. mammosum*, *S. toryum*, *S. sisymbriifollium*, and *S. toxicarium*. Crosses were compatible within and between the first and second groups, but were otherwise incompatible.

1. Selection –

(i) Pusa Purple Long – Extra early maturing variety.

(ii) Pusa Purple Cluster – Tolerant to bacterial wilt.

(iii) Pusa Purple Round.

(iv) Arka Shirish – Green type fruits are produced by this variety.

(v) Arka Kusumkar- Green type fruits are produced by this variety.

(vi) Arka Sheel.

(vii) Arka Nidhi.

(viii) Punjab Bahar – Round type fruits are produced by this variety.

(ix) Punjab Chamkila.

(x) Punjab Neelum.

(xi) Pant Samrat – Tolerant to bacterial wilt, less infected by fruit and shoot borer, it is a pure line selection variety.

(xii) Azad Kranti – Less seeded variety.

(xiii) Manjri Gota – Bicolour round type.

(xiv) Bhagyamati

(xv) Annamalai – Resistant to aphid.

(xvi) Jamuni Gole.

(xvii) BR 112.

(xviii) Type 3 – Round Type.

(xix) Co1.

(xx) Co 2.

(xxi) MDV – 1- Round type.

(xxii) PKM – 1.

2. Hybridization –

1. Pusa Kranti – Pusa Purple Long × Hyderpur × Wynad local – Long type.
2. Pusa Bhairav – Resistant to phomopsis blight.
3. Pusa Anupam - Resistant to phomopsis blight (Pusa Purple Long × Pusa Kranti), fruits long.
4. Pusa Uttam – GR × Pant Rituraj.
5. Pusa Upkar- GR × PB 91-1
6. Pusa Bindu – GR × Pant Rituraj.
7. Punjab Neelum – Jamuni Globe × Pant Rituraj.
8. Punjab Barsati – Pusa Purple Cluster × PH-4, Long type.
9. Punjab sadabahar – Tolerant to fruit borer.
10. Hisar Shaymal.
11. Hisar Jamuni – Suitable for both kharif and rabi season.
12. Pant Rituraj – T3 × Pusa Purple Cluster, round type.
13. Vaishali – Arka Kusumkar × Manjri Gota, Bicolour.
14. Pragati – Vaishali × Manjri Gota.

Public sector Hybrid –

(a) Pusa Anmol.
(b) Pusa Hybrid 5 – Long type, early cultivar.
(c) Pusa Hybrid 6 – Round type, early cultivar.
(d) Pusa Hybrid 9 – Round type, early cultivar.
(e) Arka Navneet - IIHR – 22-1 × Supreme – Highest yielding variety.
(f) Vijay.
(g) Azad.
(h) ABH 1, 2 – Oval, Early, purple type.

- Annamalai – Aphid resistant variety.
- Pusa Kranti – Can grow both in spring and autumn season.
- Black Beauty – Introduced variety.
- Private sector Hybrid – Vardan,Nisha, Shiva, Vaishali, Bicolour, Suphal.
- **Bacterial wilt resistant -**
  1. Pusa Anupam
  2. Pusa Purple Long
  3. Arka Keshav
  4. Arka Nidhi
  5. Pant Rituraj
  6. Neelkantha
- **Phomopsis blight resistant -**
  1. Pusa Bhairav (Highly resistant)

2. Pusa Anupam
3. Florida Market

- Phomopsis blight and bacterial wilt resistant variety-Pant samrat.
- NDB – 25 – It has soft joint and is very easy to harvest.
- Pusa Ankur – New variety of brinjal (small fruits) and is early fruiting (45 days).
- Pusa Purple Cluster (IARI, Katrain) is an early maturing variety, developed by selection from the indigenous material
- Pusa Kranti (IARI, New Delhi) has been developed by selection from a three-way cross (Pusa Purple Long × Hyderpur) × Wynad Giant.
- Pusa Hybrid 9 is an early maturing thornless hybrid with upright growth.
- Some of the early bearing varieties are *ABH 1 & 2*, *Arka Navneet*, *Arka Nidhi*, *Azad Kranti*, Mysore Green and *Pant Samrat.*
- The medium-bearing varieties are *Annamalai*, *Arka Keshav*, *Arka Sheela*, *Pusa Anupam* and *Pusa Kranti.*
- Late bearing varieties are *Manjri Gota* and *Punjab Bahar*.
- Little leaf resistant varieties- 1. Arka Sheel 2. Manjarigota
- Application of "Cu" favours increased yield in Brinjal crop.
- Application of "Zn" favours increased weight in Brinjal crop.
- Parachlorophenoxy acetic acid (4-PCPA) and 2, 4 – dichlorophenoxy oxalic acid improves fruit set, induces parthenocarpy, increases fruit set and advances maturity in Brinjal.
- Application of 2, 4 – D (2 ppm) at flowering stage increases total yield of Brinjal.
- Seed rate – 200 g/ha is sufficient.
- Approximately 30,000 – 45,000 seedling/ha is required.
- In North India sowing is done in June-July for autumn winter (Takes 4 week for seeding to become ready) and November for spring summer (Takes 6 week).
- To raise seedlings for one hectare, 500 g seed sown in 150 m^2 area is sufficient. The seedlings are ready for transplanting in four weeks during summer (autumn crop) and 6-8 weeks during winter (spring crop).
- Spraying of B, Mn, Cu and Zn stimulate photosynthesis and increases sugar level.
- Yield –350 –500 q/ha in true breeding varieties and about 500 – 625 q/ha in hybrid varieties.
- Storage – At 10-11°C and 92% relative humidity for 2-3 weeks.
- **Resistance source –**
  (1) *S. xanthocarpum* – Resistant to phomopsis blight.
  (2) S. *indicum* - Resistant to root knot nematode.
  (3) *S. sisymbrifolium*–Resistant to root knot nematode.
- Sandy loam soil – Early crop production can be taken.
- Silt loam or Clay loam soil – Most suitable for heavy crop production.
- Brinjal green fruit have longest shelf life (4 weeks).

- Brinjal have bi-directional root system.
- Fruit and shoot borer *(Leucinodes orbonalis)* is the most destructive pest of brinjal.
- Eggplant is highly sensitive to root knot nematodes (Meloidogyne species). The affected plants show formation of galls on roots.
- Shoot and fruit borer (*Leucinodes orbonalis*) It is the most serious pest of eggplant. Drooping of shoot tips is an indication of the infestation. The growing point is killed. Fruits bear holes and the damaged fruits eventually rot.
- In brinjal Jassid (*Amrasca biguttula*) transmit mycoplasma disease 'little leaf' and viral diseases like 'mosaic'.
- Little leaf (Mycoplasma like organism) caused by Mycoplasma like organism is transmitted by jassid is more severe in ratoon crop.
- Mosaic virus is highly contagious in brinjal and transmitted while performing inter-cultural operations and by aphids.
- Bacterial wilt (*Ralstonia solanacearum*) of brinjal is soil borne in nature. A characteristic symptom of the disease includes drooping of young top leaves and shoots followed by sudden wilting of the plants.
- Oozing of bacterial mass is seen when affected stem is cut and immersed in water.
- Phomopsis blight (*Phomopsis vexans*) is a serious problem of eggplant under north Indian conditions and is more serious in seed crop. The pathogen survives in the infected plant debris in the soil and in seed.
- Weeds in eggplant are controlled by using a pre-plant application of Fluchloralin @ 1.5 litres per hectare or by Pendimethalin @ 1.0 litre per hectare + one hoeing 45 days after transplanting.

5. "CHILLI" (HOTPEPPER)

B.N. – *Capsicum annum.*

1. *Capsicum Annum* L. (Fruits are long or small, but pungency is high)
2. *Capsicum frutescence* L. (Large and inflated variety with very little pungency)

Family – Solanaceae or nightshade family.

2n = 24.

Origin – Mexico.

Fruit type – Berry and edible Part – Fruit.

- Chilli was introduced into India from Brazil in the 16th century by Portuguese (1584).
- India is major producer, consumer and exporter of chilli in the world.
- Karnataka (45910 ha) is having maximum area under chillies cultivation followed by Bihar (42910 ha).
- In chillies production, Karnataka (673000 MT) is the leader followed by Madhya Pradesh (669000 MT).
- Andhra Pradesh, Karnataka, Maharashtra, Orissa, Tamil Nadu, Madhya Pradesh, West Bengal and Rajasthan account for more than 80% of the total area and production.

- The word chilli is a variant of 'chil' derived from the Nahuatl (Aztec) dialect which referred to plants now known as Capsicum.
- It is photo-insensitive and day length affects neither flowering nor fruit set.
- It requires a frost-free period of about 130-150 days.
- The seeds germinate rapidly at 25-30°C.
- Temperature range of 21-30°C is optimum for growth, 20-25°C for fruit setting and 18-24°C for fruit colour development.
- Chilli fails to set fruit when the night temperature is above 30°C or below 15°C
- Capsicum also known as Bell pepper is restricted to cooler region of the country like Banglore, Mysore, Nilgiri, Tamil Nadu, Pune, Ranchi, Darjeeling, Himachal Pradesh and Jammu and Kashmir.
- In Spain, capsicum is referred to as 'pimientos' to describe the red colouration and sweet taste.
- Major Bell Pepper (Capsicum) producing country is China followed by USA.
- Haryana (4560 ha) is having maximum area under capsicum cultivation followed by Karnataka (3820 ha).
- In capsicum production, Karnataka (65000 MT) is the leader followed by Himachal Pradesh (58000 MT).
- Chilli is mainly used in culinary, adding flavour, colour, Vitamin-A, C, E, P, and pungency.
- Chilli is rich in vitamin C (ascorbic acid content 120-130 mg/100g fresh) and appreciable amount of vitamin A (280-290 IU/100 g fresh).
- Curry powder, curry pastes are also prepared from Chilli.
- In food and beverage industries, Chilli is being used in the form of oleoresin which permits better distribution of colour and flavour in food.
- Capsaicin ($C_{18}H_{27}NO_3$) is pungent principle of chilli, which is used in preparation of balms.
- Capsaicin a condensation product of 3-hydroxy 4 methoxy benzylamine and decylenic acid produces a highly irritating vapour on heating.
- Capsanthin is principle pigment of chilli, which is responsible for red colour and it is used in high quality cometic preparations like lipstick.
- A non conventional use of chilli is in self defence spray which is gaining popularity in USA.
- Indian chillies are mainly exported to Sri Lanka, USA, Canada, UK and Saudi Arabia etc.
- Chilli is basically self-pollinated crop, but a high degree of cross pollination (even up to 68%) which makes as facultative cross pollinated crop.
- To maintain the purity of the varieties, isolation distance of 400-500 m is safe for seed production.
- Chilli cultivars are classified commercially on the basis of concentration of capsaicinoids.

- Paprika types have 10-30 ppm capsaicinoids, chilli peppers have 30-600 ppm, and red peppers have capsaicinoids content of 600-13000 ppm.
- **Other Chilli Species –**

  (a) ***C. frutescens:*** Grown where extreme pungency is required such as oleoresin extraction, large type of fruits of *frutescens* include the Tabasco Pepper. In India grown in Kerala, Karnataka, Maharashtra, Vizianagar district of Andhra Pradesh and North Eastern states.

  (b) ***C. chinense:*** Most variable of domesticated species and most widespread within Latin America. In India grown in backyard of Kerala and North Eastern states.

  (c) ***C. baccatum:*** Grown on garden and field scale in South America – Peru, Bolivia and Brazil. It has useful gene mainly for powdery mildew and anthracnose fruit rot resistance. In India this species is not known.

  (d) ***C. pubescens:*** High Land pepper does not set fruit satisfactorily, pungent, fruits, thick pericarp and contains lot of moisture.

- Genus capsicum is divided into three sections by Hunziker (1956).
  1. Monotypic
  2. Tubocapsicum
  3. Pseudoacnistus
- **White Flower –**

  (1) *C. annum .*

  (2) *C. chinensis.*

  (3) *C. frutescens.*

  (4) *C. baccatum. var – baccatum.*

  (5) *C. baccatum var – pendulum.*

- **Purple Flower –**

  C. pubescens.

  C. eximinum.

  C. cardenasii.

Five cultivated species of capsicum can be identified on the basis of corolla, anther, calyx, peduncle and seed characters

S. No.	Species	Corolla	Anther	Calyx	Peduncle	Seed
1.	*C. annum*	Milky White	Blue	Inconspicuous calyx teeth	One or two	Straw coloured
2.	*C. chinensis*	White, milky-greenish	Blue	A constriction on calyx	Two or more	Straw coloured
3.	*C. frutescens*	Pale-greenish or yellowish white	Blue	No calyx teeth	Ttwo or more	Straw coloured

4.	*C. baccatum*	Cream, yellow, brown spot	Yellow brown	Prominent calyx teeth	One or two	Straw coloured
5.	*C. pubescens*	Purple	Purple	Conspicuous calyx teeth	One	Black coloured

Recommended Varieties of Chilli

S.No.	Variety	Features
1	Pusa Jwala	Plants dwarf, light green fruits, 9-10 cm long, thin, turning red at ripening, dry fruits wrinkled, suitable for both the seasons in plains and March-April in the hills. Mildly pungent. Maturity 130-150 days. Yield 75-80 q/ha fresh and 7-8 q/ha (Dry).
2	Pusa Sadabahar	Fruits 6-8 cm long, erect bearing habit in clusters of 6-14 fruits/cluster. Highly pungent, resistance to leaf curl virus. Maturity 150-170 days. Yield 110-125 q/ha. Fresh and 15-20 q/ha (Dry).
3	Bhagyalakshami (G-4)	Fairly tolerant to diseases and insects. Suitable for sowing in both the seasons. Maturity 150-170 days. Yield 110-130 q/ha. fresh and 10-13 q/ha (dry).
4	Sindhur (CA-960)	Fruits long, mildly pungent, suitable for both the seasons. Maturity 130-180 days. Yield 120-140q/ha. fresh and 9-11 q/ha. (dry).
5	Pant C-1	Plant short statures, more primary branches, fruit erect bearing small in size, highly pungent, 5.5 cm long. Light green fruits turn red at maturity. Tolerant to leaf curl virus. Maturity 90-130 days. Yield 100-120 q/ha. fresh and 9-11 q/ha. dry fruits.
6	Punjab Lal	Plant dwarf, bushy, fruits erect, dark green turning red at ripening. Rich in capsaicin content suitable for both the seasons. Maturity 120-180 days. Yield 100-120 q/ha. fresh and 9-10 q/ha. (dry).
7	Kashi Anmol	Plants determinate, fruits are pungent (0.5% capsicin), suitable for long distance transport, harvesting 50-55 DAT. Yield 250 q/ha. (Fresh) recommended for UP, Karnataka, MP, Uttranchal, Jharkhand, Chatthisgarh.
8	Kashi Vishwanath	Red ripe fruits in 220 days and yield 200 q/ha. It is a CMS based hybrid sutiable for dry fruit production in UP, Bihar, MP, Jharkhand, Uttranchal, Karnatakam AP, Delhi and Haryana.

9	Kashi Early	For both green and dry fruits, yield 300 q/ha in 240 days. Suitable for AP, MP Jharkhand, Uttranchal, Karnataka, Delhi, Punjab and Haryana etc.
10	Arka Lohit	Plant tall and highly pungent suitable for irrigated and rain fed cultivation. Duration 180 days, yield 25 t/ha. (Fresh) and 37 t/ha (Dry), Suitable for Orissa, Maharashtra, Karnataka.
11	Arka Suphal (PMR 57 / 88 K)	Fruits green, medium long, pungent, suitable for irrigated and rain fed cultivations. Tolerant to powdery mildew, Duration 180 days. Yield 25 t/ha green chilli and 3 t/ha dry chilli. Suitable for Karnataka, Kerala, AP.
12	Arka Meghna (F_1 hybrid)	Plants tall and spreading, fruit length 10 cm. very early type (24 days for 50% flowering after transplanting). Fruits green turn deep red on maturity. Duration 120-140 days. Tolerant to powdery mildew and viruses. Green yield 33.5 t/ha. and dry 5 t/ha.
13	Arka Harita (F_1 hybrid)	Plants tall, spreading, fruits dark green (10 cm long.) Duration 150-160 days, resistant to powdery mildew and viruses. Green yield 33.5 t/ha. and dry 5 t/ha.
14	Arka Abhir	Plants tall, fruits have low pungency (0.05%), suitable for Oleoresin extraction. Duration 160-180 days, yield 2 t/ha (Dry).
15	Phule Jyoti	It is released by MPKV, Rahuri. It is tolerant to thrips, mites and wilt. Average yield of green fruits 250 q/ha and red dry fruits is 24 q/ha.
16	Phule Mukta	Ripe fruits are dark red in colour. Duration is 210 days. Resistant to powdery mildew and Fusarium wilt. Average yield of dry chilli is 23 q/ha.
17	Parbhani Tejas	It is released by MAU, Prabhani (MH) for rain fed as well as irrigated conditions. Fruits are longer (maximum 22 cm). Yield of dry red fruit is 18 q/ha.
18	Konkan Kirti	It is released by BSKKV, Dapoli (MH). Average yields 12-14 t/ha of green chilli. Fruits are dark green, good keeping quality and good export potential.
19	Arka Neelanchal Prabha	Fruits are medium long, first harvesting starts at 65-70 days of planting. It is moderately tolerant to anthracnose and yields 10-12t/ha.
20	Arka Khyati	High yielding hybrid with 35-38 t/ha fresh yield and 5-5.5t/ha dry yield.
21	Arka Sweta	High yielding hybrid with 28-30 t/ha fresh yield and 4.5 t/ha dry yield.

[A] Crop improvement in Chilli -

- Chilli is often cross – pollinated crop. Cross pollination is reported upto 63%.
- Flower opens at 5 AM.
- For making cross, a bud one day prior to anthesis is ideal for emasculation.

1. Selection –

(i) NP 46-A-Pure line selection tolerant to virus.
(ii) G1, G2, G3, G4- Pure line selection.
(iii) NP – 41.
(iv) K1 – Tolerant to thrips.
(v) CO1.
(vi) Patna Red.
(vii) NP 34.
(viii) Sindhur- Mild pungent variety.
(ix) Bhagyalaxmi (G4)-Fairly tolerant to disease and pest.
(x) Arka Lohit.
(xi) Sabour Angar.
(xii) Kalyanpur Yellow.

2. Hybridization –

(i) Andhra Jyoti - Cherry type fruit.
(ii) Pusa Jawala - NP 46 A × Puri Red – Tolerant to thrips, mites and aphid.
(iii) Pusa Sadabahar – (Pusa Jwala × IC 31339), Virus resistant.
(iv) Pant – C1 (NP46-A × Kandhari), Tolerant to viruses.
(v) Pant – C2 (NP46A × Kandhari), Tolerant to virus.
(vi) Punjab Lal – (Perennial × Long Red).
(vii) Bhaskar – (G4 × Yellow Anther mutant).
(ix) X 235 – (G4 × Anther mutant).

3. Mutation –

Alberto Pirovano (1922) - First man to use X-rays for inducing mutation in capsicum.

I. MDU – 1 Through gamma-rays.
II. Ankur – 1 Gamma rays.

4. Private Sector hybrid –

(i) Tejashwini
(ii) Champion
(iii) Gayatri
(iv) Gni
(v) Delhi hot
(vi) Skyline

- CH – 1 – A public sector hybrid, related by Punjab Agriculture University, Ludhiana by using Genetic Male Sterile Line (GMS).
- In private sector first – F1 hybrid of tomato "Karnataka Hybrid" and of sweet pepper "Bharat" was released for commercial cultivation in 1973 by a private seed company "INDO-AMRICAN" Hybrid seed company, Banglore (Karnataka).

Recommended Varieties of Green Pepper (Capsicum)

S.No.	Variety	Features
1	California Wonder	Upright Plants, fruits smooth, blocky, 3-4 lobed, dark green turning red on ripening, 6.5-8.5 cm long and 6.5-7.5 cm in diameter. Suitable for sowing in plains from Sept. to Oct. and March to April in hills. Maturity 90-120 days. Yield 150-180 q/ha.
2	Yolo Wonder	Dwarf bushy plant, fruits 6.5-8.5 cm long and 4.5 to 6.5 cm in diameter, thick flesh, and deep green turning red on ripening. Maturity 110-150 days. Yield 130-180 q/ha.
3	Arka Mohini	Medium to large, drooping fruits, dark green turning red on ripening, suitable for both the season in plains and March-April in hills, Maturity 125-140 days. Yield 180-250 q/ha.
4	Arka Gaurav	Fruits dark green, thick fleshed, medium size, fruits turn to orange yellow on ripening. Foliage covers the fruits to avoid sunscald. Good for both seasons. Maturity 125-150 days. Yield 180-250 q/ha.
5	Arka Basant	Fruits waxy (creamish white), conical, erect bearing, thick fleshed prolific continuous bearing suitable for both the seasons. Maturity 125-140 days. Yield 150-200 q/ha.
6	Bharat	Plant tall, prolific bearing, fruits dark green, blocky, 3-4 lobed, thick fleshed, large size, suitable for both the seasons. Maturity 125-140 days. Yield 150-200 q/ha.
7	Pusa Deepti	Plants erect, fruit erect bearing, 9-11 cm long and 3-5 cm in diameter, very early fruiting and suitable for both the seasons. Maturity 90-105 days. Yield 300-350 q/ha.
8	Arka Athulya	High yielding F_1 hybrid with powdery mildew tolerance. Suitable for fresh green market and yields 45-50t/ ha in 140-150 days. Fruits are firm, blocky with 3-4 lobes and medium large (100-120g). Suitable for *kharif and rabi* season cultivation under open field conditions.

Crop Improvement in Capsicum:

1. Introduction –

(i) California Wonder.

(ii) Yolo Wonder.

(iii) Golden Wonder.
(iv) Bullnose.
(v) Sweet Banana.
(vi) World Beater.
(vii) Chinese Giant.

2. Selection –

(i) Arka Mohini.
(ii) Arka Gaurav.
(iii) Arka Basant.

3. Public Sector Hybrid –

(i) Pusa Deepti-IARI Katrain, F_1 Hybrid, Light green in colour.
(ii) Green Gold.

4. Private sector hybrid –

(i) Bharat.
(ii) Indira.
(iii) Early Bounty.
(iv) Hira.
(v) Lario.

- Fruit rot resistant variety – K2.
- Leaf Curl resistant variety –
  (i) Pusa Jawala. (ii) Pusa Sadabahar. (iii) Pant C1.
- Leaf curl, TMV, wilt and dieback resistant variety – Punjab Lal.
- Multiple disease resistant variety – Punjab Surakh.
- Bacterial wilt resistant variety – Utkal Rashmi, Arka Gaurav.
- Jawalamukhi – Suitable for high density plantation, low pungency and resistant to viruses.
- Arka Lohit – Tolerant to powdery mildew.
- Bhaskar – Yellow anther type, resistant to thrips and mites.
- Pusa Sadabahar- Variety of *C. fruitescence* which is perennial in nature.
- Arka Abir – Paprica (*C. fruitscence*), suitable for colour extraction.
- Punjab Lal – Used for colour extraction.
- Himachal Pradesh is leading state in supplying capsicum in off season.
- Green to dry chilli ratio - 10:1.
- Drying is done for 3-10 days under sun, lowering moisture to 8 – 10%.
- 15 Irrigation are required in capsicum.
- India share in total export of chilli in world - 4%.

- Fruit drop – NAA and triconitol is used to control fruit drop.
- 7g. good quality seed provides 100 plants.
- Most extensive cultivation of chilli can be seen on vertisols covering the state Andhra Pradesh, Karnataka, Maharashtra and Tamil Nadu.
- **Sowing time –**

  Chilli can be grown in West Bengal, Maharashtra and South India almost throughout the year. However, chilli is grown in three distinct seasons *viz.*

  1. May-July (For autumn-winter crop)
  2. September-October (For winter-spring crop)
  3. January-February (For summer-rainy crop)

  In the Hills chillii is sown in February-April.
- **Seed rate –**

  Chilli – 1.5 kg/ha.

  Capsicum – 250-300g/ha.
- 50-60 thousand seedlings are sufficient to cover 1 ha.
- Damping off (*Pythium aphanidermatum*) is normally soil borne in nature and spreads by planting infected seedlings and irrigation water.
- The disease normally occurs in patches in nursery beds and is responsible for causing pre-and post-emergence death of seedlings
- **Yield –**

  7.5-10 tons/ha. Green chilli.

  12-20 tons/ha. Bell pepper.
- Long dry period – Flower shedding and fruit fall occurs.
- Dry chilli contains about 6% stalks, 40% pericarp and 54% seed.
- Excess of 'N' with water stress leads to BER of capsicum (Blossom end Rot).
- Variety having thin pericarp, less seed and strong spike is suitable for drying.
- Punjab Guchedar – Highly resistant to mosaic and tolerant to LCV.
- Punjab Sadabahar – Resistant to Chilli mosaic virus, Tomato mosaic virus, Leaf curl virus.
- Ujjwala – Resistant to bacterial wilt.
- Recently two varieties of Chilli ("Swarna Praphulya"; "Swarna Tejashwi" for pickle purpose) and one variety each of Capsicum ("Swarna Atulya") developed by ICAR-Research Complex for Eastern Region, Patna (Bihar).
- Green chillies can be stored in the cold storage up to 40 days at 0^0C and 95-98 % relative humidity.
- Green chillies treated with 200 ppm GA_3 significantly increase the shelf life.
- The moisture content of the dry pods is to be kept at 8-10 %.
- Aphids (*Aphis gossypii* and *Myzus persicae*) damage the crop by sucking cell sap and by spreading the mosaic virus.
- Thrip (*Scirtothrips dorsalis*) - Both nymphs and adults feed on young leaves and tender shoots.

- It also damages the crop by spreading leaf curl virus and is more serious during dry weather.
- Bacterial wilt (*Ralstonia solanacearum*) is soil borne and is a serious problem under protected cultivation as the prevailing temperature is higher than the outside and proper crop rotation is not followed due to economic reasons.
- Bacterial leaf spot (*Xanthomonas vesicatoria*) is primarily seed borne and is the most serious bacterial disease of chilli in India.

6. "OKRA" (LADY'S FINGER)

B.N. – *Abelmoschus esculentus.*

Family – Malvaceae.

Chromosome no. : 2n = 126, 130 132, 144.

Origin – West Africa/ India.

Type of fruit – Capsule.

Edible part – Fruit.

- India is largest producer (72 %) of Okra in the world.
- West Bengal (77550 ha) is having maximum area under okra cultivation followed by Gujarat (75270 ha).
- In okra production, Gujarat (921000 MT) is the leader followed by West Bengal (914860 MT).
- The root and stem are used for clearing cane juice in preparation of 'Gur" and "Jaggary".
- High Iodine content of fruit helps to control goitre while leaves are used in curing inflammation and dysentery.
- The fruit also helps in cure of renal colic, leucorrhoea and general weakness.
- It is a good source of minerals particularly calcium (66 mg/100 g fresh) and phosphorus (56 mg/100 g fresh).
- Dry seed – Contains 13-22% good edible oil and 20-24% protein.
- Okra accounts for 70% of the (30%) exchange earning other than onion from export of vegetables.
- For fruit export Okra fruit should be tender, 6-9 cm long, 4-5 ridged with pointed tips.

  (1) Pusa Sawani
  (2) Prabhani Kranti — All of these are suitable for export.
  (3) Versha Uphara
  (4) Pusa A – 4
- Before the release of Prabhani Kranti, Selection-2 released earlier, was the major exported variety.
- Major okra exporting area – Nasik, Ozar, Saikheda, Dindori, Kolhar, Sholapur (Maharashtra).

- The fruits can be stored at 7-9 ^{0}C temperature with 80-90% relative humidity for 7-10 days.
- Dutta and Naug (1968) suggested that the chromosome number for *A esculentum* was 2n = 72, 108, 120, 132, 144 are an indication of a series of polyploids of × = 12.
- The polyploidy nature of *A.esculentus* was also suggested by Charrier (1984).
- Okra variety Pusa Sawani resistant to YVMV was the first example of successful breeding for disease and pest resistance at IARI, New Delhi.
- Pusa Swami has revolutionized the production of okra in all okra growing areas but now it is susceptible to YVMV.
- Major breeding objective in Okra is to develop variety resistant to YVMV.

Recommended Varieties of Okra

S.No.	Variety	Features
1	Pusa Sawani	Early, fruits 12-15 cm long, 5 edged, suitable for spring summer cultivation. Less sensitive to temperature fluctuations but sensitive to virus. Maturity 50 days (first harvest), yield 120 q/ha.
2	Pusa A-4	Plants with short inter-nodes. Fruits dark green, 12-15 cm long. Excellence shelf life, resistant to yellow vein mosaic virus, tolerant to aphids, jassid and borer. Responds to pruning to extend the crop period for additional harvests. Maturity 45 days (first harvest), yield 140 q/ha.
3	Punjab Padmini	Plants tall (180-200 cm), fruit 15-20 cm long, 5 ridges, remain tender for a long period (3-4 days). Resistant to YVMV and tolerant to jassid and cotton boll warm. Suitable for kharif and summer of North and winter season of Southern states. Maturity 50-55 days. Yield 110-125 q/ha.
4	Parbhani Kranti	Plants single stemmed, resistant to YVMV. Maturity 50 days. Yield 85-90 q/ha. during summer and 110-115 q/ha. during kharif season.
5	Punjab – 7	Plant medium tall (105 cm) and fruits medium long, green, tender, 5 ridged, slightly furrowed. Resistant to YVMV. Maturity about 54 days. Yield 80-100 q/ha.
6	Pusa Makhamali	Excellent fruit quality, susceptible to virus, suitable for cultivation in hills and virus free areas. Plants hairy, erect, 180-200 cm tall. Maturity in 50-55 days. Yield 80-100 q/ha.
7	Co-1	Plants 90-120 cm tall, 5 ridged, scarlet red (colour not persistent on cooking) and maturity in 50 days. Yield 85-90 q/ha. Year round cultivation in Tamil Nadu.
8	Arka Anamika	Released by Central Variety Release committee. Excellent yield in southern but lower in north India, Resistant to YVMV. Maturity in 55 days. Yield 120 q/ha.

9	Punjab 8	Plants 150 cm tall, resistant to YVMV and tolerant of fruit borer. Maturity 50 days. Yield 105 q/ha. Variety developed by mutation at PAU, Ludhiana.
10	Hisar Unnat	Fruits green, attractive, 5 ridged, 15-16 cm long. Resistant to YVMV. Maturity in 55 days and yield about 120 q/ha. Variety developed at HAU, Hissar.
11	Varsha Uphar	Fruit attractive dark green fruit, 18-20 cm long, long tapering up. Resistant to YVMV, field tolerant to borers. Maturity 45-50 days. Yield 100 q/ha. Variety developed by HAU, Hissar following pedigree selection.
12	Azad Kranti	Plants are fast growing occasionally branched spares pigmentation. Fruits 5 ridged, smooth, shiny, green, long beaks. Highly tolerant to YVMV. Maturity 50-55 days. Yield 125 q/ha. Variety developed by CSAUAT, Kanpur.
13	Arka Abhay	It is a sister line of Arka Anamika, resistant to YVMV, tolerance to fruit borer and may suit prunning for ratoon crop. Fruits free from spines having 5-6 ridges. Maturity in 120-130 days and yield 18 t/ha.
14	Kashi Vibhuti	Resistant to YVMV, Okra Leaf Curl Virus, yield 170-180 q/ha. A plant bears 18-22 fruits having 8-10 cm length.
15	Kashi Pragati	Fruits are 8-10 long. Yield 180-190 q/ha. during rainy and 130-140 q/ha during summer season. Recommended to Chattisgarh, Orissa and A.P.
16	Kashi Satdhari	Resistant to YVMV, Okra Leaf Curl and yield 110-140 q/ha. Recommended to Punjab, Bihar and Jharkhand,
17	Shitta Jyothi	Suitable for warm, humid climate. Yield 180-200 q/ha. Resistant to YVMV and OLCV. Recommended for Gujarat, Haryana, Delhi, Chattishgarh, Orissa and A.P.
18	Kashi Bhairav	Fruit length 10-12 cm. Yield 200-220 q/ha. Suitable for all the states.
19	Kashi Mahima	Fruit length 12-14 cm. Yield 200-220 q/ha. Suitable for all the states.
20	Kashi Mohini	Plants are tall, height 110-140 cm, flowers at 4-5 nodes during summer and 5-7 nodes during rainy season. Yield of 130-150 q/ha. It is tolerate to high temperature during summer season and resistant to YVMV under open field conditions.
21	Hybrid- Kashi Bhairav	Plants are medium tall with 2-3 branches; fruits are dark green with 10-12 cm length; yield 200-220 q/ha. This is resistant to YVMV and OLCV under field conditions.
22	Pusa Bhindi- 5 (DOV-66)	A high yielding variety (18.0 t/ha) of okra with superior fruit quality and Yellow Vein Mosaic Virus resistance for Kharif season for northern plains.
23	Arka Nikitha	It is developed through GMS line. Rich in iodine content (33.31μ g/kg).Yields 21-24 t/ha in 125 -130 days duration.

Crop Improvement

1. Introduction –

(a) Velvet Green.
(b) Ghana Red.
(c) Perkins Long Green.
(d) Clemon's Spineless.

2. Selection –

(a) Pusa Makhmali -Pure line selections susceptible to YVMV.
(b) Gujarat Bhindi No. 1 Pure line selection.

3. Hybridization –

(i) **Pusa Sawani:** (Pusa Makhmali × IC 1542) - Developed by Dr. Harbhajan Singh through selection from a cross. Earlier (1962), it was resistant to YVMV but now resistance had been broken down. It is a day neutral variety which is spineless.

(ii) **Punjab Padmini:** (*A. esculentus* × *A. manihot sp. manihot*).

(iii) **Prabhani Kranti:** Interspecific cross (1955), (*A. exculentus* cv. Pusa Sawani × *A. manihot*).

(iv) **Arka Anamika:** (*A esculentus* × *A tetraphyllus*).

(v) **Arka Abhay:** An interspecific hybrid between Abelmoschus esculentus(IIHR 20-31) × A.manihot spp. Tetraphyllus (Res. To YVMV) followed by backcross.

(vi) **Varsha Uphar:** (Lam sel. 1 × Parbhani Kranti).

(vii) **Punjab 7:** *A. esculentus* cv. (Pusa Sawani) × *A. manihot* sp. *manihot*, resistant to YVMV.

(vii) **Hissar Unnat (HRR55):** (Selection 2-2 × Prabhani Kranti), Released by Central Variety Release committee and notified in 1996, resistant to YVMV.

(viii) **Selection 2-5:** (Pusa Sawani × Best one) × (Pusa Sawani × IC 7194), It was once very popular in Maharashtra for fresh fruit export.

4. Mutation –

(i) MDU-1 – Evolved at TNAU, Coimbatore in 1978 as an induced mutation from Pusa Sawani.

(ii) Punjab-8 (EMS8) Induced mutant from Pusa Sawani by treating seed with EMS 1%, resistant to YVMV and tolerant to fruit borer.

5. Other Varities –

- Perkins Long Green – Suitable for slightly cooler area.
- Hissar Barsati – Resistant to YVMV.
- Pusa A-4 – Released in 1994 by IARI as a substitute for Pusa Sawani, resistant to YVMV and tolerant to jassid and shoot and fruit borer, it has excellent shelf life.

- Co. 1 - Developed in 1976 as single plant selection from a heterogeneous population of Red Wonder, tolerant to YVMV, resistant to fruit borer and powdery mildew.
- Harbhajan Bhindi – The variety Perkins long Green was released in 1983-84 as Harbhajan Bhindi from College of Agriculture, Solan to commemorate the memory of Dr. Harbhajan Singh, resistant to YVMV.
- Azad Kranti – Tolerance to YVMV.
- Punjab no. 13 – (*A. esculentus* (Pusa Sawani) × *A. Manihot* var. Manihot) Susceptible to YVMV.
- Higher temperature – 40-42° C flower may desiccate and drop, causing yield loss.
- Pusa Sawami has tolerance to salt and also to large variation in pH range.
- **Seed rate –**

  18-28 kg/ha Spring (summer).

  8-10 kg/ha Rainy (Kharif).
- **Sowing Time –**
  1. January - February : Spring- summer crop in estern region.
  2. February - March : Spring- summer crop in north and western region.
  3. May – July : Rainy season in most area.
  4. November : Winter crop in southern region.
- Export market requires tender, dark green, straight, short 6-8 cm fruits.
- Vector of YVMV is white fly (*Bemicia tabaci*).
- Jassid (*Erias* sp.) – Serious pest of Okra.
- Arka Abhay and Pusa A-4 give quick branching after pruning.
- Private sector hybrid – Resistant to YVMV.
  1. Vijay – Indo American Hybrid Seed Company.
  2. Adhunik and Panchali – Century Seed.
  3. Hybrid No. 6 7 – Mahyco.
  4. Nath Sobha – Nath Seed Company.
  5. Sungro 35 – Sungro seed company.
- Okra gives 300-350% crop land use efficiency as an intercrop in cassava and cucurbits.
- Dr. Harbhajan Singh started systematic work on okra.
- *A. manihot* is also used as leaf vegetable and it considered being a polyspecies complex.
- Okra flowers are self – fertile but usually up to 10% cross-pollination occurs by insect, thus it is classified as an often cross – pollinated crop.
- A flower bud takes about 22-25 days from initiation to full bloom.
- Anthesis starts from 7.00 am, continues up to 10.30am.
- Pollen remains viable at 25°C temperature and 50% relative humidity for 30 days.
- 50000 plants/ha are recommended for branching type (Spacing 60 × 30 cm).
- 66000 plants/ha are recommended for non-branching type (Spacing 45 × 30 cm).

- **Average yield**

  100-120 q/ha during spring-summer season.

  150-180 q/ha during kharif/rainy season.
- **Resistant source –**

  (1) *A. manihot* sp- *manihot*

  (2) *A. tuberculatus* – All species are resistant to YVMV.

  (3) *A. pungens* –
- Okra Amphidiploids 2x = 192

  A. esculentus × A. manihot

  (2x =124) (2x =68)

  Resistant to YVMV
- **Diversity of wild species of *Abelmoschus* in India -**

S. No.	Species	Area
1.	*A. ficulneus*	Deccan plateau, Punjab and West Bengal
2.	*A. manihot*	West Bengal, Southern Karnataka, Kerala
3.	*A. moschatus*	Eastern and western peninsular tracts
4.	*A. angulosus*	Western ghats and Nilgiri hills of South India
5.	*A. pungens*	North-eastern tracts
6.	*A. tuberculentus*	North and north-western plains
7.	*A. tetraphyllus*	North-western terai region

- Species which are found both in wild and cultivated forms are *A. esculeutus*, *A. manihot* and *A. moschatus* and these species are distributed in Africa, Southeast Asia, India and other okra growing areas.
- The wild species *A. angulosus*, *A ficulneus*, *A crinitus* and *A. tuberculentus* are confined in Asia and Australia.
- The wild species particularly *A. manihot* and *A. tetraphyllus* have been extensively used in breeding okra resistant to yellow vein mosaic virus.

7. IMPORTANT PEST AND DISEASE OF SOLANACEOUS AND OKRA CROPS

Sr. No.	Pest and Disease	Symptoms	Management
1.	Fruit Borer *(Helicoverpa armigera)*	• Young larvae feed on tender foliage • Thrust only a part of its body into fruit and eat the inner content.	• Grow simultaneously American marigold and tomato seedling at 1:16 rows • Setup pheromone trap with Helilure at 10-12/ha • Spray 1 ml/lit of Quinolphos/ Prophenophos.

2.	Serpentine leaf miner *(Liriomyza trifolii)*	• Leaves with serpentine mines • Drying and dropping of leaves	• Collect and destroy mined leaves • Spray NSKE 5%
3.	Leaf eating caterpillar *(Spodoptera litura)*	• Young larva scrap leaves on ventral surface Grownups defoliate crops	• Install Pheromone traps (Pherodin SL) @15/ ha to attract male moths and spray chlorpyriphos 20 EC 2 lit/ha or dichlorovos 76 WSC 1 lit/ha
4.	Whitefly (*Bemisia tabaci)*	• Downward curling and drying of leaves • Vector of tomato leaf curl disease	• Use yellow sticky traps at 12/ ha to attract and kill insects. • Spray Dimethoate 30 % EC @ 1.0 ml/lit or Thiamethoxam 25 % WG @ 4.0 ml/10 lit.
5.	Mealybug (*Ferrisia virgata)*	• Presence of white, cottony mealy bugs on the leaves and twigs	• Spray Phosphomidon 40 SL 2ml/lit or Chlorpriphos 20 EC 2ml/lit
6.	Tobacco cutworm (*Spodoptera litura)*	• Come out from soil at night and cut the small plant at filed level	• Grow castor as a trap crop • Spray Emamectin benzoate 5 % SG @ 4 g/10 lit or Flubendiamide 20 WDG @ 6.0 g /10 lit.
8.	Nematode *(Meloidogyne spp)*	• Root galls formation • Stunted growth due to slow debility in roots function for nutrient as well as water uptake.	• Soil solarisation of the nursery beds with 25 micron polythene sheet for 15 days during May/ June along with carbofuran (soil application) 1 kg a.i./ha at 45 DAT. Planting of marigold flower near to irrigation channel.
9.	Shoot and fruit borer *(Leucinodes orbonalis)*	• Withering of terminal shoots/dead hearts • Bore holes on shoots and fruits plugged with excreta	• Avoid continuous cropping of brinjal crop • Spray Azadirachtin 1.0% EC (10000 ppm) @ 3.0 ml/ lit or Chlorpyrifos 20 % EC @ 1.0 ml/lit or Emamectin benzoate 5 % SG @ 4 g/10 lit or Quinalphos 25 % EC @ 1.5 ml/lit.
10.	Aphid *(Aphis gossypii)*	• Development of black sooty mould due to the excretion of honey dew	• Spray Imidacloprid 200 SL at 100 ml/ha or Phosphamidon 40 SL 600 ml/ha.

11.	Damping off (*Pythium aphanidermatum*)	• In the pre-emergence the phase the seedlings are killed just before they reach the soil surface and the post-emergence phase is characterized by the infection of the collar at the ground level.	• Seed treatment with fungal culture *Trichoderma viride* (4 g/kg of seed) or Thiram (3 g/kg of seed) is the only preventive measure to control the pre-emergence damping off and drench with Copper oxychloride 0.2% or Bordeaux mixture 1%.
12.	Early Blight (*Alternaria solani*)	• Early blight is first observed as small, black lesions mostly on the older foliage. • Spots enlarge; concentric rings in a bull's eye pattern can be seen in the centre of the diseased area.	• Practicing crop rotation helps to minimize the disease incidence. • Spray the crop with Mancozeb 0.2 % for effective disease control.
13.	Fusarium Wilt (*Fusarium oxysporum*)	• In young plants, symptom consists of clearing of veinlet and dropping of petioles. • At later stage, browning of vascular system occurs.	• The affected plants should be removed and destroyed. • Spot drench with Carbendazim (0.1%) • Crop rotation with a non-host crop such as cereals.
14.	Bacterial Leaf Spot (*Xanthomonas campestris*)	• Moist weather and splattering rains are conducive to disease development. • Infected leaves show small, brown, water soaked, circular spots surrounded with yellowish halo	• Spraying with a combination of copper and organic fungicides in a regular preventative spray program at 5 to 10 day intervals or Spraying with Agrimycin-100 (100 ppm) thrice at 10 days intervals effectively controls the disease.

8. COLE CROPS

Family – Cruciferae/Brassicaceae

Origin: Mediterranean region.

- Cole crops consist of a large group and have originated from a common parent wild cabbage or colewart (*Brassica pleracea* var. *sylvestris*).
- The term cauliflower is derived from two Latin words caulis means cabbage and floris means flower.
- All species of genus Brassica contain gluconsinolates which in crushed leaves is broken down by myrosinase enzyme to give bitter taste and goitrogenic substance.

- **Floral Parts:-**
  1. Calyx - 4 Sepals erect.
  2. Corolla - 4 Petals, Poly petalous, pale yellow.
  3. Androecium - 6 Stamens, Tetradynamous, Filaments straight
  4. Gynoecium - Superior ovary of 2 joined carpels, with a single short style.
- Fruit type - Siliqua
- The flowers open in the morning. The stigma is receptive for few hours earlier than the dehiscence of anthers & release of pollens (Protogyny).
- Crucifer vegetables *viz.* Cauliflower, Cabbage, Broccoli and Brussels sprouts contain a substance called indole-3-carbinol that affects metabolism of estrogens in a way that might protect against breast and other female cancers.
- The diploid species *Brassica campestris (syn. Brassica rapa)* (2n = 2x = 20, A genome) and *B. oleracea* (2n = 2x = 18, C genome) have close evolutionary relationship between them which indicates that deletions and insertions may have occurred after divergence of the two species.

S.N	Common Name	Botanical Name	2n = 2x	Edible Part
1.	Cabbage	*B.oleracea* var. *capitata*	18	Head
2.	Cauliflower	*B.oleracea* var. *botrytis*	18	Curd
3.	Knol-Khol (Kholrabi)	*B.oleracea* var. *gongylodes* *(Syn. B. Caulorapa)*	18	Knob/Tuber/ Flower bud
4.	Sprout Broccoli	*B.oleracea* var. *italica*	18	Unopened Flower bud
5.	Brussels Sprout (Mini cabbage)	*B.oleracea* var. *gemmifera*	18	Sprout/Buttons
6.	Chinese Cabbage	*B. compestris* var. *pekinensis* *B. compestris sp. Chinensis*	20	Leafy Head
7.	Kale/Collard	*B.oleracea* var. *acephala*	18	Leaves/ Shoot
8.	Rutabagas	*B. napus* var. *napobrassica*	38	Root and Leaves
9.	Red cabbage	*B.oleracea* var. *capitata f. rubra*	18	Head
10.	Savoy cabbage	*B.oleracea* var. *capitata f. sabauda*	18	Leaves

- According to the Triangle of U theory, *B. oleracea* is very closely related to five other species of the genus Brassica. The cultivars of *B. oleracea* are grouped by developmental form into seven major cultivar groups, of which the Acephala ("Non-Heading") group remains most like the natural wild cabbage in appearance:
  1. Acephala group – Kale and collard greens
  2. Alboglabra group – Kai-lan (Chinese broccoli)
  3. Botrytis group – Cauliflower, Romanesco broccoli and broccoflower (*Brassica oleracea* var. *botrytis* × *italica*)

4. Capitata group – Cabbage
5. Gemmifera group – Brussels sprouts
6. Gongylodes group – Kohlrabi
7. Italica group – Broccoli.

Seed rate of Cole Crops

S. No.	Crop Name	Seed rate/ha.
1.	Cauliflower	500-700 g Early type 300-500 g Mid and Late type
2.	Cabbage	500- 600 g Early type 400 g-Mid and Late type
3.	Knol-Khol	1-1.5 Kg
4.	Sprout Brocoli	400-500 g
5.	Brussel's Sprout	500 g
6.	Kale	350-400g

Sowing Time

Cauliflower	In North India Plains	May – July – Early
		July-Aug – Mid
		September – October (Snowball -Late)
	In Hills	August – Sept
Cabbage	In North India Plains	July-Aug – Early
		September-Mid
		October-November Late
Knol-Khol	In North India Plains	August - September Late
Sprout Brocoli	In North India	August – Oct
	In Hills	March and July-Aug
Brussel's Sprout	In Hills	Aug – October
Kale	In Hills	October – November

- Genus brassica has three monagenic and three amphidiploids digenomic species.

Species	Cultivated type	Ch. No. (n)	Genomic constitution
1. Monogenomic –			
B. nigra	Black mustard	8	B
B. oleracea	Cole group	9	C
B. compestris	Turnip group	10	A
2. Digenomic –			
B. carinata	Abyssinian Mustard	17	Bc
B. juncea	Brown mustard	18	Ab
B. napus	Swede group	19	Ac

- All the amphidiploids digenomic sp. have been developed from the crosses between the monogenomic sp. as reported by U.N. (1953) and illustrated by U.S. triangle.
- The cytogenetic relationship of the species is given by the U triangle (1935) in which the three vertices are represented by *B. nigra* (n=8), *B. oleracea* (n=9) and *B. compestris* (n=10), which by intercrossing gave rise to *B. carinata* (n=17), *B. juncea* (n=18) and *B. napus* (n=19). As a consequence of these occasional intercrosses new amphidiploid species originated. The genome A was attributed to *Brassica campestris* L., the genome B to *B. nigra* L. and genome C to *B. oleracea* L.

Fig.

Among the amphidiploid species, *B. carinata* (Ethiopian mustard), derived from the union of the BB and CC genomes, is widespread in Abyssinian Plateau, *B. Juncea* (mustard), derived from the union of AA and BB genomes, is grown in Asia and has several forms, and *B. napus* (oilseed rape), derived from the union of the AA and CC genomes, is widely grown in Asia, Europe and North America.

(A) Male sterility

Male sterility is characterized by non-functional pollen grains, while female gametes function normally. It occurs in nature sporadically, perhaps due to mutation. The phenomenon of male sterility is of great significance to produce cost effective hybrid seeds. The different types of male sterility involving GMS, CMS and CGMS, respectively are outlined. GMS is commercially utilized in chilli and cucurbits crops, while CMS is utilized in cole crops and root crops.

1. **Genetic Male Sterility (GMS):** Reported in calabrese and purple heading broccoli by Cole (1959) and Brocher (1966) in brussels sprout by Johnson (1958).
   - GMS is governed by recessive male sterility gene (MS) mutated from male fertile gene (MF).
   - In broccoli several male sterile genes MS1, MS2, MS4, MS5 and MS6 have been identified which are not allelic.
2. **Cytoplasmic Male Sterility (CMS):** The CMS was discovered by Ogura (1968), which is now popularly refered to by his name, Ogura CMS. It is not found apparently in *B. oleracea.* It was transferred into *B. oleracea* from *B. nigra* by "Pearson" through repeated back crosses and embryo rescue technique. This CMS is used worldwide in F_1 breeding of *B. napus*, *B. juncea*, *B. oleracea* and *R. sativus.*

   CMS can be maintained by crossing a male sterile line (A line) with the pollinator strain (maintainer line) used as a recurrent parent in the back cross programme since the nuclear genotype of the pollinator is identical with that of the new male sterile line. However, CMS has been successfully exploited at IARI Regional Research Station, Katrain to develop the cabbage hybrid "H 64"
3. **Cytoplasmic-genic male sterility (CGMS):** This type of male sterility is reported in carrot, onion, sugar beet, chilli, capsicum and *Brassica napus*.

(B) Self-incompatibility

All cole crops are interfertile (i.e., they can be crossed) and many are self-incompatible (i.e., flowers cannot be fertilized by pollen from the same plant). These characteristics have made it easy to select for new types of cole crops. Self-incompatibility also makes hybrid seed production economical.

(a) Gametophytic: Pollen pistil interaction is genetically controlled by haploid genome of each pollen grain and diploid genome of pistil tissue.

(b) Sporophytic: Pollen pistil interaction is genetically controlled by the genome of somatic tissue in which pollen grain is formed.

- Broccoli, Brussels sprouts and Knol-Khol have sporophytic self-incompatibility controlled by single gene with multiple alleles known as "S" alleles.
- Self-incompatibility in these crops can be studied by counting seed set a few weeks after self-pollination or by counting pollen tube in the style after 24 hours of pollination (The level of self-incompatibility can be assessed by the number of seed produced after a self or cross-pollination).
- Break down of Self incompatibility - Bud pollination treatment with CO_2 (3-5%) for 8-24 hours at 100% relative humidity and spraying the crop at full bloom with 5% sodium chloride solution.
- Self-incompatibility is used to develop hybrid varieties.

(C) Biotechnological Innovations

- **Micro-propagation and haploid culture:** Homozygous lines through haploid culture followed by diploidization of the haploids by colchicine treatment have been developed for onward use in breeding and hybrid seed production.
- **Somatic hybridization:** Somatic hybridization by protoplast fusion is effective in producing cybrids in interspecific crosses using Brassica oleracea and other Brassica species.
- **Molecular markers:** Different molecular markers have been found useful in studying genetic variability and tagging of specific gene in a new genetic background. The dominant male-sterility gene, located on chromosomes 1 and 8, was observed to be linked to RFLP marker pBN11.

9. "CAULIFLOWER"

In England in 1586, cauliflower was referred to as 'Cyprus coleworts', suggesting its recent introduction from Cyprus. Cauliflower was first grown in North America in late 17th century. It was introduced into India by Dr. Jemson, Incharge Company Bagh Sharanpur (U.P.) a botanist from Kew Gardens, London in 1822.

- In India two separate group of Cauliflower are commonly grown.
  1. Indian or Tropical type (Originated in India during the last 175 years).
  2. Annual temperature type known as "Erfurt" or "Alpha" or "Snowball type".
- Cauliflower is a thermosensitive crop i.e. temperature play an important role influencing vegetative, curding and reproductive phase.
- India is the second largest producer of Cauliflower in the world.
- West Bengal (74430 ha) is having maximum area under cauliflower cultivation followed by Bihar (62040 ha).

- In Cauliflower production, West Bengal (1939000 MT) is the leader followed by Madhya Pradesh (1008000 MT).
- Cauliflower thrives best in a cool and moist climate.
- Tropical type requires 20-25° C temperature for curd formation.
- Temperate type requires 10-16° C temperature for curd formation.
- Optimum monthly average temperature for its growth and development is 15-20°C with the average maximum at 25°C and the minimum 8°C.
- Curd is a prefloral fleshy apical meristem.
- Snowball type sets seed only in the hill region.
- Seeds are sown in raised beds, which are sterilized by drenching the soil with 4.0-5.0 litres of 1.0-1.5 per cent formalin solution per square metre area.
- For early crop 600-700 g per hectare seed is required and for the late group 400-500 g per hectare seed is sufficient.
- Sowing is done in lines at a spacing of about 7-8 cm and at a depth of 1.5-2.0 cm.
- **Classification of cauliflower varieties depending upon sowing time and temperature requirements**

Group	Varieties	Time of sowing	Time of maturity	Temperature required
Group I (early)	Early Kanwari, Pusa Deepali, November, Pant Gobi 2, Pant Gobi 3	May-June	Late Aug- mid	20-27°C
Group II (early mid season)	Improved Japanese, Pant Gobhi-4, Pusa Sharad, Pusa Kartik Sankar	July-Aug	Late Nov- end December	16-20°C
Group III (mid season)	Pusa Synthetic, Giant Snowball, Pusa Shubhra, Pant Shubhra, Kashi Agahani	Aug.-Sept	Mid January- mid Dec	12-16°C
Group IV (late season)	Snowball-16 Pusa Snowball K-1 Pusa Snowball-1	Mid Sept- October	Mid January- March	10-15°C

Recommended Varieties of Cauliflower

Variety	Features
Early Kunwari	Curds are semi-spherical with even surface. Suitable for early July transplanting. Maturity 80-85 days. Yield 80-100 q/ha. A very early variety for growing in Punjab, Haryana, H.P. and Delhi.
Pusa Early Synthetic	Plant erect with bluish green leaves, curds small to medium in size. Suitable for transplanting in early July. Adaptable to non-traditional areas like Karnataka and Kerala. Yield 110-120 q/ha.

Pusa Meghna	Curds semi dome shaped, white, weighing about 350-450 g. suitable for planting in the early July, Maturity 90-100 days. Yield 125 q/ha.
Pant Gobhi-3	A synthetic hemispherical compact, non-ricey curd. Suitable for planting in the early July, Maturity 80-85 days. Yield 120 q/ha.
Pusa Deepali	Plants with bluish green and waxy leaves. Suitable for transplanting in the first fortnight of July. Maturity 100-120 days. Yield 200 q/ha. Recommended for entire North Indian plains.
Pant Gobhi –2	Curds are medium compact and yellowish. Suitable for transplanting in the first fortnight of July. Maturity 95-100 days. Yield 100 q/ha.
Arka Kranti (IIHR, Bangalore)	It is an early maturing variety developed by selection from a local material collected from Bihar. Curds are compact and attain maturity 60 days after planting. Average yield is 22-25 t/ha.
Pusa Sharad	The plants are semi-erect with small stalk. Leaves are bluish-green, along with narrow apex and wavy margins. The curds are dome shaped each weighing 0.750-1.0 Kg. Average yield is 25 t/ha.
Pant Gobi 4 (GBPUA&T, Pantnagar)	The plants have medium long stem and semi erect leaves. The curds are medium sized, semi-spherical, creamy white and medium compact. Average yield is 15 t/ha.
Pusa Kartik Sankar	It is a F_1 hybrid suitable for June-July sowing. Mean temperature of 22-25°C is optimum for curd formation. Average curd weight is 475 g and yield is 15 t/ha.
Pant Shubhra	It is suitable for cultivation in West Bengal, Assam and Satluj Ganga Alluvial plains. Its curds are compact, slightly conical and creamy-white. Average yield is 25-27.5 t/ha.
Pusa Shubhra	It is developed from the hybrid population of triple cross involving MG 2-3, 15-1-1 and D 96. Each curd weighing 700-800g. Average yield is 25 t/ha. It has field tolerance to black rot and curd blight.
Kashi Agahani	The curds are white, compact, slightly dome shaped and take 80-90 days to attain marketable maturity. Average curd wieight is 850-900 g and yield is 30-35 t/ha.
Pusa Snowball 1 (IARI, Katrain)	It is developed from a cross between EC 12012 and EC 12013. The leaves are straight and upright. Inner leaves cover the curd. The curds are compact and pure white. Average yield is 20-25 t/ha.
Pusa Snowball K1 (IARI, Katrain)	It is developed by selection from an exotic material. The curds are dome shaped, compact and pure white. Average yield is 20-22 t/ha. It has field tolerance to black rot.

Snowball 16 (IARI, Katrain)	It is developed by selection from the exotic material. The curds are compact and pure white. Average yield is 15-20 t/ha.
Punjab Giant-26	Curds solid, white, medium-sized. Ready for harvesting from mid November to December. Average yield is 17 t/ha.
Dania Kalimpong	Commonly grown in eastern parts of India. Late season variety. Curds are attractive and white. Less sensitive to fluctuations of environment. Harvesting in Jan to April. Average yield is 25-30 t/ha.
Pusa Pausija	A mid-season variety release in year 2008. Ready to harvest after 75 days of transplanting and average yield 30-35 t/ha.
Pusa Shukti	A mid-season variety release in year 2009. Ready to harvest after 75 days of transplanting and average yield 40-44 t/ha.

Varieties according to maturity time:

[A] Early maturing (September-October)

(a) Early kunwari – Developed by PAU, Ludhiana, Sensitive to riceyness.
(b) Pusa Katki – Selection.
(c) Pusa Deepali – Cultivation in entire North Indian plains.
(d) Pusa Early Synthetic.
(e) Improved Japanese - Introduced from Israel.
(f) Pant Gobhi - 1.
(g) Pant Gobhi - 3 – A synthetic variety developed by crossing eight inbred lines of Katki, free from riceyness.
(h) Pant Gobhi – 2 – For North Indian plains.

[B] Mid season varieties – (November – December)

- Pusa Synthetic – Synthesized from seven lines with good general combining ability and resistant to curd blight.
- Pusa Hybrid 2- First public sector hybrid of Cauliflower made by crossing self incompatible line and selection 1-3-18-19, resistant to downy mildew.
- Pusa Aghani – Selection.
- Pusa Subhra – Selection from a triple cross. Free from riceyness, resistant to black rot and inflorescence blight.
- Pusa Himjyoti – Only variety which can be grown in spring summar (April-July) in hills, it is a self blanched variety.
- Pant Shubhra – Selection.
- Pant Gobhi – 4 - Free from riceyness.
- Punjab Giant – 26, 35.

[C] Late season variety (December- January)

- Snowball – 16 (Introduced from Holland)
- Pusa Snowball - developed by crossing (EC 12012 × EC 12013) – 1, Self blanching and tightly covering the curd.
- Pusa Snowball-K1 (Self blanched, resistant to black rot and inflorescence blight).
- Pusa Snowball – 2 - Selection from an introduction EC 120-12.
- Lata Dania – Sensitive to fluctuation in environment.
- Most of late types (snowball) have self blanched habit.
- The first illustration and description of Cauliflower was presented by the herbalist Dodens in 1544.
- Light soil is suitable for early crop.
- Loamy and clay Loam- Suitable for mid and late season crop.
- It can be grown in saline soil provided the C - Index is within tolerance limit.
- Cauliflower is sensitive to high acidic conditions.
- Blanching – Preventing the curds of Cauliflower from direct sunlight that make the curd white and improve flavour and tenderness.
- If cauliflower curds are exposed to sun during development, peroxidase enzyme activity is stimulated and curds develop yellowish-white colour instead of pure white
- C2 and C4 – Early cultivar (Nieuwhof 1969).
- C6 – Late Cultivar
- Very low temperature during curd development causes elongation of internodes and open heads.
- High temperature after the curd has formed causes a reversion of vegetative growth leading to production of bracts on the head.
- During curd initiation phase, which extends from 10-15 days, temperature should not fluctuate too much as this will enhance occurrence of physiological disorders like riceyness, fuzzyness, etc.
- If soil pH is higher than 7.0, availability of boron is reduced resulting in browning of curds.
- Cauliflower is only crop in group of cole crops in which the intermediate stage of curding lies between vegetative and reproductive stage.
- Scooping – Removal of central portion of curd for easier initiation of flower stalk.
- Application of GA_4+GA_7 @ 80 ppm shorten the periods from transplanting to harvest.
- After harvest, the produce is segregated into three grades depending upon size viz. large (more than 1.5 kg), medium (1-1.5 kg) and small (less than 1 kg). Medium sized heads are most acceptable.
- Maturity in cauliflower is indicated by curd size and its compactness. At maturity, the curd is at least 15 cm in diameter.
- Early varieties if sown late produce button heads and late varieties if sown early will keep on producing vegetative growth and will produce curds very late. .

- Cauliflower is a highly cross-pollinated crop and an isolation distance of 1000 m for certified seed and 1600 m for Breeder/foundation seed is required between two varieties to produce genetically pure seed.
- **Difference between Indian and European Cauliflower**

Indian type	**European type**
Tolerate to heat	Not tolerate to heat
Curd formation at an above 20°C.	Curd formation at 5-20°C.
Annual and early type	Biennial and late type
More self-incompatible.	Less self-incompatible.
No need of vernalization.	Need to vernalization.

Physiological disorders of cauliflower and broccoli

Disorders	Symptoms	Cause
Buttoning	The term buttoning is applied to the development of small curds called buttons. Plants do not develop normally and the leaves remain small.	Due to deficiency of nitrogen, selection of improper varieties, temperature stress (too low or too high), water stress, poor quality seedlings, crowding in nursery and transplanting of over-aged seedling.
Blindness	In blindness lack of terminal bud. The leaves are large, dark green and leathery. Sometimes, the axillary buds develop but the plant fails to produce a marketable head.	Prevalence of low temperature reaching subzero °C when the plants are still small or due to damage to terminal bud by mis-handling or injury caused by insect-pests.
Leafiness or Bracting	This is a condition in which the bracts grow out of the curd.	Excessive nitrogen application and if temperature is higher at the time of curd development.
Riceyness and Fuzzyness	The pedicles or peduncles grow out giving full velvety or partially velvety appearance at the curd.	Poor seed stock and growing of varieties at temperatures considered unfavourable.

Mineral deficiency -

Deficiency	Symptoms	Cause
Browning (Brown rot or red rot)	Water soaked brown spot formed on stem and curds, which leads to the formation of hollow stem and brown colour of curd.	"Boron" deficiency is main cause of browning; apply 10-15 kg per hectare Borax as a soil application or two sprays of 0.3 per cent Borax on the seedlings.

Whiptail	The leaf blades do not develop properly. When the deficiency is severe, only the midribs develop.	Deficiency of "Molybdenum" in acidic soils. Application of Sodium Molybdate 1.0-1.5 kg per hectare. Application of lime increases the availability of molybdenum.
Chlorosis	Interveinal yellow and mottling of lower older leaves.	Deficiency of "Mg" when grown on high acidic soils.
Hollow stem	The thick, fleshy stem core splits due to an uneven growth rate and an elongated cavity forms. When this occurs, infections by fungi and bacteria are common.	It can also arise from excessive N fertilization. Deficiency of boron or its unavailability in limed soils is the cause of this condition.

10. "CABBAGE"

- The word cabbage was derived from the French word 'caboche', meaning head.
- The Latin name Brassica was derived from the Celtic word 'bresic' also meaning cabbage.
- In India cabbage ranks IInd in area and production.
- Cabbage covers 4% area under vegetable crops in our country.
- Hybrid – 22.2% area, Fruit type – Siliqua.
- Among Cole crops cabbage rank Ist in production in world as well as in India.
- Cabbage is the fourth most widely grown vegetable crop of India.
- West Bengal (79460 ha) is having maximum area under cabbage cultivation followed by Bihar (37940 ha).
- In cabbage production, West Bengal (2280000 MT) is the leader followed by Orissa (1058000 MT).
- Nieuwhof (1969) recognized the three cultivated forms of cabbage:
  White cabbage (*Brassica oleracea* L. var. *capitata* f. alba),
  Red cabbage (*Brassica oleracea* L. var. *capitata* f. rubra),
  Savoy cabbage (*Brassica oleracea* L. var. *capitata* f. sabauda).
- Flavour in cabbage leaves is due to glucoside sinigrin.
- Cabbage has an anticancer property, it protects against bowel cancer due to the presence of Indole -3- Carbinol.
- Cabbage is an important source of vitamins, especially A, C and B, and minerals, especially calcium and potassium.
- It provides the roughage that prevents constipation. Per 100 g edible part of cabbage contains 91.9 percent moisture, 4.6 g carbohydrates, 1.8 g protein, 0.1 g fat, 1200 μg β-carotene, 0.06 mg thiamine, 0.09 mg riboflavin and 124 mg vitamin C.
- Cabbage juice was used as remedy against poisonous mushroom and as a gargle against hoarseness.

- Cabbage is highly cross-pollinated (as high as 73%) crop.
- In cabbage the stigma remains receptive for five days before anthesis due to protogynous nature of the flower and four days after anthesis.
- Cytoplasmic and genetic male sterility found in Cabbage is also helpful for the production of F_1 hybrids.
- Myers (1915) classified cabbage depending on maturity size, shape and colour of head into eight district groups -
  1. Wakefield on Winningtadt group
  2. Copenhagen Market group
  3. Flat Dutch or Drumheaded group
  4. Savoy group.
  5. Danish Ballhead group
  6. Alpha group
  7. Volga group
  8. Red cabbage group
- Mass selection has been found to be effective in cabbage breeding for disease resistance and ascorbic acid content.
- Cabbage requires specific low temperature for chilling 4-8° C for 40-60 days (Vernalization) so that plants grow to reproductive stage of flowering.
- A detail breeding programme has been given by Dickson and Wallace (1990) on back cross breeding.
- Back cross method of breeding is used for transferring the disease resistance governed by one or two genes, from the donor parent to an otherwise desirable cultivar.
- About seven generations of back crossing are needed to transfer a dominant or recessive gene from the recurrent parent.

Recommended Varieties of Cabbage

Research for cabbage improvement was started in the late fifties and the main emphasis in the initial stages was on standardization of seed-production technology to meet the requirement of seed which was being imported till early fifties. Cabbage cultivars can be divided into 2 groups *i.e.* early and late. Varieties of early group take 55-70 days from transplanting to head formation which weigh 0.8-2.5 kg, whereas those of late group mature in 85-130 days and their head size is 3-7 kg.

S.No.	Variety	Features
1	Golden Acre	It is a selection from Copenhagen Market developed by IARI, New Delhi. Head small, round, average wt. 1.2 kg. Fewer cup shaped outer leaves. Maturity 60-70 days. Yield 25-30 t/ha.
2	Pusa Mukta	Developed through inter-varietal hybridization. Flattish round, medium sized compact head, average wt. 1.4 kg. Resistant to black rot (*Xanthomonas compestris*). Puckered outer leaves. Maturity 70-80 days. Yield 27-32 t/ha.

3	Pride of India	It was developed at Solan (H.P.) by a selection from Copenhagen Market. Small, round, compact head, average wt. 1.2-1.5 kg. Maturity 70 days. Yield 27-37 t/ha.
4	Copenhagen Market	Large sized, round, compact head, average weight 1.5-3.0 kg. Maturity 75-85 days. Yield 35-40 t/ha.
5	Pusa Drum Head	It is a selection from Japanese material (EC 6774). Plants with larger frame, large size flat shaped less compact head. Resistant to black leg (*phoma lingam*) disease. Maturity 100 days. Yield 40-50 t/ha.
6	September	It is an introduction from Germany and is popular in Nilgiri hills. Bluish green foliage and large sized frame. Resistant to cracking, maturity 110 days. Yield 40-50 t/ha.
7	Late Large Drum Head	Wide frame, large sized, medium compact head, weighing 3.5kg, maturity 115-120 days. Yield 35-50 t/ha.
8	Pusa Ageti	First tropical variety can grow and produce marketable head at a temperature range of 15-30° C, weighing 0.6-1.2 kg and maturity 70-90 days. Suitable for plating in March and August-September. Yield 11-30 t/ha. Seed also produce under subtropical condition of North Indian plains.
9	Sri Ganesh Gol	Round, compact, bluish green head, weight 1.5-1.7 kg. Maturity 85-95 days. Yield 50-70 t/ha.
10	Nath Laxmi 401	Round, very compact uniform sized heads, longer shelf life, and weight 1.5-1.7 kg. Maturity 85-95 days. Yield 50-70 t/ha.
11	BSS-32	Very compact, round large sized heads weighing 1.5-1.8 kg. Longer shelf life. Maturity 90 days. Yield 70-80 t/ha.
12	Quisto	Large sized frame, compact, weighing 1.6-1.8 kg. Suitable for early sowings. Maturity 110-120 days. Yield 70-80 t/ha.
13	Drumhead Savoy	It has a dark green foliage which is much blistered or wrinkled, the quality is superior but it is not commercially acceptable in our country.
14	Red Cabbage	All the red cabbage cultivars are tolerant to diamond back moth insect which is serious pest of cabbage. It has distinct coat of wax and produces a head of 1-2 kg, the yield is usually low.
15	Pusa Sambandh	A synthetic variety from Regional Station, IARI, Katrain, has wide adaptability with round, compact heads, early, narrow frame, suitable for close planting.
16	K-1	Heads are large sized with inner leaves white in colour, tolerant to black rot disease, Average yield is 20-30 t/ha.
17	KGMR-1 (Hybrid)	It is hybrid resistant to black rot disease and tolerant to high temperature, developed by IARI, New Delhi. Its heads have better staying capacity in the field (non-splitting habit) Yield: 35-40 t/ha.

Varieties

(a) On the basis of type of Head –

Round head or ball head	Conical Head	Flat Head or Drum Head
(i) Golden Acre. (ii) Pride of India. (iii) Green express. (iv) Copenhagen market. (v) Mammoth Rock Red. (vi) Pusa Sambandh.	Jersey Wakefield	Pusa Drum Head Savoy – Chieftain

(b) On the basis of maturity group –

Early varieties – (50-70 days)

- Golden Acre (Introduced from Denmark) – Selection from Copenhagen market.
- Pride of India – Selection from Copenhagen Market, 1-1.5 Kg head weight.
- Pusa Synthetic – Synthetic variety.
- Pusa Ageti –
- Pusa Sambandh – Suitable for close spacing (H.D.P.) (has wider adaptability). It is a synthetic variety.
- Copenhagen Market - Exotic variety, popular in West Bengal, Head 2-3 Kg in weight.
- Shri Ganesh Gol- Mahyco, Maharashtra Hybrid Seed company Ltd. It has much better staying ability.
- Green Boy – Introduce from Japan.
- Early Drum Head.

Mid varieties –

(a) Pusa Mukta – (EC 24855 × EC 10109), Resistant to black rot (*Xanthomonas compestris*).

(b) September – Introduced from Germany, popular in Nilgiri hills, better staying power, Weight of head 4-6 kg., highly susceptible to black rot disease.

Late Varieties (100 – 120 days)

1. Pusa Drum Head – Pure line selection variety, Resistant to black leg. *Phomalingum* or dry rot common in hills, earliest of this group.
2. Late K 1 – Resistant to black rot.
3. Late Large.
4. Drum Head – Introduced from Denmark.

Private sector Hybrids

1. Questo.
2. Shree Ganesh Gol.
3. Uttam.

4. Bajrang.
5. Green Boy.
6. Green Express – Can tolerate hot weather (30-35^0 C temperature).
7. Sudha.
8. Stone Head.
9. Green Challenger.
10. Green Cornet.
11. Savarna.
12. Vishesh.

- In cabbage round variety matures first (earlier) followed by conical variety.
- Cabbage is grown mainly as a winter crop in plains and lower hills. At higher altitudes, it is grown in a cool summer and rainy season.
- In north Indian plains, sowing for early crop is done from August-September and for late season crop from September to October.
- In hills of the same region, sowing is done from early March to end of June.
- In eastern India, sowing is done from mid to end of September. Cabbage is sown throughout the year in hills of south India.
- Seed rate of cabbage varies from 400-500 g per hectare. Higher seed rate is required for early season crop.
- A still higher seed rate of 2.5 kg per hectare is required when the crop is raised by direct seeding as is the practice in some parts of north India.
- Direct seeding not only economizes labour but the crop also matures earlier than the transplanted one.
- The early cultivars are planted at a spacing of 45 × 45 cm/ 60 × 30 cm while the main and the late season cultivars are planted at a spacing of 45 × 60 cm and 60 × 60 cm.
- Spray of 50 ppm boric acid at flowering enhances the seed yield.
- Sauerkraut – Value added product prepared from white cabbage used to cure scurvy, very popular in European countries.
- Spray of CCC (2500-5000 ppm) or SADH increase the low temperature resistance.
- Cabbage grown in saline soil is more prone to black leg disease.
- Cabbage produces seed in the temperate areas only.
- Cabbage bear seed in special kind of bicapillary pod called as "Siliqua".
- Popularity of F1 hybrid cultivars in cabbage is due to comparatively easy and cheaper method of hybrid seed production due to presence of genetic mechanism of self-incompatibility.
- Speedling – Sowing of seed in theremacole plates, followed in cabbage.
- Cabbage is harder than cauliflower and can tolerate frost and extreme cold weather. It grows better in cool and moist climate.
- Germination is optimum when temperature is 20-25°C. The optimum temperature for growth and head formation ranges from 15-20°C.
- When young plants are exposed to temperature of around 2-7°C for extended periods,

the plants develop seed stalks instead of heads.

- However, some varieties like Pusa Ageti can form heads even if the prevailing temperature is 30°C.
- White cabbage is more tolerant to salts in the soil than the other two types.
- Early varieties take 60-80 days to reach maturity while the late varieties take 100-120 days.
- Pusa Ageti (IARI, New Delhi) it is an early maturing tropical variety developed by selection from an F_1 hybrid TKCBH 28 introduced from Taiwan and is recommended for cultivation under **high temperature** conditions.
- Pusa Mukta (IARI, Katrain) is **tolerant to salinity and resistant to black rot**.
- Pride of India (IARI, Katrain) it is an early maturing variety developed by selection from Copenhagen Market.
- Pusa Drumhead (IARI, Katrain) it is a late maturing variety developed by selection from the material introduced from Japan.
- Bajrang F_1 (Beejo Sheetal seeds pvt Ltd). It is early maturing (65 days) hybrid, tolerant to high temperature and resistant to Fusarium wilt.
- Swarna (Beejo Sheetal seeds pvt Ltd) is a late maturing hybrid and takes 110 days from sowing to reach maturity.
- Splitting of heads -Splitting of heads may occur due to heavy rains after prolonged drought, excessive fertilizer application and delayed harvesting.
- Early varieties are more prone to head splitting than the late maturing ones.
- Variety – Nath Laxmi has very good shelf life.
- Yield of cabbage varies from 125-150 quintals per hectare in early season varieties and 200-300 quintals per hectare in main and late season varieties.
- In grading the medium sized heads weighing around 1.0 kg are preferred. The heads are uniformly trimmed to equal number of wrapper leaves before marketing.
- In cold stores storage temperature is 0 - 1.7°C with 90-95% relative humidity for 4-6 weeks for early cabbage, whereas late cabbage can be stored for 12 weeks at same temperature.
- Cabbage is highly cross-pollinated crop and requires an isolation distance of 1000 m from other varieties and also from other cole crops.
- Seed of cabbage except the heat tolerant variety Pusa Ageti is produced in temperate climates of mid and higher hills. Pusa Ageti seed can be produced in sub-tropical climates of Indian plains.
- It requires temperature of 5°C for 6-8 weeks to initiate seed stalk formation.
- There are two methods of seed production *viz.* in-situ (seed-to-seed) method and head-to seed (transplanting) method.
- In-situ method is commonly followed under Indian conditions where crop is allowed to over-winter. The heads are selected when they are fully developed.
- After plant selection either the whole head is chopped off and only the stump with outer whorl of leaves is retained or the head is sliced from all sides without damaging the central core or the whole head is retained but a cross cut is given at the top of head to facilitate seed stalk emergence.

11. "KNOL-KHOL"/ KHOL RABI/ GERMAN TURNIP/TURNIP CABBAGE

- Very popular in Kashmir, West Bengal and Karnataka.
- Knol-khol or kohl rabi is known as ganth gobi in Hindi.
- Whole Afghanistan is well suited for knol-khol cultivation.
- Kohlrabi is a combination of German words, 'kohl' meaning cabbage and 'rabi' meaning turnip. The word kohlrabi, therefore, means cabbage with a turnip like appearance.
- Edible part of knol-khol is the swollen stem called knob and is formed from thickening of stem tissues just above the ground. The leaves emerge from surface of the knob.
- Tubers are produced above the ground, consumed mostly as salad.
- Excess use of cole crops results in swelling of thyroid glands, a disease known as goiter in human being and other animals.
- Goitrin, thiocynates, isothiocynates and glucosinolates derived product interfere with the functioning of thyroid glands causing the disease.
- Knol-Khol was known to the Romans.
- Inflorescence – Racemose.
- The species of *B. oleracea* is a triple tetrasomic with six basic genome and genomic formula is AABBCCDEF.
- Highly cross pollinated crop (up to 91%).
- Main pollinating agent – Honey bees and flies.
- Male sterility – Absence of functional pollen grain (Male gamets).
- Knol-khol is an important source of minerals, protein and fibre. Per 100 g edible part, knol-khol contains 92.7 percent moisture, 3.8 g carbohydrates, 1.1 g protein, 0.2 g fat, 0.05 mg thiamine, 0.09 mg riboflavin, 85 mg vitamin C, 21 mg calcium and 0.4 mg iron.
- It contains sulforaphane, which has anticancer properties
- Optimum temperature for Knol-khol growth is 15-25°C and it is tolerant to frost.
- If temperature falls below 15°C for longer periods, early maturing varieties develop flowering stalks without forming the knobs. At higher temperature, the knobs become fiberous and taste bitter.
- In North India sowing is done from September to November. In lower and midhills, sowing is done from July to October and in higher hills from March to July.
- To raise nursery for one hectare 1-1.5 kg seed is sufficient.
- The seedlings are ready for transplanting 30-40 days after sowing.
- Row to row spacing is maintained at 30 cm and plant-to-plant spacing at 10 cm.
- The early varieties are harvested when knobs attain 6-8 cm diameter and late varieties when they attain 10-12 cm diameter.
- Most of the varieties cultivated in India attain marketable maturity 45-70 days after transplanting.

- However, the late varieties grown in Europe and other temperate regions take 70-100 days to reach maturity.
- **Pusa Virat (IARI, New Delhi)** its knobs are round and large, each weighing 800-900 g and 13-14 cm in diameter. The knobs remain non-pithy and without fiber for longer periods. Average yield is 190 quintals per hectare.
- **Recommended Varieties of Knol-khol:** There are mainly two types of knoll-khol's varieties exist i.e. white and purple. However, internally both varieties have similar cream-yellow coloured edible flesh. Some of the popular varieties are:

S.No.	Variety	Features
1	White Vienna	An early variety, plants dwarf with light green foliage and stem. Knobs are globular round and green with fewer leaves. Flesh creamy white, tender and crisp. Maturity 55-60 days. Yield 120-150 q/ha.
2	Early White Vienna	An early variety, knobs are globular-round and green with fewer leaves. Flesh creamy white, tender and crisp. Maturity 50-55 days. Yield 110-130 q/ha.
3	Large green	A late variety. Plants vigorous with dark green foliage, knob flattish round, dark green. Maturity 60-65 days. Yield 140-170 q/ha. Adapted well in Himachal Pradesh.
4	King of North	It is dark green, flattish round knobs. Dark green leaves are well spread over the knobs. Maturity 60-65 days. Yield 130-150 q/ha.
5	Early Purple Vienna (Verma Pocha)	Early variety. Foliage purple, knobs globular-round, large with purple skin and green flesh. Maturity 55-60 days. Yield 120-140 q/ha.
6	Palam Tender Knob	Better shelf life, early variety with light green knobs and average yield of 250-275q/ha.

Other varieties:

1. Golith White (Sadashiv).
2. Suttons Earliest purple.
3. Purple Vienna.
4. Purple speck-Earliest variety.

- Knol-Khol

  For flowering and fruiting - Annual

  For seed production - Biennial.
- Early varieties are more prone to premature bolting.
- Purple varieties are more susceptible to bolting.
- If harvesting is delayed the knobs become fibrous. Before sending to the market, the roots are removed and the knobs are tied together in bunches.
- Yield in knol-khol varies from 125 to 200 quintals per hectare depending upon planting time, locations and the variety.

- Like other cole crops, knol-khol is a cross-pollinated crop and requires 1000 m isolation distance from other varieties and crops of the cole group.
- Seed is produced by both seed-to seed and transplanting methods
- Knol-khol is stored at 0°C and 90-95 per cent relative humidity. The knobs with leaves can be stored for about two weeks. De-topped knobs can be stored for upto three months under ideal conditions.

12. "SPROUTING BROCOLI"

- The word broccoli comes from Italian word broccolo, which means 'flowering crest of a cabbage'.
- Among cole crop it has the highest nutritive value.
- Iron – 205 mg/100g.
- Vitamin A – 9000 IU/100g.
- Thiamin – 300 mg/100g.
- Vitamin C – 137 mg/100g it has 130 times more Vitamin A than Cauliflower and 22 times more than cabbage. It is the richest source of protein among cole crops.
- Due to its cauliflower like appearance, broccoli in the local parlance is known as 'hari gobi', meaning green cauliflower.
- There are three commonly grown types of broccoli i.e. Calabrese, sprouting and purple broccoli.
- **Calabrese broccoli** has large (10 to 20 cm) green heads and thick stalks. It is also referred simply as 'heading broccoli' grown as cool season annual crop.
- **Sprouting broccoli** has a somewhat branching cluster of green flower buds at top a thick, green flower stalk. Small axillary heads/ sprouts arise from the stem at leaf.
- **Purple cauliflower** has a head shaped like cauliflower, but consisting of tiny flower buds. It sometimes, but not always, has a purple cast to the tips of the flower buds.
- Broccoli is very popular in hotels in Mumbai, Calcutta, Delhi and Chennai.
- Mainly cultivated in hilly areas of Himachal Pradesh, Jammu and Kashmir and Uttar Pradesh.
- USA – Largest producer of sprout brocoli in world.
- Early type- Annual.
- Late type – Biennial.
- Rich source of sulphoraphane – compound associated with reducing risk of cancer.
- In dry condition crop become fibrous.
- Inflorescence – Cymose.
- Broccoli also contains certain amount of goitrogens, the naturally occurring substances that can interfere with the functioning of thyroid gland.
- Broccoli is grown as a winter crop in most parts of India. However, in higher hills it is grown as a spring-summer crop.
- Exposure to high temperature during development stage of heads induces bracts and deteriorates quality of heads.

- A temperature of 20-22°C is considered optimum for proper growth and development.
- In north Indian plains, sowing is done from mid August to mid September; in mid hills from July to August and in higher hills from March to April.
- About 500-600 g seed is sufficient to raise nursery for one hectare. Transplanting is done when seedlings are 4-6 weeks old. Broccoli can also be grown by direct seeding.
- A line × plant spacing of 45 × 45 cm is followed to maintained optimum plant population.
- Sowing Time :-

  In plain- Mid September to early November

  In hills- September-October
- Mass Selection – Most common method used in cross-pollinated crop for improvement of qualitative characters governed by single or few genes.
- **Varieties –**

S.No.	Variety	Features
1.	Punjab broccoli	The sprouts are compact, attractive and succulent. It is ready for harvest in about 65 DAT. Its average yield is 175 q/ha.
2.	Fiesta	Mid season produces dense, well-domed heads with thick stems on stout plants. Has limited heat tolerance at maturity, but is ideal for late summer sowing.
3.	Palam Vichitra	Medium size, heading broccoli and bears dark green leaves with purple tinge stem. This variety is suitable for cultivation under low hill conditions. Its average yield potential is 225 q/ha. Other varieties that are suitable for cultivation under low hill conditions are **Palam Haritika** and **Palam kanchan**.
4.	Pusa KTS 1	Medium-tall variety (65-70 cm), dark green waxy foliage with slightly wavy margins and weight 350-450 gm respectively, matures in 90-105 DAT.
5.	Palam Samridhi	Mainly for Sub-Tropical condition and a HYV with large terminal head weighing about 300-400 gm each.
6.	Gypsy	It is an early broccoli variety with a strong root system for good productivity in poor soil. This variety is tolerant to heat.
7.	Arcadia	It produces uniform and purplish-green heads. Strong cold tolerance and resistant to head rot and downy mildew.
8.	Express	It is a mid-season variety, produces uniform plants with deep blue-green heads.
9.	Italian Green, Green Head	From University of Horticulture and Forestry (UHF), Solan.

- Purple Sicilian broccoli is also known as purple cauliflower.
- Green type cultivars are more commonly cultivated than white type.
- Calabrese type broccoli developed from Italian green sprout broccoli are becoming more popular with the growers as well as consumers for both fresh market and processing.
- In India mainly early and mid-type of broccoli are grown and these are susceptible to frost.
- Late type cultivar – Requires vernalization for the development of head and initiation of flower stalk.
- The heads are harvested as soon as they reach marketable size but the buds are still tight and well before yellow petals begin to appear.
- Marketable mature heads are of 15-25 cm diameter weighing 250-600 g each.
- Most varieties reach marketable maturity 55-75 days after transplanting. A total of 4-6 harvestings at 2-3 day intervals are possible.
- Seed production of early varieties of broccoli is possible under conditions of north Indian plains, whereas, that of main and late season varieties is possible under conditions of mid and higher hills.
- Broccoli is a cross-pollinated crop and requires an isolation distance of 1000 m from other broccoli varieties and other cole crops.
- Yield of broccoli varies from 100-150 quintals per hectare.

13. "BRUSSEL'S SPROUTS"

- Brussels sprouts got its name from Brussels, Capital of Belgium.
- The crop is the present from appeared in Brussels (Belgium) in 1750 and reached England and France in 1800.
- Brussels sprouts require a long cool and moist growing season.
- Germination is optimum when soil temperature is around 10°C.
- The optimum temperature for plant growth and sprout quality is 16-20°C.
- At higher temperature, the sprouts produced are loose and taste bitter.
- Brussels sprouts can tolerate short spell of frost and continue to grow even at 0°C.
- Edible part – Axillary bud known as sprouts (small heads that resemble miniature cabbages).
- Brussels sprout contain good amount of vitamin A and C, folic acid and dietary fibre. Moreover, it contains sulforaphane, which has anticancer properties. Likewise, other Brassicas, Brussels sprouts is also a good source of indole-3-carbinol, which boosts DNA repair in cells and block the growth of cancer cells.
- Sprouts after cutting vertically into two halves or fried with besan (gram flour) and paste are used for preparation of "Pakoras".
- Genetic Male Sterility was reported by – Johnson (1958).
- **Varieties** –
  1. Hilds Ideal.

2. Rubine – F_1 hybrid.
3. Amager Market.
4. Jade cross.
5. Danish Prize.
6. Long Island.
7. Dwarf Improved.
8. Kvik.
9. Frontier Zuerg.
10. Early Mom.
11. Catskill.
12. Red Vein.
13. Royal Marvel.
14. Bubbles
15. Oliver.
16. Prince Marvel.

Sowing of seed and harvesting of sprout:

S. No.	Part of India	Sowing time	Harvesting
1	North Indian plains	September to October	January-February
2	East India.	Second fortnight of September	February-March
3	In mid hills	August to mid September	November-March
4	In higher hills	March to April	July-September

- Late planting results in pre-mature bolting.
- The seed rate is about 500-600 g per hectare for transplanted crop and one kilogram for directly seeded crop.
- Transplanting is possible 5-6 weeks after sowing seed in nursery.
- Seed priming is get uniform stand of seedlings in directly seeded crop.
- When temperature is low, plastic mulch is helpful in facilitating germination.
- Row-to-row × plant-to-plant spacing is maintained at 60 × 45 cm.
- Brussels sprouts is a long duration crop and requires heavy doses of fertilizers.
- Topping is as an important operation followed in Brussels sprouts in which plants are pinch-off from growing point (topping). This is usually done to stimulate growth of sprouts at the top of plant and provide uniform development for harvesting.
- Brussels sprouts take 85-90 days from transplanting to reach marketable maturity.
- About three weeks before harvesting, the plants are de-topped to divert photosynthates towards developing sprouts.
- The first harvesting is done as soon as lower sprouts on the main stem attain proper size (3-4 cm diameter) and before the lower leaves turn yellow.
- Leaves are also used for green when they are tender. The crop yields about 200- 300 q/ha.
- **Storage:** 0°C for 16 weeks and 20°C for 11 days.
- Sprout in sealed polythene packs at 0-1°C in modified atmosphere, 2.5% CO_2 and 14.5-16.8% O_2 reduces yield losses and improves quality.

- Seed production of Brussels sprouts is possible only under temperate conditions of mid and higher hills. It requires a temperature of 5-8°C for 40-60 days to facilitate seed stalk formation.
- It is a cross-pollinated crop and requires an isolation distance of 1000 m between two varieties.
- A similar isolation distance is required from other crops of the cole group.
- Seed crop is inspected at formation of sprouts and flowering and pod development stages to rogue out diseased and off-type plants.

14. "CHINESE CABBAGE"

- *Brassica rapa*, subspecies *pekinensis* and *chinensis* can refer to two cultivar groups of Chinese leaf vegetables: the Pekinensis Group (napa cabbage) and the Chinensis Group (bok choy). It is a non heading strain of cabbage.

Varieties –

1. Palampur Green.
2. Pusa Sag – Wongbok (Suttons) × Turnip.
3. Solan Selection –
4. Chinese sarson No. 1 – Non Heading type, resistant to Alternaria leaf spot.

15. "KALE"

- In India introduced in 19th century and very popular in Kashmir valley.
- Hardiest crop (as low as 0-15° C).
- Seeds are sown in October-November.
- Leaves are used for vegetable purpose. It is one of the best low-calorie foods, loaded with vitamin and minerals. It is high in Vitamin K and Manganese. Kale is also a source of the carotenoids – lutein and zeaxanthin.

Varieties –

1. Karamsag – Tall variety, mostly grown in Jammu and Kashmir.
2. Dwarf Green.
3. Hamburger Market.
4. Scottish – For temperate region.
5. Dwarf Moss.
6. Siberian – For temperate region.

16. IMPORTANT PEST AND DISEASE OF COLE CROPS

Sr. No.	Pest and Disease	Symptoms	Management
1.	Diamond backmoth (*Plutella xylostella*)	Scrapping of epidermal leaf tissues producing typical whitish patches on leaves	• Grow mustard as trap crop at 2:1 ratio (Cabbage: Mustard) to attract DBM for oviposition at least 10 days ahead of planting of main crop. • Install Pheromone traps @12/ha. • Crop rotation with cucurbits, beans, tomato and melons. Insecticide — Dose Lufenuron 5.4 % EC — 1.2 ml/lit. Spinosad 2.5 % SC — 1.2 ml/lit. Trichlorofon 50 % EC — 1.0 ml/lit.
2.	Cabbage borer (*Hellula undalis*)	The larva webs the leaves and bore into the stem, stalks or leaf veins.	• Collect and destroy of caterpillars. • Spray *Bacillus thuringiensis* @ 2g/lit at primordial stage. • Spray Cartap hydrochloride @ 500g/ha.
3.	Cabbage butterfly (*Pieris brassicae*)	• Defoliation • Bores into the heads of cabbage	• Collect and destroy of caterpillars in the early stage. • Conserve parasitoids like *Cotesia glomeratus* • Spray insecticides like quinalphos 25 EC @1000 ml.
4.	Leaf Webber (*Crocidolomia binotalis*)	• Young larva feeds and webs within the leaves.	• Set up light traps@1/ha. • Encourage the activity of parasitoid: • Spray malathion 50 EC @ 0.1% or carbaryl 0.2%.
5.	Tobacco caterpillar (*Spodoptera litura*)	Damage leaves and heads of cole crops.	• Install Pheromone traps (Pherodin SL) @ 15/ha to attract male moths. • Spray chlorpyriphos 20 EC 2lit/ha or dichlorovos 76 WSC 1 lit/ha.

6.	Cabbage aphid *(Brevicoryne brassicae)*	• Yellowing • Crinkling and cupping	• Install yellow sticky trap @12/ha. • Spray any one of the following insecticide Insecticide — Dose Dimethoate 30 % EC — 2.0 ml/lit. Phosalone 35 % EC — 1.5 ml/lit.
7.	Damping off (*Pythium debaryanum*)	Black and reddish brown lesion on the collar region. Finally infected seedling is toppled on the ground surface.	• Drench with COC 0.2% or Bordeaux mixture 1%. • Seed treatment with *Trichoderma viride* (4 g/kg of seed) or Thiram (3 g/kg of seed) is the only preventive measure to control the pre-emergence damping off.
8.	Black leg (*Phoma lingam*)	Affected stem when split vertically, shows black discoloration sap. Whole root system decay from bottom upwards.	• Seed treatment with Captan or Thiram 4g/kg of seed, followed by seed treatment with *Trichoderma viride* 4g/kg.
9.	Downy mildew (*Peronospora parasitica*)	Purplish or yellow brown spots on the upper surface of the leaf, while fluffy downy fungus growth is found on the lower surface.	• Seed treatment with Metalaxyl (Apron 6g/kg of seed) • Foliar spraying with Metalaxyl (Ridomil) 0.4%.
10.	Root rot (*Rhizoctonia solani*)	Young plants show soft, water soaked lesion on the stem near soil level.	• Seed treatment with Captan/Thiram 4g/kg, followed by seed treatment with Trichoderma viride 4g/kg.
11.	Black spot (*Alternaria* sp.)	Damage heads and curds during seed production stage.	• First foliar spraying with Tridemorph 0.1% followed by spraying with Mancozeb 0.25% a month interval.
12.	Clubroot (*Plasmodiophora brassicae*)	Club like swelling of root and root lets	• Soil fumigation with 1kg/10m ^{2} Methly bromide followed by covering with plastic film. • Lime application @ 2.5 t/ha or soil drenching with COC 0.25%.

13.	Powdery mildew (*Erysiphe polygoni*)	Initially, white tufts of mould arise on the upper surface of the leaves and later the entire leaf becomes covered with greyish white mycelium.	• Spary inorganic sulphur 0.25% or Dinocap 0.05%.
14.	Black rot (*Xanthomonas campestris pv. campestris*)	The infection of the foliage results in yellow 'V' shaped spots arising along the margin which extend in the direction of the midrib.	• Seed treatment with Aureomycin 1000ppm for 30 min • Nursery drenching with formaldehyde 0.5%. • Application of bleaching powder at 10.0 to 12.5 kg/ha controls the disease.
15.	Wire stem (*Rhizoctonia solani*)	This disease makes the seedlings unsuitable for transplanting and affected plants will die or grow poorly.	• Sterilize soil and seedbed drenches with Copper oxychloride 0.25% will give good disease control

17. ROOT CROPS

S. No.	C. N.	B. N.	Family	2x (2n)	Edible part
1.	Radish	*Raphanus sativus*	Cruciferae	20	Fusiform
2.	Carrot	*Daucus carota*	Umbelliferae	18	Conical form
3.	Turnip	*Brassica rapa*	Cruciferae	20	Napiform
4.	Beetroot	*Beta vulgaris*	Chenopodiaceae	18	Napiform

18. "RADISH"

- Origin - It is believed to have originated from Europe and Asia, while it is widely distributed as a weed crop in Europe.

  Indian (red tail) radish–India.

  Japanese radish – Japan.

  European radish – Mediterranean region.
- Radish is a cross-pollinated vegetable due to sporophytic self- incompatibility.
- West Bengal (40960 ha) is having maximum area under radish cultivation followed by Haryana (33060 ha).
- In radish production, West Bengal (539000 MT) is the leader followed by Haryana (531000 MT).
- Rat-tail radish *Raphanus sativus* var. *caudatus* (pods are used), which is similar to common radish, is grown for its long slender pods and used as salad or cooked vegetable. It will not produce fleshy root as in radish.
- Pollination – Entomophilous (Honey Bees).
- Radish show considerable inbreeding depression upon selfing.
- Red coloured roots are high in ascorbic content.
- Potassium application increases the vitamin C content in radish.
- Pungency – Due to Isothiocynates.
- Red colour- Anthocyanin is responsible.
- Roots are used in treating urinary, liver troubles, jaundice and piles complaints.
- **Autopolyploids:** Raphanobrassica (Radish cabbage) – Cross done by Karpenchenko (An allotetraploid). Karpenchenko wanted to synthesis a plant with root like radish and shoot like cabbage but unfortunately "Raphanobrassica" has root like cabbage and shoot like radish, hence it was not applicable.
- **Recommeded Varieties of Radish:** The varieties can be divided into three groups –

(A) Asiatic or tropical types

(B) European or temperate types

(C) Indian types (An indigenous type, Jaunpuri Giant, cultivated Jaunpur in Uttar Pradesh, has roots of 75- 90 cm length, 50-60 cm girth and weight up to 5-15 kg).

S.No.	Variety	Features
Asiatic/Tropical type (Annual): More pungent, long duration (45-55 days), large size roots, Yield high (250-300q/ha.) produce seeds in plains.		
1	Pusa Desi	Roots 30-35 cm long, tapering, white with green stem end and mildly pungent in taste. Suitable for sowing from August to October. Maturity 50-55 days. Yield 300 q/ha.
2	Pusa Reshmi	Tolerant to slightly high temperature. Roots 30-35 cm long, tapering, mildly pungent in taste. Suitable for sowing from Oct. to Dec. Maturity 60-65 days. Yield 35-40 t/ha,
3	Japanese White (Introduction, Japan)	Developed at IARI Regional Station, Katrain; roots 20-30 cm long, cylindrical, mildly pungent and white with green shoulder; yield 150-300 q/ha in 60-65 days; suitable for sowing between Oct. to Dec. in plains and Sept. in hills. Seed produced only in hills, less bolting.
5	Pusa Chetki (Introduction from Denmark)	Suitable for sowing in hotter months, i.e., from March to August when no other variety is available in North Indian Plains; duration 40-45 days, Yield 200-250 q/ha.
8	Arka Nishant (By mass selection)	Resistant to pithiness, premature bolting, root branching and forking (Multiple resistant). It yields 200-300 q/ha and root and shoot length ratio is 1:1.
9	Punjab Safed	Roots are pure white, tapering, smooth, free from forking and 30-40 cm long. Developed at PAU, Ludhiana.
10	Punjab Pasand (PAU, Ludhiana)	It has been developed by selection from Pusa Chetki and suitable for sowing in October-November. It is quick growing (45 DAS). Average yield is 540 q/ha.
11	Kashi Sweta (IIVR, Varanasi)	It also developed by selection from Pusa Chetki and suitable for sowing in all seasons. It is a quick growing variety (30-35 DAS). Average yield is 500 q/ha.
12	Mino Early (Introduction, Japan)	It is a quick growing variety and roots attain harvesting maturity 40 days after sowing. Roots are pure white, 30-40 cm long, slightly tapering and mildly pungent.
13	Early 40-Days (Introduction, Japan)	Roots are about 30 cm long, white, cylindrical with semi-blunt end and mildly pungent. It is suitable for sowing in late spring. Attain marketable maturity 40 days after sowing.
14	Palam Hriday (CSK HPKV, Palampur)	Developed by selection from the exotic material. Roots are oblong with stump end. The roots attain marketable maturity 50-60 DAS. Average yield is 300 q/ha.

15	Punjab Agethi	Roots medium long, red skinned at top and white skinned at lower half, sowing from April to August.
16	Kalyanpur No.-1	Roots 22-23 cm long, white with green shoulder and tapering.
17	CO.1	Roots milky white, 23 cm long and cylindrical; yield 9-10 t/ha in 45 days.
18	Naduani	A popular variety of Himachal Pradesh and developed through selection from local material. Roots are long, slim, light pink with tapering to a point.
19	Chinese Pink	It has 12-15 cm long pink coloured roots. The lower end of roots is white. Roots mature in 50-55 days.
20	Pusa Mridula	Roots globular in shape, bright red in colour, white flesh with soft texture. Maturity in 20-25 DAS. Suitable for winter crop (September-February). Yield : 13.5 t/ha
21	Pusa Jamuni	First purple fleshed unique trait, nutritionally rich radish variety. Distinct advantage in root size, shape, yield and consumer preference over the existing varieties. Higher anthocyanins and ascorbic acid.
22	Pusa Gulabi	First entire pink fleshed unique nutritional rich radish variety. Medium root size, cylindrical shape, optimal yield and consumer preference over the existing varieties. High carotenoids, anthocyanins and optimal ascorbic acid.
European or temperate types (Biennial): Short duration (25-30 days), produce good quality roots with less pungent, small size, low yield (75-100 q/ha.).		
1	White Icicle	Roots small (8-10 cm long), pure white, tapering, sweet in taste. Suitable for sowing in kitchen garden or container garden from Oct. to Jan. Maturity 30 days. Yield 50 q/ha.
2	Rapid Red White tipped	Roots small, round or globular, upper portion of the roots is red and lower portion is white, sweet in taste. Suitable for sowing in kitchen garden or container garden from October to January. Maturity 25 days. Yield 50 q/ha.
3	Pusa Himani (IARI, Katrain) Black × Japanese white.	Suitable for sowing from December to February in plains and In hills it is only variety which can be grown throught the year barring three winter months viz. November –January. Maturity 50 to 60 days. Yield 340-370 q/ha.
4	Pusa Mridula (IARI, New Delhi)	The roots are globular, red skinned and white fleshed with soft texture, quick growing variety (20-25 DAS) and sowing from Sept. to Feb. Average yield is 135 q/ha.

5	Scarlet Globe	Roots bright red, round, small, flesh white, takes 25-30 days from sowing to harvest.
6	Scarlet Long	Roots long, tapering with red skin and white flesh.
Indian types	Jaunpuri Mooli	A local variety of Jaunpur (U.P.). Roots are much long (40-45cm), thick (7-10cm), less pungent, sweet and white. On an average, each root weighs 2.5-3.0 kg.

Sowing Time:

Variety	**Sowing time**	**Root availability**
Pusa Himani Early Mino	January to February	February to March
Punjab Safed, Punjab Pasand	March	April
Kashi Sweta	April to August	May to September
Punjab Pasand, Japanese White	Mid September to October	October to December
Japanese White	November to December	December to January

In North India plains, radish can be grown throughout the year. Since temperate radish tolerates frost, it is can grown between Sept. and January in plains.

- Develop best flavour, texture and root size when the temperature is between 10-15°C.
- In temperate types, maximum root growth occurs initially at 20-30°C and later at 10-14°C.
- More foliage growth is favoured at temperature above 25°C.
- In hot weather roots become pungent and tough before attaining marketable maturity.
- Asiatic varieties if late planted in the season start bolting without forming edible roots.
- Long photoperiod and warm temperature favour early bolting even before proper root development. Normally plants bolt when days are 8-10 h long.
- Cultivated Radish – Originated from *Raphanus raphanistrum.*
- Fodder type – *Raphanus sativus variety mongri.*
- Seed rate and Spacing –
  For Tropical type: 8-10 Kg/ha and 45 × 7.5 cm
  For Temperate type: 10-12 Kg/ha and 30 × 7.5 cm
- For sowing in the off-season, high seed rate of 15-20 kg per hectare is recommended.
- Seed production of Asiatic type varieties of radish can be under-taken in north Indian plains where as that of European types can only be under-taken in temperate regions comprising higher hills.
- All root vegetables are highly cross-pollinated. Insects, mainly honeybees, do pollination and introduction of beehives in seed production blocks increases seed yield and quality.
- A minimum of 1000 m isolation distance is required to produce true-to-type seed.

- Seed in root crops is produced by two methods viz. root to seed method and seed to seed method.
- In seed to seed method, the plants are allowed to bolt in-situ. However, root selection cannot be exercised in this method.
- In root to seed method, plants are uprooted and roots selected for re-planting in seed production block. Uprooting is done when the roots have attained marketable maturity and before they develop pithiness.
- Stecklings are prepared by chopping off 2/3rd of the root and retaining only a new leaf growth. Chopping of roots is done for convenience in re-planting.
- One hectare of root crop produces stecklings sufficient for planting four hectares of seed crop. Spacing is maintained at 60 × 25 cm.
- The seed is dried to 6 per cent moisture level or less, cleaned and stored. Average seed yield is 5.0-6.0 quintals per hectare.

Physiological disorders –

1. **Akashin:** Induced by high day and night temperature between 30°C and 20 °C, as well as by low soil moisture.

   Caused by Boron deficiency in which root growth is checked.
2. **Pithiness:** More in summer crop than spring or autumn.

   It Caused due to high temperature before harvesting.

   The excess application of N, P, K and stress also causes pithiness.
3. **Hollow root wart:** High temperature during 16-30 days of sowing.

   Soil moisture stress causes wart disorder
4. **Forking:** Splitting of roots is called forking and is a common problem in radish.

   It caused due to hard soil pan, delayed harvesting and water stress followed by excessive water application.
5. **Brown heart:** Dark spots that first appear on roots characterize it. The plants remain stunted due to retarded growth rate. Leaves show variegated appearance with yellow and purplish red blotches. Roots remain small showing distorted and grayish appearance.

   Commonly occurring disorder in radish and turnip caused by boron deficiency.

Brown heart can be controlled by soil application of borax @ 15-20 kg per hectare or foliar application of 0.1 per cent boron.

- Storage – 0-1°C at 90-95% R.H. for 2 months.
- Yield - European varieties yield 5-7 t/ha while in Asiatic cultivars, it varies from 15-20 t/ha.
- Viability of seed 4-5 years.
- Application of fresh undecomposed manure will result in forking of roots in radish.
- White rust is caused by *Albugo condida.*
- Phyllody – Seed disease of radish.

19. "CARROT"

- Origin – Southern Western Asia (especially Afghanistan).
- Punjab and Kashmir are also considered its primary centers of origin with secondary centre in Ethiopia and North America.
- Annual – For root production.
- Biennial – For flowering and seed set.
- Haryana (26560 ha) is having maximum area under carrot cultivation followed by Bihar (14000 ha).
- In carrot production, Haryana (445000 MT) is the leader followed by Punjab (197000 MT).
- Taste of carrot is due to the presence of glutanic acid.
- Carrot juice is a rich source of carotene.
- Per 100 g carrot root has 86 g moisture, 10.6 g carbohydrates, 0.9 g protein, 0.2 g fat, 1890 μg β-carotene
- A special type of beverage known as Kanji is prepared from black carrot which is used as appetizer.
- A sweet preparation called *Gajar halwa* is very famous dish in North India.
- Outer core consist of a thin periderm (a layer of cork cell) and a relatively wide band of secondary phloem. In this region only the sugar is stored.
- Inner core has the secondary xylem and pith.
- Inflorescence - A compound "umbel".
- Flower – Protandrous.
- In temperate countries it is grown in spring, summer and autumn and in tropical and sub-tropical countries carrot is grown in winter.
- Most of the tropical varieties are red coloured due to the presence of the pigment lycopene.
- Orange coloured (Temperate) carrot contains more carotene than Asiatic type and is rich in thiamine and riboflavin.
- Orange colour of the European types is due to β-carotene pigment, a precursor of vitamin A.
- Purple and yellow colour of carrots is due to the pigments anthocyanin and xanthophylls respectively. Absence of all these pigments results in white carrots grown in eastern and western Europe as cattle feed.
- The western carotene carrot (European type) had been derived from the eastern anthocyanin carrot (tropical type) and the later had probably been developed from anthocyanin containing forms of *D. carota ssp. carota* such as those found in Afghanistan
- Carrot green leaves – Contain appreciable amount of protein, minerals, and vitamins.
- Breeding Objective – To develop varieties with high root yield and carotene content, self coloured core, cylindrical, uniform tapering, non-branching.

- The Temperate types form root under both tropical and temperate conditions but set seed only under temperate condition. They need low temperature of 5-8° C for 40-60 days before throwing out the seed stalk (break dormancy).
- Asiatic type are high yielding and produce seed under tropical condition, but they are poor in carotene content and other quality attributes.
- Carrot juice is used for colouring butter and other foods.
- Highest accumulation of carotene occurs in older cells of the phloem.
- Cytoplasmic male sterlity (CMS) is reported in carrot.
- Carrot varieties are divided into two major groups viz. Asiatic or tropical types and European or temperate types.
- European types are firm textured, sweet, highly flavoured, yellowish orange to deep orange, slow bolting and adapted to cool growing conditions.
- Asiatic types are soft textured, less sweet, moderately flavoured, red or reddish orange in colour, quicker to bolt and are adapted to warm growing conditions.
- The temperate types are biennial whereas tropical types are annual.
- Heavy tops, large roots and big core characterize the Asiatic types. These are sown earlier than the temperate types. The Asiatic types have more of anthocyanin pigment and less of carotene.
- The varieties of carrot are divided into two groups namely, Asiatic type and European type:

Asiatic or tropical type	European Type or Temperate Type
1. It produces seed in plains.	1. It produces seed in hills
2. It does not require low temperature treatment for flowering.	2. It requires chilling (4.8-10°C) for 40-60days for flowering.
3. Roots are long and red in colour with white or creamy core.	3. Roots are medium in size and orange in colour with centre core.
4. Rich in lycopene	4. Rich in carotene

- **Recommeded Tropical Varieties of Carrot :**

S.No.	Variety	Features
1	Pusa Kesar (Selection of Local red × Nantes half long)	Red coloured roots and self coloured core, early, rich in carotene 38mg/100g (7753 IU/100g). Most desirable features of this variety are sets seeds freely in plains. Suitable for early sowing (mid August to early October). Maturity 90-100 days. Yield 300 q/ha.
2	Pusa Meghali (Selection of Pusa Kesar × Nantes)	Orange coloured roots, self coloured core and short top. Suitable for early sowing (mid August - early October). Maturity 100-110 days. Yield 250 q/ha. Highest vitamin A (carotene) content (11574 mg/100g.)

3	Arka Suraj	The variety was developed by pedigree selection between Nantes and IIHR 253. Deep Orange roots with self coloured core, conical shape; root length is 15-18 cm with TSS 8-10%, tolerant to powdery mildew and nematode. It flowers and sets seeds under tropical condition.
4	Pusa Rudhira	It has self core red coloured with delayed bolting. Maturity 85-90 days. Average yield 330 q/ha. Gives more than 50% higher yield over check. Pusa Kesar.
5	Pusa Asita	It has self black coloured and is late bolter. Maturity 95-100 days. Average yield 300 q/ha. Gives more than 30% higher yield over check Pusa Kesar.
6	Pusa Vrishti	It is a new heat tolerant tropical carrot variety. It is suitable for early sowing beginning in July under North Indian plains. Maturity 85-90 days. Average yield 250 q/ha.
7	Pusa Vasuda	First public sector tropical carrot hybrid developed using CMS system. Self red coloured carrot hybrid. High in total carotenoids, lycopene, TSS and minerals.
8	Kashi Krishna	Attractive black coloured roots along with self coloured core. Tolerant to bolting, suitable for salad, juice, halwa, kanji (fermented juice) and nutraceutical purpose. Rich source of anthocyanin (285 mg/100g FW), phenolics and antioxidants. Root yield (20-22 t/ha).

Recommeded Temperate Varieties of Carrot

S.No.	Variety	Features
1	Nantes Half Long (IARI, Katrain)	Cylindrical, stumpy, orange coloured roots, self coloured core suitable for sowing from mid October to early December. Maturity 100-120 days. Yield 100-125 q/ha.
2	Pusa Yamdagni	Roots with orange flesh and self coloured core. Suitable for sowing from mid October to early December. Maturity 100-110 days. Yield 150-200 q/ha.
3	Early Nates	Roots with orange flesh and self coloured core. Almost cylindrical roots, terminating abruptly in small thin tall. Suitable for sowing from mid October to early December. Maturity 90-110 days. Yield 100-120 q/ha.

4	Chantaney (Introduction, France)	Roots with orange coloured flesh and self coloured core, thick at the shoulder with gradual tapering. Suitable for sowing from mid October to early December. Suitable for canning, Maturity 100-120 days. Yield 125-150 q/ha. It is suitable for canning and storage.
	Ooty 1 (TNAU,Ooty)	Developed by selection from the local type DC 3. Roots are long, deep orange, slightly tapering and with thin self-colour central core. Harvesting is possible 100-110 days after sowing. Average yield is 490 quintals per hectare. It is resistant to powdery mildew and leaf spot.
5	No : 29	Selection from a local variety, Roots light red, long tapering and thin. Developed by IARI.
6	Sel. 233	Roots are orange, semi cylindrical, maturity duration 90-100 days.
7	Sel 21	Roots red with light orange core, quick growing, suitable for pickles, candy and juice.
8	Imperator (Nantes × Chantaney)	Deep orange coloured roots, self coloured core, suitable for sowing from mid October to early December. Maturity 100-120 days. Yield 120-150 q/ha.
9	Zeno	Introduction from Germany. Roots long orange. Medium tops maturity 110-120 days and suitable for Nilgiries.
10	Danvers	Suitable for fresh market and processing.

- The optimum temperature is 7-24°C for germination and 18-24°C for growth.
- The best root colour develops at 15-20°C. Temperature higher than 30°C, particularly in later stages of development, induces undesirable strong flavour and coarseness in the roots.
- Shape of the roots is not affected by the increase in temperature.
- The tropical types are sown early in the season and sowing continues till early September.
- Sowing of temperate types begins from early October. In hills carrot is sown from March to July. In south and central India, sowing is done in January-February, June-July and then in October-November.
- Successive sowings at 10-15 day intervals ensure continuous supply of roots.
- Sowing is done on ridges spaced at 45 cm for tropical types and 30 cm for temperate types.
- Shallow sowing about 1.5 cm deep gives good germination.
- Germination is slow and usually takes 10-20 days for seedlings to emerge.
- Seed rate – 5-6 kg/ha.
- Excess "N" – Effects quality of root like sugar content, dry matter, carotene, vitamin-C, storage quality etc. In field will increase the rotting of roots during storage

- Un-decomposed FYM should be avoided as it induces forking and excessive hairiness.
- High temperature in carrot cause bolting.
- Roots harvested with top are called bunch carrot while those without the tops are called bulk carrots.

Physiological disorder –

1. **Root splitting:** Splitting or cracking of carrot roots is a major problem.

   Possible reasons:

   - Wider spacing as larger roots tend to split more.
   - Dry weather followed by wet weather is conducive to cracking of roots.
   - High nitrogen application
   - Early cultivars tend to split more readily than late ones.
2. **Cavity spot:** It appears as a cavity in the cortex. In most cases, the subtending epidermis collapses to form a pitted lesion.

   Possible reasons:

   - Calcium deficiency associated with an increased accumulation of K and decreased accumulation of Ca.
3. **Forking:** It is a common disorder in carrot and radish formed by the enlargement of secondary root growth.

   Possible reasons:

   - Excess moisture during the root development, it also occurs on heavy soils due to soil compactness.

Yield –

Asiatic – 25-30 tons/ha.

European – 10-15 tons/ha.

- Storage – 0-4.4 ° C temperature and 93-98% relative humidity for 6 months.

Seed Production

- Seed production of Asiatic type varieties of carrot can be under-taken in north Indian plains whereas that of European types can only be under-taken in temperate regions comprising higher hills.
- A minimum of 800 m isolation distance is required to produce trueto-type seed.
- Seed is produced by two methods viz. root to seed method and seed to seed method.
- In seed to seed method, the plants are allowed to bolt in-situ. However, root selection cannot be exercised in this method.
- Root to seed method is preferred for seed production.
- In root to seed method, plants are uprooted and roots selected for re-planting in seed production block.

- Stecklings are prepared by chopping off lower 2/3rd of the root and retaining only a new leaf growth.
- One hectare of root crop produces stecklings sufficient for planting four hectares of seed crop.
- Seed is harvested from primary and secondary umbels, as the tertiary umbels produce poor quality seed.
- The seed is dried to 8 per cent moisture level before storage.
- Average seed yield of carrot is 5.0-6.0 q/ha.
- Other carrot family related crops are-

Crop	**Botanical Name**	**2n (2x)**	**Remarks**
Parsnip	Pastinaca sativa	22	It is a root vegetable closely related to carrot and parsley.
Chervil	Anthiriscum cerefolium	32	A multifunctional nutritive herb.
Celeriac	Apium graveolum var. rapaceum	22	It is also known as turnip-rooted celery or knob celery.
Skirret	*Sium sisarum*	12	Skirret is a minor European root crop with a wild distribution across Asia.

- All four plants belong to Umbelliferae family.
- Presence of polyacetylenes, contributes to the unpleasant odour in carrots.
- Carrot seed contains germination inhibitors known as carrotal.
- Propazine – Excellent residual and selective weed control in carrots.
- In India carrot was introduced from Persia, Probably carrot was first grown in Arne, the Salem gardens about 1620 AD.
- Dietary supplementations of a combination of carrot and orange juice have been found to control the oxidation of low density lipoproteins in habitual cigarette smokers.
- Carotene reduces the risk of cataract formation (clouding of the lens of the eye which leads to a decrease in vision).
- **Diseases:** The important disease are leaf blight, leaf spot or Cercospora blight, powdery mildew, watery soft rot, black rot, and bacterial soft rot.
- **Insect-pests:** The serious pests are rust fly and turnip moth.

20. "TURNIP"/ CRUCIFERAE ROOT VEGETABLE

- The Mediterranean area is thought to be the primary centre of origin of European types.
- Eastern Afghanistan and the adjoining areas of Pakistan are considered to be the primary centre of origin of Asiatic types.
- Asia Minor, Transcaucasus and Iran are considered its secondary centre of origin.

- Inflorescence is terminal raceme.
- Young leaves are rich source of vitamin–A (15660 IU/100g.), vitamin–C and Iron.
- It is a biennial in nature.
- The fleshy thickened underground portion of turnip is actually the hypocotyl.
- A distinct tap root and secondary roots arise from the lower part of the swollen hypocotyl.
- Mainly grown in Bihar, Punjab, Tamil Nadu, Jammu and Kashmir, Himachal Pradesh and Haryana.
- It can be grown at an elevation of 1500 M. at MSL or above, but it is not suitable for growing in low lands of wet tropics.
- Mass Selection – Most adopted breeding method for crop improvement.
- Turnip has sporophytic self-incompatibility; hence there is no need for emasculation.
- Pollination – Entomophilous (flies) and Honey bees (main source).
- Sodium chloride ($NaCl_2$) and Carbon dioxide (CO_2) can also be used for overcoming self-incompatibility.
- Selfing can be achieved by application of fresh pollen on decapitated stigma.
- Thinning –Important operation to maintain optimum plant to plant spacing (10-15 days after germination).

Recommended European or Temperate type Varieties of Turnip:

S.No.	Variety	Features
1.	Pusa Swarnima (IARI, Katrain)	It is an early maturing variety and has been developed by selection from the cross Japanese White × Golden Ball. Roots large, flattish round, pale amber coloured, rich in carotene content and sweet in taste. Suitable for sowing from October to December. Maturity 65-70 days. Yield 350-375 q/ha.
2	Pusa Chandrima (IARI, Katrain)	It is suitable for sowing from October to December in plains and developed by selection from the cross Snow Ball × Japanese White. Flesh is sweet and tender. Roots are ready for harvesting 50-60 DAS. Average yield is 400 q/ha.
3	Golden ball (Introduction, England)	It is suitable for sowing late in north Indian plains. Roots are round, medium sized with yellow skin and flesh. It reaches marketable maturity 70 DAS. Average yield is 175 q/ha.
4	Purple Top White Globe (Introduction, USA)	This is suitable for sowing late in north Indian plains. Plant top is medium heavy. Roots are large, round, smooth with bright purple top and white base. Maturity 60 DAS. Average yield is 200 q/ha. Best for dehydration.

5	Snow ball	It is an early variety with medium sized roots. Roots are round, smooth with pure white skin, whereas the flesh is white, finely grained, sweet and tender. It has a small erect top with medium yellow green and cut leaves.
6	Early Milan Red Top	Extra early (45 day) variety. The roots are deep flat with purplish red tops and white underneath. The flesh is pure white, well gained, crisp and mildly pungent. The tops are very small with only 4 to 6 sessile leaves.

Recommended Asiatic Tropical Type Varieties of Turnip

S.No.	Variety	Features
1	Pusa Sweti	Very early maturing variety. Roots pure white, round and slightly flat with rat tail habit. Suitable for sowing from early August to September. Maturity 45-50 days. Yield 250-300 q/ha.
2	Pusa Kanchan (IARI ,New Delhi)	Developed by selection from the cross Local Red Round (Asiatic type) × Golden Ball (Temperate type). It produces seed satisfactorily in the plains, ready for harvesting 50-55 DAS. Yield from 250-400 q/ha.
3	Punjab Safed – 4 (PAU, Ludhiana)	A selection from local cultivar. Early maturing, roots pure white, round, medium size and mild taste.
4	L-1 (PAU, Ludhiana)	Developed from the cross PTWG × 4-White. Roots are round, pure white, smooth, rat-tailed and crisp with mild flavour. It takes 45-60 days to reach marketable maturity. Average yield is 260 q/ha.

European or temperate type	Asiatic or tropical type
The temperate varieties are sweet, tender and palatable	Asiatic types are pungent with tough texture and are suitable for pickle making.
Temperate types set seed only in hills.	The Asiatic types set seed freely in plains
The temperate varieties can withstand cold temperature and can be sown in late winters also.	The tropical types are tolerant to high temperature and therefore, sown early in the season.

- **Sowing time –**
  July-September-(Asiatic Types)
  October - December-(European Types) In North Indian Plains
  In Hills - July-September.
- Root develops best flavour, texture and size at a temperature of 10-15°C.
- In hot weather, roots become fibrous, tough and pungent.

- The Asiatic types can tolerate fairly high temperature but flourish under cool weather.
- Short day length and cool weather favours proper root development and long day and high temperature induces early bolting.
- Row × plant spacing is maintained at 45 × 7.5 cm.
- Sowing is done on ridges for proper development of the roots.
- Seed germination takes place in less than a week.
- Seed rate – 3-4 kg/ha.
- Seed sowing depth-1 .5-2.5 cm.
- Yield – 200-400 q/ha.
- Seed production of Asiatic type varieties of turnip can be under-taken in north Indian plains where as that of European types can only be under-taken in temperate regions comprising higher hills.
- It is highly cross-pollinated and insects, mainly honeybees, do pollination.
- A minimum of 1000 m isolation distance is required to produce true-to-type seed.
- Turnip is cross compatible with Chinese cabbage, rutabaga, mustard and rape. Therefore a similar isolation distance is required from these crops.
- Seed production is done by two methods viz. root to seed method and seed to seed method.
- One hectare of root crop produces stecklings sufficient for planting four hectare of seed crop. Spacing is maintained at 45 × 15 cm.
- At least three inspections at different plant growth stages viz. vegetative growth, root development and flowering are required to rogue out off-type and diseased plants.
- Seedpods turn reddish-brown at seed maturity.
- Average seed yield is 6.0-8.0 q/ha.

Physiological disorder –

1. Whiptail – Due to "Mo" deficiency, more common in acidic soils.
2. Brown Heart – "B" deficiency, more common in acidic soils.

- Turnip yellow mosaic virus disease- transmitted through flea beetle, cabbage may serve as an alternate host for this virus.
- Crinkle – Viral disease.
- Seed viability – 5-6 years.

21. "BEET ROOT" (CHUNKANDAR)

- Origin from Mediterranean region.
- Fruit type – Capsule.
- Inflorescence – Spike (develops in 2nd year).
- Probable progenitor of garden beet is *Beta vulgaris sps. maritina.*
- Beet root, sugar beet and Palak (*B. vulgaris*) are cross compatible.

- After sugar cane it is the second most important crop for the preparation of sugar. The sugar content of beetroots varies between 6 to 10 per cent on fresh weight basis (high sucrose contents in the roots).
- Beet root is the rich source of protein, carbohydrate, Ca, P and vitamin C.
- The colour of root is due to the presence of red-violet pigment Beta-cyanins and yellow pigment Beta-xanthins.
- Upper portion of root develops from the hypocotyl and the lower portion from the tap root, from which the secondary roots arise.
- Beet crop can tolerate freezing to some extent.
- In some European countries, it is also used as an animal feed.
- Under high warm weather "zoning" is caused which is marked by appearance of alternating light and dark-red concentric circles in the roots.
- The seeds usually contain one or more seeds and are called "Seed ball" which is viable for 5-6 years under normal storage condition.
- Thinning is practiced to maintain the intra-raw spacing and optimum population.
- Thinning is not practiced in the crop grown for processing, as small roots are preferred for processing.
- Seed ball's rate – 7-9 kg/ha.
- **Spacing –**
  40-60 cm Row × Row
  8-10 cm Plant × Plant
  2.5 cm Depth of seed
- **Sowing time –**
  March -July — In Hills region
  September - November — In North Indian Plains
  July-November — In South India.

Recommended Varieties of Beetroot

S.No.	Variety	Features
1	Crimson Globe (Introduction, USA)	Roots are globular to flattened, medium red skin, flesh medium dark, crimson red with indistinct zoning. Heavy yielder. Maturity 80-90 days. Yield 180 q/ha.
2	Detroit Dark Red (Introduction, USA)	Roots are round with smooth, uniform, deep red skin. Flesh dark blood red with light red zoning. Heavy yielder. Maturity 80-100 days. Yield 180 q/ha.
3	Crosby Egyptian (Introduction, USA)	Roots are flat globe and smooth outer surface. The flesh colour is dark purplish red with some indistinct zoning. Maturity 55-60 days. Yield 170 q/ha.
4	Early Wonder (Introduction, USA)	Roots are flattened globe with a smooth, dark red skin. Flesh dark red with some light red zoning. Maturity 55-60 days. Yield 160 q/ha.
5	Ooty 1 (Udhagamandalam)	Suitable for growing in South Indian Hills. Its skin is thin and flesh colour is blood-red. Maturity 120-130 days. Average yield is 310-450 q/ha.

- Madhur, Ruby Queen and Ruby Red are a few of the varieties marketed by private seed industry. It is a cool season crop and can tolerate freezing temperature.
- During sowing, if temperature is high, the roots produced would be coarse with woody-flesh and dull colour.
- Unlike other root crops, beetroot does not have tropical type varieties.
- Beet root seed is multigerm which produce 3-4 seedlings per seed ball hence thinning is an important operation.
- Temperature of 4.5-10°C for 15 days induces premature bolting.
- All varieties are temperate types and biennial in nature.
- Based on the shape of roots, beetroot varieties are classified into five groups viz. flat, short top, round or globular, half long and long.
- Most of the varieties cultivated in India are the exotic ones.
- Beetroots develop best colour, sugar content, root quality and shape when temperature remains between 15-19°C.
- The germination is slower and erratic if temperature is below 7°C.
- The probable origin of beet root is Mediterranean region for *B. vulgaris sp. maritima* possibly by hybridization with *B. patula.*
- Beetroot is sensitive to high acidity and low temperature.
- It can grow in saline and alkaline soil up to 9-10 pH.
- Raking is done to break the crust as seeds sprout poorly in crusted soil.
- Optimum temperature for growth is 18.3-21°C.
- Harvesting is done when they are about 3.5 cm in diameter.
- Exposure of crop to temperature between 4-10°C for 15 days or more induces premature bolting.
- Beets for bulk marketing or canning are topped in the field when they attain a diameter of 6.5-8 cm.
- At temperature above 25°C, the root colour, texture and quality are adversely affected.
- Roots become stingy and tough in response to high temperature.
- Yield - 250-300 q/ha.
- Storage in 0°C with 90% R.H up to 4-6 months.
- Beet mosaic yellow a viral disease is transmitted by aphid.
- Curly top disease is transmitted by hoppers.
- High temperature in beet causes poor colour development.
- Seed production is done in hills above 1200 meter elevation.
- Beetroot is a biennial and requires temperate climate for seed production.
- The plants require exposure to 4.4-7.7°C for 6-8 weeks for initiation of seed stalks.
- The seed is produced by root-to-seed and seed-to-seed methods. The root crop for seed production is raised the same way as for the vegetable crop.
- One ha of root crop is sufficient for planting 10-12 ha of the seed crop.

- Beetroot is highly cross-pollinated (by wind) and a minimum isolation distance between two varieties is 1000 m, similar isolation distance is required from other crops, like Swiss chard, sugar beet etc., cross-compatible with beetroot.
- The crop is harvested when 70-80 per cent of the seed balls become hard and those at base of the shoots turn brown.
- Average seed yield is 10-12 quintals per hectare.

Physiological Disorder

1. Internal Black Spot / Brown Heart/ Heart rot/ Breakdown of beet – Hard and corky black spots are scattered throughout the root.
   - Caused due to "B" deficiency in neutral or alkaline soils.
   - "B" deficiency first affects young cells and tissue.
   - Detroid Dark Red variety of beet show comparatively less symptoms of "B" deficiency.
2. Speckled yellow is caused due to deficiency of "Mn". "Mn" deficiency is very severe in sandy and very alkaline soils.

   Symptoms – leaves with yellow blotches between veins tend to curl up in summer.
   - "B" deficiency is common in alkaline soil or on soils relatively high in "Ca".

Viral disease

- **Beet mosaic virus:** The virus is transmitted by aphids. The affected plants remain stunted. The disease reduces photosynthetic activity, thus adversely affecting root size, sugar content and root quality.
- **Curly top:** The virus is transmitted by leafhoppers. The leaves remain under-sized, curl upward and leaf veins become transparent.
- **Yellows:** The virus is aphid transmitted. The affected leaves become chlorotic, thickened, leathery and brittle.

22. IMPORTANT PEST AND DISEASE OF ROOT CROPS

Sr. No.	Pest and Disease	Symptoms	Management
1.	Leaf hopper *(Empoasca Pujabensis)*	Both Adult and Nymph suck sap.	Apply Monocrotophus or Phosphamidon.
2.	Cut Worm *(Agrotis ipsilon)*	Seedlings are cut in the base at night by cut worm.	Soil drenching with Chloropyriphos.
3.	Carrot Fly *(Psila rosae)*	Larva feeding upper mid portion of the top root.	Insecticides used are organophosphates or carbamaetes.

4.	Aphids (*Myzus persicae, Brevicoryne brassicae*)	It attacks autumn season root crop of radish and seed crop of both radish and turnip. Both nymphs and adults of the tiny insect damage the crop by sucking cell sap and by spreading the virus diseases.	Spray the crop with Malathion 50 EC @ 0.05 percent or Rogor 30EC (dimethoate) @ 0.03 percent.
5.	Leafminer (*Pegomya hyoscyami*)	It affects the crop by mining the leaves and feeding on them. The leaves show blistered appearance and plant growth is checked	Spray systemic pesticides like Rogor 30 EC (dimethoate) @ 500 ml per hectare.
6.	Alternaria leaf spot (*Alternaria raphani*)	Symptoms first appear on leaves as small, yellowish and slightly raised lesions. The spots enlarge as they become older. Later on the lesions also appear on stem and pods. The infection spreads rapidly in humid conditions.	Follow long term crop rotation, use disease free seed, treat the seed with hot water at 50°C for 30 minutes and then with Captan or Thiram @ 2 g per kg seed.
7.	White rust (*Albugo candida*)	The disease is more serious on radish seed crop. The symptoms appear as white blisters on leaves, small stems and floral parts. Infected plant parts become deformed and bear only malformed flowers. Pod formation, seed setting and seed yield is drastically reduced	Follow crop rotation with non-crucifer crops. Spray the crop with Dithane Z-78 or Ridomil MZ @ 0.2-0.3 per cent. Repeat the spray after 10-15 days.
8.	Mosaic Virus	The virus is aphid transmitted. Symptoms first appear as a small circular to irregular chlorotic lesions inbetween and adjacent to the veins. Little or no leaf distortion is noticed. Radish plants may develop enations.	Remove and destroy the diseased plants and alternative host plants from vicinity of radish and turnip fields. Check aphid population as recommended under insect-pests.
9.	Cercospora leaf spot (*Cercospora carotae*)	It is an important foliage disease of carrot. The symptoms first appear as elongated lesions along the edge of the leaf segment resulting in a lateral curling.	Spray the crop with Dithane Z-78 @ 2.0 g per litre of water when 50 per cent of middle leaves show symptoms. Subsequent sprays are applied at 7-10 day intervals.

10.	Downy mildew (*Peronospora farinosa*)	Light-green spots of varying sizes appear on upper surface of the leaves. The fungal growth (mildew) is seen on underside of the leaves. In dry weather, the margins of leaves may show pale red pigmentation. In humid conditions, the disease appears on all above ground plant parts.	Treat the seed with Captan or Thiram @ 2-3 g per kg seed. Grow resistant varieties and spray the crop with Indofil Z-78 @ 0.3 per cent.
11.	Bacterial Blight (*Xanthomonas campestris* pv. *carotae)*	The bacterium causes irregular brown spot on leaves, dark brown streaks on petioles and a blighting of floral parts.	Spraying early with Copper Oxychloride 0.25 %
12.	Bacterial soft rot (*Erwinia carotovora* sp. *Carotovora)*	Cells become water soaked, the middle lamella is destroyed and the cells collapse into a soft, watery slimy mass.	Dipping in a solution of 1:500 of sodium hypochlorite before storage or transits reduce the disease.

23. BULB CROPS

Family - Amaryllidaceae

Common Name	Botanical Name	2n	Edible Part
Onion	*Allium cepa*	16	Bulb
Garlic	*Allium sativum*	16	Cloves
Leek	*Allium ampeloprasum* Var. *porrum*	32	Stem and Leaves
Multiplier / Underground Onion	*Allium cepa* Var. *aggregatum*	16	Leaves and bulbs
Perennial onion / Shallot	*Allium cepa Var. ascalonicum*	16	
Tree Onion	*Allium cepa* Var. *viviparum/ proliform*	16	
Japanese Onion	*Allium fistulosum*	16	
Chive	*Allium scheonoprasum*	16, 24, 32	

- All bulb crops belong to monocotyledonae vegetables and primary center of origin is Central Asia.

24. "ONION"

- Economically most important bulb vegetable.
- Harvested organ: swollen base of the stem (bulb) with numerous fleshy leaves.
- India rank IInd in area and production in the world after china and third in export (12% of total world export).
- In export Netherland (21% of world export) ranks first and Spain IInd.
- In India Maharashtra, Karnataka, Gujarat, Orissa, Andhra Pradesh, Uttar Pradesh, Tamilnadu, Rajasthan and Bihar are major onion producing states.
- Maharashtra (507960 ha) is having maximum area under onion cultivation followed by Karnataka (195280 ha).
- In onion production, Maharashtra (8854000 MT) is the leader followed by Madhya Pradesh (3701000 MT).
- Average yield per unit area is very low 10.6 tonnes/ha when compared with world (16.5 tonnes/ha., USA 43 t/ha., Japan 42.6 t/ha.).
- Onion accounts for 77% of total foreign exchange earning among fresh vegetables.
- In India– Yield is highest in Gujarat (23.37 t/ha.) followed by Punjab (20.13 t/ha.).
- Per capita availability of onion in Netherlands (32.99 kg/year) is highest followed by Turkey (29.82 Kg/year).
- In India – The availability of onion per capita is 3.94 kg/year.
- Onion is divided into 3 main groups:
  1. The common onion bulbs are formed as single plants ,inflorescence does not form bulbet, biennial, often grown as an annual ,most important commercially propagated in the main from true seeds ,sometimes planted as transplants, small bulbs (sets) are produced from thickly planted seed when sets are replanted, these bulbs reach maturity rapidly, cultivars vary for bulb shapes, dry scale colour (white, yellow, and red), pungency (ranging from mild and sweet to pungent), soluble solids, adaptation to different day lengths when harvested early they produce spring or bunching onions, mature bulb onions are usually dried before marketing .
  2. The aggregatum group with many lateral bulbs or shoots inflorescences, lacking bulbets, sterile seed production, propagation by vegetative means includes the potato onion or multiplier onion, ever-ready onions, and shallots, shallots are sometimes called scallions, a source of confusion because *A. fistulosum* is also called scallions.
  3. The proliferous (proliferum) group – ground bulbs are sometimes poorly developed, infloresence bears bulbets, true seed is usually lacking, reproduction is by vegetative infloresence and bulbets not commercially cultivated almost exclusively for home gardening, includes tree onions, top-set onions, and Egyptian onions.
- The Jews loved onion so much that they build a city near the Gulf of Suez in 173 BC and named it 'Onion'.

- Mostly the Indians preferred and pungent type of onion while in Japan, Europe and America yellow coloured varieties are preferred.
- Per 100 g edible part of onion bulb contains 86.6 percent moisture, **11.1 g carbohydrates, 1.2 g protein,** 11 mg vitamin C, 47 mg calcium and 0.7 mg iron
- Alliums are characterized by the presence of remarkable sulphar containing compounds, which gives their distinctive smell and pungency.
- Oil yields methyl and propyl compounds in concentration 0.002-0.03g/100g. of fresh bulb by steam distillation.
- Fructans, the most abundant soluble carbohydrate in onions, are a type of dietary fiber that helps to reduce chances of colorectal cancers.
- Onion is rich source of carbohydrate, proteins, minerals (calcium) and vitamins (riboflavin).
- Red colour varieties are more pungent followed by brown, yellow and white.
- Pungency is due to – Allyl propyl -di-sulphide.
- Red colour is due to – Anthocyanin.
- Yellow colour is due to –Quercetin.
- Antifungal activity-Phenolic factor i.e. "Catechol".
- Inflorescence – Umbel.
- Flower colour – White or blueish.
- Number of seed stem – 1 to 20 or more.
- Onion is a highly cross pollinated crop due to its protandry.
- Pollination is chiefly by honey bees.
- Flies are excellent pollinators as they go from flower to flower.
- Onion (Allium) is known to benefit in the prevention and treatment of atherosclerosis and coronary heart disease.
- Onion is a cool season crop.
- For good vegetative growth 12.8-23°C temperature before bulbing is best.
- For bulb development 15.5 - 25°C temperature is best.
- In onion very low temperature in early stage favours bolting.
- Sudden rise in temperature causes early maturity in rabi crop resulting in small bulb size.
- Almost all cultivars grown in plains in India are short day cultivars.
- "Brown Spanish" – A long day variety, require day length of more than 14 hours and low temperature for proper bulb development.
- Kharif onion varieties require day length of 10-11 hours.
- Rabi onion varieties require day length of 12-13 hours.
- Long day varieties do not bulb under short day condition where as short day varieties if planted under long day condition will develop early bulbs.
- Onion is a cool season and photosensitive crop.

- Short day varieties require 11-13 hour day length where as long day varieties require day length of 13-15 hours.
- Short day varieties if grown under long day conditions will develop small bulbs.
- All varieties adapted to north Indian plains are short day varieties.
- Seed longevity in onion is poor, compared with most of the other vegetables. To obtain optimum germination, it is important to use fresh seed.
- In North and North-East India, rabi onion nursery is sown from mid-October to mid-November. Sowing is done in September-October in South and hilly regions and November-December in Maharashtra.
- Prepare 15-20 cm raised nursery beds in 200 sq m area.
- Sowing is done in lines approximately 5 cm apart and at a depth of 1-2 cm depending upon the soil type.
- First irrigation is applied 1-2 days after transplanting.
- Delayed planting results in reduced bulb size and pre-mature bolting.
- Planting is normally done on flat beds. However, planting on raised beds improves bulb size and quality.
- Bed spacing is maintained at 60 cm. Each bed accommodates three rows.
- Where the season is sufficiently long to provide early pre-bulbing growth, onion can also be raised by direct seeding, as is the practice in some parts of Maharashtra.

Bulbset technique

- Nursery for kharif onion production often fails due to hot weather conditions.
- 'Bulbset technique' is a sure method of raising kharif onion crop.
- To produce bulb sets, sowing is done in mid-March the same way as for the main (rabi) season crop.
- Bulbsets are uprooted when they attain pencil size thickness by end of June and are stored in baskets in a ventilated room.
- The bulbsets are transplanted in the field in second fortnight of August.
- The crop is ready for harvesting in end November or early December. However, the tops remain green and do not fall over due to low temperature.

Seed rate

10-12 kg/ha for rabi season.

12-15 kg/ha for Kharif season.

- Seedlings are ready for transplanting in 8-9 weeks in rabi where as 6-7 weeks in kharif season.
- The best time for sowing kharif onion nursery is middle of June.

Sowing time -

August - September for kharif season crop.

October - November for rabi season crop.

March- April for Hilly areas.

- Spacing 4-5 cm apart in rabi season.
  5-7 cm apart in kharif season and 2-3 cm deep.
- Planting by bulb – To meet the demand of green onion for sale in early winter, planting by bulbs is done.
- Bulbs planted in August-Sep, become ready in October – December.
- If planting is done by sets, 1.5-2.0 cm of sets amounting 10 q/ha are required.
- "Zn" application (1-3 ppm) increases yield and improves quality.
- For direct sowing of the variety "Agrifound Rose", 22.5 kg/ha seed is required.
- In case of multiplier onion sets 1-1.2 tonnes of bulbils are required for planting in a hectare.
- "Cu" and "B" @ 13.4 kg and 1.8 kg/ha respectively are beneficial in increasing the yield.
- Onion responds very well to organic manures. Therefore, the soils for onion should be liberally manured and fertilized.
- To produce 30 tonnes of bulb yield per hectare, it removes 85, 42 and 130 kg of nitrogen, phosphorus and potash respectively.
- Deficiency of nitrogen enhances maturity but reduces bulb size and ultimately bulb yield.
- Higher doses of nitrogen result in excessive vegetative growth, increased bulb size with large thick neck and poor storage quality.

Crop Improvement

- Jones and Clarke (1943 and 1963) were first to use cytoplasmic male sterility (CMS) for producing hybrid seed in onion.
- Selfing and massing technique – Suggested by Jones and Mann (1963), improvement in cultivar can be affected by selfing followed by massing.
- Onion seed production industry has moved away from open-pollinated types to hybrid production systems.
- Cytoplasmically inherited male sterility factor (*S*) and a nuclear encoded gene (*Ms*) which can normally override the male sterility or "restore" the plant to a normal fertile phenotype.
- When this "restorer" gene is in the homozygous recessive condition only in the *S* cytoplasm does male sterility result.

  S msms (= sterile)
  S Msms (= fertile)
  S MsMs (= fertile)
  N msms (= fertile) (*N msms* plants are known as maintainers)
  N Msms (= fertile)
  N MsMs (= fertile)
- Onion is insect (bee) pollinated.

- In hybrid production it is important to have both parental lines flowering at the same time.
- Ratio of 8 female: 2 male rows are commonly used.
- Three-way hybrids is prepared by crossing an *S msms* F_1 hybrid (male sterile) to a third inbred parent.
- Similar male sterility system occurs in chives (*A. schoenoprasum*).
- Five major genes determine onion bulb colour:
    - *i*-inhibits the production of colour and the bulbs are white (the expression of red, yellow or chartreuse pigments requires that the plant be *ii*).
    - *C* locus conditions the production of coloured bulbs. All *iicc* bulbs are white.
    - Plants with a dominant allele at the *C* locus and homozygous recessive *rr* at the *R* locus are yellow to brown in colour.
    - Red bulbs are conditioned by dominant alleles at both the *C* and *R* loci (*iiC-R-*).
    - A dominant allele at the *G* locus conditions a golden colour without pink. Chartreuse bulbs are homozygous recessive at the *G* locus.
    - *L* conditions a red pigment when the plant has the genotype *iiC-rrG-L-*.
- According to their photoperiod requirement, they are classified as long day, short day and day neutral cultivars.
- Onion cultivars are also classified according to their growing season as rabi season and kharif season cultivars.
- **Recommended Varieties of Onion**

S.No.	Variety	Features
1.	Pusa Red	Medium sized, flattish round, light red coloured bulbs. Bulb weight 90 g. TSS 12-13%. Very good storage. Maturity 140-145 days after transplanting. Yield 250-300 q/ha.
2.	Pusa Riddhi (IARI, New Delhi)	Medium sized, round with tight neck and with good keeping and export quality. Maturity 125-130 days after transplanting. Bulb weight is 70-100 g, yield 315 q/ha. Suitable for both rabi and kharif season.
3.	Early Grano	Large globular yellow bulbs, good for salad and for green onion purposes. Bulb weight 100-120g. TSS 6-7%, very poor in storage. Maturity 95-110 days. Yield 400-500 q/ha.
4.	Pusa Madhvi (IARI, New Delhi)	Developed from a local material collected from Uttar Pradesh by selfing and massing. Light red, roundish flat bulbs. Bulbs weight 90-100 g. TSS 11-12%. Good in storage. Maturity 130-135 days. Yield 300-350 q/ha.
5.	Pusa White Flat	Flattish round, white bulbs. Bulb weight 95-100g. TSS 11-12%. Drying ration 9:1. Suitable for dehydration purposes and also good for green onion purpose. Good in storage, maturity 130-135 days. Yield 325-350 q/ha.

6.	Pusa White Round	Roundish flat, white bulbs. Bulb weight 90-95g., TSS 12-13%, drying ratio 8: 1. Suitable for dehydration and green onion purposes. Maturity 130-135 days. Yield 300-325 q/ha.
7.	Pusa Ratnar (NBPGR, New Delhi)	Developed by selection from segregating population of an F1 hybrid 'Granex', introduced from the USA. Roundish flat, deep red bulbs. Bulbs weight 100-110g. TSS 9-10%. Fair in storage. Maturity 125 days. Yield 300-400 q/ha.
8.	N 2-4-1	Roundish flat, deep red bulbs. Bulb weight 80 g. TSS 12%, good in storage. Maturity 130-140 days. Yield 200-250 q/ha.
9.	N-53	Roundish flat, deep red coloured bulbs. Bulb weight 80 g. TSS 9-10%. Poor in storage. Suitable for Kharif season. Maturity 100-110 days. Yield 150-200 q/ha.
10.	Agrifound round	Roundish flat, light red bulbs. Bulb weight 90 g. TSS 12%, good in storage. Maturity 135-140 days. Yield 250-300 q/ha.
11.	Agrifound Light Red	Bulbs are light red and globular with TSS content of 13 %. It has less incidence of pre-mature bolting and good keeping quality. Maturity 140 days. Average yield is 300-325 q/ha.
12.	Agrifound Dark Red (NHRDF, Nasik)	Bulbs are medium large with tight skin, globular, deep red and moderately pungent with TSS content of 12-13 %. Among the kharif season varieties, it has the best storage quality. Maturity 95-110 days. Average yield is 300 q/ha.
13.	Arka Niketan	Developed through mass selection from IIHR-153 (sel.-13). Globe shaped medium large bulbs with an average weight of 100-180 g. TSS 12-14% Duration 145 days. Suitable for rabi season. Yield 37 t/ha.
14.	Arak Bindu	Developed by 5 generation of mass selection for IIHR-402 (local collection) (Rose onion selection). Bulbs small (2.5-3.5 cm diameter), deep red, highly pungent, flattish globe shape, high TSS (14.16%) with high dry matter content. Suitable for export market. Duration 90-100 days. Yield 25 t/ha.
15.	Arka Pitamber (IIHR, Bangalore)	Developed through pedigree selection from the cross U.D 100 × IIHR-396 (Yellow onion). Bulbs yellow, medium sized (5.2-6.0 cm diameter). Less pungent with TSS 11% and total sugar 9.81%. Suitable for export market and kharif and rabi seasons. Duration 140 days. Yield 35 t/ha.

16.	Arka Kirthiman	An F_1 hybrid between MS-65 and Sel. 13-1-1 (Hybrid). Bulb red in colour with weight 120-130g. Tolerant to purple blotch, basal rot and thrips, long storage (4-5 months). Suitable for kharif and rabi seasons. Duration 125 days. Yield 47 t/ha.
17.	Arka Lalima	An F_1 hybrid between MS-48 and sel. 14-1-1 (Hybrid 5). Bulbs red in colour with weight 120-130g. Suitable for kharif and rabi seasons. Duration 130-140 days. Yield 50 t/ha. It is tolerant to purple blotch, basal rot and thrips
18.	Brown Spanish	Developed by IARI, Long day type, can grow in hills, maturity 160-180 days, yield 25 t/ha.
19.	Arka Kalyan	Globe shaped bulbs with deep-pink coloured outer scales, high TSS (10.5 to 12%) suitable for Kharif season having a yield about 470 q/ha. It takes 140 days from seed to bulb.
20.	Arka Pragati	Globe shaped thin necked and deep pink bulb. It has fleshy scales with high pungency. TSS (10.5-12.5%). The yield potential is 450 q/ha. Suitable for both the seasons. It takes only 130 days from seed sowing to bulb harvesting.
21.	Kalyanpur Red Round (CSAUAT, Kalyanpur)	Bulbs are bronze-red, globular, and moderately pungent with TSS content of 13-15 percent. It takes 120 days from transplanting to harvesting. Average yield is 250-300 q/ha.
22.	Punjab White (PAU, Ludhiana)	It has been developed by mass selection from exotic material English White Spanish. The bulbs are white with tight neck and less pungent. It has TSS content of 15 % and is suitable for dehydration. Average yield is about 337 q/ha.

- Onion is a shallow rooted crop and therefore requires light but frequent irrigation throughout the crop season.
- Fluctuations in water supply cause splitting of bulbs.
- Excessive irrigation produces bulbs with thick neck.
- To improve storage life of bulbs, stop irrigation at least 15 days before harvesting.
- Depending upon soil and weather conditions, total number of irrigations would be 15-18 in rabi season and 8-10 in kharif season.
- Irrigation at bulb formation is a must and moisture stress at this stage results in poor bulb development and consequently low yield.

Varieties classification on basis of colour:

1. **Red coloured varieties – High yielding, mild to high pungent and good keeping quality.**

(a) Agrifound Dark Red – Suitable for kharif season.

(b) Agrifound Light Red – Recommendation for rabi season.

(c) Agrifound Red – Multiplier onion variety.

(d) Agrifound Rose – Suitable for kharif, export purpose mainly grown in Kolar an Banglore district of Karnataka and Chuddaph district of Andhra Pradesh.

(e) N.53- Most suitable for kharif, developed by MPKVP Rahuri (Maharashtra) yield 250-300 q/ha.

(f) Arka Kalyan – Kharif season, resistant to purple blotch.

(g) Arka Niketan – Rabi season, good for storage.

(h) Arka Bindu – Recommendation for Rabi season, suitable for export to Malaysi and Singapore.

(i) Arka Pragati – Suitable for both rabi and kharif season.

(j) Pusa Red – Rabi season, very good keeping quality less bolting and less pungent short to intermediate day length type.

(k) Pusa Ratnar – Rabi season, fair in storage quality.

(l) Pusa Madhvi – Rabi season, good keeping quality.

(m) Punjab Naroya: Tolerant to purple blotch disease and insect thrips (*Bemisia tabaci*).

(n) Punjab Red Round – Rabi season, early maturing.

(o) Punjab Sel. – Rabi season, comparatively more tolerant to purple blotch and thrips, suitable for dehydration.

(p) Kalyanpur Red Round – From Chandrasekhar Azad University of Agriculture and Technology, Kanpur, suitable for rabi season.

(q) Udaipur 101 – Rabi season, good for salad.

(r) Udaipur 103 – Rabi season.

(s) Hisar -2 – Rabi season.

(t) N 24-1 – Kharif season, tolerant to alternaria blight and thrips.

(u) Banglore Rose – Suitable for export.

(v) Basant -780: Kharif season, good for storage, developed at MPKVP Rahuri (Maharashtra).

(w) VL-3 – Suitable for Rabi in Hills.

2. **White coloured Varieties – Best suitable for dehydration purpose, TSS 11-12%.**

(a) Pusa White Flat – Rabi season, suitable for dehydration, drying ratio – 9:1, good for green onion. Developed in 1975.

(b) Pusa white round – TSS 12-13%, dry ratio 8:1, suitable for dehydration and green onion purpose, very good for storage.

(c) Punjab White – High TSS (15%), suitable for dehydration.

(d) Punjab -48 (S48) – Suitable for dehydration, Rabi sowing, Identified in 1975 for sub humid plains of Sutlej Ganga, good keeping quality.

(e) Udaipur 102 - Suitable for dehydration and rabi season, TSS 12%.

(f) N 257 9-1 – Suitable for dehydration and Rabi season, Developed at MPKVP Rahuri (Maharashtra.).

3. **Yellow coloured varieties**
   (a) Arka Pitamber – Short day, developed by pedigree selection by crossing between two short day local genotypes UD 102 (White) × IIHR -396 (Red).
   (b) Phule Swarna – Developed by MPKV Rahuri (Maharashtra), less pungent, suitable for export to Europe, Australia and America.
   (c) Early Grano – Good for salad, green onion purpose, Introduced from USA.
   (d) Brown Spanish - Long day type variety, excellent storability.

   Multiplier onion varieties:
   (a) CO1 – 7-8 bulblets / plant.
   (b) CO2 - 7-9 bulblets / plant, resistant to purple blotch.
   (c) CO3 -8-10 bulblets / plant, resistant to thrips.
   (d) CO4 -8-10 bulblets / plant, resistant to thrips.
   (e) MDU 1 – Resistant to thrips, tolerant to lodging due to thick erect leaves.
   (f) Agrifound Red – Light red in colour.

- Delay in irrigation may result in seedling mortality upto 80-90%.
- Critical stage of irrigation is bulb formation.
- Excess moisture induces incidence of purple blotch especially when vegetative growth is over.
- Green onions are harvested when they are just ready for earthing.
- Yield –
  Common onion – 250-300 q/ha.
  Pickling type -160-200 q/ha.
  Multiplier onion – 150-180 q/ha.
- Onions are cured when neck is tight and dried scales rustle.
- Curing is done to remove excess moisture from the outer skin which helps to reduce the infection of disease.
- Delhi market prefers big sized bulbs.
- Jute bags are used for onion packaging for sending to distant market by trucks.
- Onion should be packed in 14-15 kg capacity cane baskets for export purpose (Bangalore Rose and Multiplier onion).
- Temperature is more important than day length in seed production.
- Day length is more important that temperature for bulb production.

Crop Improvement:

(1) Introduced Varieties –
   (i) Early Grano from USA.
   (ii) Brown Spanish.
   (iii) Barmuda Yellow.

(2) Selection Varieties –

Most of varieties in India had been developed through mass selection which is common in cross pollinated crops.

(i) Pusa Ratnar
(ii) Pusa Red
(iii) Pusa White Round
(iv) Pusa White Flat
(v) N-53
(vi) Arka Kalyan
(vii) Arka Niketan
(viii) Arka Bindu
(ix) Arka Pragati
(x) Nasik Red
(xi) Patna Red
(xii) Agrifound Dark Red
(xiii) Bangalore Rose

(3) Selfing – Pusa Madhvi

Brown paper bag and three rings muslin cloth bags are used for selfing.

(4) Hybrid –

(i) Arka Pitamber – White × Red
(ii) Arka Kirtiman
(iii) Arka Lalima – Developed through heterosis breeding (MS based).

- Male Sterile lines have been isolated in Pusa Red at IARI (Good storage and wide adaptability).

(5) Small onion varieties –

1. Bangalore Rose
2. Arka Bindu
3. Agrifound Rose

- Largest cultivated onion varieties – Orange and yellow coloured.
- Onion is used against sunstroke (*Loo* in North India).
- Hybridization is done when we want to introduce character from other varieties.
  1. Intervarietal Hybridization - Very common.
  2. Interspecific Hybridization – Very rare.
     Utilized in *A cepa* × *A fristulosum.*

(6) Long day type

(1) White Spanish
(2) Cream Gold
(3) Early Lockier
(4) Brown Spanish–Best in areas at 1000 MSL, 28 tons/ha yield.

- Seed sowing – September-November.
- Transplanting – March.
- Harvesting – August.

Long day varieties crop can only be grown in hills.

- Rabi onions are harvested one week after 50-70% neck fall.
- Kharif crop is harvested when leaves colour changes to slightly yellow and tops starts drying.
- The varieties having drying ratio of 6:1 are more suitable for dehydration.
- Storage – 0°C temperature and 65-70% R.H. for 6-8 months.
- Storage models had been created by National Horticultural Research and Development Foundation (NHRDF) and National Agricultural Cooperative Marketing Federation of India Ltd. (NAFED).
- For Kharif season onion spraying of M.H. 25 ppm 75-90 days after transplanting check the sprouting up to satisfactory level.
- MH should not be used in rabi crop; it would not work at higher temperature.
- At BARC, Trombay irradiated bulb with gamma rays at very slow dose level (4,000-9,000 Krads) offered an effective physical method to control sprouting of onions.
- Export is canalized through the NAFED and its associate's shippers.
- From India large onion is being exported to Malaysia, Dubai, Oatar, Kuwait, Singapore, Srilanka and Mauritius.
- Small onion- Exported to Bangladesh, Singapore and Malaysia.
- Big onions are exported from Maharashtra, Gujarat, and Tamil Nadu ports.
- Non bolting cultivars are – (1) Early Grano, (2) Texas Early Grano.
- Export to Bangladesh and Nepal is from U.P. and Bihar by road.
- 90% Export varieties - N-53, N-2-4-1, Bellary Red, Patna Red, Agrifound Dark Red, Agrifound Large Red, Pusa Red, Pusa Madhvi, Arka Niketan.
- Export potential small onion variety – Agrifound Rose, Arka Bindu and Bangalore Rose.
- Export potential multiplier Onion varieties – CO4, CO3, and Agrifound Red (10%).
- Bangladesh prefers 3-4 cm diameter of bulbs for import.
- Gulf countries and Far East countries prefers 4-6 cm of bulbs.
- European and Japan prefers 6-7 cm.
- Malaysia and Singapore prefer 2.5-3.0 cm.
- Bulb colour preferences vary from region to region.
- Middle East countries, Bangladesh and Sri Lanka prefer light red to dark red
- Malaysia dark red.
- The European Union, Japan and America prefer yellow and brown coloured onions with mild pungency.
- Major diseases of onions: pink rot, *Fusarium* basal rot, *Botrytis* leaf blight, neck rot, white rot, smut, downy mildew, *Alternaria* purple blotch, black mold, *Stemphyllium* blight, smudge and bacterial soft rots.
- Insect pests of onions: onion thrips, onion maggot, aster leafhopper and vegetable leafminer.

- Seed in onion is produced by two methods viz. seed to seed and bulb to seed method.
- Bulb to seed method is preferred to produce true to type seed.
- The bulbs for seed production are produced the same way as for the table crop. Bulbs are stored for replanting in next crop season.
- Poor seed germination is the main constraints in onion because seeds lose their viability within a year under ordinary storage condition.
- Bulbs are planted at a spacing of 45 × 30 cm to raise a seed crop.
- Due to protandry, onion is a highly cross-pollinated crop and requires an isolation distance of 500 m from other varieties.
- For initiation of flowering low temperature (less than 12°C) and low humidity are desirable. Flowering is, however, little affected by day length.
- Onion seed has a short storage life. Under ambient conditions seed maintains acceptable level of germination for one season only.
- Average seed yield is 10-12 q/ha.

Physiological disorder –

1. Sprouting of bulbs – In white cultivar more common than purple and yellow.
   Causes –
   1. Excess soil moisture at maturity.
   2. Supply of excess "N".
2. Bolting/ Pre-mature bolting - Pre-mature bolting is a condition when the plants develop seed stalks before forming proper bulb size. This could be attributed to early planting, poor seed stock and unstable weather conditions.
   (1) Low Temperature.
   (2) Poor "N" Supply.
   (3) Late seedling transplanting.
3. Splitting of bulbs - It is a condition when the bulb instead of developing into a single unit splits into two or more than two units.

Causes: It could be attributed to water stress during bulb development stage, early planting, heavy soil texture, excessive fertilizer application, especially of nitrogenous fertilizers and poor seed stock. Large sized varieties are more prone to splitting than the small sized ones.

25. "GARLIC"

- Next to common onion most widely used of the cultivated allium is probably garlic.
- Frosty hardy crop.
- China rank Ist in area and production.
- India rank IInd in area and IIIrd in production.
- Egypt rank Ist in productivity (21.5 t/ha.) followed by USA (18.6 t/ha.).
- Indian productivity – 4.1 t/ha.

- Per capita availability of garlic is highest in Korean Republic (11.4Kg/Year).
- In India – 0.340 kg/ capita/ year.
- In India Madhya Pradesh is the leading state accounting 22.2% area and 24.40% production with average yield of 3.91 t/ha.
- Gujarat has the highest percentage share of production i.e. 28.98%.
- In India yield/ha. is highest in Punjab (17.20 t/ha.) followed by Haryana (12.80 t/ ha.) and Maharashtra (6.85 t/ha.).
- Garlic has higher nutritive value than other bulb crops.
- Great-headed garlic (*Allium ampeloprasum*) also known as Oriental garlic or elephant garlic is not true garlic.
- *Allium longicuspis* is believed to be wild ancestor of garlic.
- Alliin – A colourless, odourless, water soluble amino acid present in uninjured garlic.
- Alliicin - On injury of cell, an enzyme, allinase comes in contact with alliin and causes its breakdown into a sulphar containing product allicin which is the antibacterial substance of garlic and has typical odour of fresh garlic.
- Diallyl disulphide – Responsible for pungency in garlic which is found in alliin in an allyl radical form.
- Garlic contains 0.1 – 0.4% essential oil.
- Chief constituents of oil are diallyl – disulphide (60%), diallyl trisulphide (20%) and allyl-propyl disulphide (6%).
- True Garlic odour is due to Diallyl di-sulphide.
- Garlic oil is an item of commerce.
- Per 100 g of garlic contains 62.0 per cent moisture, 6.3 g proteins, 0.1 g fat, 29.8 g carbohydrates, 0.06 mg thiamine and 0.23 mg riboflavin.
- For recovery of oil fresh garlic is the best.
- Fresh garlic bulb yields 0.1-0.2% essential oil on steam distillation.
- Whole plant yields 0.05-0.9% essential oil on steam distillation.
- Diallyl, dimethyl and allyl methyl sulfide, disulphide and trisulfide and minor –n–propyl group- All formed by thermal decomposition of allicin and its homologues.
- Garlic is a perennial species.
- Garlic differs from *A. cepa* as the leaf base dose not store food.
- Cloves – Well developed axillary bulbs within the thin foliage leaves.
- Garlic is a sexually sterile diploid.
- Galic is propagated by vegetative method i.e. by cloves and in some cases inflorescence bulbils (Jones 1983).
- *Allium longicuspis* – Wild ancestor of garlic.
- Garlic is mainly used for flavouring and seasoning.
- 50% production of garlic comes from Madhya Pradesh and Gujarat.
- Garlic is considered to possess antibacterial, antibiotic, antitumour, antiviral, antifungal, anti-cordidal, antimycotic, antithrombotic, fibrinolytic, antirosclerosis,

crosis, platelet aggregation, inhibitory, hypoglycemic, cytotoxic and lipid lowering properties. Eg. Allicin, ajoen, diithins and essential oils are principle contributors in this direction.

- Indore, Mansore (MP), Angul (Orissa), Junnar Taluka (Maharashtra), Manipur, Etah and Etania (UP), Sikar, Jhunjhunu, Jhalawar, Ajmer and Udaipur (Raj.) are major garlic growing pockets.
- In India small cloved garlic varieties are being grown in plains, recently large garlic cultivation has been introduced in North Indian plains.
- Garlic is a frost hardy crop, requires cool and moist period during growth and relatively dry period during maturity of bulbs.
- The critical day length for bulbing is 12-hours. It must make most of the vegetative growth before the bulbs begin to enlarge.
- Bulb formation in response to warming of temperature and lengthening of days.
- The optimum temperature for vegetative growth is 10-15°C and for bulb development is 25-30°C.
- Garlic can be grown well up to 1000-1300 MSL.
- In India mostly short day types are grown.
- Garlic should be planted early to promote vegetative growth under short day condition and cool temperature.
- Critical day length for bulbing is 12 hours.
- Expose of dormant cloves or young plants around 20° C or lower for 1-2 months hastens subsequent bulbs.
- Garlic requires medium black, fertile loamy soil with good content of potash.
- Sandy or loose soil results in bulbs lighter in weight.
- Highly alkaline and saline soils are not suitable for its cultivation.
- **Recommended Varieties of Garlic**

S.No.	Variety	Features
1	Yamuna Safed (G-1) (NHRDF, Nasik)	The Bulbs are compact with attractive white creamy sickle shaped. 25-30 cloves/plant. Tolerant to insect pest and disease like thrips, purple blotch and stenophyllium blight, TSS 38%, dry matter 39.5% Maturity 150-160 days. Yield 150-175 q/ha. Recommendation for cultivation in all India.
2	Yamuna Safed-2 (G-50)	Bulbs diameter 3.5-4.0 cm suitable for North India, tolerant to purple blotch and stenophyllium disease. The bulbs are compact with attractive white creamy flesh. Number of cloves 35-40/plant. TSS 38-40%. Maturity 150-160 days. Yield 150-200 q/ha.
3	G-282	This is a local variety from Dindigul (Tamil Nadu). The leaves are wider than other varieties, bulbs creamy white and bigger sized 5-6 cm having 15-16 cloves/plant. Dry matter 39-43%, TSS 39-43%. Maturity 125-130 days. Yield 175-200 q/ha. Suitable for export.

4	Agrifound White (NHDDF, Nasik)	First garlic variety released by central variety release committee for growing in western, southern and central parts, TSS 41%, dry matter 43% bulb diameter 3.5-4.5 cm, Number of cloves/bulb is 20-25. Selection from a local cultivar of Bharsharif of Bihar. Bulbs are silvery white.
5	Agrifound Parvati (NHRDF)	It is a selection from a local variety from Hong Kong. Bulbs are white with pinkish tinge. Long day type, as such suitable for North Hilly areas, Bulbs bigger in size 5-6 cm in diameter, 10-16 cloves/bulbs, suitable for export.
6	Godavari	Selection from Jamnagar collection. Bulbs pink coloured medium size. Tolerant to mites. Duration 130-140 days.
7	Sweta	Selection from Gujarat collection. Bulbs white duration 120-130 days.
8	Ooty 1	Bulbs bigger in size, dull white, duration 120-130 days.

Other Varieties –

- G-51 – Tolerant to purple blotch and stenophyllium disease.
- Godavari
- Sweta
- HG-1
- HG-6
- Pusa Sel. 10– IARI New Delhi
- LCC 1 – PAU Ludhiana
- ARU 52- UPKAS, ALMORA
- Other local varieties – Madrasi, Tabiti, T56-4 and Jamnagar Fawari and Rajalle Gaddi grown in Bellary District.
- Varieties released from NHRDF.
  Agrifound White,
  Yamuna Safed (G1)
  Yamuna Safed – 2 (G50)
- All these varieties are small bulb size and have number 20-30 of smaller cloves.
- G 282 and Agrifound Parvati – Bigger bulb and bigger cloves.
- Propagation – By cloves, sometimes bulbils are also used.
- Cloves are dibbled 5-7 cm deep, keeping growing ends upwards.
- Spacing – 15 10 cm – Bigger size cloves.
  10 × 7.5 cm – Small size cloves.
- Spacing in garlic varies as per the bulb size.

- To improve bulb size, planting is done on beds spaced at 35 cm. Each bed accommodates two rows 7.5 cm apart.
- Planting time – September – November – Main season
  March – April – Hills
- About 500 kg cloves of 8-10 mm diameter are required to plant in one ha.
- Excessive 'N' results in thick neck and sprouting before harvest.
- Borax up to 10kg/ha. Increases bulb size and yield.
- $MnSO_4$ 0.1%, boric acid 0.2%, $CuSO_4$ 0.02% and $ZnSO_4$ 0.02% stimulate dry matter accumulation in cloves.
- Garlic is produced only in one season – Winter (Rabi).
- **Harvesting** – March-April in Plains.
  July – September in South Hills and North Hills.
- Close spacing cause purple blotch disease.
- In India harvesting is done manually by hand digger.
- Yield – 100-200q/ha.
- 25-50% losses take place in garlic if it is not stored properly.
- Curing is done to remove excess moisture and to allow the bulbs to become compact and go into dormancy.
- Curing is a process of drying to remove the excess moisture and to allow the bulbs develop firmness and dormancy.
- Curing enhances storage life of bulbs. Harvested bulbs are allowed to cure in the field for about a week.
- Tops provide cover to the bulbs to prevent their direct exposure to sun. Thereafter the bulbs, with or without the tops, are cured in shade for another week.
- Garlic must be graded for market to sort out diseased and damaged bulbs and to ensure proper size. There are three grades viz. large (30 mm diameter), medium (25 mm diameter) and small (<20 mm diameter).
- Garlic is stored under a wide range of temperatures. However, bulbs must be kept dry and if stored at relative humidity more than 70 per cent at any temperature results in growth of molds.
- Properly cured bulbs can be stored for 6-8 months in well-ventilated, cool ordinary store.
- The storage life is 3-5 months under cool (16°C), dry and dark conditions.
- For extended period of storage, bulbs are stored at 0-2.2°C and 60 per cent relative humidity.
- Storage life is prolonged and loss in weight is reduced by spraying 3,000 ppm M.H on crop before harvest.
- For export quality 40-60 mm size diameter and 10-15 cloves in each bulb are required.
- Longer day length – Higher yield.
- Physiological disorder –

1. Sprouting
2. Rubberization – Nilgiri hills of Tamil Nadu

- Storage life of garlic is prolonged and weight loss is reduced by spraying MH 3000 ppm, 3 weeks before harvest.
- Irradiation with 6 Krad of cobalt 60 gamma rays has been recommended for successful storage of garlic.
- Garlic is a condiment crop.
- Garlic being vegetatively propagated, its seed production is not much different from bulb crop.
- Large sized and healthy bulbs are selected for seed bulb production.
- A minimum of 5 m isolation distance is required between two varieties of garlic to avoid physical mixtures.

26. "LEEK"

- Non-bulb forming tetraploid plant of onion family.
- Grown only in kitchen garden in India.
- Used as salad and in flavouring the soups.

Varieties –

1. PPL –1-Selection, Good replacement for green onion.
2. London Flag.
3. American Flag.

Seed rate – 5-6 Kg/ha. in nursery during September-October.

- Blanching is an important cultural operation in leek cultivation which improves the quality of crop.
- Yield – 250-300q/ha.

27. IMPORTANT PEST AND DISEASE OF BULB CROPS

Sr. No.	Pest and Disease	Symptoms	Management
1.	Onion Thrips (*Thrips tabaci*)	Small white silvery patches are seen all over the leaves and in severe infestation, leaves dry from tip to downward.	Spray insecticides dimethoate @ 0.06% or profenofos @ 0.05%
2.	Onion Fly/ Maggot (*Delia antique*)	Onion leaves get rotten, shredded and rolled up Growth of onion tubers gets hampered, tissues get decayed and becoming unfit for marketing	Adopt proper crop rotation Spray malathion 1 lit/ha

3.	Ear wig (*Euborellia annulipes)*	Both nymphs and aduts bore into the onion bulb and feed.	Soil application of Carbofuran 3 G at 30 kg/ha on 20 DAS and earthed up.
4.	Basal Rot (*Fusarium oxysporum* f. sp. *cepae)*	The entire plant shows complete drying of the foliage and the bulb of the affected plant shows soft rotting and the roots get rotted.	Soil drenching with Copper oxychloride 0.25 %.
5.	Downy mildew (*Peronospora destructor)*	White downy growth appears on the surface of the leaves.	Three spraying with Mancozeb 0.2 % is effective.
6.	Purple blotch (*Alternaria porri)*	Circular to oblong concentric black velvety rings appear in the chlorotic area.	Three foliar sprayings with Copper oxychloride 0.25 % or Chlorothalonil 0.2 % or Mancozeb 0.2 %.
7.	*Pythium* root rot (*Pythium aphanidermatum, P. debaryanum and P. Ultimum)*	This disease causes seed rotting, pre-emergence damping off. The disease appears in circular patches in the field here and there. The disease also occurs after establishment of the crop *i.e* after 15 to 30 days of sowing or planting. This stage is called post-emergence damping off.	Seed treatment with Thiram or Captan @ 4g/kg. The bulbs may be dipped in Thiram solution 0.25%. After sprouting, soil drenching with Copper oxychloride 0.25%.

28. LEGUMES (PEAS AND BEANS)

Family – Leguminaceae/Fabaceae

Common Name	Botanical Name	2x	Origin
Pea	*Pisum sativum*	14	Central Asia, Abyssinia, Mediterranean
Cowpea	*Vigna unguiculata*	22	Africa
Cluster Bean	*Cyamopsis tetragonoloba*	14	India
French Bean	*Phaseolus vulgaris*	22	Central and South America
Lima Bean	*Phaseolus lunatus*	-	Guate mala
Broad/Faba/Horse Bean	*Vicia faba*	12	Central Asia, Mediterranean and South America

Indian /Lab-Lab Bean	*Dolichos lablab / Dolichos purpureus*	20, 22, 24	India
Winged Bean	*Psophocarpus tetragonolobus*	18, 22	Madagascar in Africa
Soyabean	*Glycine max*	22	China
Jack Bean	*Canavalia ensiformis*	22	Central America
Sword Bean	*Canavalia gladiate*	22,44	-

29. "PEA"

- In India grown mainly as a winter vegetable in the plains of North India and as a summer vegetable in Hills.
- Rich source of digestible proteins (7%) along with carbohydrates and proteins and minerals like Ca and P.
- Green pea is second only to Lima bean as fresh vegetable source of protein.
- Fresh pea contains 75 g water, 6.2 g protein, 0.4 g fat, 17 g carbohydrates, 2.4 g crude fibre, 1.0 g ash, 32.0 mg calcium and 102 mg phosphorus.
- It is used as a fresh vegetable or in soup, canned, processed or dehydrated.
- Average yield in India – 6.12 t/ha.
- Average yield of world– 6.1 t/ha.
- Uttar Pradesh (221000 ha) is having maximum area under peas cultivation followed by Madhya Pradesh (94990 ha).
- In peas production, Uttar Pradesh (2511000 MT) is the leader followed by Madhya Pradesh (961550 MT).
- *Pisum elatius* – Wild species or ancestor of P*isum sativum*.
- Pea is a diploid (n = 7, 2n = 14).
- Two sub species :
  - *Pisum arvense* – Field pea, coloured flower.
  - *Pisum sativum* – Garden pea, sweet pea, horticultural pea.
- *P. arvense* is used as pulse, seed round or little angular, hard and whitish, plants are very hardy, resistant to drought and frost.
- *P. sativum* – Used for vegetable (table purpose) and canning purpose, seeds wrinkle, sweet and green, tender pods are used.
- Grown as early as the Stone Age.
- Harvested organs: Mature or immature seeds and/or young, tender pods grown either for the fresh market or processed for canning or freezing.
- Edible-podded peas sometimes referred to as a separate variety (*P. sativum* var. *macrocarpum*) (e.g. snow and snap peas).
- The field pea, *P. sativum* var. *arvense*, is grown for drying, green manure and as a forage crop.

- The anti-nutrient factor in pea is cyanide containing compounds.
- Pea is a self-pollinated, herbaceous winter annual.
- Green pea – 7.2 g protein and dry pea – 19.7 g/100g.
- Pea protein – 25% amino acid and sugar 12%.
- *Pisum sativum var – hortense* contribute 67% of total production.
- Varieties differ for pod and seed colour (light and dark green), plant height tall (indeterminate) and short (determinate), wrinkled (sweet) or smooth (starchy) seed surface.
- Varieties of pea are classified on the basis of seed surface (smooth or wrinkled), growth habit (dwarf, medium tall and tall) and maturity (early and main season).
- Early varieties take 50-60 days to reach marketable maturity, whereas, mid-season varieties take more than 80 days.
- Early varieties are usually dwarf and tolerant to high temperature.
- Mid-season varieties are medium tall or tall and require staking.
- The wrinkled seeded varieties are sweeter than the smooth seeded varieties.
- Plant type of the edible-podded pea or snow pea is more-or-less the same as that of garden pea. However, the pods of snow pea are flatter, broader, more tender, more fleshy and free of hard parchment layer on the inner side of the shell.
- Plant type of snap pea varieties is also similar to that of green pea varieties but the pods are stringless and the whole pod is consumed.
- Gregor Mendel conducted experiments with pea plants, including analysis of the inheritance of traits such as wrinkled and smooth seed, led to the science of genetics.
- Direct seeded sowing is done.
- **Recommended Varieties of Peas**

S.No.	Variety	Features
1	Arkel	Introduced from UK, dwarf (50cm), pods dark green about 8 cm long, slightly curved and well filled, 100 seed weight 17g, in early sowing escapes the attack of powdery mildew but in late planting highly susceptible to powdery mildew, also susceptible to collar rot, Grains very sweet and wrinkled when dry. Suitable for sowing in mid-October. Maturity 55-60 days. Yield 100-125 q/ha.
2	Early Badger (Wisconsin)	Early dwarf, wrinkle seeded variety, pods pale green, 7.5 cm long pods. Suitable for early October sowing. Maturity 60-65 days. Yield 50-60 q/ha.
3	Meteor (England)	Plant dwarf 5-60 cm tall, round seeded variety suitable for early October sowing, maturity 55-60 days. Yield 90-100 q/ha.
4	Pusa Pragati	Early, powdery mildew resistant variety. Avg. pod length 10 cm with 9 seeds per pod. Maturity 60-65 days. Yield 67 q/ha.
5	Jawahar Matar-4	Plant height 70 cm green pods, stem and foliage, pods well filled mature seeds wrinkled, green. Maturity 65-70 days. Yield 60-70 q.ha.

6	Early December	Plants bushy 70 cm tall with white flowers. Pods round to oval shaped, 5 cm long, 1 cm broad and five seeds per pod. Maturity 65-70 days. Yield 60-65 q/ha.
7	PM-2	Cross of Early badger and IP-3, Plant height 50-60 cm and 6 seeds per pod. Fruits setting starts from 6th node. Escapes powdery mildew. Maturity 55-60 days. Yield 100-120 q/ha.
8	Arka Ajit	A mid-season variety. Pods 8-9 cm long; seeds bold green, sweet shelling percentage 55, resistant to both powdery mildew and rust, duration 90 days and yield 10 t/ha.
9	Arka Sampoorna	The plants are bushy erect. Pods are flat, perchmentless, crisp sweet, resistant to powdery mildew and rust. Duration 80 days and yield 8 t/ha.
10	Arka Karthik	Fruits bushy erect and resistant to both powdery mildew and rust. Pods are long (11-12 cm) with 8-10 green, sweet seeds. Crop duration 90 days. Pod yield 11 t/ha.
11	Kashi Nandini	This is the earliest variety in the seed chain of the country and grown in the northern plains (25 Oct – 25 Nov.), Plants erect (60 cm), pods 8-9 cm long. Shelling quality 47-48%, duration 55 days. Yield 110-120 q/ha. Tolerant to leaf miner and pod borer.
12	Kashi Udai	Early maturing, 60-65 days duration, yield 110 q/ha.
13	Kashi Mukti	Early maturing 60-65 days duration, yield 110-120 q/ha.
14	Kashi Shakti	High yielding medium maturing sown in November in plains shelling quality 48-49%, yield 100-120 q/ha.

Crop improvement –

- Arka Ajit was developed by IIHR by crossing Bonneville to powdery mildew and rust resistant lines and then again back crossing to Bonneville. It is resistant to rust and powdery mildew.
- Jawahar Pea 15 (JNKV, Jabalpur) was developed by selection from a three way cross (JMI × R 98 B) × JP 501 A/2. It is resistant to Fusarium wilt and powdery mildew.
- Arka Karthik (IIHR, Bangalore) It has been developed by selection from the cross Arka Ajit × IIHR 554. Plants are bushy and erect. Pods are 11-12 cm long with 8-10 sweet grains. It is resistant to rust and powdery mildew.
- Arka Sampoorna (IIHR, Bangalore) it is a snow pea variety developed by selection from a multiple cross involving Bonneville, IIHR 209, Freeza 656 and Manao Sugar. Pods are flat, sweet and crisp with the parchment layer. It attains marketable maturity 80 days after sowing. Average yield is 80 quintals per hectare.
- Though 25% heterosis is reported for yield in pea, F_1 hybrid seed production is almost restricted because of the high cost involved in manual emasculation; pollination and availability of less number of seeds per pod unlike the fruit of solanaceous crops.

Classification of varieties according to maturity –

1. **Early Maturing** (The early varieties are more popular in India because of better economic return. They are not high yielders but give high initial remuneration to the growers. Two pickings of green pods are generally done in these varieties, the first being after 60 days of sowing and second after 15-20 days of first picking)-
   - (a) Arkel – Early wrinkled variety suitable for fresh market and dehydration.
   - (b) Asauji – Smooth seeded variety.
   - (c) Alaska – First early canning variety.
   - (d) Meteor –Round seeded early variety.
   - (e) Early Giant – Suitable for hilly areas, seeds sweet in taste.
   - (f) Harbhajan – Suitable for dry areas, extra early variety, first picking 50 days, 100 seed weight 12 g, highly susceptible to powdery mildew.
   - (g) Jawahar Matar-3 – 100 seed weight is 18 g, susceptible to powdery mildew but may escape due to earliness.
   - (h) Jawahar Matar- 4 - Susceptible to powdery mildew but may escape due to earliness.
   - (i) Jawahar Pea – 54 - (JP-54) Double cross of (Arkel × JM-5) × (46C × JP-501), resistant to powdery mildew.
   - (j) Kanwari – Tall, double podded variety, a mucilage excretion on the pod is a characteristic feature of this variety, largely grown in hills and plains.
   - (k) Little Marvel – Introduced from England (Chelsea × Sutlon Alaska) dwarf variety.
   - (l) Matar Ageta – 6 Long pods, 12-15 in number, 50% pods can be harvested in first picking.
   - (m) Pant Matar -2 – This variety suffer badly due to powdery mildew attack.
   - (n) Pant Sabzi Matar - 3 – Arkel × GC 141, susceptible to powdery mildew.
   - (o) VL Ageta Matar- by VPKAS, Almora (Pant Uphar × Arkel), free from the incidence of powdery mildew.
   - (p) Early Superb – From England.
   - (q) Early Badger – Good for canning of green pea, resistant to Fusarium wilt and tolerant to heat and drought.
   - (r) Yates Early crop – Australian variety.
   - (s) Kelvendon Wonder – Dwarf plant, flowering 40 days after sowing.
   - (t) Lucknow Boniya – First variety to come to Delhi market from the plains in the season.
2. **Main Season Variety** (These are high yielding varieties capable of giving 3 pickings. The first picking can be done 90 days after sowing and the subsequent two at an interval of 15 days) –
   1. Bonneville – Highly susceptible to powdery mildew disease.
   2. Jawahar Matar -1 - Bigger pod, highly susceptible to powdery mildew.

3. Jawahar Matar 2 – Greater progress × Russian 2, Highly susceptible to powdery mildew disease
4. Jawahar Pea 83 (JP-83) – (Arkel × JP-829) × (46C × JP–501), resistant to powdery mildew.
5. Jawahar Matar 71- (Arkel × SP-827) × (JP 501 × JM1), resistant to powdery mildew.
6. Pant Uphar (IP-3) – Highly susceptible to powdery mildew disease.
7. P-88 – Highly susceptible to powdery mildew.
8. VL-3 – Susceptible to powdery mildew, resistant to wilt.
9. Lincon – Better suited for hilly regions of North India also suitable for late sowing.
10. **JP-19- Whole pod are consumed, no need of removing the peelings.** This quality of green pod is due to lack of sclerenchymatous tissues. This character of pod not only saves yield loss but also the burden of house wives.
11. JP4 (JM6) - Resistant to powdery mildew.
12. FC-1 - Resistant to powdery mildew.
13. PRS-4- Moderately resistant to powdery mildew.
14. Arka Ajit- Resistant to powdery mildew and rust.
15. UN 53 (6) – Snap pea (Whole pod edible).
16. Ooty 1- Pure line selection, resistant to white flies.
17. Sylvia – Whole pod edible.
18. Perfection New Line.
19. Thomas Laxtan – Most suitable for freezing.
20. Alderman – Most suitable for kitchen garden.
21. Palam Priya – Tolerant to powdery mildew.
22. Solan Nirog – Resistant to powdery mildew.
23. Duke of Albany – Also known as Telephone, need stacking, produce Ist flower on 12-14th node.
24. Khaper Kheda – Popular cultivar of M.P.
25. Delwiche Commando – Resistant to wilt.
26. N.P 29 - Suitable for dehydration purpose.
27. Oregon Sugar Pod (Introduction, USA) it is a snow pea variety.
28. Oregon Sugar Pod II (Introduction, USA) it is also a snow pea variety. The plants are bushy and about 75-90 cm tall. The pods are 10-12 cm long, tender, crisp and suitable for stir-frying.
29. Sugar Snap (Introduction, USA) It is a snap pea variety. The plants are about 120 cm tall and require staking.

Edible podded variety (snow pea or sugar pea or snap pea)

Whole pods of this type can be eaten as the pod walls contain less fibre and high sugar.

1. **Sylvia.** It is an introduced variety from Sweden. Plants are tall; first blossom appears at 14th to 16th node after 60 days of sowing. Pods are borne singly, yeelowish in colour, long (12 cm) and curved without parchment type pericarp. Pods are sweet and have general appearance of a medium sized French bean pod.
2. **UN 53 (3).** Developed at IIHR which gives 8-9 ton pod yield/ha in a crop duration of 90 days.

- Optimum temperature for germination – 22°C.
- Soaking seeds in GA_3 (10 ppm) for 12 hours gives highest yield.
- Hot weather interferes with pollination and seed setting, affect the number of pods /plant, pod weight and lowers the quality of pod produced due to the conversion of sugars into hemicellulose and starch.
- Wrinkle seeded cultivars are more sensitive to high temperature than the round seeded ones.
- Snap pea has wider adaptability and tolerates to higher temperature than the garden pea.
- Sowing time-

  October - November - North Indian Plains.

  March – Hills for summer crop.

  May – Hills for Autumn crop.
- Seed rate –

  Early - 100-120kg.

  Mid and late- 80-90kg.
- A closer spacing of 30 cm between rows and 5-10 cm between plants is maintained for early planting.
- For main season planting, spacing is maintained at 45 × 10 cm. Seeds are placed about 2.5 cm deep.
- Phosphatic fertilizer enhances yield and quality of pods.
- Prolonged application of DAP causes sulphar deficiency.
- Gypsum or pyrite – Complete the deficiency of sulphar.
- Excess moisture causes - Yellowing of crop.
- Pea sown earlier or later suffers from wilt or powdery mildew attack.
- Critical stage of irrigation is flowering, fruit setting and grain filling.
- Depending upon the soil texture and weather conditions, a total of 3-5 irrigations are sufficient.
- Staking is an important operation especially in tall growing varieties.
- The most appropriate time for staking is when plants are about 15 cm tall.
- The plants are trained on bamboo sticks, wire netting, trellis etc.
- Staking increases yield, improves quality, decreases incidence of diseases and facilitates inter-cultural operations
- Foliar application of 0.1% ammonium molybdate at the time of flowering showed favourable effect on yield and quality of peas.

- Spraying 15 ml of 10 M solution of growth regulators CCC at the five-node stage of development has favourable effect on the growth and yield.
- Low tenderometer reading is an index of high quality and vice-versa.
- Usually 2-3 pickings in early maturing varieties and 4-5 pickings in main season varieties at 10-day intervals are possible.
- Snow peas (sugar peas) are harvested when the pods have acquired the characteristic length of the variety but are thin and have just begin to develop grains.
- The optimum stage is reached 10-days after flower anthesis.
- Snap peas are harvested when the grains develop to nearly full size; the pod walls are thick, fleshy and crunchy.

  Harvesting – Early (50 Days after sowing).

  Mid (60-65 Days after sowing).

  Late (70-75 Days after sowing).
- In processing industry, the maturing of pea is tested with the help of tenderometer.
- If harvest is delayed surface become coarse which brings down the market value of product due to the conversion of sugar into starch, the quality deteriorates especially when the temperature is very high.
- Yield – Early 2.5-4.0 t/ha.

  Mid 6-7.5 t/ha. Green pod.

  Late 8-10t/ha.
- The shelling percent ranges from 35-50%.
- Yield of snow pea varieties range from 50-70 q/ha.
- Maturity – Measured by Tenderometer (tenderness).
- Peas are blanched in hot water at 95°C for 1-2 minutes to prevent enzymatic activity responsible for undesirable flavour that may develop during processing and storage.
- For freezing, the optimum tenderometer reading is 105 at harvest.
- Storage at 0°C with 90-95% relative humidity for two weeks.
- Peas meant for processing should have uniform maturity, sweet taste, fine texture and dark green colour.
- Snap peas are frozen but not canned since high temperature destroys the structure of the pods.
- As pea is highly self-pollinated crop, an isolation distance of 20 m is sufficient to prevent physical mixtures for seed production.
- The crop is monitored twice, first at the time of flowering and the second at the time of pod maturity.
- Marsh spot – Due to Mn deficiency.
- *Rhizobium leguminosarum* – Application advisable to reduce the quantity of nitrogenous fertilizers.
- Major diseases of pea: *Fusarium* wilt and root rot.
- Major insect pests of pea are: –

- Loopers (contaminant).
- Alfalfa caterpillars (contaminant).
- Stink bugs (contaminant).
- Ground beetles (contaminant).
- Pea aphids (virus vectors).

30. "COWPEA" (BLACK EYE PEA)

- Cowpea is a warm season, annual and herbaceous legume having Indian center as the origin.
- Known by various names such as black-eye bean, China pea, southern pea and lobia.
- Used for tender pods and dry seeds used as pulse for culinary purpose.
- Pods are rich in protein, vitamin B and minerals.
- Cowpea is low in saturated fat and very low in cholesterol and sodium.
- Its protein is rich in tryptophan and lysine but is deficient in methionine and cysteine.
- It is used as fodder or green manure crop.
- Most probable progenitor of cowpea is *var menensii.*
- Cowpea belongs to subfamily fabaceae.
- Flowers are white coloured.
  1. Common Cowpea – 20-30 cm long pods and small seed size.
  2. Catjang bean (*Vigna catjang*) – 7-13 cm long pod, 5-6 mm kidney shaped seeds.
  3. Asparagus bean or yard long bean (*Vigna sesquipedalis*) – 30-90 cm fleshy pod/ yard long bean with elongated kidney shaped 8-12 mm long seeds.
- It is grown as catch crop, drought tolerant and sensitive to water logging.
- Frost is harmful to this crop.
- Cowpea is a warm season crop and is adapted to hot and dry conditions.
- Germination of seed is rapid when the soil temperature is above 18-20°C.
- Growth is optimum when day temperature ranges between 27-30°C and night temperature between 17-22°C.
- **Recommended Varieties of Vegetable Cowpea**

S.No.	Variety	Features
1	Pusa Phalguni	Bushy dwarf variety suitable for sowing in February-March in the northern plains. Pods dark green, erect and 10-20 cm long. Maturity 60 days (first picking). Yield 70-75 q/ha.
2	Pusa Barsati	Suitable for growing during rainy season. Pods light green, 26-28 cm long and pendant. Maturity 45 days (first picking) Yield 70-75 q/ha.

3	Pusa Dofasli	Bushy, suitable for both in spring-summer and rainy seasons. Pods light green and about 18 cm long. Maturity 45-50 days (first picking). Yield 75-80 q/ha. It is a photo-insensitive variety.
4	Pusa Komal	Bushy, suitable for spring-summer and rainy seasons. Pods light green, and 20-22 cm long. Resistant to bacterial blight disease. Yield 120-120 q/ha. It is a photo-insensitive variety.
5	Pusa Rituraj	Bushy, 40-50 days for first picking, light green pods, seeds brown, Photo-insensitive variety.
6	Cowpea -263	Bushy, dwarf, medium green pods, 45-50 days for first picking, developed by PAU, Ludhiana, tolerant to moasaic virus.
7	C-152	Resistant to root knot nematode, tall, erect plants, 90-100 days for maturity, oblong light brown seeds. It shows better adaptation In Southern states of India.
8	Narendra Lobia 1	Bushy, 45-48 days for first picking, green pods and bold seeds.
9	Kashi Shyamal	Dwarf bushy variety, suitable for both kharif and rabi season. Yield 48 DAS. Pods yield 70-80 q/ha. and tolerant to golden mosaic virus.
10	Kashi Gauri	Dwarf bushy type, photo-insensitive suitable for spring summer and rainy season. Resistant to golden mosaic virus and leaf spot diseases. Yield 10 -12 t/ha. Duration 40-48 days.
11	Arka green	Plants are tall, vigorous and photo insensitive. Leaf colour light green, flowers purple, pods light green, long thick, fleshy and stringless. Suitable for vegetable purpose. Tolerant to heat and moisture stress. Duration 90 days. Pods yield 18 t/ha.
12	Arka Suman	Plants, erect bushy, photo-insensitive, pods medium long, crisp without parchment. Duration 70-75 days, pod yield 18 t/ha.
13	Arka Samrudhi	Plant erect, bushy, photo insensitive, early with pods above the canopy. Pods green, fleshy without parchment. Good cooking qualities. Duration 70-75 days. Yield 19 t/ha.

Five distinct sps. of cowpea:

A. Dekinotiana

B. *Mensenensis* Wild

C. *Ungiculata*

D. *Cylindrica* Cultivated in India

E. *Sesquipedalis*

Crop Improvement

1. Introduction –

(1) Philippines Early – Trailing growth habit.

(2) Yard Long Bean- Species *sesquipedalis*, late maturing variety.

2. Selection

(a) Pusa Barsati – Early variety, Suitable for rainy season.

(b) Pusa Phalguni – Dwarf variety, most suitable for February – March (Spring).

(c) Arka Garima- Bushy, pole type and photo-insensitive, evolved by back cross (*unguiculata* × *sesquipedalis*), more suitable for rainy season.

(d) Arka Samrudhi – Bushy, photo-insensitive.

(e) Arka Suman – Bushy, Photo insensitive.

3. Hybrid

(a) Pusa Dofasli – Pusa Phalguni × Philippines bush , Bushy suitable for both summar and rainy season, photo insensitive.

(b) Pusa Rituraj – Bushy and prolific bearer, being highly photo thermo insensitive, can be grown during rainy as well as summar season, suitable for dual purpose.

(c) Pusa Komal – Pusa Rituraj × P-246, dwarf and bushy, Resistant to bacterial blight, Suitable for sowing summer as well as rainy season.

(d) Pusa Sukomal

Bushy Types

Pusa Dofasli (IARI, New Delhi) It was developed by selection from the cross Pusa Phalguni × Philippines Bush.

Pusa Komal (IARI, New Delhi) It was developed by selection from the cross P 85-2 × P 426

Cowpea 263 (PAU, Ludhiana) It is a selection from the material collected from Bangalore.

Arka Samrudhi (IIHR, Bangalore) It was developed by selection from the cross Arka Garima × Pusa Komal.

Bidhan Barbati 1 (BCKV, Mohanpur) It is a selection from the cross EC 243954 (V. unguiculata) × EC 305827 (V. unguiculata ssp sesquipedalis).

Trailing Types

Pusa Barsati (IARI, New Delhi) It is a selection from an exotic material introduced from Philippines.

Lola (KAU, Vellanikkara) It is developed by selection from a local material.

Arka Garima (IIHR, Bangalore) It was developed from the cross T.U.V. 762 × V. unguiculata ssp. sesquipedalis.

Other Varieties

(a) Cowpea 263 – Suitable for rainy as well as summar season, comparatively resistant to mosaic and free from bean mosaic virus.

(b) Narendra Lobia – Being photo insensitive suitable for rainy as well as summar season.

(c) Sel. 1552 – Resistant to bacterial blight.

(d) Bidhan Barbati 1 – Resistant to cowpea mosaic and cowpea golden mosaic virus.

(e) Bidhan Barbati 2 – Semi-determinate.

(f) IIHR 220 – Resistant to rust.

- **Seed rate –**
  - ➢ Rainy Season – 12-15 Kg/ha.
  - ➢ Summar Season – 20-25 Kg/ha
- **Sowing time –**
  - ➢ In north Indian two crops viz. spring season sown in January-February and rainy season sown in May-June are grown.
  - ➢ In South India where the seasons are moderate, cowpea is grown almost throughout the year.
  - ➢ In hilly regions, sowing is done in April-May.
- A narrow spacing of 30-45 cm between rows and 20 cm between plants within the rows is required for bush type varieties where as a wider spacing of 2.0 × 2.0 m is required for trailing types.
- In trailing types, 2-3 plants are raised per hill.
- A higher seed rate of 15-20.0 kg/ha is required for bush types where as a seed rate of 4.0-5.0 kg/ha is sufficient for trailing types.
- Spacing – 45-60 RXR
15 cm PXP Rainy Season
30-30 cm RXR
15 cm P × P Summer season
- Cowpea has strong tap root system.
- The pole type varieties need support; these varieties are often restricted to kitchen garden.
- The most critical water requirement period is just prior to and during bloom.
- Application of water first after 15 days of sowing and then at flowering or pod setting improves the pod yield considerably.
- Spraying of MH (50-200 ppm) before flowering improves pod set and yield.
- The pods are ready for harvesting 60-80 days after sowing and 16-10 days after flowering.
- Picking of pods 15 days after pod setting is ideal for vegetable purpose.
- Compared with other picking stage the pod at this stage are longer and less fibrous with more protein content and are more succulent.
- Yield – 50-80 q/ha. Green pod.

12-15 q/ha. Seed yield.

- Being a self-pollinated crop an isolation distance of 50 m between two varieties of cowpea is sufficient to produce genetically pure seed.
- For seed production of mosaic virus resistant varieties, the crop is raised in rainy season for proper screening.
- Cowpea mosaic virus is seed borne and also transmitted by aphids.

31. "CLUSTER BEAN"

- Grown for green pod as vegetable, dry seed, forage and green manure. Some varieties are used for extraction of gum.
- Diploid 2n =2x = 14.
- *C. senegalenis* – Ancestor of guar.
- Guar is one of the hardiest among legume vegetables.
- Seed contain gum like mucilaginous substance called "guar gum" or "galactomannon" which is used in textiles, paper industry and cosmetic industry.
- Green pods are rich in vitamin A and Iron.
- It is drought tolerant and is also grown in arid zones as a fodder crop.

Recommended Varieties of Cluster Bean

S.No.	Variety	Features
1	Pusa Mausami	Pods attractive smooth bright green and 10-12 cm long. First picking in 65-80 days after sowing. Suitable for rainy seaon/ kharif season.
2	Pusa Sadabahar	Selection From local variety "Jaipur" of Rajasthan, Non-branching type. Pods green and 12-13 cm long, tender and fiberless. First picking in 45-55 days after sowing. Suitable for both summer and rainy seasons.
3	Pusa Navbahar	Pusa Mausami × Pusa Sadabahar, Pods 15 cm long and of better quality. Non-branching type. Suitable for both summer and kharif seasons, susceptible to bacterial blight and lodging.
4	P-28-1-1	Smooth green and long pods. A photo insensitive variety. Flowers in 30-32 days in summer and 40 days in kharif. Suitable for both summer and kharif seasons.
5	P28 1	Developed by NBPGR, Pusa Navbahar × IC11521, suitable for both rainy and summar, photosensitive.
6	Sharad Bahar (NBPGR, New Delhi)	Developed through selection from local cultivar collected from Maharashtra, profuse branching, single plant produce about 133 pods.
7	Giant type	Good quality seeds with high content of guar gum and grown in Gujarat.
8	Dwarf type	Good quality pod which can be used as vegetable, Grown in Punjab and Uttar Pradesh.

- It prefers long day conditions for growth and short-day conditions for flower induction and pod development.
- In north Indian plains, sowing is done in February-March for summer crop and in beginning of July for the rainy season crop.
- The crop is sown during December to January in south Indian plains.
- Spacing is maintained at 45-60 cm between rows and 15-20 cm between plants.
- Seed rate – 25-30 kg/ha
- Sowing Time – June – July – Main Season.

  Feb - March – Summer crop.
- Flowering and pod formation are the critical stages for irrigation application.
- Excessive irrigation leads to luxuriant vegetative growth resulting in poor pod setting and yield.
- First picking of green pods is possible 45 days after sowing in summer crop and after 55-60 days in rainy season crop.
- Pod yield in cluster bean varies from 30-40 quintals per hectare.
- Cluster bean is a self-pollinated crop and an isolation distance of 50 m between two varieties is sufficient to prevent physical mixtures.

32. "FRENCH BEAN" (RAJMAH, FARASH BEAN)

- It is also known as kidney bean, haricot bean, snap bean, navy bean, string bean, garden bean.
- The English first used the name 'kidney bean' in 1551 to distinguish common bean from Old World beans.
- It is an annual herb grown for green tender pods and dry seeds (rajmah)
- Three types of French bean are there.
  1. Bush type with short internodes.
  2. Semi pole type with longer internodes.
  3. Pole type – Longer internode then semipole type.
- Edible portion is 94% of the pods which contains 1.7% protein, 0.1% carbohydrates, 4.5% fat, 1.8% fibre and 0.5 mineral.
- Dried beans are rich in protein and closely compared with meat.
- In India grown for tender vegetable while in USA grown for processing in large quantities.
- *Phaseolus oborigineus* – Progenitor of French bean.
- Four cultivated species.
  1. *P. vulgaris* - Common red kidney bean
  2. *P. coccineus* - Scarlet runner bean
  3. *P. lanatus* - Lima bean
  4. *P. acutifolius* - Latifolius-Moth bean

- All the species are self-pollinated except *P. coccineus* which is generally cross-pollinated.
- **Recommended Varieties of French Bean**

S.No.	Variety	Features
1	Giant Stringless	A bushy variety, pods green, long, slightly curved, meaty and stringless.
2	Contender	Plants bushy bearing light green, fleshy oval and thick pods. High yielder than Giant stringless in the plains. Maturity 50-55 days (first picking) yield 80-95 q/ha. Resistant to powdery mildew.
3	Pusa Parvati	It is an early maturing variety developed by X-ray irradiation of Wax Pod, an introduction from the USA. A bushy variety bearing light green round and meaty pods. Maturity 45 days (first picking) Yield 80-85 q/ha. Resistant to powdery mildew.
4	Tender Green	Bushy type bears light green, round and oval pods.
5	Kentucky Wonder (Introduction, USA)	Its pods are long (22.5 cm), flat, curved, stringless and borne in clusters. A pole type with green, fleshy round and slightly curved pods, Maturity 60-65 days (first picking). Yield 100-120 q/ha. The variety is tolerant to wilt.
6	Pusa Himlata	It was developed by selection from an exotic material. Pole type pods medium long, round straight light green meaty and stringless. Seeds white, Maturity 60 days (first picking).
7	Kashi Param	Determinate plants, yield 120-140 q/ha. Suitable for J&K, HP, Uttarakhand, MP, Maharashtra.
8	Arka Komal (IIHR, Bangalore)	It has been developed by selection from an exotic material introduced from Australia. Has a green, fleshy, tender long and straight pod. Produces 200-250 q/ha. green pods in 70-75 days.
9	Arka Anoop	A pedigree selection from Arka Bold × Arka Komal. Plants bushy photosensitive, resistant ot both rust and bacterial blight. Pods long (17-18 cm) flat and straight. Duration 70 days and yield 20 t/ha.
10	Jampa	A Mexican variety reported to be outstanding in performance in Maharashtra. The plant has a slight tillering habit. It flowers in about 47 days and the first picking is obtained after two months of sowing. In all, eight pickings within 40 days are obtained. The pod is flat, smooth, non stringy and pale green. The seeds are black, smooth small and flat. It is resistant to wilt.

11	Pant Anupama	Plants are upright, 45 cm tall and bushy. The pods are round, 12-15 cm long and stringless. First picking is possible 55-65 DAS. Average green pod yield is 90 q/ha. Resistant to angular leaf spot and common mosaic virus disease.
12	Pant Bean 2	Pods flattish round, stingless, both for green pods and dry seeds, cross between Turkish Brown × Cotender. It is resistant to rust and common bean mosaic. Developed by GBPUAT, Pantnagar.
13	YCD 1	Selection from local type Shevroy hills. Pods slightly flat, green dual prupose developed by TNAU Coimbatore.
14	VL Boni 1	It is a rust resistant variety. Pods medium long, stringless, Pale green. Developed at Almora, UP.
15	Lakshmi	Hybridization between Contenders × Local Pole variety. Pods round, green stringless. It is resistant to angular leaf spot.
16	Phule Surekha (MPKV, Rahuri)	It is a semi-determinate variety developed by selection from an exotic material. Its pods are straight, flat and 9.2 cm long. It is photo-and thermo-insensitive, suitable for cultivation in kharif, spring and summer seasons. It is resistant to anthracnose, leaf crinkle, bean mosaic virus and wilt. Average yield is 150 q/ha.

Varieties

(A) On the basis of fiber

(1) String type – Pod contain fiber, indigenous varieties are string.

(2) Stringless type – Pods are free from fibre. Most of improved varities are stringless type.

(B) On the basis of growth habit

(1) Pole type – Tall, Indeterminate in growth and requires support.

(2) Bush type–Plants dwarf and bushy, popular because of compactness, easy harvest and short duration.

1. Pole type –

1. Pusa Himlata
2. Kentucky wonder – Introduced from USA.
3. Lakshmi - Tolerant to angular leaf spot disease.
4. SVM 1 – Hybridization between ***P. vulgaris*** var **Contender** × ***P. Multiflorus*** var. PBL 257, resistant to angular leaf spot.
5. VL Lata Bean 12
6. VL Lata Bean 17
7. Tweed wonder

2. Bush Type

1. Pusa Parvati – Developed through X-rays irradiation of an American variety, wax pod, resistant to mosaic and powdery mildew disease.
2. Pant Anupam – Early, resistant to angular leaf spot and moderately resistant to mosaic virus.
3. Arka Komal – Very good for transport and good keeping quality.
4. Arka Suvidha -It has been developed by selection from the cross Blue Crop × Contender.
5. IIHR 220- Resistant to rust, Pure line selection variety.
6. IIHR 909 – Contender × Blue crop.
7. Jampa (Maxican variety) – Highly resistant to wilt and withstand warmer conditions.
8. Contender – Early, tolerant to powdery mildew mosaic.
9. Premier – Introduce from Sweden, late sowing, susceptible to mosaic and wilt, suitable for hills.
10. Bountiful – USA introduced; pods borne on main stem in clusters.
11. Pant Bean – Moderately resistant to common virus.
12. VL Boni -1
13. TKD
14. KKL - For elevation of 1800-2400 m.
15. YED
16. Ooaty – Moderately resistant to leaf spot, anthracnose and pod borer.
17. Giant stringless.

- Most of French bean varieties are day neutrals except semi-pole varieties which are short day types.
- Optimum temperature range from 15-25°C.
- The seeds will not germinate if the soil temperature is below 12°C and the seedlings will not tolerate temperature below 10°C.
- Pole types require longer growing season and are, therefore, more suitable for growing in mid-hills.
- Bush types can be grown in north Indian plains where growing season is short.
- The crop is sensitive to frost, high temperature and excessive rains.
- Plants drop/shed their blossom or young pod in very hot or rainy season.
- Pole type – Generally grown in heavy rainfall areas of chhotanagpur of Uttar Pradesh.
- It can't withstand water logging.
- Seed rate –

  50-75 kg – Bush type/ha

  25-30 kg - Pole type/ha
- Sowing time – July-September

  Jan –Feb. – In Plains

March – April - In Hills
September - Nov -Southern India
October - Northeast and Central Zone.

- Spacing –
  30 × 5 cm - Bush type (2, 50,000 plants/ha)
  90 × 7.5 cm - Pole type (80,000 plants/ha)
- Foliar application of B, Cu, Mo, Zn, and Mg @ 0.1% – Enhance pod quality and yield.
- Seed should be inoculated with strains of *Rhizobium phaseoli* for quick nodulation and high yield.
- The seed should be treated with it's culture @ 30 g. sugar solution/kg seed.
- It is sensitive to both water excess and water stress condition.
- Flowering and pod development periods are the critical stages for water application.
- Cultivation or working among plant, where foliage is wet should be avoided, because spores of anthracnose and angular leaf spot disease are easily dispersed under wet condition.
- In temperate climate, pole beans are grown in combination with corn or okra as the stalk is used for support of the vines.
- Application of GA3 (50 ppm) on 30 days old plants is effective in increasing plant height, number of leaves and pod yield.
- 2-4 foliar spray of growth retardant CCC (400 ppm) reduces vegetative growth, causes early flowering and increase yield.
- Flooding in French bean cause anoxia leading to root rot.
- Maturity standard – Determined by Alcohol-insoluble-solids (AIS) and dry matter (DM) contents.
- French bean takes 50-60 days from sowing and 7-10 days from flowering to reach marketable maturity.
- Three to four pickings of bush types and 8-10 pickings of pole types are possible.
- Yield –
  8-10 tonnes - Bush varieties/ha.
  12-15 tonnes - Pole varieties/ha.
- Storage 5-7°C for 20-25 days.
- In India bulk is consumed as fresh.
- In USA bulk is processed, more than 80% (White colour is preferred).

Physiological disorders –

1. Delayed Flower initiation and development– Occurs below 10°C where fertilization may not occur, producing small and misshaped pods.
2. Blossom drop and ovule abortion – High temperature (35°C) is the main cause.
3. Transverse cotyledon cracking (TVCC) – Occurs when dry seed of beans are sown

in wet soils. Seed moisture content less than 12% enhance cracking of cotyledon. White seeded cultivars are more prone to this malady.

- Resistant cultivars – Show 60-80% germination.
- Susceptible cultivars – Show 20% or less germination.

4. Necrosis of hypocotyl – Low "Ca" content in seed after germination and soil rich in Ca and Mg are affected by this problem.

- Bean common mosaic virus is seed borne and is also transmitted by aphid (Myzus persicae).
- Bean yellow mosaic virus is not seed borne in beans and is also transmitted by aphid (*Myzus persicae*).
- French bean is a self-pollinated crop and an isolation distance of 50 m between two varieties of French bean is required to prevent physical mixtures.
- For seed production, the beans are harvested when pods turn yellow.

"BROAD BEAN"

- Also known as faba bean, horse bean, bakla bean.
- Mainly cultivated in Latin America.
- It is the only bean which is grown as a winter crop and autumn crop.
- Pollinated by insects.
- The pollen grains and green pods cause allergy to some people which turn in serious illness known as "Favism" (Haemolytic anaemia).

Varieties –

[A] Based on seed colour –

a. White seeded varieties – Less soft.
b. Green seeded varieties – Soft and suitable for freezing.

[B] On the basis of pod size –

1. Long pod -
   Red Epicure.
   Imperial White long pod.
   Imperial Green long pod.
   Master piece Green long pod.
   Aquadule Claudin.
2. Small poded -
   (a) Imperial white windsore.
   (b) Gaint Four Seeded Green Windsor.
   (c) Imperial Green Windsor.

- It can with stand temperature as low as 4°C, mainly grown at higher elevation.
- Seed rate – 70-100kg/ha.
- Yield – 7-10 tons/ha.
- Pusa Sumeet – Improved variety of Broad Bean, developed at IARI, New Delhi.

33. "INDIAN / DOLICHOS BEAN" (SEM)

- Also known as Indian bean, hyacinth bean, lablab bean, country bean, Egyptian bean, Tonga bean, field bean, sem or field bean.
- Crop is multipurpose, can be used as pulse vegetable and green manure.
- From India it was introduced to China, West Asia and Egypt.
- The term *Dolichos* is derived from Greek language means "Long pod" and l*ablab* is an Arabic term means "Production of sound by dried seeds in the pods".
- Dolichos bean is vigorous trailing, twining herbaceous annual propagated through seed.
- Trypsin Inhibitor is present in mature and dark coloured seeds.
- On heating it is broken down into toxic cyanogenic glucoside which is soluble in boiling water. Therefore it is necessary to boil the mature seeds in water before consumption.
- Sem is of two types –

  (a) *Dolichos purpureus var- typicus* (Garden type)

  Soft edible, pod wall have less fibre.

  (b) *Dolichos purpureus var – Lignosus* (Field bean).

  Grown for fresh dry seed as pulse, pod wall has high fibre content and have characteristics aroma.
- Per 100 g of edible green pod contain carbohydrates 6.7 g, protein 3.8 g, fibre 1.8 g, minerals 0.9 g (calcium 210 mg, phosphorus 68 mg, sodium 55.4 mg, potassium 74 mg, sulphur 40 mg, magnesium 34 mg, iron 1.7 mg) and fat 0.7 g.
- The mature seeds, however, contain anti-nutritional factors such as tannins, phylate and trypson inhibitors.
- **Recommended Varieties of Dolichos Bean**

S.No.	Variety	Features
1	Arka Jay	Developed by backcross and pedigree selection from the cross involving Hebble Avare and IIHR 93. Plants dwarf, bush type, erect and photo-insensitive. Flowers are purple and pods long, light green, slightly curved without parchment. It is a vegetable type with excellent cooking quality. Crop duration is 75 days and yield 12 t/ha.
2	Arka Vijay	Plants dwarf, bushy erect, photo insensitive, dual purpose, pods with characteristic aroma, with parchment, tolerant to heat and drought with excellent cooking qualities, light aroma in pod. Duration 75 days and pod yield 12 t/ha.
3	Konkan Bhusan (KVK, Dapoli)	It is bushy (60-75 cm tall) variety that bears about 10 cm long, tender, green and tubular pods. Its plants are compact and first picking is possible 60 DAS. The average yield is 80 q/ha.

4	Hebbal Avare 4 (UAS, Bangalore)	It was developed by selection from the cross Hebbal Avare × Co 80. Its plants are compact and give first picking around 60 DAS. Average yield of 60 q/ha in five pickings.
3	Pusa Early Prolific	It is a pole type early variety. Light green pods. Suitable for autumn and spring, Duration 200-215 days. Yield 14 t/ha.
4	Pusa Sem-2	Pods dark green, first picking in 120-125 days, highly tolerant to anthracnose and viruses, insect pest, pod borer, June-July is the sowing time, yield 15 t/ha.
5	Pusa Sem – 3	Pods green, 15 cm long, first picking in 115-120 days, highly tolerant to anthracnose and viruses, insect pest, pod borer, June-July is the sowing time, 17 t/ha.
6	Blue Lake	It is early variety, bushy, photo insensitive, dual type. Spreading white flowers, 70-80 days for harvesting.
7	Konkan Bhusan	Green and tubular pods, 100-180 days, yield 6-8 t/ha.
8	Dasarawual	Green with purple tinge, yield 8 t/ha.

- Green podded varieties are more popular in south and north India, whereas, white-podded varieties are preferred in eastern parts of India.

Other Varieties –

1. Herbal Avare 3 – Determinate growth habit, photo insensitive pods are not edible but can be used in hybridization/programme for incorporation of dwarf character.
2. Wal Konkon – Resistant to yellow mosaic virus.
3. Kalyanpur Type – 2
4. Jawahar Sem -37, 53, 79, 85 - For low rainfall areas.
5. Arka Jay and Arka Vijay – Can be grown throughout the year.
6. Climbing – Rajni, Deepali, KDB – 403, 405, 71
7. Pusa Early Rajni – An improved variety developed at IARI, New Delhi.
8. CO 10 – Mutant Variety

- Dolichos bean is a cool season crop.
- It grows well under warm, humid conditions at temperatures ranging from 18-30°C and is fairly tolerant to high temperatures.
- It responds to photoperiod, is drought hardy and is grown in arid, semi-arid and humid regions.
- High temperature and high humidity favour its plant growth but fruiting starts with the onset of winters when the temperature and humidity become moderate and continues throughout spring.
- In north India, the seeds are sown in June -July while in south India, in July-August.

- Seed rate in Dolichos varies with the varieties, depending upon their growth habit.
- For a pure crop, the row-to-row spacing is maintained from 1.0 to 1.5 m and plant-to-plant 60 to 75 cm.
- Flowering and pod development are the critical stages of crop for irrigation.
- Seed rate –

  10-12 kg – Climbing type.

  20-30 kg – Bushy type.
- In North and Central India, it is usually grown as mixed crop with ragi or sorghum.
- Yield – 3-5 tonnes/ha.
- Dolichos bean is primarily a self-pollinated crop and an isolation distance of 25 m between two cultivars is sufficient to avoid mechanical mixture of seeds.
- The seed crop is inspected before flowering, during flowering and at pod maturity to rogue out off-type and diseased plants.
- The crop is harvested when pods turn yellowish and then allowed to further dry in sun for 2-3 days.

34. "LIMA BEAN"

- Lima bean, grown predominantly as an annual for its tender pods as well as mature green seeds,
- The name Lima bean obviously came from Lima, the capital city of Peru, one point at which the species was found by early European explorers.
- The name butter bean was probably derived from its cream coloured pods
- Also known as double bean or butter bean.
- It is also used to prepare wine.
- The pods are one of the principal ingredients of succotash, a dish consisting of Lima beans and corn.
- It is cross compatible with French bean.
- Per 100 g immature green pods contain moisture 71 g, protein 7.6 g, carbohydrates 19 g, fibre 1.8 g and vitamin A 280IU.
- Lima beans also contain the phytochemicals coumestrol and saponin compounds that may impart anticancer benefits.

Varieties –

1. **Pole Type –**

   (a) Florida Butter.

   (b) Challanger.

   (c) King of the Garden.

   (d) Karolina Butter.

   (e) KKL 1 Butter bean (HRS Kodaikanal, TNAU Coimbatore)

   (f) Carolina Sieva (Introduction, USA)

2. **Semi pole type –**
   (a) Hopi.
   (b) Wilbur.
3. **Bush Type –**
   (a) Handerson Bush.
   (b) Fordhook 242.
   (c) Baby Potato.
   (d) Baby Fordhook.
   (e) Early Thorogreen (Introduction, USA
   (f) Burpees Improved Bush (Introduction, USA)

- In plains, it is grown as a winter crop where as at higher altitudes, it is grown in spring and rainy seasons.
- It grows better when the temperature range is between 15-20°C.
- If temperature exceeds 26°C, both pollination and fertilization are impaired resulting in reduced pod yield.
- Seed Rate -
  20-35kg/ha Small seeded (For pole varieties).
  45-50kg/ha Large Seeded (For bushy varieties).
- Yield – 5-10 Tonnes/ha.

35. "WINGED BEAN" (ASPARAGUS BEAN, MULTIFLORUS BEAN)

- Also known as four angled bean, Goa bean, Manila bean, Princess bean.
- It is a tropical legume of high value.
- Pod - 1.2-2.9% Protein
- Leaves - 6-15% Protein
- Pod show four sides wings protruding from angles hence the name winged bean.
- It is photosensitive and requires short day for normal flower induction.
- It is a diploid 2n = 18, 22.
- Mainly grown in Tamilnadu, Kerala, Karnataka, Goa, Orissa, Maharashtra, West Bengal, and North Eastern States.
- Multipurpose legume, every part of plant like green pods, seed, leaves, flower and tuberous roots are edible.
- Main breeding object – Development of varieties for photoinsensitivity.
- **Varieties** –IIHR sel, 20, 60, 71
  WBG2 – North East Parvatiya Anusandhan Kendra, Meghalaya

36. "SCARLET RUNNER BEAN" (PHASCOLUS COCCINEUS)

- It is L.D.P. (Long Day Plant).
- Consists of Protein (17%), carbohydrates (65%) and 2% edible oil.
- In Brazil where it is called as Pig Bean.

37. "JACK BEAN" (CANAVALIA ENSIFORMIS)

- Seeds are utilized to produce urease enzyme.

38. "SWORD BEAN" (CANAVALIA GLADIATA)

- Seed contain some toxic substance like saponins therefore it is available to use the seeds after cooking.
- Sword bean **SBS 1** is an introduction and is one of the vegetables with photo-insensitivity. It matures in 110 - 120 days. It can be grown throughout the year and gives good response to irrigation. Tender pods are ready for harvest from 75 DAS. As a pure crop it gives an average grain yield of 1356 kg/ha and green pod yield of 7500 kg/ha. This can also be grown as border crop, intercrop and a shade crop.

39. IMPORTANT PEST AND DISEASE OF LEGUME VEGETABLE CROPS

Sr. No.	Pest and Disease	Symptoms	Management
1.	Gram pod borer (*Helicoverpa armigera)*	Defoliation in early stages and larva's head alone thrust inside the pods and the rest of the body hanging out.	Install Pheromone traps for *Helicoverpa armigera* 12/ha. Apply Emamectin benzoate 5%SG 220 g/ha or Quinalphos 25%EC 1400 ml/ha.
2.	Bean Aphids (*Aphis craccivora)*	Leaves, inflorescence stalk and young pods covered with dark coloured aphids. Honey dew secretion with black ant movements	Spray Imidacloprid 17.8 SL 100-125 ml/ha or Thiamethoxam 25% WG 100 g/ha.
3.	Whitefly (*Bemisia tabaci)*	Leave mottled and yellowish in colour. Vector of yellow mosaic virus.	Spray methyl demeton 25 EC 500ml or dimethoate 30 EC 500 ml or phosphomidon 85 WSC 250 ml/ha
4.	Blister beetle (*Mylabris phalerata)*	The adult feeds voraciously on buds and flowers.	Control as same as pod borer.
5.	Pod bugs (*Riptortus pedestris)*	Poorly filled pods with shrivelled grains inside.	Spray Dimethoate 30% EC 500ml/ha or Methyl demeton 25%EC 500ml/ha (Spray fluid 500 l/ha).
6.	Blue butterfly (*Lampides boeticus)*	Buds, flowers and young pods with boreholes.	Spray Emamectin benzoate 5%SG 220 g/ha or NSKE 5% twice followed by triazophos 0.05% (Spray fluid 500 l/ha).

7.	Leaf hopper (*Empoasca kerri)*	Green colour insects found under surface of leaves	Spray methyl-o- demeton 750 ml in 700 – 1000 L water per hectare.
8.	Pea Stem fly (*Ophiomyia phaseoli*)	The maggot of the insect damages the internal tissue, consequently the entire plant dies.	Mix 30 kg/ ha Carbofuran (Furadon) 3 % granules or 10 kg /ha Phorate (Thimet) 10 % granules in the soils before sowing the crop.
9.	Thrip (*Thrips indicus*)	Foliage becomes thickened and puckered.	Spray Rogor 30EC (dimethoate) 1l/ha when the attack begins and repeat the spray at 15 days intervals.
10.	Leaf miner (*Chrotomyia horticola*)	Its larvae feed by making tunnels in the leaves. These cause serious damage during December-March.	Spray Oxydemeton methyl (Metasystox) 25 EC 1l/ha when the attack begins and repeats it at 15 days intervals.
11.	Spiny Pod Borer (*Etiella zinckenella)*	The larva feeds inside green pods and then on pod surface, webbing together 2-4 pods.	Deep summer ploughing and install pheromone and light traps. Spray cypermethrin or quinalphos 25 EC @ 1000 ml/ ha.
12.	Powdery mildew (*Erysiphe polygoni*)	Powdery white fungal growth on the upper leaf surface and pods.	Spray Karathane 40 EC @ 200ml or Sulfex @ 1.5 kg/ha. and three sprays may be given at 10 days interval.
13.	Fusarium wilt (*Fusarium oxysporum f.sp. pisi)*	The first symptom of the disease in the field is drooping of the plants followed by sudden death.	Seed tretment with Carbendazim (2 g/kg of seed) at initial stage later soil drenching with Copper oxychloride 0.25%.
14.	Rust (*Uromyces fabae)*	The entire leaf blade and other affected parts give a brownish appearance even from a distance	Application of fungicide like Tridemorph 0.1 % or Mancozeb 0.25% effectively control the disease.
15.	Leaf Spots (*Cercospora sp.*)	Leaf spots cause severe defoliation and are common during humid seasons	Spraying Mancozeb 0.25 % control the disease
16.	Anthracnose (*Colletotrichum lindemuthianum)*	Bean pods with black, sunken lesions or reddish-brown blotches.	Used certified disease-free seed and spray copper based fungicide.
17.	Downy mildew (*Peronospora destructor)*	Grey brown mould spot appear on the under surface of the leaves. Incidence of the disease is only severe in very wet weather conditions.	Dithane Z-78 or Dithane M-45 at 0.2% concentration should be sprayed

40. LEAFY AND SALAD VEGETABLES

Common Name	Botanical Name	Family	Chromosome Number	Origin	Seed rate	Sowing Time
Palak/Beat leaf/ Spinach beet	B. vulgaris Var. bengalensis	Chenopodiaceae	18	Indo-China	10-15 kg winter 20-30 kg summer	August – Nov. Plains March - May Hills
Spinach	*Spinacia oleracea*	Chenopodiaceae	12	Central Asia (Iran)	37-45 kg/ha.	Sept.- Oct. Winter Aug - Oct. Hills
Amaranthus	*Amaranthus sp.*	Amaranathaceae	32	India	1.5-2.0 kg/ha. 1 kg transplant	March summer July-Kharif
Lettuce (salad crop)	*Lactuca sativa*	Compositae	18	Med. Region	400-500g (50-75m.area)	September-November
Basella(Poi)/Indian Malabar Spinach	*B. rubra var. alba* (green leaves) *B. rubra var. rubra* (red leaves)	Basellaceae	24	South Asia (India)	12-15kg/ha	March-May. plains Mar-April-Hills Oct.-Nov. S. India
Portulaca	*Portulaca oleracea*	Portulacacea	54			
Parsley	*Petroselinum hortense*	Umbelliferae	22		250-300g/ha.	
Agathi	*Sesbania grandiflora*	Leguminaceae	24			
Water leaf/kalmisag	*Ipomoea acuatica*	Convolvulaceae				
New Zealand Spinach	*Tetragonia expansa*	Aizoaceae				
Chicory Sorrel	*Cichorium intybus* *Rumex vesicarius*	Compositae Polygonaceae	18			

- Leafy vegetables also referred to as greens or potherbs, are popular around the world especially in Asia.
- They are quick growing crops that are harvested 4-6 weeks after sowing.
- These vegetables are very perishable, therefore it grown in peri-urban areas.
- Leafy vegetables are rich source of protein, vitamins A and C; and minerals like iron calcium and phosphorous.
- These vegetables provide roughage and have an important place in balanced diet.
- Dieticians recommend daily at least 116 g of leafy vegetables in a balanced diet.
- Leafy vegetables could be grouped into:
  1. **Cultivated indigenous vegetables**
  2. **Uncultivated naturally occurring leafy vegetables**
  3. **Introduced leaf vegetables**
  4. **Cultivated perennial leafy vegetables**

1. **Cultivated indigenous leafy vegetables**

   A. Amaranthus tricolor
   B. Amaranthus viridis
   C. Amaranthus dubius
   D. Atriples hortensis
   E. Anethem rowa
   F. Alterneathera sessiles
   G. Hibiscus carbis
   H. Coriandrum sativan
   I. .Portulaca grandiflora
   J. Trigonella foenum gracum

2. **Non cultivated naturally growing leafy vegetables**

   A. Alternanthera echinate
   B. Amaranthus sessiles
   C. Amaranthus caudetus
   D. Amaranthus gangeticus
   E. Amaranthus polygamus
   F. Amaranthus spinosus
   G.Centella asiatica
   H. Celosia argentia
   I. Coleus amboinicus
   J. Jatropa prostrate
   K. Oxalis carniculate

3. **Perennial greens**

   A. Subania grandiflora
   B. Moringa olerifera
   C. Sauropus androgynus
   D. Tamarindus indica

4. **Introduced leafy vegetables**

- Spinach, Celery, Lettuce. Parsely, Basella, Ceylon (Talinium triangulare) etc,.

41. "PALAK/SPINACH BEET/BEET LEAF"

- It is mostly grown in Uttar Pradesh, West Bengal, Maharashtra and Gujarat.
- It is primarily used as potherbs.
- It is a rich source of vitamin A (5862 IU/100g.) and vitamin C (70 mg/100g).
- It is rich in nutrients 3.4 g. protein, 0.8 g fat, 380 mg Ca, 30 mg P and 16.2 mg Fe.
- Spinach beet, swiss chord and garden beet has $2n = 2x = 18$.

- Leaves of this crop might have been first used in Bengal and hence it is known as *B. valgaris var. begalensis.*
- Palak leaves contains high oxalates content hence, people suffering from kidney stones are advised to restrict its consumption.

Difference between spinach beet and spinach –

S.No.	Attributes	Spinach Beet	Spinach
1.	B.N.	*B. vulgaris var. begalensis*	*Spinacia oleracea*
2.	Ch.	2x = 18	2x=12
3.	Sex forms	Bisexual	Monoecious, Dioecious Hermaphrodite.
4.	Leaves	Have entire margin	Have lobed margin.
5.	Seed	Monogerm	Multigerms.

- **Recommended Varieties of spinach beet**

S.No.	Variety	Features
1	Pusa Bharati	It has been developed by selection from the variety Pusa Jyoti. Late variety, It has higher vitamin C content and ß carotene than Pusa Jyoti, it can also be grown during kharif and summer seasons. Performs best in winter and early spring seasons. Maturity 30-40 days. Yield 500 q/ha.
2	Pusa Jyoti	It has been developed by polyploidy breeding using colchicine for doubling chromosome number of the cultivar 'All Green'. Gives 6-8 cuttings. Rich in calcium, iron and ascorbic acid, can be grown throughout the year. Maturity 40 days. Yield 490 q/ha.
3	Pusa Harit	The variety was developed from an inter-specific cross involving sugar beet and Local Palak. Suited to hills, plants upright, vigorous, late bolting habit and wide range of adaptability to varying climate, tolerant to alkaline soil. Maturity 35-40 days. Yield 425 q/ha.
4	All Green (IARI, New Delhi)	Produces uniform green, medium sized, tender leaves, yields about 123.5 q/ha of green material. Gives about 6-7 cuttings at 15-20 days interval. Maturity 40 days. Yield 120 q/ha.
5	Jobner Green	It has been developed by selection from a spontaneous mutation identified from a local material, 'Selection No. 5'. Maturity 40-45 days. Yield 300 q/ha. Evolved by RAU, Jobner.
6	Pusa Palak	Hybridization between Swisschard × local Palak. Leaves uniform green with red pigmentation late bolting.

7	Pant Composite	Composite variety, leaves green, high yielding, yield tolerant to cercospora leaf spot. Yield 27.5 t/ha. Developed GBPUAT, Pantnagar.
8	Punjab Selection	Selection from local type. Leaves light green, thin, long narrow smooth, stem with purple pigmentation. Yield 37.5 t/ha. Developed by PAU, Ludhiana.
9	Punjab Green (PAU, Ludhiana)	Selection from local type. Yield 37.5 t/ha. Through selection, leaves free from sourness and succulent, mild purple pigment on stem, Low oxalic acid content.
10	Palak No 51-16	Selection from local type. Leaves green, several cuttings, late bolting.
11	HS-13	Selection from a local type. Leaves light green, medium sized, quick growing. Developing by HAU, Hissar.
12	Arka Anapama (IIHR, Bangalore)	It has been developed by selection from the cross IIHR 10 × IIHR 8. Late bolting variety with medium large, thick, dark green leaves. Four cuttings can be taken over a period of 65-70 days. It is resistant to cercospora leaf spot under epiphytotic condition. Yield 41 t/ha.

- **Other Varieties** – Crop Improvement – By Mass selection and Hybridization.
  1. All Green – Late bolting.
  2. Jobner Green – Evolved through spontaneous mutation detected from a local collection S.I. No. 5.
  3. Pusa Palak – Hybridization (swiss chord × local palak).
  4. Pusa Jyoti – Developed by polyploidization, result of induce mutation using All Green as a source, very big leaves with dark green colour.
  5. Pusa Harit – Hybridization (Sugar beet × Local palak) heavy yielder, late bolting, has low chilling requirement.
  6. Punjab selection – Slightly sour taste.
  7. Arka Anupama – Resistant to cercospora leaf spot, regenerate faster.
  8. Banerjee Giant – Development by S.P. Banerjee through hybridization (Local Palak × Sugar beet).
  9. Pant Composite – 1 – Tolerant to cercospora leaf spot.
  10. HS-23 - Mass selection.
  11. Palak no. 51-61 -
- Spinach beet is a winter season crop.
- It can tolerant frost better than other vegetable.
- It can also tolerant high temperature but causes premature bolting.
- Palak is tolerant to slightly alkaline soil and also highly tolerant to salts.
- Superior yields of better-quality leaves are produced in neutral soils having soil pH around 7.0.

- The main sowing season in plains is from last week of August to second week of November.
- In places with mild climate, it can be grown throughout the year. In hilly regions, spinach beet is sown from March to May.
- Line sowing is more desirable as it facilitates weeding, hoeing and harvesting.
- Line spacing is maintained at 20 cm and thinning is done to maintain plant spacing within lines at about 5 cm.
- Seed rate varies from 10-15 kg for winter crop and 25-30 kg for summer crop.
- Spacing 20 × 5 cm (RXR and PXP).
- Palak is a leafy vegetable which requires more "N" for crown growth, apply "N" @ 20-25 kg/ha after every cutting as top dressing.
- During summer – Only one harvesting is recommended.
- Washing of leaves is not advocated because rotting may take place at center of bundle.
- Storage – 0°C temperature, 90-95% relative humidity for 10-14 days.
- Spinach beet is a highly cross-pollinated crop and requires an isolation distance of 1000 m from other cultivars and also from beetroot and Swiss chard.
- To rogue-out early bolters, diseased and off-type plants, the seed crop is inspected atleast twice; once when leaves are fully developed and then at bolting stage.
- Seed crop is harvested when plants have fully dried. It is then threshed, seed cleaned and dried to nine percent moisture or less and stored.
- Average seed yield is 10-12 q/ha.

42. "SPINACH" (VILAYATI PALAK)

- The name spinach is derived from Persian word 'ispanai', meaning green hand.
- It is a cool season vegetable and an annual crop.
- It is extensively cultivated in USA, Canada and Europe.
- Leaves are rich in Vitamin - A (5580 IU), Vitamin – C (28mg) folic acid (123mg), Iron (17.4mg), Ca (190mg) and P (42mg) / 100g.
- Spinach is related to Swiss chard, sugar beet, table beet, pig weed and saltbeet.
- In China it was introduce from neighbouring hilly kingdom Nepal in 647 AD.
- China is the largest producer of spinach accounting for 76 percent of the world production.
- Spinach is a long day crop/plant (LDP).
- This plant tolerates frost better than most other vegetables.
- This plant is sensitive to acidity.
- It is a dioecious plant and produce extreme male, vegetative male and female plants.
- In spinach 4 sex types exist – Tetramorphic.
    1. Male – Plants bloom early and hence are poor yielders.
    2. Vegetative Male

3. Female – Give fairly high yield because they flower late and have longer vegetative phase.
4. Monoecious

Recommended Varieties of Spinach

S.No.	Variety	Features
1	Virginia Savoy (Introduction, USA)	This is a smooth seeded variety having blistered, large, dark green leaves with round up. Plants are upright and vigorous in growth. Maturity 30 days. Yield 80 q/ha.
2	Early Smooth Leaf	This is a smooth seeded variety and produces small light green leaves with a pointed apex. Maturity 25-30 days. Yield 60-70 q/ha.
3	Long Standing (Introduction, USA)	Leaves are dark green, large, thick and conical in shape. First cutting is possible 48 DAS. Average yield is 100-125 q/ha.

- Virgnia Savoy – Prickly seeded variety.
- Early smooth leaf – Smooth seeded.
- Khara Lucknow, Khara palak, Banerjee Giant and Banarasi (Katvi Palak) are popular desi varieties.
- Being leafy vegetable, spinach require higher dose of "N" for increasing leaf yield.
- Sowing time of spinach varies from place to place and with cultivars. In plains, winter crop is sown in September-October and in hills from August-October.
- Spinach is sown on flat soil or on raised beds of varying dimensions, by broadcast method or in rows.
- Spinach seed has poor longevity and it is important to use fresh seed for plantings.
- Seed rate varies from 12-15 kg/ha.
- Seed rate in spinach is high because 50 per cent of plants are males with poor growth and it is not possible to remove them before seed stalk formation.
- Sown in lines, plant × row spacing is maintained at 30 × 5-7.5 cm. Within row plant spacing is maintained by thinning.
- Spinach requires a long cool growing season. It does not grow well during hot weather.
- High temperature and long days promote pre-mature seed stalk formation.
- Prickly seeded cultivars are best suited for the autumn winter crop in the hills.
- Smooth seeded cultivars are suitable for spring summer in hills and autumn sowing in plains it tolerates frost better than most other vegetables.
- Seed germination is poor at high temperature and is optimum when the temperature is around 4-5°C.
- The growth is best when temperature ranges from 10 to 15°C. However, it can tolerate temperature as low as 7°C.

- Thinning is done to keep plant spacing within rows at 5-7.5 cm.
- Spinach is ready for harvesting in 40-50 days after sowing at 5-6 leaf stage or just before seed stalk formation.
- Spinach has a large leaf surface ratio to weight ratio and has very high respiration rate.
- Storage 0°C – 90-95% relative humidity.
- Yield – 8-10 tons / Ha.

43. "AMARANTHUS" (CHAULAI)

- It is the cheapest leafy vegetable in tropical and subtropical parts of the country.
- Buddhist monks and the Muslim invaders took leaf amaranth from India to the neighbouring countries.
- Amaranth is also known as pigweed and used primarily as potherb.
- In India grown during summer and rainy season.
- It fits well in crop rotation because it is short duration and high yield per unit area.
- Amaranth is a rich source of protein, vitamins A and C; minerals, especially calcium and iron and dietary fiber.
- Amaranth protein is rich in lysine (5 per cent of total protein) and sulfur containing amino acids (4.4 per cent of total protein).
- High level of oxalic acid inhibits the absorption of calcium and it is prohibited by the people suffering from kidney disorders and stone.
- Amaranth prefers temperature between 25-30°C and is susceptible to frost.
- Seed germination is quicker and better when soil temperature is 20°C or above.
- The ideal soil pH range is between 5.5 to 7.5, but some strains of Amaranth can be grown in soils with pH as high as 10.0.
- Among leaf vegetable types
  1. *Amaranthus tricolor* (Badi chaulai) – Main cultivated species in India.
  2. *Amaranthus blitum* (Choti chaulai).
  3. *Amaranthus tristis* – More popular in South India.
- Amaranthus is a C4 plant which can make efficient use of Co_2 and suppress its photorespiratory loss.
- Amaranthus is susceptible to water logging.
- It can be grown in soil as high as 10 pH.
- *A dubius* is a tetraploid species having chromosome number 2n = 4x = 64.
- All other species are diploid.
- *A tristis* and A. blitum have chromosome number 2n = 2x = 32 whereas A. tricolour has chromosome number 2n = 2x = 34.
- *A. tricolour* is an ornamental and is often grown in flower gardens.
- Yellow and green dyes are extracted from *A. tricolour*.

- Classification of amaranthus

1. Leaf type

1. Amaranthus tricolor/ganetious
2. Amaranthus tristis
3. Amaranthus lividus
4. Amaranthus blitum
5. Amaranthus durbius
6. Amaranthus viridis

2. Grain Amarath species

1. Amaranthus caudatus
2. Amaranthus hypochondriacus.
3. Amaranthus cruentus
4. Amaranthus edulis.

Recommended Varieties of Amaranthus

S.No.	Variety	Features
1	Pusa Kirti (IARI, New Delhi)	It was developed through selection from the material introduced from Tamil Nadu. Leaves green with broad ovate lamina. Petiole 3-4 cm long, stem: leaf (1:3.4). Suitable for growing in spring-summer (March to June) season in the plains. Maturity 30-35 days. Yield 500 q/ha.
2	Pusa Kiran (IARI, New Delhi)	It has been developed from the naturally occurring inter-specific cross between *A. tricolour* × *A. tristis*. Pusa Kiran has predominantly characteristics of *A. tricolour*. Leaves glossy green with broad ovate lamina. The lamina 7-9 cm long and 6-7 cm wide. Suitable for growing in kharif (July to October) season in the plains. Maturity 25-30 days. Yield 300 q/ha.
3	Pusa Lal Chaulai	The upper surfaces of the leaves deep-red or magenta, lower surfaces purplish red. The stem deep-red in colour. Suitable for sowing in both spring-summer and kharif seasons in the plains. Maturity 25-30 days. Yield 500 q/ha. in spring-summer and 400 q./ha. in Kharif seasons.
	Pusa Chhoti Chulai	Its plants are erect, dwarf with thin stem; leaves are small, green in colour; responds well to cutting. It is best suited for early summer and can be grown in rainy season also.
4	Co-1	Dark green in colour, grows vigorously, high yielder, released by TNAU, Coimbatore. Developed by selection from a local type. Suitable for late harvesting. Average yield is 70 q/ha.

5	Co-2	Dark green lanceolate leaves, 20-25 days duration, 10-11 t/ha. Developed by TNAU Coimbatore through selection from a local material. This cultivar is suited for early harvest. Average yield is 100 q/ha.
6	Co-3	This is suitable for clipping of tender greens. First clipping is possible 20 DAS and total about ten clippings spread over 3 months are obtained. The leaf: stem ratio is high and this enhances the palatability of cooked vegetable. Its average yield is 307 q/ha.
7	Co-4	A dual purpose type for greens and grains. Plants dwarf 20-25 days for leaves, 80-90 days for grains. Yield 7-8 t/ha. (Green) and 2-2.5 t/ha. (Grain)
	Co. 5 (TNAU, Coimbatore)	The leaves are double shaded (green and pink) and have rosette appearance in early stages. The stem is tender and consumable. First harvesting is possible 25 DAS. Average yield is 400 q/ha.
8	Lal Sag	Grown popularly in many states, belongs to *A. mangostanus,* high yielding Indian variety. Produces seed early, has small flowers.
9	Arka Suguna (IIHR, Bangalore)	It is a pureline selection from an exotic collection from Taiwan (IIHR-47), leaves are light green, broad and succulent stem. First harvest in 25-30 DAS and 5-6 cuts in 90 days. It is moderately resistant to white rust and yield 25-30 t/ha. It is moderately resistant to white rust
10	Arka Arunima	A new mutant which is resistant to white rust and rich in calcium and iron and low in anti nutrient factors like oxalate and nitrates. It is multicut, purple amaranth variety. It becomes ready for first cutting in 30 DAS. Two subsequent cuttings can be taken at 10-12 days, interval with a total yield of 27.4 t/ha.
11	Badi Chaulai	By selection, suitable only for kitchen and container gardening. Plants tall 2-3 cuttings suitable for summer and kharif sowing. Developed by IARI, New Delhi.
12	Choti Chaulai	By selection, highly suitable for commercial cultivation, Plants dwarf suitable for spring-summer season. 6-7 cuttings. Developed by IARI, New Delhi.
13	Amt 105	Green leaves, stem green with reddish base. Suitable for central zone of Kerala. Yield 13 t/ha.
14	Amt 237	Green leaves and stem. Suitable for Kerala. Yield 15-16 t/ha. (Greens)

Other Varieties – It is a cross- pollinated (anemophilous) crop.

1. Pusa Kiran – Suitable for rainy season, stem: leaf (1-4.6).

2. CO1 – Especially suitable for late harvesting.
3. CO2 - Especially suitable for early harvesting.
4. CO3 - Especially suitable for clipping harvesting, local Name – Araikeerai.
5. CO5 – Tetraploid and very vigorous growth.

- Pusa Kiran – *Amaranthus tricolor* × *Amaranthus tristis.*
- Pusa Kirti – *Amaranthus triocolor.*
- Badi Chaulai – *Amaranthus tricolor.*
- Chhoti Chaulai – *Amaranthus blitum.*
- Pusa Lal Chaulai – *Amaranthus tricolor.*
- CO1 – *Amaranthus dubius.*
- CO2 – *Amaranthus tricolor.*
- CO3 – *Amaranthus tricolor.*
- CO4 – *Amaranthus hypochondriacus*, Dual purpose variety useful for grain and leaf.
- The seed is sown by broadcast method or by drilling in lines 20-30 cm apart, depending upon the variety.
- In the South, there is a practice of transplanting amaranth especially the variety 'Badi Chulai' either as a pure crop or along the border of beds of other vegetables.
- Seed rate is about 2 kg/ha for direct sowing and 1 kg/ha for transplanted crop.
- Dried seed with 6% moistures content are stored, yield 100-500q/ha.
- It can be stored for 6 days at 24-28°C temperature.
- For seed production, row × plant spacing is maintained at 60 × 25 cm.
- Most cultivated species of amaranth are monoecious and vegetable amaranth is largely self-pollinated due to large number of male flowers in the inflorescence.
- It requires a minimum of 200 m isolation distance between two varieties of amaranth
- Average seed yield is 2-3 q/ha.

44. "LETTUCE"

- The Latin name (*Latuca*) is derived from the Latin root word "lac" or milk. "Lettuce" is derived from the French "*Laitue*" meaning "milk" and "*Sativa*" means grown from seed.
- Lettuce is major salad crop in North America and is also important in Australia and most countries of Europe and South America.
- It occupies the maximum area among salad crop.
- Its popularity is increasing in Africa, the Middle East and Japan.
- In India it is grown in the kitchen garden and pre urban area to meet the demand of continental hotels.
- There are four important botanical varieties of cultivated lettuce.

1. **Head type (*Latuca sativa var. capitata*)** is divided into 2 groups.

   (a) Butter head type – Produces relatively small and loose head, soft leaves;

smooth texture; varieties bred for both outdoor summer conditions and greenhouse winter conditions; summer butterheads larger and firmer than the winter types; winter butterheads smaller and less compact; popular in N. Europe.

(b) Crisp head type – Popular in North America and European countries, produces brittle textured and light folded large head (up to 1 kg.) which have excellent shipping and handling abilities, brittle or crisp textured; prominently veined leaves; wrapper leaves green; inner leaves whitish-yellow; predominantly outdoor types.

2. **Leafy or Bunching type – (*Latuca sativa var. crispa*)**

   This group includes non heading cultivars cultivated in America, Australia, Europe and India.

3. **Cos or Romaine type – (*Latuca sativa var. longifolia*)**

   Produce elongated leaves to form a loaf shaped head. Elongated leaves developing into large loaf-shaped heads; slower to bolt than other lettuces; useful as a warm-weather crop; popular in S. Europe and the US.

4. **Asparagus or stem type – (*Latuca sativa var. asparaginia*)** Also known as celery lettuce, consumed after peeling.

- Harvested organ: leaves, eaten raw, often in salads.
- It is rich in "Vitamin A" (900 IU), Vitamin C (10mg), Choline (178mg) "Ca" (50mg) and "P" (28mg).
- If cooked most of the vitamin C content of leaves is lost.
- In UK and the USA, crisphead is referred to as 'iceberg' lettuce.
- Varieties of Lettuce

S.No.	Variety	Features
1	Great Lakes (Crisphead type)	Introduction from USA, Head large, firm with green leaves, outer leaves blistered. Fairly resistant to tip burn, susceptible to powdery mildew.
2	Slow bolt (Leaf type)	Leaf Broad, frilled, slightly yellowish green, crisp and tender, very late in bolting.
3	Chinese Yellow (Leaf type)	Early, leaves with light green, crisp and tender.
4	Imperial 859 (Crisphead type)	Head medium large and well covered by heavily blistered outer leaves. Resistant to tip burn but susceptible to powdery mildew.
5	White boston (Butterhead type)	Has soft, solid heads, inner leaves oily, soft greasy texture.
6	Dark Green	Narrow leaves with elongated upright heads.
7	Alamo 1 (Introduction, USA)	It is a heading type variety. Heads are compact and average head weight is around 500 g. Leaves are dark green, curly and crisp. It attains harvesting maturity 90-100 days after sowing. Average yield is 235 q/ha.

Other Varieties –

1. Punjab Lettuce No. 1 – Non heading type.
2. Slobolt – Leaf type, late in bolting, suitable for gardens.
3. Chinese Yellow – Leafy type.
4. White Baston – Butter head type.
5. Dark Green – COS type
6. Celtuce – stem type non heading.

- The Portuguese or the British most likely introduced Lettuce into India during the 16th century.
- *Latuca serriola* - Probable progenitor of cultivated lettuce.
- Lettuce is a self-pollinated, cool season crop. (13-16°C temperature).
- Sensitive to both day length and high temperature.
- Late spring days and high temperatures prematurely bring on flower initiation, causing bolting.
- Lettuce seed may go into dormancy when subjected to high temperature in the dark and its exposure to chilling at 4-6° C for 3-5 days results in breaking of dormancy (**Thermo-dormancy**).
- High temperature promotes seed stalk and causes bitter taste of leaves and induces tip-burn injury.
- Seed do not germinate when soil temperature goes above 22°C.
- It is sown normally in the nursery beds and requires a very fine seedbed due to small seed size.
- Seedlings are ready for transplanting in about 4-6 weeks.
- To raise seedlings for one hectare, 400-500 g seed is required to be sown in about 50-75 square meter area.
- Plant spacing is maintained at 30-45 cm between the rows and 20-25 cm within the rows.
- Summer grown lettuce will be ready for harvest in 50-60 days while winter grown lettuce will take about 120 days.
- Increase CO_2 enrichment (1000-1500 ppm) under glass house condition results in high yield.
- High temperature causes bolting or acceleration of seed stalk in lettuce.
- Mulching is recommended for lettuce.
- Lettuce for shipped is often kept fresh by vacuum cooling.
- Breeding objectives: resistant to diseases and pests, increased yield and uniformity, improved quality and bolting resistance.
- Big vein of lettuce is due to virus which is transmitted by olpidium brassicae fungus.
- It is highly sensitive to acidic soils. Optimum soil pH: 6.5 to 6.8.
- Harvesting – 50-60 days in summer season.
  120 days in winter season.

- Crisp head lettuce is considered mature and ready for harvest when the head is firm to hard and weighs about 1 kg.
- Storage -0°C and 90-95% relative humidity for 2-3 weeks.
- It is better not to harvest the lettuce immediately after a rain or while the dew is on it, because the leaves being crisp and brittle, break easily while handling.
- Major insect pest are aphids, loopers, leafhoppers, cutworms and leafminers.
- Pre and post-harvest spray of BA 5-10 ppm help delay senescence in storage and improves the shelf life.
- It is predominantly a self-pollinated crop and requires a minimum isolation distance of 25 m from other varieties for the production of genetically pure seed

Physiological disorders

- **Tip burn:** Burning or scorching of lateral margins of inner leaves of mature head. Deficiency of calcium is primary reason of its occurrence. The deficiency is caused due to faster growth of leaves than the rate of its translocation to the growing points. Exposure to high temperature also causes tip burn of lettuce.
- For its control grow resistant varieties and do not apply excessive fertilizers, especially the nitrogen.
- In strongly acidic soils, increase soil pH by liming to improve calcium availability.
- **Russet spotting:** It is due to Ethylene injury.

45. CHINESE CABBAGE

- It is also called tango and tip-top and a popular leafy vegetable in China.
- Head of Chinese cabbage is not round like cabbage head but is tall like romaine lettuce head with a broad central midrib.
- The outer leaves are pale yellow and the inner ones are whitish.
- Unlike other Brassicas, Chinese cabbage has no trunk.
- Chinese cabbage is classified into four types viz. loose leaved (var. dissoluta Li), semiheading (var. infarita Li), fluffy topped heading (var. laxa Tsen et Lee) and heading (var. cephalata Tsen et Lee).
- Variety cephalata is further classified into three types viz. ovate type (*f. ovata* Li), flat topped type (*f. depressa* Li) and cylindrical (*f. cylindrica* Li).
- Chinese cabbage is a native of China where it has been in cultivation since the fifth century.
- Chinese cabbage might have originated by hybridization between turnip (*B. campestris* spp. rapifera) from North China and Pak-choi (*B. campestris* L. ssp. chinensis Rupr Olsson) from South China
- Chinese cabbage is an excellent source of calcium, potassium, vitamin A and folate.
- Chinese cabbage is an important potherb vegetable. It is also stir-fried, sautéed and grated; used in soups and eggrolls and eaten raw in salads.
- Punjab Agricultural University has released Chinese Sarson No.1 in 1987.

- Chinese Sarson No. 1 (PAU, Ludhiana) it is a non-heading type with light green, broad leaves and semi-erect plant habit. It takes about 30 days for first cutting after transplanting. It is quite rejuvenating, giving 6-8 cuttings. Its average yield is 400 q/ha.
- It is field resistant to Alternaria leaf spot.
- Chinese cabbage thrives best at temperature from 15-20°C.
- High temperature and long days promote seed stalk formation and leaves develop bitterness.
- It prefers soils with pH ranging from 5.5 to 7.6 but growth is best in neutral soils with pH around 7.0.
- Chinese cabbage can be raised by direct seedling as well as by transplanting the seedlings. Direct seedling is more commonly followed, as it is believed to give a head start over transplanted crop.
- To raise one-hectare crop, 500 g seed is required through transplanting method and 2.5 kg for direct seedling.
- Generally, 60 cm × 45 cm row × plant spacing is recommended for early maturing cultivars, 70 × 50 cm for medium maturing cultivars and 75 cm × 55 cm for late maturing cultivars.
- Chinese cabbage is also highly cross-pollinated.
- To produce its genetically pure seed, an isolation distance of 1000 m between Chinese cabbage varieties is required.

46. "BASELLA"

- It is commonly called poi, Indian spinach or Malabar spinach or climbing vine.

  Basella alba – Dark green and round to oval leaves.

  Basella rubra– Red aired leaves and stems.

  Cultivars having dark green round oval leaves (Petiole and stem) with reddish petiole and stem and dark green cordate leaves are available.

Eclipse

- Producing a crop in 55 - 60 days in warm areas.
- Very early cultivar producing small
- and compact plants that can be planted close together.
- The leaves are thick and
- medium to deep green in colour.
- Yields very well under warm humid conditions.

Red

- The leaves, stems and flowers are tinged with red.
- The colour is lost when the plant is cooked and so it is best used in salads.

- *Basella alba* is a fast-growing, soft-stem vine, reaching 10 m in length.
- Its thick, semi-succulent, heart-shaped leaves have a mild
- flavour and mucilaginous texture. The stem of the cultivar *Basella alba* var.Rubra is
- reddish-purple.

Other spp.– *Basella connifolia, Basella nigra, Basella saponica,Basella cordifolia.*

Malabar Spinach (Red Vine)

This is a unique variety that has red-purple vines and dark green leaves. This variety has relatively small leaves and vines. The vines grow much faster than the Green Vine variety. Many people also like to grow this variety as the backyard decorative plants due to the beautiful vine and flowers.

Malabar Spinach (Green variety)

This variety produces large dark green leaves and vines. This vegetable is very popular in Chinatown and Vietnamese markets. Young leaves and tips are excellent for stir-fry cooking

- Inflorescence – Axillary, solitary.
- Flower – Sessile in the axil of the leaves, regular bisexual, rose coloured or yellowish white.
- Vitamin A – (3250 IU).
- It can be propagated by both seed (12-15 Kg/ha) and stem cutting.
- Succulent leaves with petioles and tender stem, which are used as vegetables.

47. "PORTULACA"

- Summer season vegetable.
- Fresh branches and leaves are used as vegetable.

48. "PARSLEY"

- It is a biennial herbaceous plant.
- Both leaves and stem are consumed (Salad, garnishing, flavouring.)
- Rich source of Iron, Vitamin A and C.
- It is a cool season, hardy crop which flourishes well both in temperates as well as in tropical climate.

  Three types of varieties are there.

  1. Plain leaf.
  2. Double curled.
  3. Moss curled – plants are dwarf, compact and bushy.

- Humburg – Plants tall, coarse and tough stem, leaves resemble those of celery shoots.

49. "AGATHI"

- It is a tropical tree valued for it's leaves and flowers.
- It is grown as a standard for black pepper and betel vines, as shade plant for the coconut seedlings and as wind break for Banana in Tamil Nadu.
- The bark contains tannin and yield good fibre and gum.
- The crop grows best in black cotton soil.
- It can tolerate seasonal water logging.

Varieties –

1. Red Flower.
2. White Flower – Suitable for kitchen garden.

50. IMPORTANT PEST AND DISEASE OF LEAFY AND SALAD CROPS

Sr. No.	Pest and Disease	Symptoms	Management
1.	Amaranthus weevil (*Hypolixus truncatulus)*	Irregular zig-zag tunnels in the pith region filled with excreta.	Spray malathion 50 EC 2 ml/lit after the harvest of leaf
2.	Leaf caterpillar (*Hymenia recurvalis)*	Webbed leaves become devoid of chlorophyll and dry up	Spray malathion 50 EC 1 ml/ lit
3.	Blue beetle (*Altica caerulescens*)	The grubs feed on tender leaves and make mines. Adults feed on leaf margins.	Spray the Metasystox 25EC (methyl demeton) @ 2.5-3.0 litres or Rogor 30EC (dimethoate) @ 1.0 l/ha.
4.	Leafminer (*Pegomya hyoscyami*)	Tiny yellowish larvae of the insect make tunnels inside the leaves.	Spray NSKE 5%
5.	Aphids (*Lipaphis erysimi, Myzus persicae*)	It sucking cell sap from young leaves and tender shoots. They infest the underside of the leaves causing leaf curl. Infested leaves look yellowish and are unfit for consumption.	Spray Imidacloprid 200 SL @ 100 ml/ha or Dimethoate 30 EC 500 ml/ha or Phosphamidon 40 SL 600 ml/ha (500 l spray fluid/ ha).
6.	Damping-off (*Pythium sp., Rhizoctonia sp.*)	Pre- and post-emergence death of seedlings	Drench the soil with 0.2 percent solution of Captan or Thiram. Follow crop rotation and avoid planting in sick soils.

7.	White rust (*Albugo occidentalis*)	White and shiny blister like pustules appear on lower side of leaves.	Spray Ridomil MZ @ 2 g per litre of water.
8.	Downy mildew (*Peronospora effusa*)	The symptoms first appear as yellowish, irregular spots on upper surface of leaves, it appears in cool and moist weather.	Spray the crop with Indofil M-45 @ 2 g/l of water or Alliete @ 2-3 kg per hectare.
9.	Leaf spot (Cercospora beticola)	Symptoms appear as small, brown lesions on leaves.	Spray the crop with Blitox @ 0.2 percent.
10.	Spinach mosaic (Virus)	Light and dark green patches and mosaic mottling of infected leaves.	Control the insect-vector (Aphid) by using systemic insecticides like Rogor 30EC or Malathion.

51. PARENNIAL VEGETABLES

Common name	**Botanical name**	**Family**	**2n**	**Edible part**	**Origin**	**Seed rate**	**Sowing time**
Asparagus	*Asparagus officinalis*	Liliaceae	20	Spears	Europe, Asia and Africa	3-4 kg/ha	July – Nov. Plains March – May Hills
Globe artichoke	*Cynara* scolymus	Compositae / Asteraceae	34	Flower bud	Med. Region	Suckers or off shoots	Feb. - March
Rhubarb	*Rheum rhaponticum*	Polygona-ceae	44 **(4X)**	Leaf stalk/ petiole	Asia (Siberia)	Crowns	-
Drum stick	*Moringa oleifera*	Moringa-ceae	28	Green pods	Tropical Africa and N.W. India	500-600 g/ha	-
Curry leaf	*Murraya koenigi*	Rutaceae	18	Leaves	Tarai tract of U.P.	-	-
Chekuramanis	*Sauropus androgynus*	Euphorbia-ceae		Green leaves	Indo-Burma region	-	-
Ceylone spinach / water leaf	*Talinum triangulare*			Leaves and tender shoots	Brazil	-	-

52. "ASPARAGUS" (SHATAWAR)

- Asparagus" is derived from Greek word "aspharagos" ("sprout" or "shoot").
- Origin: Coastal areas of Asia, Africa and Europe.
- First cultivated by Romans 2,000 years ago.
- Brought to N. America by European settlers.
- Cool season, cold hardy, dioecious vegetable (male and female flowers on separate staminate and pistillate plants, respectively).
- It is an herbaceous perennial and is grown in large area is USA, Germany and France, while Indian contribution is meager.
- Tender shoots called "spears" stem (cladophyllus) are used as vegetables and in preparation of soup.
- First harvest after 2 or 3 years. Plantings may produce for 15 to 20 years.
- Spears are harvested when they become 13 to 25 cm.
- The tender shoots contain a white crystalline substance, asparagine, which is used in medicine as diuretic in cardiac, dropsy and chronic gout.
- Asparagus is a dioecious herb with erect branches and grows to a height of 1-3 m, with winter dormancy.
- The roots are tuberous, horizontally inside the soil and serve as storage organs.
- Plant is composed of ferns, crown and root system.
- The crown is made up of roots and rhizomes.
- The fern is a photosynthetically active modified stem.
- The aerial stem (spear) arises from a bud on the rhizome.
- Male flower – Yellowish green and conspicuous.
- Female flower – Less conspicuous.
- The male and female plants are separate and the female plants produce berries with 1-3 seeds/fruit.
- Insect pollinated.
- Female flowers produce red berries.
- Male plants are more productive than female plants.
- The main objective of asparagus breeding is development of male hybrids, fusarium and rust resistance, improved spear quality and high yielding ability.
- Being a dioecious crop with male and female plants, cross-pollination is the rule and the cultivars should be grown in isolation to maintain genetic purity.
- Male plants yield more and live longer than female plants and are comparatively more disease resistant.

Varieties –

1. Green – More popular and produced mainly for fresh market.
2. Light green - Used in processing.
3. White – Used in processing.

Other varieties – Martha Washington, Marry Washington, Reading Giant, Palmetto, Colossal, Argentenil, Berr's Mammoth, Mammoth White, Jersey Queen, Jersey Giant, New Jersey, Improved perfection, UC 7266 , SL 831, Limbrass 22,126, Lucullas 1813J.

- Perfection – Improved variety released at IARI, New Delhi.
- Jersey Centennial – Vigorous, rust resistant and more productive than Mary Washington.
- Blanching is done to produce white asparagus.
- Green asparagus contains more vitamin - A then white.
- Pre cooling in asparagus is done by hydrocooling.
- Asparagus has two types of varieties, one with spears overcast with purple and the other with spear tips almost free from the purple overcast.
- Spears with open (Feathered) tips decay more rapidly, toughen and are unattractive.
- Asparagus is propagated through seeds or crowns.
- Crown becomes dorment during winter and formation of spears start in the beginning of spring.
- Storage – 0°C and 90% relative humidity for 3-4 weeks (spears).
- Mounding the soil to a height of 25-30 cm over the row is practiced to blanch the young spears and get white asparagus.
- Takes 3 year to obtain real yield.
- Major diseases are fusarium crown, root rot and cercospora leaf spot rust.
- For fusarium crown and root rot control: use certified disease-free crowns and avoid fields with a history of the disease, carefully manage soil pH, fertility and drainage.
- Fusarium crown and root rot are promoted by low pH, poor drainage and infertile soil.
- Hybrids are more tolerant to fusarium than open-pollinated varieties.
- Major insect pests are asparagus beetles, cutworms, tarnished and alfalfa plant bugs and asparagus aphids.

53.. "GLOBE ARTICHOKE"

- Common name: Globe artichoke or French artichoke.
- Latin name: *Cynara scolymus.*
- Family name: Compositae (Asteraceae), Diploid (2n = 34).
- Origin: Central and Western Mediterranean over 2,500 years ago.
- Harvested organ: Immature flower bud, surrounded by leaf-like bracts; the thickened bracts (phyllaries), heart (receptacle) and choke (flower bud) are eaten.
- Immature flower heads or buds are used as vegetable.
- Small heads are eaten raw or cooked.
- Large heads are eaten only after cooking.
- The thick receptacle known as heart is used for canning.
- Flower heads are considered useful in the dietary of diabetes.

- Leaves are bitter and considered to be useful in dropsy and rheumatism.
- The above ground portion dies every year during winter and the crown of individual plant decays after a year growth.
- It is an allogamous and entomophilous plant.
- Cynarin is the main ingredient in globe antichoke which stimulates bile and fat digesting enzymes.

Varieties –

Green Globe, Purple Globe. Both are popular variety.

Violet de Provience – Fresh Market.

Catanese - Fresh Market.

Spinosa - Fresh Market.

Bianco Tarantino – For processing.

Precoce di Jesi - Processing

Bull, Tudella, E 15, Brindisino, Romonesco are some of other popular varieties.

- It is a cool season crop that can be grown successfully at a temperature of 12-18°C. Temperature below 28°C causes tough bud and tendency to spread occurs.
- A light frost may not destroy the edible quality of bud but causes blistered appearances.
- Freezing temperature (-2°C to 1°C) prior to harvest, kills the buds and above ground growth. This will result in delayed harvesting upto 6-8 weeks.
- Cross pollinated which is assured by protandry and main pollinator is honeybee.
- Propagation –

  By seed – Less popular, inferior quality.

  By suckers or off shoots – Commercially used.
- The suckers or off shoots are removed when they are 30-45 cm in height.
- Perennial - plantings produce for 3 to 7 years. Mature plant grows to about 1.5 m.
- Leaves blue-green, sometimes spiny. Flower stalks bearing a central bud and 2 or 3 secondary buds also arise from the central crown.
- Plants grown form seeds come to bearing in about 8 months.
- Plants grown form suckers come to bearing in about 6-7 months.
- Harvest season – September - October is best.
- A mature plant will produce 12 or more stems and 45-50 edible buds.
- The buds can be stored at 0°C and 90-95% relative humidity for a period of 3-4 weeks.
- Pre cooling of buds at 5°C is practiced to achieve less weight loss, discolouration and incidence of decay than buds that are not precooled.
- Yield – 10-12 tonnes / ha.

Physiological disorders –

1. Scuffing

2. Split bracts
3. Detached epidermis
4. Bract wilting – All are marketing disorders.

54. "RHUBARB"

- Plant herbaceous, perennial, with large fleshy under ground rhizome, having a fibrous root system.
- Economic parts are large thick leaf stalk or petiole. They are used for sauces and pies.
- It was first grown in Italy.
- During late 18th century it was introduce in USA.
- Small greenish white flower are borne in panicles.
- Flower are protandrous, self-pollinated within a flower does not take place.
- Lower flowers are fertilized by pollen from upper flower.
- Rhubarb is a cold resistant plant.

  Propagation –

  1. Seed – Not recommended as the seed do not give rise to true to type plants,, only used in development of new varieties.
  2. By division of crowns – Each crown should contain good strong bud.
- Where the freezing temperature is common, the crowns are planted during spring.
- Where the temperature is non freezing, the crowns are planted in autumn.

  Planting pattern –

  Depth – 5 cm below soil surface

  1.2 - 1.8 m Row × Row

  0.6 -1.2 m Plant × Plant

  Varieties– Victoria (Oldest variety)

  Linnaeus (Oldest variety)
- Mcdonald, Ruby Valentine, Sunrise are new varieties, alternatives with red stalk.
- Strawberry and Cherry Red – Also popular variety of Rhubarb.
- Major insect – Snout beetle boring into stalk.
- Major disease – Foot rot / crown rot.
- Production of rhubarb stalks during the winter is called forcing of rhubarb in regions where the climate is suitable for production of vigorous crowns.
- Rhubarb is harvested from the second year of planting.

55. "DRUMSTICK"

- The Moringaceae is a single genus family with 14 known species.
- *Moringa oleifera* (syn. *M. pterygosperma*) – Most wildly grown and utilized species.
- It is commonly called "horse radish" tree (arising from the taste of a condiment

prepared from the roots) or drumstick (arising from the shape of a pod) or ben oil tree (arising from the ben oil extracted from seed) and has a host of other country specific vernacular names.

- Popularly known as Ganigana, Mullakkai, Murraugi, Shajan and Muringa.
- Its tender leaves and immature or halfmature fruits are eaten.
- It is most popular for its distinct appearing flavouring fruits.
- Seeds contain an oil called ben or behen oil which has been much used for illumination, soap industry and highly priced for lubricating watches, computer and other delicate machinery etc.
- Seed contain 38-40% of non-drying oil which is clear and odourless and never becomes rancid.
- The oil cake is water coagulant and used for purifying effluent water from rivers, ponds and shallow wells and it is used as organic substitute for water purifying chemicals as aluminium sulphate (alum).
- In India commercially grown in Tamil Nadu.
- It is a deciduous tree of 8-10 m height.
- Flower are white or creamy white.
- The flowers are bisexual, oblique, stalked, axillary and hermaphrodite which are highly crosspollinated due to heteromorphism.
- Zygomorphy of flower may be due to particular pollinators used in this aspect and it is akin to sub family caesalpinoideae.
- Fruits is hanging, three sided brown capsule of 20-45 cm length which bears dark brown, globular seeds.
- Anthesis 4:30 – 6:30 AM, 5:30 AM peak period.
- Anther dehiscence- 4:30 – 6:30 AM.
- At full maturity, anthers are greenish yellow and after dehiscence they turn to pale colour.
- Pollination occurs through honeybees.
- Pollen viability is 72% at anther dehiscence.
- The fruit set due to artificial pollination using fertile pollen may go even upto 100% while it is only 11-15% under natural pollination.
- It is a tropical plant, highly susceptible to frost and high temperature exceeding 40°C causes flower shedding.
- It can not grow in stiff clay soil.

Varieties –

1. Jaffna –Yazphanam type, introduced from Sri Lanka, highly suited for coastal tract of Tamil Nadu, Kerala, Karnataka, pods are long (60-90 cm length), 400 pods/tree.
2. Moolanoor Murangai -
3. Chavakacheri Murungai – Ecotype of Jaffna type, 90-120 cm, damage occurs during transportation due to 90-120 cm length of pod.

4. Chemmurungai- Pod tip red colour, grown in home gardens.
5. Palmurungai – Ecotype of Jaffna.
6. Punamurungai - Ecotype of Jaffna.
7. Kodikanal Murungai – Annual type, pods 20-25 cm in length, predominantly cultivated in betelvine garden, propagated by seed.
8. Kattu Murungai – Released from Annafarm at Kudumiyamalai (Tamil Nadu), annual type, development through pureline selection, propagation by seed, plants are like shrub and harvested easily.
9. Palamedu Murungai – Released from Tamil Nadu as seed propagated moringa.
10. KM – Selection from an annual type, propagated by seed, Developed at Annafarm, Kudumianmali of Pudukkottai district in Tamil Nadu.
11. PKM 1 – Annual type developed at Periyakulam (Tamil Nadu).
12. PKM 2 – Released during January 2000 at Periyakulam (Tamil Nadu). Cross – (MP 31 × MP 28)

- Seed Rate - 625 gm/ha
- Propagation – Perennial type by limb cutting (1-1.5m), Annual type by seed.
- The annual drumstick types come to harvest in six months after sowing while the perennials types propagated through limb cuttings take 8-9 months for bearing.
- The fruits are ready for harvest in 60 days after flowering. The period of harvest extends for 2-3 months and each plant bears 200-250 fruits in annual types.
- In perennial types, the yield will be generally low (80-90 fruits/plant/year) in the first two year of bearing. Then it increases to about 500-600fruits/plant/year in 4th and 5th year and the pods are harvested mainly in march-June.
- A second crop can be harvested in September to October. The ratooned crops will develop new shoots and will start bearing after six months.
- Fruit weight – 200- 230 g containing 10-20 seeds each.
- Pollarding is the process of pruning the plants to promote branching, fruiting and flowering practiced during November-December.
- Yield Annual moringa 200-400 fruits
- Perennial type 80-90 fruits /year to 500-600 fruits/year.

56. "CURRY LEAF"

- Belongs to Rutaceae family with origin in Tarai region of UP.
- Ground curry leaf with mature coconut kernel and spices form an excellent preserve.
- Its leaves are used mainly to improve the taste and flavour of foods, rich in "Ca".
- Its leaves, roots and bark are credited with tonic, stomachic and carminative properties.
- Leaves are reported to cure piles and allay heat of body.
- Green leaves are eaten as raw for treatment of dysentery.
- An Infusion of roasted leaves is used to stop vomiting.

- Decoction of leaves is drunk for dropsy.
- Volatile oil (curry leaf oil) is produced by steam distillation of fresh leaves, (2.6%) under high pressure (90 IB/Sq. inch).
- Oil is deep yellow in colour with a strong spicy colour and pungent clove like taste.
- A crystalline glucoside "Ksenigin" from the leaves and "murayin/muragin" from flower are extracted which are industrial products.
- It is a self-pollinated crop.
- *Murraya paniculate* – Indigenous to Burma.
- *Murraya exotica* - Indigenous to India.
- It is cultivated in Coimbatore, Salem and Tiruchirappalli district of Tamil Nadu on a commercial scale.
- Flower cymes, berries purplish black when ripe and two seeded tree bears flower from February - May.
- Grows well in warm climate.
- Yield 20000 kg/ha annually from 4th to 5th year onwards.

Varieties –

1. DWD 1 (Suvasini) – 0.1629 mg of chlorophyll content per gram of fresh leaf, highly aromatic oil 5.22%, sensitive to winter season, evolved at University of Agri. Sci. (UAS), Dharwad, clone of single plant root suckers.
2. DWD 2 – Good aroma, oil 4.09% evolved at UAS, Dharwad, seeding progeny, winter insensitive, high yield then DWD – 1.
3. Senakanbu – Mainly grown in Tamil Nadu, red petioled.

- There are two types of curry leaf viz broad leaved and small leaved, small leaved are more fragrant and hence used for extraction of essential oil.
- Propagation – By "seed"; but also by "root suckers".
- Polyembryony has been reported and about 14% of the fruit contains two seeds per fruit.
- Normally fruits for seed are available during July-August. Each fruit contains 2-3 seeds.
- Spacing 4 × 4 M.
- 1 Year old seedlings are planted in main field.

57. "CHEKUMANIS"

- It is a perennial, small shruby leafy vegetable.
- Due to high nutritive value it is called "Multivitamin Greens".
- It is occupying prominent place in almost all household kitchen garden of Kerala.
- It is reported to be first introduced into Kerala from Malaysia in 1953 and is known in Malayalam as Madura Keera.
- The inflorescence is axillary with small reddish flowers.
- Flowers are monoecious.

- The crop is highly cross-pollinated and entomophilous because of protogynous and monoecious nature of flower.
- The leaves are very rich in protein (6.8-7.4%) as compared to Amaranthus (3.2%), also rich in minerals and vitamins.
- Leaves and tender shoots – Mainly used as leafy vegetable either as salad or after frying.
- The leaves are also said to be used to give a light green colour to pastry and to fermented rice in the Dutch East Indies and Java for preparation of soup.
- When the plants are raised under shade, it produces broader leaves.
- Propagated by seed and stem cutting (Soft wood and semi hardwood) (500 ppm IBA/IAA treated before planting).
- For planting a hectare area about 1, 00,000 cuttings are required.
- It can be trained as hedge or even on bowers and on trellises.
- The annual productivity is 30-50 t/ha with per plant yield of 1-3 kg leaves per year.

58. "CEYLON SPINACH OR WATER LEAF"

- It is a soft mucilaginous leafy vegetable grown in tropics.
- Being a shade loving species it is suitable for cultivation under other crop canopy also.
- It was introduced to Java from Brazil.
- Short day condition (SDP) favours flowering and fruiting.
- The crop is highly self-pollinated.
- It grows upto an altitude of 1000 M, most humid conditions favour optimum growth and development of crop.
- Propagation – By seed or herbaceous stem cuttings.
- Flower – Pink colour.

Important Pest and Disease of Perennial Vegetables

Sr. No.	Pest and Disease	Symptoms	Management
1.	Asparagus beetle (*Crioceris asparagi)*	Chewed spear tips; brown stains; bleached ferns.	During the harvest season, apply insecticides when 1-2% of the spears damage.
2.	Leaf caterpillar (*Noorda blitealis)*	Larva feeds on the leaflets reducing them into papery structure	Spray insecticides like carbaryl 50 WP@ 1gm/lit or malathion 50 EC 2 ml/lit of water.
3.	Hairy caterpillar (*Eupterote mollifera)*	Scraping the bark and defoliation of the tree.	Spray FORS @ 25g/lit or carbaryl 50 WP @ 2g/lit.

4.	Bark caterpillar (*Indarbela tetraonis)*	Zig-zag galleries and silken webbed masses.	Clean all webbed material, plug the holes with cotton wool soaked in fumigants like formalin or petrol and seal it with mud.
5.	Bud worm (*Noorda moringae)*	Larva bores into flower buds and causes shedding.	Collect and destroy the flower and buds and spray malathion 1 litre/hectare.
6.	Crown Rot/ Seedling Blight (*Fusarium oxysporum f.sp. asparagi)*	Infected seedlings will exhibit stunting, yellowing and wilting of the foliage as the primary roots are rotted off.	The disease is seed- and soil-borne, Use strong healthy plants.
7.	Rust (*Puccinia asparagi)*	Red or brown elongated spots appear on the shoots spears or needles of asparagus.	Remove or bury crop residue in the fall to help limit infection.
8.	Damping off (*Pythium aphanidermatum*)	Disease of nursery beds and young seedlings resulting in reduced seed germination and poor stand of seedlings.	Seed treatment and soil drenching with fungicides.
9.	Brown leaf spot (*Cercospora spp*)	Brown spots appear on the leaves and then spread to cover them entirely.	Spray of fungicide like Carbendazim, Difolatan, mancozeb etc.
10.	Powdery Mildew (*Leveillula taurica*)	White to light grey powdery growth on the under surface of the leaves.	Spraying wettable sulphur 0.2 per cent at gives better control.

59. CUCURBITS

Family - Cucurbitacea

Common name	Botanical name	(2n)	Origin	Salient features
Cucumber	*Cucumis sativus*	14	India	Ethral increase female flowers 2nd most cultivated cucurbit
Water melon	*Citrullus lanatus*	22	Tropical Africa	Bitterness due to Cucurbitacin E Heavy dull sound – Maturity indices
Round melon	*Citrullus lanatus/ Praecitrullus fistulosus*	22	Tropical Africa	The fruit is a type of berry called a pepo.
Bottle gourd	*Lagenaria siceraria*	22	Tropical Africa	Processed products: Kofta and petha

Bitter gourd	*Momordica charantia*	22	Indo-Burma region	Bitter taste – Due to Momorcidin Rich in iron
Pointed gourd	*Trichosanthus diocea*	22	India (Bengal and Assam)	Propogated by vine cuttings
Snake gourd/ viper gourd/ snake tomato/ long tomato	*Trichosanthus anguina / cucumerina*	22	India	The AIDS drug Compound Q is a refined protein called trichonanthine which is derived from the trichosanthes (snake gourd) family
Musk melon	*Cucumis melo*	24	Tropical Africa (Sahara desert)	Maturity indices - Netting stage except Hara Madhu
Long melon (Kakri)	*Cucumis melo var. utillisimus*	24		Very suitable for river bed cultivation
Snape melon (Phoot)	*Cucumis melo var. momordica*	24		A source of high fruit acidity
Gherkin	*Cucumis anguira*	24		A savory pickled cucumber
Ash (wax) gourd	*Benincasa hispida*	24	Java, Japan	The famous ayurvedic 'Kooshmanda rasayana' is made of ash gourd fruits.
Ivy gourd (Scarlet gourd)	*Coccinia indica/ groundis*	24	India	It is also known as Baby watermelon or Gentleman's toes
Ridge gourd (Jhinga tori; kalitori; turiya)	*Luffa acutangula*	26	India	Seeds are eaten to expel intestinal worms, While leaf juice used externally in eczema.
Sponge gourd (Chikani tori)	*Luffa cylindrica*	26	America	Commonly called loofa, vegetable sponge, Bath sponge or dish cloth gourd.
Chow-chow (Chayote)	*Sechium edule*	24	South Mexico and Central India	Best alternative to shifting cultivation in Mizoram.
Spine gourd / kakrol / Teasle gourd / Kantola	Momordica diocea	28	Indo- Malayan region	Tea-sle gourd, Kantola /Kakrol Meetha Karela, Khekhsa, Pa-dora, Bhaat Karela Tea-sle gourd, Kantola /Kakrol Meetha Karela, Khekhsa, Pa-dora, Bhaat Karela Tea-sle gourd, Kantola /Kakrol Meetha Karela, Khekhsa, Pa-dora, Bhaat Karela Tea-sle gourd, Kantola /Kakrol Meetha Karela, Khekhsa, Pa-dora, Bhaat Karela Tea-sle gourd, Kantola /Kakro Tea-sle gourd, Kantola It contains high amount of carotene and protein.

Pumpkin (Squash)	Cucurbita moschata	40	Mexico, Peru	Specing 2.5 to 3.0m (row to row) × 100 to 150cm (plant to plant)
Summer squash (Chappan kadu)	*Cucurbita pepo*	40	Mexico	Also known as bush squash, vegetable marrow, Vilayati Kaddu/ Chappan Kaddu/Safed Kaddu.
Winter squash	*Cucurbita maxima*	40	Mexico	This species is originated from the wild *Cucurbita andreana* over 4000 years ago.
Sweet gourd (Gac Fruit/ Red Melon)	Momordica cochinchensis	28	South east Asia	Low germination capacity

Sex forms in cucurbits –

1. Monoecious – Where staminate (Male) and pistillate (female) flowers are separately produced on the same plants eg. Cucumber, Watermelon. Bottle gourd, Bitter gourd, Ash gourd, Ridge gourd, Sponge gourad, Snake guard, Pumpkin etc.
2. Androecious – Where only staminate flower are present on the plant.
3. Andromonoecious – Where staminate and hermaphrodite (perfect) flowers are produced separately on the plant. eg Muskmelon (desert), Snapmelon (Phoot) Long or Serpent melon, Kachri, Vellarikkai.
4. Gynoecious – When only pistillate (female) flowers are present on the plant, eg. Cucumber.
5. Gynomonoecious – Where pistillate and perfect flower are produced separately on the same plant. eg. Cucumber.
6. Gynoandromonoecious – (Trimonoecious) where staminate, pistillate and perfect flowers are produced on the same plant. eg. Cucumber, Muskmelon, Ridge guard.
7. Hermaphrodite – Where only bisexual or perfect flowers are produced. eg. Cucumber (original forms),"Satputia" cultivars of Ridge gourd, Muskmelon.
8. Dioecious – Where staminate and pistillate flowers are produced separately on separate plants .eg. 1. Parwal (Pointed gourd) 2. Ivy gourd 3. Kakrol.

Pollination mechanism –

- All cucurbits are highly cross pollinated crops (Entomophillus).
- Cucurbits are day neutral plant.
- Genetic male sterility is found in – Muskmelon, Water Melon.
- Cytoplasmic male sterility is found in – Cucumber.

New Cultivars –

1. Pusa Uday – Cucumber
2. Pusa Ujjwal, Pusa Shakti - Ash gourd
3. Pusa Sneha – Sponge gourd
4. Pusa Shandar – Snap melon
5. Pusa Sandesh – Bottle gourd

Seed rate, planting distance and depth of sowing in different cucurbits

Crop	Seed Rate (kg/ha)	Planting distance		Depth of seed sown (cm)
		Row to Row (cm)	Hill to Hill (cm)	
Watermelon	3.7-5.0	250-350	90-120	2-2.5
Roundmelon	3.7-5.0	180-240	60-120	1.5-2.0
Muskmelon	3.0-7.0	180-240	60-120	1.0
Snapmelon	2.4-3.7	180-240	60-120	1.0
Longmelon	2.4-3.7	180-240	60-120	1.0
Cucumber	2.5-3.7	150	60-90	1.0
Bottlegourd	3.0-6.0	200-300	100-150	2.5
Bittergourd	4.5-6.0	150-250	60-120	2-2.5
Ridgegourd	3.7-5.0	180-250	60-120	1.5-2.0
Spongegourd	2.5-3.7	180-250	60-120	1.5-2.0
Snakegourd	5.0-6.0	180-250	60-120	2.0-2.5
Ashgourd	5.0-7.0	250-300	60-120	2.0
Pumpkin	6.7-9.0	250-300	100-180	2.0-2.5
Bush Squash	8.0-10.0	60-75	45-60	2.0-2.5

Public sector hybrids of cucurbits

Name of hybrid	Source
Muskmelon	
Punjab Hybrid-1, MHL-10	Ludhiana
MHY-3, MHY-5	ARS, RAU, Durgapura
Pusa Rasaraj, DMH-4	IARI, New Delhi
Watermelon	
Arka Jyoti	IIHR
RHRWH-2	Rahuri
Cucumber	
Pusa Sanyog	IARI, Katrain
RHRWH-2	IMPKV, Rahuri
AAUC-1, AAUC-2	AAO, Jorhat
Bottle gourd	
NDBGH-4 NDBH-7	NDUAI, Faizabad
Pusa Manjari, Pusa Hybrid-3, Pusa Meghdoot	IARI
PBOG-1, PBOG-2	Pantnagar
Summer squash	

Pusa Alankar	Katrain
Pumpkin	
Pusa Hybrid-1	IARI
Bitter gourd	
Pusa Hybrid-1, Pusa Hybrid-2	IARI
RHRBGH-1	Rahuri

Anthesis –

1. Muskmelon – Morning 5.30 – 6.30 am.
2. Cucumber – Morning.
3. Sponge Gourd – Morning.
4. Bottle Gourd – Evening 5-8 pm.
5. Snake gourd – Evening 6-9 pm.
6. Ridge gourd – Evening 5-8 pm.

- I^{st} Hybrid for cucumber was developed in Japan (1940).
- I^{st} commercial hybrid in vegetable was developed by IARI.
  1. Pusa Meghdoot (1971) - By Choudhary and Singh.
  2. Pusa Manjari (1971) - By Choudhary and Singh.
- Recalcitrant seed – Pointed gourd, Chow-Chow.
- Pumpkin – 9:1 (male: female Ratio).
- Highest Cu – Pointed gourd.
- Highest Cl – Cucumber.
- Dr. Kihara (Japanese Scientist) developed Pusa Bedana (Tetra 2 × Pusa Rasal) (2n = 33 triploid seedless) in 1951.
- Ash gourd – Longest storage life among cucurbits.
- Flower Colour
  1. Pointed gourd – White
  2. Snake gourd – White
  3. Bottle gourd – White
- Except above three cucurbits all other bears yellow colour flowers.
- Punjab Hybrid – Cultivar of muskmelon, First commercial vegetable hybrid which was developed by using male sterility cross between MS1 × Hara Madhu.
- Amphidiploid – (*Cuurbita moschata* × *Cucurbita maxima*).
- Autotriploids – (In Watermelon and Sugarbeet), Triploids (3x) by crossing between Tetraploid (4x) × Diploid (2x), Developed by Dr. Kihara (1951).
- India is an important supplier of fresh vegetable to South East Asia and Gulf countries.

60. "WATER MELON"

- Watermelon is thought to have originated in Kalahari Desert of Africa.
- China is the largest producer of watermelons in the world.
- Annual, Dicotyledon, Trailing vine, Diploid (2n= 22) (but note that some cultivars are tetraploid, and seedless watermelons are triploid).
- Water melon crop is grown in garden lands and river beds in India.
- The fruit has cooling effect and is used as an expectorant, diuretic and stamachic and in allaying thirst.
- The fruit juice is consumed by after adding a pinch of salt and black pepper.
- Fruit contains 95% water, 0.2% protein, 0.3% minerals and 7% carbohydrates /100g fresh weight.
- It is also a rich source of Iron.
- The seed is also eaten as a snack after shelling and roasting.
- *Citrullus colocynthis* occurs from India to tropical Africa.
- This is also known as Tumba in Bikaner and other arid tracts.
- They are cooling, purgative, anthelmintic, antipyretic and carminative in action.
- They cure a wide range of ailments such as ulcers, asthama, bronchitis, jaundice, enlargement of spleen, dyspepsia, constipation and anemia.
- The ground roots are applied externally to control inflammation of breasts and pain in the joints and uterus.
- Water melon is indigenous to tropical Africa more specifically South Africa.
- It is truely wild or native only in sandy dry areas of South Africa chiefly Kalahari Desert.
- Cirtrullus genus belongs to the subtribe Benincasinae, tribe Bininace and sub family cucurbitaceae.
- *Cirtrullus vulgaris* is now renamed as *C. lanatus*.
- The species classification of citrullus and their relationship have been studied in relation to cucurbitacin content (bitter principle) by South African investigators.
- In *C. vulgaris, C. colocynthi and C. ecirrhosus* Cucurbitacine E. is the main bitter substance (Tetracyclic Terpenes).
- In *C. naudinianus* – Cucurbitacin-B together with E and their derivatives are the main bitter substance.
- From the results of morphological and cytogenitical studies it is observed that all these four species were cross compatible with each other and several factors like geographical isolation, difference in flowering habit, generic difference and structural changes in chromosome would have contributed to the maintenance of their identity.
- Pangalo (1930) and Goldhousen (1938) – Reported that bitter and nonbitter wild races of vulgaris should be distinguished and treated with *C. edulis* and *C. colocynthoides*.
- The citroides comprising the citron or preserving melon produce fruits with hard inedible flesh and greenish or tan seeds.
- The citron or preserving melon (var - *citroides* of Bailey) can be considered as intermediate stage between primitive bitter and cultivated sweet forms.

Recommended Varieties of Water Melon

S.No.	Variety	Features
1	Sugar Baby	Fruit round, 7-8 inches in diameter, weight 3 to 5 kg. Rind bluish green with faint black stripes, thin, hard, and tough. Flesh medium red, firm, crisp, sweet (TSS 11-13%) fine texture. Seeds relatively few, very small, dark, tan mottled with black colour. Maturity 85-90 days. Yield 250-300 q/ha.
2	Pusa Bedana	A seedless watermelon. Dark rind with faint strips, somewhat triangular shape, tough rind, red flesh with remnants of false seeds, TSS 12-13%. Fruit weight 5-6 kg, produces 3-6 fruits per vine, very good keeping quality. Maturity 115-120 days. Yield 200-250 q/ha.
3	Arka Jyoti	F_1 hybrid between IIHR 20 (a Rajasthan selection × Crimson Sweet), An early season crop. Flesh colour deep pink, very sweet (12 to 14% TSS) Average yield 800 q/ha. Duration 90-100 days.
4	Arka Manik	A selection from advanced generations of a cross between IIHR 21 × Crimson Sweet by IIHR Bangalore. Average fruit weights 6 kg excellence granular texture, high TSS (12-15%) deep crimson flesh, pleasant aroma. Average yield 600 q/ha. Good keeping and transport quality. Duration 90-100 days. Resistant to powdery mildew, downy mildew and anthrocnose and bloosom end rot (multiple resistances).
5.	Durgapur Meetha (ARS, Durgapur)	The fruits are round, light green and weighing 6-8 kg. It has thick rind, dark red flesh and TSS content of 11 per cent. It is a late maturing variety and takes about 125 days to reach marketable maturity. Average yield is 400-450 q/ha

Crop Improvement –

(1) Introduction:-

(a) Asahi Yamato – Introduced from Japan, TSS 11-13%, fruit 6-8 Kg in weight.

(b) Sugar Baby – Introduced from USA, TSS 11-13%, very sweet in taste, suitable for North India.

(c) Improved Shipper – Introduced from USA, 8-9% TSS

(d) New Hemisphire Midget – 1.5-2.0 kg weight, suitable for home gardens.

(e) Dixie

(f) Fucken

(2) Selection –

(a) Special No. 1 – Early maturing, red flesh and red seed, selection made by PAU, Ludhiana.

(b) Durgapura Meetha – Good keeping quality, TSS 11%, dark red colour, matures 125 DAS (Late cultivar).

(c) Dugapura Kesar – Late cultivar, flesh is yellow in colour, moderate sweet, large seeds.

(d) Pusa Rasal – Old cultivar.

(3) Hybridization –

(a) Pusa Bedana (Triploid) – Tetra-2 × Pusa Rasal (Seedless variety), 3-4 fruit/vine.

(b) Arka Manik – (IIHR 21 × Crimson Sweet) followed by modified back cross method.

(c) Arka Jyoti – F1 hybrid (IIHR 20 × Crimson sweet).

(4) Polyploidy breeding –

Polyploids were produced by treating seedlings at the cotyledon stage with an emulsion of 0.2% colchicine, stearic acid, morphaline and lanoline.

- Fruits produced on triploid plants had seeds devoid of embryo.
- Seedless watermelons -To get rid of seeds, Dr. Kihara in Japan conceived production of seedless watermelons. The technique involves production of tetraploid lines by application of colchicine to growing tips of young seedlings. Colchicine interferes with the normal mitotic cell division and leads to production of tetraploid plants.
- Tetraploid lines are stabilized and subjected to selection for developing inbreds to be used as female parents. Tetraploid lines are crossed with superior diploid parents to obtain triploid F_1 hybrids.
- Triploid hybrids are self-sterile and produce fruits only when pollinated by diploids. These hybrids are not entirely seedless but develop rudimentary seeds that are soft and eaten undetected in the flesh.
- Seedless watermelons are sweeter probably due to the reason that the photosynthates saved from seed production are diverted towards production of sugars.
- Triploid watermelon seeds are very costly and are not being grown in India at the moment to any considerable extent.
- In general, the triploids have some germination problems also due to their thicker seed coat. Triploids are commercially popular in the United States and many other countries. Due to higher seed cost, they are not directly seeded but transplanted.

(5) Resistance breeding –

- Arka Manik highly resistant to powdery mildew and moderate resistant to anthracnose.
- Resistance to red pumpkin beetle has been reported in lines Sl. – 120, Afghan.

Locally grown cultivars are:-

- Farrukhabadi, Moradabadi – UP.
- Mateera - Rajasthan (grown in rainy season around Bikaner).
- Katagolan – Cultivar of Jamuna river bed whose flesh is not sweet but it keeps well for 2-3 months at ambient temperature during July-September.
- Private sector hybrid – Madhu, Milan, Mohini.
- MHW 4, 5, 6, 11, 12, 15, NS 246, 295, Suruchi, Samtrupti, Amruth and Century 2- Appealing appearance, good quality flesh, transportability, resistance and tolerance to wilt.
- Arid region of Rajasthan are best suited for production of quality fruits.
- High temperature is beneficial during ripening.
- Water melon requires hot dry climate and a long growing season preferably with warm days and cool nights.
- It can't withstand frost or very low temperature.
- Temperature for seed germination – 25-30°C.
- Temperature for fruiting – 24-27° C.
- Plant growth is optimum under 28-35°C.
- Sandy loam is suitable for early crop.
- Loam – High yield potential.
- Alluvial river beds are also good for water melon crop.
- Excess humidity cause powdery mildew, downy mildew, anthracnose and virus diseases and pest such as fruit fly.
- Cool nights and warmer days – Ideal for accumulation of sugar in fruit.
- Seed rate – 3.0-3.5 Kg for small seeded types.

  5.0 kg for large seeded type.
- Spacing – 2.5 – 3.0 M (Row × Row).

  90 – 120 Cm. (Hill × Hill).
- Sugar baby- 2.0 M (Rows × Row).

  1.0 M (Hills × Hill).
- Plant to plant spacing could further be reduced to 0.6 m to accommodate around 16,600 plants/ha, if single plant is retained in place of 2.
- Spacing – 30 × 60 cm for sugar baby for Zone IV at NDUAT at Faizabad.
- Sowing time – Mid February in Northern plains.

  Nov. – January for river beds.

  December – January for Central and South India.
- Sugar Baby – (December 20 - January 20) under Maharashtra conditions gives highest yield.
- Asahi Yamato performs best during April – May in Northern Hills.
- Mateera – August - September in North-West Rajasthan, a rainy season crop of a local watermelon because of absence of heavy rainfall.

- In some areas where early crop is desired, seeds can be sown in alkathene bags (10 × 15cm) of 100-200 guage.
- 4-5 seed in a hill is sown and finally two vines are retained / hill.
- Training and prunning – The excess vine growth can be pruned manually to restrict vegetative growth and promote higher female: male flower ratio.
- Yield and fruit quality are improved with prunning.
- Fruit thinning is useful and retaining of 2 fruits /vine improves fruit size as well as fruit yield.
- 3-4 main shoots per plant are retained on Sugar Baby for optimum production.
- Yield loss upto 30% have been observed due to weeds.
- In watermelon to increase the flowering and fruit yield, exogenous application of chemicals such as TIBA 25-250 ppm, boron (3-4 ppm), Mo (3-4 ppm), Ca 20-25 ppm is recommended at 2 true leaf stage, Repeat the spray at 4 true leaf stage.
- Water stress at fruit setting and development results in reduced fruit setting and fruit size and increases incidence of blossom end rot.
- However, excessive irrigation results in fruit cracking.
- At fruit maturity irrigation is restricted in watermelon.

Maturity –

- Colour change of the fruit portion which rests on the ground from white to creamy in case of light skin colour or to yellow in case of dark skin colour.
- Metallic sound when fruit is tapped denotes immaturity.
- Heavy dull sound indicates ripening.
- At maturity, the rind becomes hard and cannot be punctured with thumbnail.
- Tendril dries and the ground spot has turns from white to yellow.
- Watermelons reach maturity approximately 40 – 45 days after blooming; 90-120 (Days after sowing), highly dependent upon cultivar.
- Immature watermelons have turgid tendrils; a completely wilted tendril can be indicative of over-maturity.
- Fruit is rich in sugar (soluble solids); a sugar content of 17% is considered excellent quality.
- Fruit should be cut cleanly from the vine to avoid stem damage and prevent stem-end.
- Yield – 200-250 q/ha.
  300- 400 q/ha. (Hybrid/high yielding varieties).
- Cylindrical or oblong shape like that of Charleston gray is suitable for transportation in trucks.
- Most of cultivars have been deep-pink or pink or pale-pink flesh colour with slightly reddish-tinge containing largely lycopene and anthocyanin pigments.
- Uniform development of pink colour from center to rind is necessary.

- More often in Indian cultivars "White heart" seen at the central portion indicating poor quality.
- Hollow heart and fibrous flesh signify over maturity.
- Average TSS for watermelon – 8-12%.
- Seed maturing and edible maturity coincide with each other.
- Water melon from Uttar Pradesh and Andhra Pradesh are transport to Culcutta market.
- Fruit is chilling sensitive and should not be stored below 7°C; optimum storage temperatures are 7° to 10°C.
- Major diseases of watermelons – bacterial fruit blotch, damping off, anthracnose, downy mildew, *Fusarium* wilt, gummy stem blight, various viruses (e.g. watermelon mosaic virus) (transmitted by aphids and cucumber beetles; effective control of these diseases involves control of the insect populations).
- As in tomato, calcium deficiency can result in a disorder called "blossom-end rot".
- Major insect pests: cucumber beetle, aphids, mites and sap beetles.
- Cucumber beetles are of lesser concern to watermelon growers than cucumber and muskmelon growers because watermelon is not susceptible to bacterial wilt.
- Protection of watermelons from cucumber beetles is necessary when plants are small and high beetle populations are feeding on stems, or when beetles are feeding on the fruit.
- Pesticides application toxic to honey bees ,during flowering can decrease pollination and fruit set.
- Due to cross-pollinated nature, a minimum isolation distance of 500 m is required between two varieties of watermelon for seed production. Since watermelon and citron are cross compatible, a similar isolation distance is required between these two crops.

61. "MUSK MELON" (CANTALOUPE, MELON CASABA, WINTER MELON)

- Musk is a Persian word for a kind of perfume. The word melon is derived from the Latin word 'melopepo' means 'apple shaped melon'.
- Mature fruits (*Cucumis melon* 2x = 24) are round in shape and 8-16 cm in diameter.
- Muskmelon is considered to be native of Persia (Iran) and adjacent areas on the west and the east.
- Central Asia comprising southern Russia, Afghanistan, Pakistan and north-west India is regarded as its secondary centre.
- Uttar Pradesh (21000 ha) is having maximum area under muskmelon cultivation followed by Andhra Pradesh (9900 ha).
- In muskmelon production, Uttar Pradesh (548000 MT) is the leader followed by Andhra Pradesh (314000 MT).

- Nevertheless some oblong, flat and oval fruit shaped varieties are also available.
- In Ayurveda and Unani medicines muskmelon have cooling effect on the body system.
- The roots have emetic and purgative properties.
- The fruits are sweet and musky in flavour. Fruit is rich in Vitamin A, B and C, Ca, P and Fe.
- Immature melons are used fresh in salads, cooked (soup, stew, curry, stir-fry) or pickled.
- Asia produces more than half of the world's melon crop. China alone produces about 25 per cent of the world's crop.
- Mature fruit eaten fresh as a desert fruit, canned or used for syrup or jam; dehydrated slices (lightly processed) for short-term or moderate storage can be reconstituted; and the pressed juice can be canned.
- Melon seeds are a dietary source of unsaturated vegetable oil and protein, and may be lightly roasted and eaten like nuts.
- Salmon-orange flesh contains carotenoid pigment, green fleshed ones contain chlorophyll and the white fleshed have none.
- Orange fleshed varieties have higher quantities of carotene than green and white-fleshed varieties.
- Sweetness in muskmelon is due to the sugars such as glucose, fructose, sucrose, mannose etc.
- Muskmelon is considered beneficial to the people suffering from heart diseases because they contain an anticoagulant called adenosine.
- Due to their high β-carotene content, muskmelons rank high as an anticarcinogenic.
- It is helpful to the people suffering from high blood pressure due to their high potassium content
- Muskmelon provides a wholesome food.
- For export the fruits should have good keeping quality.
- Jaffrey (1980) classified the entire genus cucumis in two subgenera.

1. Subgenus *cucumis* x=7, 3 or 4, Sino – Himalayan species including *C. sativur* (*C. hystric* × *C. muriculatus*).
2. Subgenus *melo* x=12 species mostly in Tropical and South Africa.
   (a) **Anguria group:** Dioecious, monoecious and andromoenocious perennial or annual with yellowish or brown fruit, closely related and partially fertile interspecific hybrids. *C. anguria, C. dispaceus, C. prophetarum. C. myriocarpus, C. sacleuxii.*
   (b) **Metuliferus group:** Monoecious, annual with red spiny fruits, *C. metuliferus.*
   (c) **Melo group:** Monoecious, andro monoecious perennial or annual with smooth fruit. *C. Melo, C. sagittatu, (syn. C. anglolensis), C. dinteri, C. humifructus.*

Types of Melons

- Honeydew melons. The fruits are silvery-white, smooth and round each weighing 1-2 kg, flesh is green or white.
- Casaba melons. The fruits are large each weighing 2.0-3.5 kg, tapering towards the stalk end. The flesh is green and the rind is yellow and ribbed.
- Crenshaw melons. The fruits are large, smooth with dark green rind that turns yellowish green when ripe. The conventional types have greenish flesh but the new types may have orange flesh.
- Some other groups of lesser economic importance are:
- C. melo var. momordica. Snapmelon or phut of north and central India. The fruits are oval to oblong, the rind may crack or split open on ripening. The rind is yellow; flesh is dry and tastes sour.
- Cucumis melo var. chito. Known as chate of Egypt, mango melon, lemon melon, orange melon and apple melon. The fruits are small i.e. orange sized.
- C. melo var. conomon. Known as Chinese white cucumber, Oriental pickling melon and sweet melon. The fruits are smooth with variable shapes, glabrous and without musky flavour.
- C. melo var. dudain. Known as fragrant melon, pocket melon, dudain melon and is grown for ornamental purposes. The fruits are small as that of var. chito, brown, shining and fragrant. The immature fruits are variegated but turn yellow and then whitish when fully mature.
- C. melo var. makuwa. Known as Japanese cantaloupe
  1. *C. anguria* – Resistant to nematode.
  2. *C. ficifolius* - Resistant to nematode.
  3. *C. metuliferus* – Resistant to nematode.
- Melons may be andromonoecious (perfect and staminate flowers), gynoecious (pistillate flowers), gynomonoecious (perfect and pistillate flowers), hermaphrodite (perfect flowers), or monoecious (pistillate and staminate flowers).
- Monoecious and andromonoecious are the most common.
- Since muskmelon is highly cross-pollinated crop and honeybees do the pollination, introduction of 6-7 beehive colonies per hectare significantly improves fruit yield.
- Hybrid gynoecious melons (with multiple disease resistances, earliness, uniformity of harvest and fruit quality) are becoming increasingly popular.
- Two major genes control sex expression in melon (*A* and *G*):

  A-G- = monoecious

  A-gg = gynomonoecious

  aaG- = andromonoecious

  aagg = hermaphroditic

  gy interacts with *A* and *G* to produce a stable gynoecious phenotype.

Recommended Varieties of Musk Melon

S.No.	Variety	Features
1	Pusa Sharbati	An early maturing variety developed from a cross between Kutana and PMR 6, an introduction from the USA. Fruits round, straw coloured, netted and striped skin, flesh salmon orange, firm, thick with small seed cavity. TSS 11-12 %. Maturity 80-85 days. Yield 150-170 q/ha.
2	Pusa Madhuras	Fruits flattish round, sparsely netted, yellowish green, smooth skin with salmon orange flesh. Very sweet, TSS 12-14%. Maturity 95-100 days. Yield 180-200 q/ha.
3	Durgapur Madhu	Fruits oblong, light green flesh with big seed cavity, TSS 12-13%. Maturity 85-90 days, Yield 160 q/ha.
4	DMDR 2	Fruits oval-round with green strips, orange flesh and average fruit weight 950-1000g. TSS 11-12% Resistant to CGMMV and downy mildew diseases. Maturity 85-90 days. Yield 160 q/ha.
5	Hara Madhu (PAU, Ludhiana)	It is an inbred selection from the local material collected from Uttar Pradesh. Fruits round, tapering towards the stalk end with prominent green sutures, rind thin, smooth, turn pale yellow at maturity. Flesh green, thick, very juicy, melting and sweet. TSS 12-15% Maturity 110-120 days. Yield 125-150 q/ha.
	Punjab Sunehri (PAU, Ludhiana)	It has been developed from the cross between Hara Madhu and Edisto, an introduction from the USA. Sutures are absent. It is 10-12 days earlier than Hara Madhu. The fruits develop full slip stage at maturity and good keeping qualities. Average fruit weight is 800 g and yield is 160 q/ha.
6	Pusa Rasraj	The first monoecious hybrid, Fruits slightly oval, skin smooth, non striped, dull white with green and sweet flesh. TSS 11-12%, Maturity 70-75 days. Yield 200-250 q/ha.
7	Punjab Hybrid	It is an F_1 hybrid developed from the cross MS1 × Hara Madhu. Fruits round with distinct sutures. TSS 12-13%. High keeping quality, moderately resistant to powdery mildew and fruit fly. Maturity 75-80 days. Yield 160-170 q/ha.
8	Kashi Madhu	Fruits round. Harvesting starts from 90 DAS - 105 days. Tolerant to powdery and downy mildew and mosaic virus. Medium maturity, yield 250 q/ha. in 105 days.
9	Arka Rajhans	An early-medium variety, medium-large oval fruit, weight 1 to 1.5 kg dirty white, fine nets, excellent keeping and transport quality. TSS 20 to 14%. Highly resistant to the powdery mildew. Yield 250-300 q/ha. 85-90 days duration.
10	Arka Jeet	Fruits small, attractive orange yellow, round weight 400 to 800g. Very sweet, TSS 15 to 17% with high vitamin C (41.6mg/100g fresh weight), Yield 140-150 q/ha.

Crop Improvement –

1. Selection –

(a) Durgapura Madhu – Very early cultivar, Oblong, 500-600 g. /fruit weight, very sweet, 13-14% TSS, seed cavity big.

(b) Hara Madhu – Late maturing variety, round, 1 Kg fruit weight, TSS 13%, keeping quality poor, never reaches at full slip stage.

(c) Pusa Madhuras – Mid season variety, 12-14% TSS, poor keeping quality, suitable for growing in North Indian plains.

(d) Arka Rajhans – Mid season variety, large oval fruits (1kg), fruit transportable quality, keeping quality is excellent, tolerant to powdery mildew.

(e) Arka Jeet – Early cultivar similar to Lucknow safeda of Uttar Pradesh, dwarf habit 300-500 g, flat and small, orange brown skin, white flesh, big seed cavity, TSS 12-14%, Excellent flavour and high vitamin C content.

2. Hybridization

(a) Pusa Rasraj – F_1 hybrid (Monoecious 3 × Durgapura Madhu), developed by using monoecious sex forms as female parent, 11-12% TSS, 1-1.2 kg fruit weight, yield 25 ton/ha.

(b) Pusa Sharbati – Kutana × Resistant (contaloups) No. 6 of USA, 11-12% TSS, early cultivar.

(c) Punjab Rasila – WMR 29 × Hara Madhu, Resistant to downy mildew and powdery mildew, TSS 11%, fruit wt. 600g.

(d) Punjab Sunehri – (Hara Madhu × Edisto), 11-12% TSS, early maturing.

(e) Hisar Madhur – Pusa sharbati × 75.

(f) Punjab Hybrid – F_1 hybrid between a male sterile line (MS1) and Hara madhu, vine long 2-2.5 m.

Private sector hybrid – Shweta and Swarna

Heterosis Breeding –

- Expression of quality character (TSS %) is highly influenced by environmental factors.
- The andromonoecious nature of muskmelon involves the tedium of emasculation of perfect flower in the female parents for hybrid seed production.
- Further more monoecious line reduces the extent of self pollination in the female parent when seed production is done under open pollinated condition with adjacent row of male and female parents.
- Three stable gynoecious lines 86-104, 105 and 118 lines have been developed at IARI, New Delhi.
- Till now four true breeding homozygous monoecious lines (M1, M2, M3 and M4) have been developed.
- Andromonecious varieties are used to study the extent of heterosis in muskmelon.

Two commercial hybrids have been developed.

1. Pusa Rasraj – Using monoecious sex forms as female parent at IARI.
2. Punjab Hybrid – Using male sterile stock as female at PAU.

Polyploidy breeding – Attempt was made to induce polyploidy in muskmelon as early as in 1953 by using colchicine.

- Quality of tetraploid strains was superior to that of diploid but other character like productivity, size, fertility and number of seed, the tetraploid strains were inferior to the diploid so polyploidy has no scope for muskmelon improvement.

Resistant breeding –

- Cantaloupe, PMR-5, 6, FM 5 (Cornell 83-275-61), Gulf Stream, Jacumba, Perlita - Resistant to powdery mildew under natural epiphytotic condition.
- PMR6 and campo have same locus for resistance.
- Gulf Stream differs from campo by two factors. Budama type 1 3 and 3, Phoontee Goomuk Nakkados. Kaya, Ex 1, Annamalai are resistant to downy mildew.
- In cucumber a gynoecious sex form WI 998 have been isolated in a stable condition in muskmelon in USA.
- Harela, Durgapura Madhu and Punjab Sunehri – Resistant to both *fusarium oxysparum* sp. *melonis* and *F. solani* (fusarium wilt).
  1. *Cucumis figareic* (GBNR 1804).
  2. *C. zeyheri* 1 (73252 Hg).
  3. *C. zeheri* 2 (GBNR 1053).
  4. Smiths perfect.
  5. VAC 1.
  6. *C. callosus*.
- All above six species are resistant to *Fusariun oxysporum species melonis*.
- Florida 67, PMR 6, Jacumba, PMR, Honey Dew, Bulgariams-10, Bacharacha, Durgapura Madhu, Gulf stream, Kabul Melon and Ludhiana Col, 2 – Resistant to Cucumber Mosaic Virus.
- *Cucumis figarei*, *C. myriocarpus* 1, *C. myriocarpus* 3, *C. africanus* 1, *C. africanus* 2, *C. meeusii*, *C. ficifolius* and *C. zeyheri* 2 were resistant to virus to CGMMV (Cucumber green mottle mosaic virus).
- Immune nature of resistance was later confirmed by back inoculation technique and, transmission electron microscopic study of *C. figarei*, *C. ficifolius*, *C. meeusii*, *C. africanus* 1 and *C. zeyheri* 2.
- Phoot (*C. melo* var. *momordica*) and Kachri (non□dessert types), FM 1 and FM 2 (Cornell breeding lines) and *C. myriocarpus* 2 were also resistant to CGMMV with very mild symptoms.
- By using phoot (*C. melo var. memordica*) as resistant source, five GMMV resistant lines namely VRM 5-10, 29-1, 31-1, 42-4, 43-6 have been developed.
- Cassba and PI 70683 – Resistant against red pumpkins beetle

- *C. melo var callosus* – Resistant against fruitfly.
- Long tap root system is adapted to the growth of cucurbits in riverbeds.

  Local Cultivar –

  1. Lucknow Safeda.
  2. Baghpat Melon of Meerut.
  3. Jaunpuri Netted.
  4. Mau Melon of Azamgarh.
  5. Tonk Melon – from Rajasthan.
  6. Kharri Melon – Hoshangabad (Madhya Pradesh).
  7. Goose Melon and Kavit Jam or Neel (Maharashtra).
  8. Sharbat – e - Anar (Andhra Pradesh).
  9. Bathesa – A.P.
- Non desert type –
  1. Phoot (Snap melon)
  2. Kachri or Kachauri
  3. Dosakaya.
  4. Budamkaya – Cucumber and vallerakai (a salad cucumber type) of Tamil Nadu.
- Muskmelon is susceptible to frost.
- Excess humidity will promote disease like powdery mildew, downy mildew, anthracnose, virus disease and fruit fly.
- For good fruit quality and sweetness in muskmelon dry weather during fruit development is necessary.
- Cool night and warm days are ideal for accumulation of sugar in the fruits.
- Maturity hasten if nights are warm.
- It requires long days and dry weather with abundant sunshine.
- It does not tolerate even mild frost. The seeds fail to germinate if the soil temperature is below 10°C.
- Optimum temperature for seed germination 24-29°C.
- Temperature for growth and fruit development should be fairly high, the optimum range being 20-32°C.
- Temperature above 45°C causes vines to wilt temporarily resulting in sunburning of fruits and reduced shelf life
- Seedlings become ready for transplanting at 2-4 true leaf stage.
- **Seed rate** - 1 kg by dibbling method.

  2.5 - 3.75 kg by direct sowing in kera and pora method.
- 2-3 seeds/hills are sown.
- High temperature at the time of fruit maturity and ripening increases the sweetness.
- Desert muskmelons are grown only in those seasons when the weather is dry and warm during fruit development.

- During rainy season non-desert forms of muskmelon are grown primarily in most parts of the country.
- **Sowing time –**
  North India - November - January (Riverbed).
  South and Central India - October - November.
- November transplanted crop take significantly more days to fruit maturity than March transplanted.
- In the mainlands, sowing is done in late February.
- In muskmelon insitu sowing is practiced.
- For obtaining an early crop, sowing of muskmelon and other cucurbits is done in polythene bags of 15 × 10 cm size and 100 gauge thickness, in end of January.
- The seedlings are transplanted into the field in end February when they have developed 2-3 true leaves.
- To obtain highest yield of good quality fruits in cultivar Hara Madhu, treat the seed for 8 hours with a solution containing nitrophenolate sodium + nitroguaiacol sodium at 0.5%.
- **Spacing –** 1.5-2.5 m. (Row × Row)
  60-120 cm (Hill × Hill)
- 350 cm Row × Row 60 cm Plant × Plant (for cultivar Durgapura Madhu, Rajasthan) for zone VI by AICVIP.
- A close spacing of 30 cm is recommended for Hara Madhu under Punjab condition.
- Soaking of seed in ethaphon at 480mg/l. for 24 hours improves germination in muskmelon at low temperature.

Training and Pruning –

- In muskmelon the highest yield and good fruit quality are obtained with prunning (as the first hermaphrodite flower is borne on secondary branch arising from the eighth node the secondary branches are pinched off upto the seventh node) plus stacking.
- Removing the secondary shoots upto the seventh node on the main stem is found optimal to improve plant growth and fruit set and induce early flowering.
- Maximum number of fruits with highest TSS is produced between 9th and 12th nodes on the main stem.
- Application of 128 – 180Kg Nitrogen/ha significantly reduce the perfect: staminate flower ratio.
- N and K as foliar spray @ 1.5% increases the number of hermaphrodite flower in cultivar Hara Madhu. Potassium treatment also increases TSS, ascorbic acid, flavour, taste and juiciness of fruits.
- Irrigation at 0.6 – 0.8 PEC (Pan Evaporation Coefficient) results in good vegetative growth and higher yield of musk melon cultivar "Hara Madhu".
- Muskmelon is a climactric fruit.
- Among cucurbits muskmelon is most suitable for daria cultivation.

Effect of Plant Growth Regulations (PGR) –

- In muskmelon application of Ethrel (250 ppm) increases the fruiting and the yield.
- Exogenous application of silver thiosulphate (300-400 ppm) induces the male flower.
- These PGR should be applied twice at 2nd true leaf stage and secondly at 4th true leaf stage.

Harvesting

- In muskmelon fruits attain marketable maturity 30-40 days after fruit set and 65-70 days after sowing.
- Muskmelon is a climacteric fruit that ripens during transit and storage.
- *Cucumis melon – var momoridica* (phoot) takes about 27-28 days from anthesis to harvesting.
- Yield – Pusa Rasraj (F_1) - 25t/ha.
  Pusa Madhuras -12-16t/ha,
  Pusa Sharbati - 10-13t/ha.
- Muskmelon is harvested at full slip stage except Hara Madhu cultivar.
- Full slip stage – Lucknow Safeda, Durgapura madhu, Cantaloupes of USA.
- In netted cantaloupe green colour of vein tract between the netting changes into yellow and netting become pale-or-dirty white. These cantaloupes can be picked at ¾ slip stage for long distance transportation.
- Honey Dew and Casaba – Picked on the basis of external colour because netting do not separate at maturity.
- Long or serpent melon – Carpel separation in cross section.
- More juicer the flesh, the less will be the keeping quality of the whole fruit.
- The sweetness is entirely due to reducing and non reducing sugar in flesh chiefly glucose, fructose, sucrose, mannose etc.
- Salmon- Orange colour flesh contains carotenoid pigment.
- Green colour contains chlorophyll, white flesh have none.
- Half slip stage – Harvest for distant market to avoid losses from over ripening and decay.
- Flesh of muskmelon is used in ice-cream in western countries.
- Candy can be prepared from muskmelon.
- Pickling melon is used in making pickles.
- Jagger and Scott (1937) developed powdery mildew resistant contaloupe 45, PMR-45 of muskmelon.
- Muskmelon should have 10% TSS for Indian palate.
- TSS range – 11 -17% in most of cultivars.
- Muskmelon fruits are perishable and can be stored for 2-4 days at room temperature.
- In cold store at 2-4°C and 85-90 per cent relative humidity, the fruits can be stored for 2-3 weeks.

- Honeydew, Casaba and Crenshaw types are stored for one month at relatively higher temperature of 7-8°C as they are susceptible to chilling injury.
- Muskmelon is a highly cross-pollinated crop, therefore, an isolation distance of 500 m between varieties and 1000 m between hybrids is required to maintain genetic purity of the seed.
- Since muskmelon, long melon, snap melon and wild melon are cross compatible; a similar isolation distance from these cultivated and wild crops is also maintained.

62. "CUCUMBER" (GHERKIN)

- It is the second most widely cultivated cucurbit after watermelon.
- Origin: India.
- Latin name: *Cucumis sativus* L. , Family: Cucurbitaceae.
- Fruits are used for salad and pickling; leaves are used as vegetable.
- In temperate countries grown in glasshouse.
- Fruit and seed have cooling properties.
- Seed oil is used as antipyretic.
- Fruits are good for people suffering from constipation, jaundice and indigestion.
- Fruit contains – 0.4 % Proteins/100g fresh wt.

  2.54 % Carbohydrate/ 100g fresh wt.

  1.5 mg Fe/100g fresh wt.

  2 mg Vitamin C/100g fresh wt.
- India is considered to be the home of cucumber.
- Haryana (17390 ha) is having maximum area under cucumber cultivation followed by Madhya Pradesh (9460 ha).
- In cucumber production, Haryana (274000 MT) is the leader followed by Madhya Pradesh (154000 MT).
- The immature fruits of *Cucumis sativus* L. cucumber, *(gherkin)* are used as salad and for pickling.
- The open pollinated cultivars of cucumber are monoecious in sex form.
- Generally slicing cucumber have white spine colour and pickling cucumber have black spine colour.
- Cucumber prefers slightly lower temperature than watermelon and muskmelon.
- *Cucumis hardwickii* – Small bitter cucumber with sparse and stiff spines found wild in the foothills of Himalayas which is the progenitor of cucumber.
- Free hybridization with cultivated sativus with no reduction of fertility in F_2 generation indicates that *C. hardwickii* might be a feral or more likely proginator of the cultivated cucumber.
- In cucumber, the yield could be doubled by using tropical gynoecious F_1 hybrids.
- Used for fresh consumption (slicing cucumber), or for preservation, marinated with vinegar, salt, or spices (picking cucumber).

- Warm season, frost-susceptible vine crop having separate staminate and pistillate flowers.
- Optimum temperature: 20° to 30°C.

Recommended Varieties of Cucumber

S.No.	Variety	Features
1	Poinsette	Fruits dark green, 20-25 cm long, resistant to mildew diseases, heavy yielder, suitable for summer season. Maturity 40-50 days. Yield 180 q/ha.
2	Pusa Uday	Early, 13-15 cm long, light green, non-wrinkled, suitable for both spring summer and rainy season. Maturity 48-52 days yield 140-155 q/ha.
3	Pusa Sanyog (F_1 hybrid)	Very early, fruits 28-30 cm long, cylindrical, attractive very heavy yielder. Maturity 50 days. Yield 200 q/ha.

Crop Improvement –

1. **Introduction** –
   (a) Japanese long green – Suitable for temperate region, extra early, by IARI, Katrain, suitable for hill region.
   (b) Straight Eight – Early cultivar, suited to hills.
   (c) Poinsett – From USA, Multiplied by (NSC) National Seed Corporation, developed originally at Charleston (S. Carolina) USA, resistant to powdery mildew, downy mildew, anthracnose and angular leaf spot (multiple resistance)
   (d) China Long
2. **Selection.**
   (a) Sheetal – Recommended for Kankan region, Developed by KKV, Dapoli (Maharashtra), suitable for high rainfall areas.
   (b) Khira 75 – Developed at Dr. VSPUHF, Solan HP.
   (c) Khira 90 – Developed at Dr. VSPUHF, Solan HP.
3. **Hybridization** –
   (a) Himangi – Poinsett × Kalyanpur Ageti, developed by MPKV, Rahuri (Maharashtra), resistant to bronzing, fruit white in colour.
   (b) Phule Shubhangi – (Poinsett × Kalyanpur Ageti), green skin, tolerant to powdery mildew, developed by MPKV, Rahuri (Maharashtra).
   (c) Pusa Sanyog – It has gynoecious female parents (Japanese Gynoecious line × Green long (Napel).

- All of them are mostly restricted to the temperate region of the country, developed at IARI, Katarain.
- Private Sector Hybrid – Priya.
- The open pollinated cultivars of cucumber are monocecious in sex-form.

- The first F_1 hybrid, Pusa Sanyog developed in India in 1971 has a gynoecious female parent.
- It is suited to both temperate and subtropical regions of the country.
- This hybrid has shown heterosis for early yield, total yield and number of fruits per plants.
- Presence of gynoecious sex forms has made possible phenomenal progress in exploitation of F_1 hybrid under glasshouse condition.
- Temperate gynoecious lines are unstable for gynocey under high temperature and long photoperiodic condition hence three tropical gynoecious lines had been developed namely.

  Gyc1 (87-304)
  Gyc2 (87-3/9)
  Gyc3 (87-338)
  All three developed at IARI, New Delhi
- Gyc4 (PKG -1) Developed at MPKV, Rahuri (Maharashtra) having parthenocarpic gene in it.
- DCH 1, 2 PH-9, 18 having involved tropical gynoecious lines as female parents are being tested in multilocational trials of PDVR.

4. Resistant Breeding –

The Indian variety Banglora was the starting point for developing downy mildew resistance in cultivar in the United States, Korea, Dutch and GY 598 – All these are less susceptible to red pumpkin beetle.

- *Meloidogyne incognita* resistant variety – GY 5937-587.
- Three different flower types namely staminate (male), pistillate (female) and hermaphrodite (both male and female) occur in cucumber.
- Main sex types in cucumber:
- Monoecious plants; separate staminate and pistillate flowers on the same plant.
- Androecious plants; staminate flowers only.
- Gynoecious plants; pistillate flowers only.
- Hermaphroditic plants; hermaphrodite flowers only.
- Andromonoecious; separate staminate and hermaphrodite flowers on the same plant.
- Dioecious; male flowers only on one plant and female flowers only on another.
- Three major genes determine sex in cucumber (*M*, *Acr* and *a*):

  Gynoecious = *MM, AcrAcr*
  Hermaphrodite = *mm, AcrAcr*
  Monoecious = *MM, acracr*
  Andromonoecious = *mm, acracr*
  Androecious = *acracr, aa*
- *a* is responsible for androecy and has an effect in *acracr* plants only.
- The symbol *F* (*female*) is used synonymously with *Acr*.

- *F* locus has recently been shown to be very closely linked to a gene encoding ACC synthase, a key enzyme of ethylene biosynthesis.
- Identification of a 1-aminocyclopropane-1-carboxylic acid synthase gene linked to the *female* (*F*) locus that enhances female sex expression in cucumber.
- *Acr (F)* may be modified by an intensifier gene *In-Fe (In-F)*.
- *Tr* determines formation of trimonoecious plants (with staminate, pistillate and hermaphoditic flowers on the same plant).
- If a purely female variety is grown, there is need to provide an appropriate pollinator.
- Parthenocarpy is the ability to develop fruits without pollination.
- Parthenocarpy is controlled by an incompletely dominant gene *Pc*.
- Generally slicing cucumber have white spine colour and pickling cucumber have black spine colour.
- Balam Khira of Sharanpur (UP) regions is akin to pickling type of west.
- Pickling type – Fruits small, seed cavity develops more slowly, an important character affecting suitability for storage in brine.
- In pickling type thin skin with a lighter colour is preferred.
- K 90, K 75, Solan hybrid – Developed from Dr. YSPUHF, Solan.
- Poona Khira - Produce small size pale green fruits.
- Balam Khira – Similar to pickling type of USA.
- Cucumber is a thermophilic and frost susceptible crop.
- It is essentially a warm season crop.
- Excess humidity promotes disease like powdery mildew, downy mildew, anthracnose and pest such as fruit fly.
- Cucumber prefers slightly low temperature compared to melon and hence it is grown extensively in subtropical regions.
- Optimum temperature – 26.4°C for growth.
- Optimum temperature – 13-18°C for anthesis and dehiscence of flower.
- If the temperature goes above 30°C the female flower production is reduced considerably.
- **Sowing Season** – November – Jan in North India (River beds)
  June – July for rainy season
  Mid February (Garden crops)
  April – May (Hills)
- In North India seeds can be sown in alkathene bag (15 × 10cm) of 100-200 gauge to avoid low temperature.
- Transplanting is done at 2 leaf stage.
- Seed rate – 2.5 - 3.5 kg/ha.
- Spacing - 1.5-2.5m (Raw × Raw)
  60-90cm (Hill × Hill)

- Cultivar Shinko Ago – Spacing of 100 × 60 cm and pruning of plants to a single stem gives the highest yield.
- Cultivar Poinsett (Monoecious) requires spacing of 60 × 60 cm under greenhouse.

Effect of Plant Growth Regulators –

- Application of Ethrel (150-200 ppm) increase the number of female flower, fruit set and in turn increase the fruit yield.
- In cucumber, chemicals/plant regulators may be applied at 2-true-Ieaf stage.
- GA (1500-2000 ppm), AgNo3 (200-300 ppm) and aminoethoxyvinylglycine induce the male flowers on gynoecious cucumber.
- This enable maintenance of the gynoecious line in pure form and by using this gynoecious line, phenomenal heterosis has been achieved in cucumber.
- These PGR are applied at 2 true leaf and spray is repeated at 4 true leaf stage.
- Fruits of cucumber attain edible maturity with in a week from anthesis of female flower.
- Thermosensitivity – Sensitive to low temperature of January - February and high temperature of April-May.
- In salad or slicing cucumber dark green skin colour should not turn brownish-yellow or russeting, white spine colour is also a useful indication for their edible maturity.
- For commercial purpose, cucumber is harvested at immature condition 5-7 days after pollination.
- Crook-neck-shaped fruits are produced because of imperfect pollination and fertilization and also due to fly attack.
- Storage – 10°C. with 85% relative humidity.

  Physiological disorders –

  1. Chilling injury – When fruits are exposed to temperature below 10° C for prolonged period.
  2. Pillow – Due to "Ca" deficiency.

- White spine colour indicates maturity in slicing cucumber.
- Black spine colour indicates maturity in pickling cucumber.
  1. Japanese Long Green.
  2. Straight Eight.
  3. Pusa Sanyog. All three varieties are suitable for temperate region.
- Light intensity and day length influence anthesis more than temperature. Temperature influence another anther dehiscence and pollen fertility.
- GA at higher concentration induces maleness but at the low concentration (10-25 ppm) increase the number of female flower.
- Pollinated by bees: 2 to 3 bee colonies per acre are recommended.
- Bacterial wilt is transmitted by striped cucumber beetles.
- Open pollinated cultivars of cucumber are monoecious.

- Pusa Uday – Suitable for both spring and Kharif season, developed by selection.
- Sex expression in cucumbers is modified by low fertility, higher temperature; longe light period. All these induce maleness.
- MH @ 100 ppm, NAA @100 ppm, Ethrel @ 150-200 ppm and "B" and Mo @ : ppm once at 2 leaf and second at 4 leaf stage increase the number of female flower fruit set and in turn increase the fruits yield.

  Palmetto –
  PR - 27 – Resistant to downy mildew.
  Santee –
  Palmer –
  Tokoyo Long Green
  Chinese Long All these are resistant to Cucumber Mosaic Virus (CMV
  Wisconsin
  Table Green
- Cucumber is characterized by monoecious sex expression but andromonoecious gynoecious and trimonoceious are also found.
- Gynoecious lines are used in hybrid seed production because these lines do no require rouging of monoecious plants.
- Bitter principle in cucumber is cucurbitacin (tetra – cyclic-tri-terpenes).
- Bitterness in cucumber is due to metaxenic effect, when bitter pollen fertilizes non-bitter ovules, the resulting fruit become bitter, this phenomenon is called metaxenia
- Preservation is done by brining cucumber in solution containing 5-8% Nacl during fermentation.
- During brine storage bloating, softening, bleaching and development of off-flavours and odours are the major defects which usually occur.
- Chilling injury as a physiological storage disorder of cucumber is noticed wher fruits are exposed to temperature below -10°C for prolonged period.
- Bloating occurs due to production of excess CO_2 in cucumber.
- Cucumbers undergo lactic acid fermentation during storage which offers the advantages of acid fermentation.
- *C. foetidissima* – A xerophytic perennial sps; origin from Northern Mexico, alsc called buffalo gourd. Seeds rich in oil and protein.

63. "BOTTLE GOURD"

- Schwein furth has identified remains of *Lagineria siceraria* from Egyptian tombs dated about 3000-3500 B.C.
- A theory of trans-oceanic diffusion or drift of bottle gourd has been suggested to explain pre-columbion distribution in Tropical America.
- The wild species occurs only in East Africa and Mada gascar.
- *L. siceraria* (Syn *L. vulgaris*, *L. leucantha* (Bottle gourd) calabash, white flowered Zucca melon.

- *L. sphaerica* (Perennial wild Sp.) – Used for hybridization. The hybrid is sterile and possibility of transfer of desirable traits to cultivated species requires investigation.
- It is commonly grown in India, Ethiopia, Africa, Central America and other warmer regions of world.
- The fruits of bottle gourd are used for making sweets halva, kheer, burfi and pickles.
- Kofta is most popular preparation.
- A decoction made from the leaf is a very good medicine for curing jaundice.
- As a vegetable it is easily digestible therefore used even by patients.
- Fruit has cooling effect, cardiatonic and diureatic.
- Pulp is used for overcoming constipation, cough, nightblindness, and as an antidote against certain poisons.
- Seed are used in dropsy.
- The dry hard shell of the fruits have been used for making a wide range of articles of common use, including bowls, bottles, ladles, containers, floats for fishing nets, pipes and musical instruments.
- It is good for people suffering from biliousness and indigestion.
- Fruit pulp is very good source of fibre-free carbohydrate and fruit pericarp for crude fibre.

Recommended Varieties of Bottlegourd

S.No.	Variety	Features
1	Pusa Summer Prolific Long	Fruits long club shaped, curved thin 'neck', 40 to 45 cm long 20 to 25 cm thick at edible stage. Green in colour. Maturity 60 days. Yield 250q/ha. in summer and 275 q/ha. in Kharif.
2	Pura Naveen	Fruits short, cylindrical and straight without curve or neck green colour, 25-30 cm long, 15-20 cm thick. Average fruit weight 800g. Suitable for sowing both in summer (February-March) and Kharif (June-July). Maturity 55 days. Yield 275 q/ha. in summer and 300 q/ha. in kharif.
3	Pusa Sandesh	Fruits round globular, medium size, green colour, Average fruit weight 500 g. suitable for sowing in both summer (February – March) and Kharif (June-July) Maturity 55 days. Yield 270 q/ha. in summer and 300 q/ha. in kharif.
4	Pusa Hybrid -3	Fruits long, slightly club shaped, straight without neck green colour, 35-40 cm long and 15-30 cm thick. Average fruit weight 1.2 kg. Maturity 55 days. Yield 425 q/ha. in summer and 470 q/ha. in Kharif.
5	Pusa Meghdut	A hybrid from cross between Pusa S.P. Long and S1-2. Fruit long. Light green, Yield about 258q/ha.
6	Pusa Manjri	A hybrid from cross between Pusa S.P. Round and S1-11. Fruits round, light green. Yield 253 q/ha.
7	Arka Bahar	Economy-sized, medium long fruits (43cm), 26 cm girth, each weight about one kg. Skin light green, white tender flesh, yield 400-450 q/ha. Maturity 115-120 days in both the seasons.

Crop Improvement –

1. Selection

(i) Pusa Summar Prolific Long.

(ii) Pusa Summar Prolific Round- Prolific bearer.

(iii) Pusa Naveen - Fruit free from crookneck, weight- 850 g/fruit, cylindrical shape.

(iv) Punjab Long.

(v) Punjab Round.

(vi) Arka Bahar – Devoid (without) of crookneck, weight 1 kg/fruit, pleasant aroma.

(vii) Samrat – Released in 1992 by MPKV, Rahuri (Mh.) good for box packaging, highest yield 40 t/ha.

2. Hybridization –

(i) Pusa Meghdoot – 1st F_1 hybrid by public sector (Pusa Summar Prolific Long × Sel 2) released in 1971, fruits long, suitable for both spring and summer.

(ii) Pusa Manjari – (Pusa Summar Prolific Round × Selection 11), released in 1971, It has given 48% higher early yield over Pusa Summar Prolific Round.

Other varieties –

1. Pusa Sandesh – New variety suitable for spring, summer and rainy season.
2. Punjab Komal – Early variety, oblong fruit, 10-12 fruits/vine, tolerant to cucumber mosaic virus.
3. Pusa Hybrid – 3 – Recently released 1kg /fruit weight, fruits long, suitable for packaging in card board boxes for distant marketing, suitable for both spring summer and Kharif, very early in maturity, 40-45% higher yield than Pusa Naveen.
4. Pusa Sneha – Improved variety over hitherto, very popular.
5. Pusa Supria – Suitable for spring and summer, kharif, high yielding variety.
6. Gutka –
7. Harit –
8. Varad –
9. NDGB -2- Highly preferred for diara land, long type convenient for packaging and transportation due to less bending, recommended for spring summar.
10. PBOG –Suited for Northern plains.
11. NDGB -4
12. Kalyanpur Hari Lambi – Suitable for summar and rainy season.
13. Phule BTG-1 – Cultivated in Maharashtra. Fruit long and Bottle shaped.
14. Rajendra Chamatkar-bottle shaped.
15. Kalyanpur Long Green – Developed by CSAUAT, Kanpur.
16. CO1 – Fruits typical bottle shaped.

- Sex form – Sex expression (Wide ranging) favours out breeding.
- Cucurbits do not suffer much from inbreeding depression.

- Breeding goal for bottle gourd are high yield, greater fruit number, great fruit wt., earliness, high female: male flower ratio, round long cylindrical fruits, sparse hairs present on skin, non fibrous flesh at edible stage, non bitter fruit, attractive green fruit, resistant to disease and pest.
- The bottle gourd is a typical tropical plant which requires a hot and humid climate for best growth.
- It is highly sensitive to photoperiod (24 - 27°C).
- Higher temperature induces emergence of male flower and wide sex ratio.
  Temperature requirement – Night Temperature – 18 - 22°C and
  Day Temperature – 30-35° C. (For proper growth and development)
- Sowing Time – Dec-Jan in summar and June-July in rainy season.
- Seed rate – 3-5 kg/ha. (450-500 seeds /100g)
- Spacing – 2-3 M (Raw × Raw)
  1.0 M – 1.5 M (Hill × Hill)
- Brownish or whitish brown seed germinates well.
- Nitrogen deficiency – Yellowing of veins and foliage occur.
- Higher "N" produces more number of male flowers which is undesirable.
- Potassium deficiency – Reduce plant height, flower drop and checks fruits completely.
- In Maharastra bottle gourd is trained on bower system which increases yield.
- The Samrat cultivar of bottle gourd gives highest yield potential of 80 tonnes/ha on bower training system.
- Bower system prevents fruits of bottle gourd from rotting because fruits remain away from soil.
- High productivity on bower system is due to better use of sunlight by maximum number of leaves of side branches resulting in better assimilation of chlorophyll.
- PGRS – Ethrel (100-150 ppm), MH (400 ppm), Tri iodobenzic acid (50 ppm), "B" (3-4 ppm) increase fruit set at 2 leaf and 4 leaf stage. MH. (400 ppm) along with 100kg N/ha promotes the female flower production and increase fruit set.
- Tenderness and edible maturity are judged by pressing the skin and little pubescence persisting on skins.
- For hybrid seed production male flower buds of female parent are pinched before flower opening.
- The rows of female and male parents are kept in the ratio of 2:1 or 3:1.
- Processed product – Tutty fruity.
- Rainy season is very good for bottle gourd.
- Nail Test – Done for maturity.
- Summer season is suitable for seed production.
- **Yield-** 42.5 t/ha. Spring /Summar Season.
  47 t/ha. Kharif Season

64. "BITTER GOURD" (BALSAM PEAR OR BITTER CUCUMBER)

- Indo-Burma Centre of origin has been reported by Garrison 1977.
- Bitter taste of fruit is due to momordicin.
- Bitter gourd is also known as a balsam pear in the United States.
- Fruits turn to orange – yellow colour when ripe, rich in Iron, vitamin A, B, C and are inexpensive source of protein and minerals.
- The fruit is reported to have herbicidal effect and is laxative and easily digestible.
- Good for curing blood disease, rheumatism, diabetes and asthma.
- Cucurbitacin (bitter glucoside) – Helps in prevention of spoilage of cooked vegetable of bitter gourd.
- *Momordica dioica* - Spine gourd (Kakrol) 2x = 28.
- *M. cochinchinesis* – Sweet bitter gourd of Assam.
- *M. charantia* is incompatible with *M.diocea and M. balsamina*.
- White coloured varieties are less bitter in taste and preferred in South India.
- Fruits are exported to Gulf and European Countries.

Recommended Varieties of Bittergourd

S.No.	Variety	Features
1	Pusa Do Mausami	Fruits medium long (15-18 cm), green, club shaped, unbroken ridges on the surface. Average fruit weight 80-100 g. suitable for sowing in February-March for summer crop and June - July for Kharif crop. Maturity 55-60 days. Yield 110-110 q/ha.
2	Pusa Vishesh	Fruits medium long, smooth broken ridges on the surface. Average fruit weight 100-150 g. suitable for sowing only in Feb-March for summer season crop. Maturity 60 days and yield 125-130 q/ha. Suitable for pickle and dehydration.
3	Pusa hybrid-1	Fruits medium long, small to medium thick, glossy green, smooth broken ridges on the surface. Average fruit weight 100 g. Suitable for growing in both summer and kharif seasons. Maturity 55-60 days. Yield 150-170 q/ha.
4	Arka Harit	Fruits short to medium long, thick, grayish green, smooth broken ridges on the surface. Suitable for growing in summer season. Maturity 60-75 days. Yield 75-90 q/ha.
5	Coimbatore	White medium long thin and broken ridges. Average fruit weight 60-75 g. Suitable for kharif season. Maturity 60-65 days. Yield 70-75 q/ha. Released by TNAU, Coimbatore.
6	VK-1 Priya	Fruits are white and 35-40 cm long, heavy bearing. First picking at 60 days and yield 50 fruits/plant. Developed by Kerala Agricultural University.
7	Pride of Gujarat	Fruits are small, roundish, whitish, green in colour, weights 8-10 g.
8	Pant Karela-1	Highly resistant to red pumpkin beetle. Developed by GBPUAT, Pantanagar.

Crop Improvement –

A. Selection

1. Pusa –Do-Mausami – Fruits dark green, grow well in summar and rainy season, 12-15 t/ha yield.
2. Pusa Vishesh – For summer season.
3. Konkan Tara – Has good keeping quality (7-8 days under ambient temperature), suitable for export market.
4. Arka Harit – Spindle shaped, good for stuffing purpose (kharif and summer), rich in Vitamin C and Iron.
5. Arka Sumeet – Less seed per fruit.
6. Arka Sujat – Tolerant to downy mildew.
7. Priyanka.
8. Harkani.
9. Preethi.
10. VK 1 Priya-Fruit 40 cm long.
11. Coimbatore Long white.
12. Co1.

B. Hybridization

1. Phule Green Gold – (Green Long × Delhi Local) Fruit dark green, tolerant to downy mildew.
2. Pusa Hybrid 1 – Early maturing, suitable for spring summer, for dehydration.
3. RHR BGH 1 – Tolerant to downy mildew, suitable for kharif and summer season.

C. Mutation

MDU 1 – Induced mutant with greenish white fruits, 30-40 cm fruit length, developed by irridiation of MC – 103 by gamma rays.

Other varieties – Kalyanpur Baramasi, Solan Green, Solan Safed

- If temperature goes above 40°C. more male flowers are produce and plant become more prone to mosaic disease.
- Short day treatment – Reduces production of staminate flowers and increase number of pistil flowers.
- Low temperature (20°C) – Increase production of female flower under short day condition.
- Application of azospirillium and phosphobacteria is also recommended to improve yield and quality of fruit.
- MH 50-150 ppm and CCC at 50-100 ppm increase female: male ratio but at higher concentration of 200 ppm CCC it is reduced.
- Ethrel at 25 ppm increase female flowers.
- GA at 60 ppm reduces the ratio of male: female flowers.
- MH (150-200 ppm) when applied at 2 leaf stage in variety Pusa Do Mousami induced the formation of pistillate flowers at the lowest node.

- Seed treatment with B9 at 3-4 ppm for 20 hours gave the highest number of female flower/plant.
- Seeds soaking in ethrel 20 ppm or GA3, MH, Ag No3 at 3-4 ppm gave the highest number of female flowers and fruits per plant.
- Seed rate – 4-5 Kg/ha.
- **Sowing time** – April – July (In hills)
  June – July (Rainy season)
  Jan – Feb (North India)
  December - January (South India)
- Spacing – 1.5 - 2.0 m (R × R)
  1.0 m. (Hill × Hill)
- Immature tender fruits are harvested (60-70 DAS).
- Fruits can be packed in poly propylene bags for extended shelf-life.

65. "PUMPKIN"

- It has high carotene content and good keeping quality. It is considered as vegetable of immature value.
- Yerusseri – Prepared from immature fruits, is very popular in Kerala.
- The flowers of pumpkin are more nutritive than fruits.
- This crop is especially known for its low cost of production and long keeping quality.
- Medicinal use to reduce tape worm infection and its use as a diuretic are yet to be explored on large scale.
- The genomic structure of *Cucurbita moschata* is AABB which indicates that it is an amphidiploid.
- Both wild and cultivated species have same chromosome number 2n = 40.
- Five cultivated species
  1. *Cucurbita moschata* (Pumpkin).
  2. *Cucurbita pepo* (Summar squash).
  3. *Cucurbita maxima* (Winter squash)
  4. *Cucurbita mixta pang* (*Cucurbita angyrosperma*).
  5. *Cucurbita ficifolia* (Big leaf gourd or Malabar gourd)

  (a) Xerophytic forms – *Cucurbita digitata, Cucurbita palmata, Cucurbita cylindiata, Cucurbita foetidissima.*

  (b) Mesophytic forms – *Cucurbita ficifolia, Cucurbita sororia, Cucurbita lundelliana.*
- Pumpkin (*Cucurbita moschata)* appears to be more closely related to xerophytic forms and is considered as the axis through which the species under xerophytic forms are related to each other.
- The mesophytic forms are mostly cross compatible and *Cucurbita lundelliana* is seen as the key species in this group.

- *Cucurbita lundelliana* has wide spectrum of compatibility with the cultivated species particularly *Cucurbita moschata*.
- Among five cultivated species *Cucurbita moschata* (pumpkin), *Cucurbita pepo* (Summer squash) and *C. maxima* (Winter squash) are of great economic importance.
- Plant is mònocecious.
- Flowers are unisexual, solitary or fasciculate and are lemon yellow to deep orange in colour.
- Isozyme studies in *Cucurbita moschata* revealed that there are 28 loci conditions in the isozymic phenotypes and this results together with the high chromosome number in the genus (2n=40) are taken as proofs for the allotretraploid origin of cucurbita.
- 85% area and 90% production of pumpkin is contributed by the state Orissa alone.

Recommended Varieties of Pumpkin

S.No.	Variety	Features
1	Pusa Vishwas	Fruits spherical and medium. Average fruit weight 5.0 kg, fruit flesh thick and golden yellow in colour. Recommended for all over the country. Yield about 400 q/ha in 120 days.
2	Pusa Vikas	Fruits small and round flat, average fruit weight 2.0 kg. Fruit flesh deep yellow. Plant type dwarf (average length 2.0 m), Maturity 110 days, yield 250-300 q/ha.
3	Pusa Hybrid-1	Fruits flesh thick, golden yellow in colour and rich in vitamin A. Avg. fruit wt. 5 kg and yield about 500 q/ha in 110 days.
4	Arka Suryamukhi (*C. maxima*)	Fruit small, economy sized (1 kg), round with flat ends, ring colour deep orange yellow, with discontinuous thin cream lining, rich in vitamin C, highly resistant to fruit fly and bears 8-10 fruits per plant. Yield about 335 q/ha in 115 days.
5	Arka Chandan	Fruit medium sized (2.5 to 3.5kg) round with slightly pressed blossom end, rind colour light brown with creamy patches at maturity. Have excellent flavor and orange flesh, rich in carotene content. Bears 2-3 fruits per plant and yield about 325 q/ha in 115-130 days.
6	Co-1 (*C. moschata*)	A late variety, matures in 175 days, fruits, attractive, globular, medium large (7 to 8 kg) each vine produces 7-9 fruits, yield 288 q/ha.
7	Co-2 (*C. mosachata*)	Fruit small sized, weight about 2 kg with orange coloured flesh. Yield about 226 q/ha. developed at Coimbatore.
8	Kashi Harit	Short vines, fruits are spherical, green, weight 2.5-3 kg, 5 pickings can be taken in 70-75 days, Yield about 36-40 t/ha.

Crop improvement –

1. **Selection**
   (a) Arka Suryamukhi (1-1.5kg wt)– Variety of *Cucurbita maxima* (Winter squash), resistant to fruit fly (D*aucus cucurbita*), Keeping and transport quality are good.
   (b) Arka Chandan (2-3 kg) – Rich in carotene (3331 IU/100g) TSS 8-10%, pleasant aroma, keeping quality good, suitable for Western ghat and Karnataka.
   (c) Pusa Biswas/Vishwas.
   (d) Pusa Vikas – Suitable for spring summer in North India.
   (e) CO2 – Fruit colour green.
   (f) Ambili – Pure line selection.

- The most popular cultivar In *Cucurbita moschata* is the butter nut squash with fruits relatively small, cyclindrical with pronounced bulb surrounding the seed cavity and bright orange flesh.
- No breeding depression has been observed due to the fact that the original form of cucurbits is hermaphrodite.

For F_1 hybrid seed production –

- Male and female parents are planted in the ratio of 1: 10 in isolation.
- Insect pollination is reported to produce hybrid seed. Two MS gene viz MS1 and MS2 have been reported in cucurbits.
- Genetic male sterility has been reported in pumpkin.
- Good flesh quality character can be transferred from *Cucurbita maxima* to *Cucurbita moschata* and resistance to squash bug and squash vine borers from *Cucurbita moschata* to *Cucurbita maxima*.
- Resistance breeding is mainly done by back cross method.
  1. Cucurbita martinezzi.
  2. *Cucurbita foetidissima* - Both are used to transfer resistance to CMV in *C. moschata.*
- A single gene when homozygous in *Cucurbita moschata* confers a high level of resistance to Yellow vein mosaic virus (YVMV).
- Short days low night temperature and high relative humidity is best for pumpkin production.
- Sowing time: - April – May (In Hills)

  June – July and January – Feb (In Plains)
- Channel and hill system of cultivation is most useful for higher yield (Scientific method).
- Seed rate – 4-5 kg/ha.
- In general high "N" under high temperature condition promotes maleness in flowering resulting in low fruit set and high yield.
- Excessive vine growth can be pruned manually to promote higher female to male ratio.

- PGRS – Ethrel 250 ppm (2.5 ml of ethrel in 10 litres of water) can be applied to increase the female flower production.
- Honeybees are chief insect pollinators for pumpkin.

Seed Production –

- The cross compatible species like *Cucurbita pepo* and *Cucurbita mixta* should be planted in isolation from *Cucurbita moschata*.
- The pollen grains of non crossable species sometimes stimulate the ovary of pumpkins that leads to development of partheocarpic fruits.
- Isolation distance – 1000 M.
- Male and female parents can be grown in a 1: 3 ratio.
- Matures 75-180 days after seed sowing.
- Full mature fruits have a long storage life.
- Yield 20-30 tonns/ha.
- Storage 2-3 weeks at 15-20°C.
- Well mature fruits can be stored for 4-6 months is well ventillated chamber.

66. "SUMMER SQUASH" (VEBETABLE MARROW OR FIELD PUMPKIN)

- It is also called Vilayati Kaddu/Chappan Kaddu/ Bushy or ornamental cucurbits fruits resemble as a cover of earthern pitcher.
- In Hindi it is called Kumra, Safed kaddu.
- Immatures fruits are used as vegetables.
- It is used as bathing sponge.

Recommended Varieties of Summer Squash

No.	Variety	Source	Remarks
1.	Punjab Chappan Kadoo-1	Punjab Agricultural University, Ludhiana (1982)	It is an early maturing variety which is ready to harvest in 60 DAS. The plant of this variety has thick foliage, green color leaves, disc shaped fruits which are green in color. The variety is resistant to downy mildew and is tolerant to virus, red pumpkin beetle and powdery mildew. It is high yielding gives about 20-25 tons/ha.
2	Patty Pan	IIHR, Bangalore	An Introduction from USA and recommended by IIHR for cultivation during 1972. A bushy-type and fruits are disc-shaped, chalky white, tender and very attractive at edible stage. A short-duration variety (80-90 days), yield 54 tons/ha.

3	Early Yellow Prolific	IARI, Katrain	Early bush-type varieties, Fruits are light yellow skin turns to orange-yellow on maturity. Flesh is tender at the stage it is consumed as vegetable.
4	Australian Green	IARI, Regional Station, Katrain	An introduction, Very early bush-type variety. Green fruits are dark-green with longitudinal white strips all over, 25-30 cm long. 15-20 fruits/plant and yield 15-16 tons/ha.
5	Pusa Alankar	IARI, Regional Station, Katrain	An F_1 hybrid early maturing having uniform dark green fruits with light coloured stripes. The flesh is tender delicious and fruits mature in 45-50 days. High yield 20-30 tons/ha.

Crop Improvement

1. **Introduction** – Patty Pan, Australian Green.
2. **Selection** - Punjab Chappan Kaddu 1.
3. **F_1 hybrid** – Pusa Alankar (EC 207050 and Selection 1).

- F_1 hybrid can be obtained from most inter specific mating between *Cucurbita pepo, Cucurbita mixta, Cucurbita moschata* and *Cucurbita maxima* but such hybrid are normally sterile.
- *Cucurbita moschata* was used as a bridge to transfer disease resistance gene from *Cucurbita martinezii* to *Cucurbita pepo.*
- The three ways cross between *Cucurbita pepo* and (*Cucurbita moschata* × *C. martinezii*) is being used to transfer resistant to powdery mildew and CMV from *Cucurbita martinezii* to *Cucurbita pepo.*
- Washek (1982) achieved a cross of *Cucurbita pepo* with the multiple disease resistant *Cucurbita ecuadorensis* through embryo culture.
- **Sowing Time** –

  Jan – March and October- November (In Plains)

  April – May (In Hills)
- Fruits are harvested at 1/3 maturity.

67. "WINTER SQUASH"

- Monoecious.
- Tolerant to frost.
- Commonly grown in hills variety.
  - (i) Arka Suryamukhi – Resistant to fruit fly.
  - (ii) Pink Banana.

68. "RIDGE GOURD" (RIBBED GOURD/KALI TORI)

- Both the species (*Luffa acutangula and L. cylindrica)* contain a gelatinous compound called "luffein".
- The genus derives its name from the product "*Loofah*" which is used in bathing sponges, scrubber pods, doormats, pillows, mattresses and for cleaning utensils.
- The sponge gourd fruits contain higher protein and carotene than ridge gourd.
- Among the gourd vegetables only bitter gourd, bottle gourd and pointed gourd have export potential.
- Monoecious, young fruits are used as vegetables.
- Anthesis – Start in evening and remain open throught the night and are ready for selfing and pollination in the next morning.
- Ridge gourd flower – Pale Yellow.
- Sponge gourd flower – Deep yellow.

Recommended Varieties of Ridge gourd

S.No.	Variety	Features
1	Pusa Nasdar	Early maturing, suitable for summar and rainy season, produce club-shaped fruit, monoecious. Maturity 55-650 days. Yield 100-155 q/ha.
2	Arka Sujat	Lush green and tender medium longer (35-45 cm) fruits much prominent ridges and delicate aroma, good transport and cooking qualities. Duration 100 days and yield 53 t/ha.
3	Arka Sumeet	Lush Green and tender long fruits (50-65cm) with prominent ridges and delicate aroma, good transport and cooking qualities, Duration 110 days and yield 50 t/ha.

Other Varieties:

1. Punjab Sadabahar – Rich in protein.
2. Satputia – Hermaphrodite in sex form (Bisexual flower).
3. Konkan Harita.
4. PKM -1 (Kharif and summer).
5. Co-1, Co-2.
6. Hybrid – Surekha, Arka Sujat.

- Sowing time as mentioned in other crops.
- Seed rate – 3.5-5 kg/ha.
- Commercial crop are trained on a kniffen system.
- The sex ratio can be regulated by exogenous application of growth regulators.
- NAA 200 ppm increases the female flower production.
- Harvesting is done when fruit is tender and still immature.
- Fruit 14-15 cm long, club shaped and surface is distinctly ribed.

"SPONGE GOURD"

- Fruit of sponge gourd are smooth and is called ghiatori.
- *L. cylindrica* (Syn. to *L. aegyptiiaca*) is also known as sponge gourd, towel gourd, smooth luffa, veg sponge (dish cloth gourd).
- It is monoecious climber.
- Anthesis occurs at 4.0-8.0 am, suitable for selfing almost throught the day.
- Fruits are 20-50 cm long, almost cylindrical in shape.

Recommended Varieties of Sponge gourd–

S.No.	Variety	Features
1	Pusa Chikni	Fruits smooth, dark green, suitable for both spring summer and rainy, early maturing high yielding, poor yielder. Maturity 50-55 days.Yield 110-120 q/ha.
2	Pusa Supriya	Short vine growth, fruits light green, smooth, tender, 12-14 fruits per vine. Gives a 2% more yield than Pusa Chikni, suitable for growing in both seasons. Maturity 40-45 days. Yield 100-120 q/ha.
3	Pusa Sneha	Fruits dark green, 20-25cm long, straight tolerant to high temperature. Suitable for cultivation in spring-summer and rainy seasons. Maturity 45-50 dyas. Yield 110-130 q/ha.

Other Varieties:

1. Phule Prajakta – Suitable for summer and kharif season.
2. Harita – F1 hybrid

- Seed Rate – 2.5-3.5 kg/ha.
- Application of Ethrel 250 ppm has been found to be beneficial.

69. "ASH GOURD"

- Ash gourd or wax gourd and white gourd also known as petha kaddu.
- It is used in the preparation of Petha sweet.
- Agra Petha is famous all over India.
- Ash gourd is considered good for people suffering form nervousness.
- Ash gourd is an annual hispid (rough white bristle like hairs) climbing, herbaceous and monoecious plant.
- It is also known as Chinese preserving melon.
- Also used to make sugar candies.
- The surface of the fruit is covered with a bluish white "Waxy" ash.
- Ethephon (250 ppm) spray increase production of female flowers.

Recommended Varieties of Ashgourd

S.No.	Variety	Features
1	Pusa Shakti	Fruits oblong-ellipsoid with greenish-white flesh. Average fruit weight 7.0 kg. Long storability, more than 6 months at room temperature. High TSS, Vitamin C and minerals. Suitable for cooking as vegetable and Petha (sweet candy). Maturity 140 days, Yield 250 q/ha.
2	Co1	Fruits globular, light green with ash coating. Suitable for growing in South India. Maturity 150 days. Yield 250 q/ha.
3	Co2	Fruits light green, small sized, average weight 2.5 to 3.0 kg also suitable for South India. Maturity 120 days. Yield 340 q/ha.
4	Kashi Dhawal	Fruits oblong, harvesting in 100-105 DAS. Most preferred variety for sweet preparation. 55-60 t/ha. For UP, Punjab and Bihar.

Other Varieties:

1. Pusa Ujjwal.
2. APAU-Fruit Long.
3. Mudliar - By TNAU.
4. Pusa Shakti – New variety.

- Short day, low night temperature and humid climate are good for production of female flower.
- Planting time – June-July.
- Seed rate – 5-7 kg/ha.
- Fruit at maturity have white waxy surface.
- The fruit is mainly purchased by the confectioners who use them to prepare "Petha".
- Ash gourd can be stored longer than any other cucurbits.
- Disappearance of ash wool on the fruit surface is an indication of maturity.

70. "POINTED GOURD" PARWAL

- It is a perennial cucurbit cultivated mostly in Bihar, WB and MP.
- Immature fruit of this plant is a popular vegetable containing 2% protein, 0.03 % fat, 2.2% carbohydrate and 153 IU Vitamin A.
- Fruits are used in making curry and also fried.
- Parwal is a dioecious climber with perennial rootstock.
- Fruit is particularly recommended during convalescence (regain health after illness).
- Parwal remains dormant during winter.

Varieties –

1. Swarna Rekha – High yielding.

2. Swarna Alaukik – High Yielding.
3. Shankolia
4. Dandali
5. Chhota Hill
6. Hilli
7. CHES Elite line – Resistant to fruit fly.
8. CHES hybrid – First parwal hybrid developed in country.
9. Faizabad Parwal 1.
10. Faizabad Parwal 3 - Excellent for culinary purpose.
11. Faizabad Parwal 4 - Recommendation for bower system.
12. Damodar.
13. Kazil.
14. Bihar Sharif

- Require humid and hot climate since it is susceptible to frost and cold. It can withstand water but not water logging.
- It remains dormant in winter and grows vigorously during spring.
- Rainfall is considered very well for this crop.
- Pointed gourd is vegetatively propagated through vine cuttings and root suckers.
- About 2000-2500 (1m. length) cuttings are required.
- Seed propagation is avoided in pointed gourd because of poor seed germination and since it is dioecious in nature produces male and female plants in equal proportion if they are planted from seed.
- In other word, 50% plants are none fruiting (i.e.male plants). In order to ensure maximum fruit set and yield only 10-12% male plants are maintained in the garden to ensure the source of pollinizer and remaining plants must be female.
- In bower system 8000 cuttings are required to raise one hectare crop with in a spacing of 1.0x1.5m.
- In flat system – 4500-5000 cuttings, 1.0 × 2.0m spacing.
- In Riverbeed system – 3500-4000 cuttings.
- February - March – Ideal time in upland for sowing.
- Mid. November – In diara land for sowing.
- Lunda or lachhi method – Cuttings folded into a figure of "8" commonly known as lunda or lachhi.

  Moist lump method
  Straight vine method
  Ring Method — Other methods
  Rooted Cutting
  Root suckers
- Minimum of 10% male plants in the field are essential to ensure the availability of enough pollen grains required for fertilization and fruit setting.
- For every 15 female plants one male plant is must to get highest yield.
- Fruit availibility – February-October.

71. "SNAKE GOURD"

- Fruits are rich in carbohydrates, proteins, minerals and vitamins A, B and C.
- It is monoecious, annual and herbaceous climber.
- Immature fruits are eaten.
- The fruit improves the apetite, acts as a tonic and stomachic and cures bilousness.
- The roots and seeds are used in the treatment of diarrhoea.
- Flowers opening occur in afternoon.
- India is native home of snake gourd.
- It's widely cultivated in South and South East Asia.

Recommended Varieties of Snakegourd

No.	Variety	Features
1	Co 1	160-180 cm long fruits with white stripes on dark green skin, flesh greenish. Each vine gives 10-12 fruits, average weight 4-5 kg. Maturity 135 days. Yield 180 q/ha.
2	Co 2	Suitable for HDP, 30-35 cm long stout fruits, greenish, stripeless skin, ashy bloom. Vine length short, can accommodate more number of plants per hectare at shorter planting distance. Maturity 125-130 days. Yield 250-300 q/ha.
3	MDU – 1	A hybrid, fruit skin green with white stripes and medium length around 60 cm. Maturity 125 days. Yield 200 q/ha.
4	APAU Swetha	Long fruits, white skin, green fingures (stripes) maturity 125-130 days. Yield 280-300 q/ha.

Other Varieties –

1. CO 4 – Early maturing.
2. TAL – 19 -- Developed by KAU, Kerala.
3. Konkan Sweta 1 – Fruits have good flesh if harvested timely otherwise is becomes hollow.
4. MDUI – F_1 hybrid.
5. PKM-1- Extra long (200 cm).

- Seed rate – 5-6 kg/ha, spacing 1.5-2.5 m. R × R, 60-120 cm P × P.
- Training is essential to this crop. The bower system of training is best for this crop.
- Sowing time –
  Summer Season (December – January) and Rainy Season (June-July).
- Fruits are hand picked when they are still tender and about 1/4 –1/3 of their full size.
- Fully mature fruits are lighter, fibrous and hard.
- Just after harvesting fruits are sent to market as they do not have shelf life more than two days.

72. "IVY GOURD" (LITTLE GOURD/COCCINIA/KUNDRU/TONDLI/ SCARLET GOURD)

- It is most commonly used as vegetable in Southern and Central India.
- The fruits are rich source of carbohydrate, protein, vitamin - A and C.
- IVY gourd is dioecious and perennial in nature.
- The immature fruits are used as vegetable.

  Propagation – By stem cutting (25 cm long).

  Planting Time – June-July and February – March.
- To get good yield 5-10% of the male type vines should be planted in the field.
- For a commercial crop, plants are trained on a bower system constructed to suit the growth of vine.
- Harvesting is determined by change in colour from dark green to bright or light green.
- Semi hard wood cutting (5-20 cm long) 1.5 cm diameter collected from female parent are planted @ 2-3 cuttings per basin, 5-6 cm deep.
- *Coccinia grandis* is a dioecious; about 10% of the total population should be male for commercial crop.
- There are certain clones which do not bear male flowers but set fruit parthenocarpically.
- Parthenocarpy – Vegetative type is found.
- In parthenocarpic clones the fruit set is without pollination hence there is no need of separate male and female clones.
- During winter plants go in dormancy under North India, however under South India plants grow throught the year.
- Yield 200-300 fruits, weight 3-4kg/vine /year.

73. "ROUND MELON" (ROUND GOURD/ INDIAN SQUASH/SQUASH MELON/TINDA)

- *Citrullus vulgaris* var. *fistulosus*.
- It has cooling effect to the body.
- The fruits are also considered good for curing dry cough and for improving blood circulation in body (Vitamin A).
- 27-30°C temperature is optimum for seed germination.

Varieties –

S.No.	Variety	Features
1	Arka Tinda	Recommended for Karnataka and grown in northern states. An early summer variety. Fruits round, light green with soft hairs. Tender flesh at marketable stage. Maturity 90-100 days. Yield 95-100 q/ha.
2	Punjab Tinda	Fruits medium in size, flat round and light green. Fruit surface shining and pubescent. Maturity 60 days. Yield 45 q/ha.
3	Bikanery Green	Fruits round, tender and green in colour. Suitable as spring summer crop in Rajasthan.
4	Hisar Selection -1	Fruits oval, green in colour with hairs, very attractive.
5	Tinda Ludhiana	Flesh White, less seed, Good keeping quality.
6	Tinda Tonk	Grown both as solo and as an inter crop.

- Seed rate – 3.5-7.0 Kg/ha.
- Picking should be done when these are still immature and small in size.
- M.H. (50 PPM) aqueous solution sprayed at 2 leaf and 4th leaf stage stimulates vine growth, giving more femaleness and enhance female flowering at lower nodes.
- Due to high water content fruits are spoiled early.

74. "CHOW-CHOW" (CHAYOTE, CHOCO, ASKAS)

- It (*Sechium edule*) is an herbaceous, perennial climbing, monoecious vine (13m. length) having large tuberous roots.
- Seed are delicacy.
- Chow-Chow is a single seeded cucurbit.
- Single seed is located near the broad end of the pear shaped fruit.
- High "Ca" content.
- Viviparous germination, highly sensitive to frost.
- Propagation - By sprouted fruit.
- Chow Chow is grown for its large tuberous roots.
- Principally grown for its pear shaped fruits.
- It is rich source among gourds in nutritive value particulary in carbohydrate (6 %) and calorific value 29 Cal/100g.

75. "PERENNIAL CUCURBITS"

- Kakrol or spine gourd (*Monordica diocia*) is dioecious.
- Remains dormant during winter, fruit ovate in shape (10-12 cm long).
- Rich in protein and Vitamin C.

- Propagation – Tuberous roots (5 lakh/ha).
- Seed – 3-4 kg.
- Being dioecious in nature about 10% of male plants should be planted for good fru set.
- Seed oil – Used as an illuminant.

76. "KARTOLI" (SWEET GOURD) (MOMORDICA COCHINEHENSIS

- Dioecious in nature.
- Propagation – Tuberous roots.

77. "SNAP MELON" (PHOOT) (CUCUMIS MELO VAR MOMORDICA

- Still grown during summer and rainy season in North India.
- The fruits brust when snap melon is ripe therefore it is popularly called phoot.
- Immature fruits having light green to dark green colour are harvested for vegetab and salad purpose.

78. "LONG MELON" (CUCUMIS MELO VAR UTILLISSIMUS)

- Commonly grown in North India.
- Two types of long melon are very popular in Lucknow.
  1. Lalia ki ungalian (Fingers of Lalian) – Fruits small and thin.
  2. Manju ki Pasalian (Ribs of Manju) - Long thin fruit
- Variety – Arka Sheetal an improved variety.

S.No.	Variety	Features
1	Arka Sheetal	Fruit medium long, light green skin, shallow furrows, fruit weigh 80-90 g. fresh crisp excellent flavour. Yield 350 q/ha. Duration 100-110 days. A selection from the local strains of Lucknow develope at IIHR. Recommended for Karnataka.

79. IMPORTANT PEST AND DISEASE OF CUCURBITS CROPS

Sr. No.	Pest and Disease	Symptoms	Management
1.	Fruit flies (*Bactrocera cucurbitae)*	Maggots feed on the pulp of the fruits Oozing of resinous fluid from fruits Distorted and malformed fruits	Use fly trap and attractant like citronella oil, eucalyptu oil, vinegar (acetic acid), an lactic acid to trap flies.

2.	Red Pumpkin beetle (*Aulacophora foveicollis)*	Grubs feeds on the roots, stem and fruits touching the soil.	Collect and destroy the damaged plant parts with larvae and spray malathion 50 EC @ 500 ml or dimethoate 30 EC 500 ml or methyl demeton 25 EC@ 500 ml/ ha.
3.	Stem borer or clear winged moth (*Melittia eurytion*)	Larva bores into the stem of snake gourd and produces galls	
4.	Stem gall fly (*Neolasioptera falcate)*	Maggots bore into the distal shoot and form galls	
5.	Snake gourd semilooper (*Plusia peponis)*	The caterpillar cuts the edges of leaf lamina, folds it over the leaf and feeds from within leaf roll.	
6.	Pumpkin caterpillar (*Diaphania indica)*	It webs the leaves and feeds within; It also feeds on flowers and bores into fruits.	
7.	Bottle gourd plume moth (*Sphenarches caffer)*	Larva feeds on leaves making small holes.	
8.	Leaf miner (*Liriomyza trifolii)*	Leaves with serpentine mines.	Spray NSKE (Neem seed kernel extract) @ 3%.
9.	Hadda Beetle (*Epilachna Implicate*)	Both adults and larvae feed voraciously by scrapping the chlorophyll of the leaves.	Foliar applications of pesticides like malathion, lambda-cyhalothrin, pyrethrin, spinosad methoxychlor, and rotenone.
10.	Aphids (*Aphis gossypii*)	Downward curling and crinkling of the leaves.	Spray of Cypermethrin (0.01%) or acetamiprid (0.01%) or bifenthrin (0.01%) can be used to control aphids.
11.	Whitefly (*Bemisia tabaci*)	Whiteflies colonize the underside of leaves. It can affect the crop directly by its feeding and by acting as a vector of viruses.	A soil application of neonicotinoid insecticide (imidacloprid or thiamethoxam) at planting effectively controls whiteflies.
12	Root-knot nematodes (*Meloidogyne incognita*)	Small leaves and bear fewer flowers. "Giant cells" within the root.	Crop rotation and apply of nematicides (carbofuran) and soil fumigation (Methyl bromide) before or at planting.
13.	Powdery mildew (*Erysiphe cichoracearum)*	White or brown mealy growth will be found on upper and lower surfaces and stems.	The disease can be controlled by spraying Wettable sulphur 0.1%.

14.	Mosaic (*PRSV/CMV*)	New growth is cupped downward, and leaves are severely mottled with alternating light green and dark green patches.	The virus is readily transferred by aphids and survives on a wide variety of plants. Spray any systemic insecticide.
15.	Downy mildew (*Pseudoperonospora cubensis)*	Symptoms resembling mosaic *viz*, pale green areas separated by dark green areas appear on upper surface of leaf.	Seed treatment with Apron SD 35 @ 2 g/kg. Followed by spraying with Mancozeb 0.2%.
13.	Alternaria Blight (*Alternaria cucumerina*)	Yellow spots which enlarge to form concentric rings on the upper leaf surfaces.	Spray the crop with Mancozeb @ 2 g /lit and use disease-free seed in fertile, well-drained soil and practice crop rotation.
14.	Gummy Stem Blight (*Mycosphaerella melonis)*	Stem lesions on melons exude a gummy, red-brown substance which may be mistaken for a symptom of Fusarium wilt.	Periodic applications of fungicide like mancozeb @ 0.2% can help limit secondary infections.
15.	Bacterial Wilt (*Erwinia tracheiphila)*	Leaves adjacent to the wilting leaves will also wilt, and eventually the entire lateral is affected. In general, more bacterial wilt is seen on the edges of fields where beetles first encounter plants.	Carbaryl, Malathion, or rotenone insecticides or combination products are registered to treat cucumber beetles.
16.	Fusarium Wilt (*Fusarium oxysporum* f. sp. *Melonis*)	Wilting and chlorosis (yellowing) of older leaves also causes vascular browning that is visible in stem cross-sections.	Planting resistant cultivars and followed crop rotation.
17.	Angular Leaf Spot (*Pseudomonas lachrymans)*	Small, angular, water-soaked lesions on the leaves. Water-soaked spots may also appear on the fruit and are frequently followed by soft rot bacteria.	Rotating with unrelated crops, planting disease free seed and Spray 400ppm Streptomycin sulphate.
18.	Anthracnose (*Colletotrichum orbiculare)*	Sunken, elongated stem cankers are most prominent on muskmelon, though leaf and fruit lesions also occur.	Seed treatment with Carbendazim 2g/kg of seed and spray of Mancozeb 2g or Carbendazim 0.5g/lit.

80. POTATO AND TUBER CROPS

Potato and Tuber Crops

Common Name	Botanical Name	Family	Ch. No	Edible part	Origin
Potato	*Solanum tuberosum*	Solanaceae	48 (4x)	Stem Tuber	South America (Peru-Bolivia)
Sweet Potato	*Ipomoea batatas*	Convolvulaceae	30, 40, 90 (6x)	Tuber	South America
Yam	*DIoscorea sps*	Dioscoreaceae	20, 30, 40, 50, 60, 70, 80, 81	Tuber	South East Asia
Cassava	*Manihot esculenta*	Euphorbiaceae	36	Tuber	Brazil
Colocasia (Taro)	*Colocasia esculenta*	Araceae	24, 28	Corms	South Central Asia
Elephant Foot Yam	*Amorphophallus compinulatus*	Araceae	26	Corms	India
Yam Bean	*Pachyrrhizus crosus*	Fabaceae	-	Corms	South America
Tannia	*Xanthosoma sagittifolios*	Araceae	26	Corms, leaves and petioles	South America
Queensland Arrowroot	*Canna edulis*	Cannaceae	-	Rhizome	South America
Jerusalem Artichoke	*Helianthus tuberosus*	Compositae (Asteraceae)	-	Tuber	Eastern United State
Coleus/ Chinese potato	*Solenostemon rotundifolius* (syn. *Coleus parviflorus*)	Lamiaceae	-	Tuber	Tropical Africa, Asia and Australia
Arrowroot/ West Indian Arrowroot	*Maranta arundinacea*	Marantaceae	-	Rhizome	West Indies
Giant Taro	*Alocasia macrorrhiza*	Araceae	-	Rhizomes and shoots	Asia
Chinese Water Chestnut	*Eleocharis dulcis*	Cyperaceae	-	Tuber	China

- Central Tuber crop Research Institute (CTCRI) was established at Trivantpuram (Kerala).
- Project Directorate of Vegetable Research is at Varanasi (UP).

- Potato contains dimethyl pyrazine (aroma containing compound).
- Cassava is monoecious while Yams are dioecious.
- Respiration rate of potato is very low.
- Yam Bean – Also known as Misrikand, Potato Bean.
- Potato and Sweet Potato are C3 plants.
- Fuyu – Product made from yam.
- Sago – Important food product from Cassava starch.
- Sweet potato and yams- Fruit type is capsule.
- Inflorescence of Cassava – Panicle.
- Highest Carbohydrates – Tapioca.
- Sweet potato is cheapest source of calorie.
- High fiber- Potato.
- Diosgenin containing yam is useful in manufacture of cortisone and contraceptive drugs.
- Cercospara leaf spot of sweet potato was 1st reported in Africa.
- Shree Shaya – Multiple hybrid of Cassava.
- Central Potato Research Institute (CPRI) – (Under ICAR) Established in 1949 with headquarters at Patna which was later shifted to Shimla in 1956.
- AICPIP (All India coordinated Potato Improvement Project) was started in 1971 with headquarters at CPRI, Shimla (HP).
- It has 22 centers.
- International Potato Centre (CIP) – Autonomous Scientific institution in Lima (Peru) established in 1972 by agreement with Govt. of Peru.
- It is one of the 16 International research and training center supported by consultative Group on International Agriculture Research (CGIAR).
- The CGIAR is sponsored by FAO of the United Nations.
- UNDP – United Nations Development Programme.
- UNEP – United Nations Environment Programme.
- TPS – True Potato Seed.

81. "POTATO"

- About 50% of potato produced in world is utilized as human food.
- Potato is used as staple diet in many of the countries especially in the west.
- It is first ranked vegetable crop of the world in area and production.
- The word potato has been derived from the word 'papa' or 'patata', the South American Indians' name of the plant.
- The edible part of the potato plant is the tuber, the swollen ends of its underground stems (stolons).
- The potato was introduced in India from Europe in early 17th century, only fourty years after its introduction in Europe by Portuguese.

- Potato is grown in India in almost all the states except "Kerala".
- In total production, India ranks second after China.
- Potato is grown during short days of winter (October-March) in about 22% area in the plains.
- Potato remained staple food of the Irish people till 1845 when late blight appeared in an epidemic form resulting in the 'Potato Famine of Ireland'.
- Potato is grown during long days of summer (April- September) in about 10% area in hills.
- Potato is a dicot plant.
- Genus Solanum contains nearly 2000 species.
- All the cultivated sp. has been grouped under the series Tuberosa of sub section potatoe.
- About 72% of the species are diploid (2x=24) and nearly 12% tetraploid (2x=48), rest triploid (2x=36), pentaploid (2x=60) and hexaploid (2x=72).
- The widely cultivated potato belongs to tetraploid sp. (2x=48) *Solanum tuberosum.*
- First mention of potato in India (1615) began on the account of a bouquet given by Asaf Khan to Sir Thomas Roe in Ajmer (Raj.).
- Probable ancestor of all cultivated potatoes is *Solanus stenotonum.*
- The annual per capita consumption in the country is 15 kg as against the world average of 24 kg.
- Uttar Pradesh (614780 ha) is having maximum area under potato cultivation followed by West Bengal (427500 ha).
- In potato production, Uttar Pradesh (15555000 MT) is the leader followed by West Bengal (12782000 MT).
- Potato breeding station, Kufri (Shimla) was established in 1935 under IARI.
- Central Potato Research Institute, Shimla was established in 1949.

Improved varieties:

Early Season	
Kufri Chandramukhi (CPRI, Shimla)	Suitable for cultivation in north Indian plains. It matures in 80-90 days. Plants are medium-tall and spreading with grey-green foliage. It has good keeping quality and attractive white, large and smooth tubers with fleet eyes. Average yield is 250 q/ha.
Kufri Ashoka	It takes 75-80 days to reach maturity. Plants are medium tall and spreading with glossy-green foliage. The tubers are oval to oblong, large sized with fleet eyes. Its yield is 275 q/ha.
Kufri Pukhraj	It takes 75 days to reach maturity. Plants are tall, erect with grey green foliage and white flowers. Its tubers are medium-sized, oval, slightly tapered with fleet eyes. Resistant to early blight, moderately resistant to late blight and has an average yield is about 325 q/ha.

Kufri Surya	Recommended for commercial cultivation in 2000. It is tolerant to high temperature and is suitable for early (September) planting. It matures in 90-100 days after planting. Tolerant to late blight and average yield is 250-310 q/ha.
Mid Season	
Kufri Jyoti	Suitable for planting in the spring season. It matures in about 100 days. Tubers are large, oval, and white with fleet eyes and white flesh. It is resistant to wart disease. Average yield is 300 q/ha.
Kufri Jawahar	Released in 1996. It takes 80 days to reach maturity. This is moderately resistant to late blight. It is suitable for inter-cropping and mechanized cultivation. Average yield is 350 q/ha.
Kufri Sutlej	Matures in about 90-95 days. It is moderately resistant to late blight. Average yield is about 350 q/ha.
Kufri Lalima	The variety was released in 1982 for Indo-Gangetic plains. It matures in about 90-100 days. Average yield is 220 q/ha. It has good keeping quality and floury texture.
Kufri Bahar	Matures in about 90-100 days. The tubers are large, round to oval with white flesh and medium deep eyes. Average yield is 315 q/ha.
Kufri Giriraj	Matures in 100 days and produces tubers that are medium to large, white and oval with fleet eyes. The variety is resistant to late blight and average yield is 200-250 q/ha.
Kufri Pushkar	Released in 2005. It matures in 90-100 days under short day conditions. It is moderately resistant to late blight and early blight. Average yield varies from 400-425 q/ha.
Late Season	
Kufri Sindhuri	Medium late variety and takes 110-120 days to mature. It is moderately resistant to early blight. It possesses very good keeping quality and average yield is 300 q/ha.
Kufri Badshah	Matures in about 110-120 days.Tubers tend to develop faint purple colour on exposure to light. It is moderately resistant to late blight. Average yield is 325 q/ha.
Kufri Swarna	The variety was released in 1988 for cultivation in eastern hilly regions. Plants are tall, erect, medium compact and vigorous with white flowers. It matures in about 120 days, highly resistant to late blight and cyst nematodes. Average yield is 270-280 q/ha.
Varieties Suitable for Processing	
Kufri Chipsona 1	Takes 100 days to reach maturity. Tubers are white, medium sized, oblong and low in reducing sugars, low phenols and high dry matter. It is resistant to late blight. Average yield is 250-300 q/ha.

Kufri Chipsona 3	Released in 2005. It is a dual purpose variety suitable for both processing (chips and flakes) and table purposes. Its tubers are round-oval shaped with dry matter content of more than 20%. It has excellent chip colour and is free from browning. The variety is highly resistant to late blight and average yield is 333 q/ha.
Kufri Frysona	Recommended for commercial cultivation in 2010. It matures in 90-100 DAP. The tubers are medium sized, long to oblong shaped with smooth white skin, shallow eyes and white flesh. Tubers have good keeping quality and high dry matter. It is resistant to late blight and is suitable for making French fries. Average yield is 400-425 q/ha.

Crop Improvement –

1. **Introduction** – Upto date from N. Ireland, most of popular varieties in India.
2. **Clonal Selection** – Kufri Red from Darjeeling.

   Red Round

   Kufri Safe both are selection from var Phulwa
3. **Hybridization** –
   - Kufri Sindhuri – K.Red × K. Kundan.
   - K. Badshah – K. Jyoti × K. Alankar, resistant to late blight, early blight and Potato virus Z.
   - K. Jyoti – 3069 d (4) × 2814 a (1) resistant to late blight, immune to wart disease.
   - Kufri Bahar – K. Red × Gineke.
   - K. Lalima – K. Red × AG 14.
   - K. Swarna – K. Jyoti × VI n2, doesn't show cracking, highly resistant to late blight and cyst nematode.
   - Kufri Jawahar – K. Neelam × K. Jyoti, moderately resistant to late blight.
   - K. Sutlej – K. Bahar × K. Alankar, moderately resistant to late blight.
   - K. Pukhraj – Craigs Defiance × JEX / B-687, resistant to early blight and moderately resistant to late blight.
   - K. Chipsona 1 – MS/78-79 × CP 2416, resistant to late blight, suitable for processing / producing light – colour chips and finger fries.
   - K. Chipsona 2 – Novazo × QB/B- 92-4, resistant to late blight and tolerant to frost, suitable for processing giving light colour chips and frying.
   - K. Alankar, K. Jevan, K. Moti, K. Deva, K. Kundan, K. Chamtkar, K. Naveen, K. Khasigaro, K. Luvkar.
   - K. Sheetman -Phulwa × Craig's Defence, resistant to frost.
   - K. Chandramukhi - Sd – 4485 × K. Kuber, early maturing.
   - K. Griraj – SLB/J132 × EX / A 680-16, resistant to late blight.

4. **Mutation breeding** – R-8- through × rays.
   - Triploid potato (Solanum × Chaucha) cultivated at Peru and Bolivia.
   - Pomato – Potato × Tomato, Developed with the help of protoplast fusion by Melcher (1978).

Other varieties

Early Maturing –

- K. Chandramukhi (75 days)
- K. Lavkar suitable both for Rabi and Kharif season.
- K. Ashoka (75 days) – Replaced K. Chandramukhi.
- K. Jawahar (80 days).

Mid Maturing –

- K. Jyoti.
- K. Lalima.
- K. Badshah.
- K. Sutlaj.
- K. Bahra.
- K. Pukhraj.
- K. Chipsona 1 (100 days).
- K. Griraj (100 days).
- K. Muthu – For Southern Hills.
- K. Lalima.
- K. Sutlej replaced K. Badhsah and K. Bahar.

Late Maturing

- K. Sindhuri (120-140 days) – Popular in Northern India.
- K. Swarna (120 days).
- K. Deva.
- K. Megha – for Meghalaya
- Varieties resistant to Late blight –K. Badshah -Also resistant to early blight, and Potato virus x.

Late blight resistant varieties

K. Jyoti – Also immune to wart disease.

K. Swarna – Also resistant to cyst nematode.

K. Jawahar.

K. Pukhraj.

K. Chipsona 1 and 2.

K. Megha, K. Jeevan, K. Kuber, K. Moti, K. Naveen, K. Kundan.

K. Badshah – also resistant to early blight, Potato virus (Multiple Disease Resistant).

Varieties for plains –

K. Ashoka, K. Pukhraj, K. Anand, K. Alankar.

K. Kuber - Resistant to black scurf.

K. Badshah – Multiple resistances: Late blight, early blight and potato cyst nematode and potato virus.

K. Jyoti – Have tendency to cracking and immune to wart disease.

K. Swarna – Resistant to cyst nematode and late blight.

K. Sherpa – Resistant to cyst nematode and late blight.

K. Pukhraj- Resistant to early blight.

K. Sindhuri – Resistant to early blight.

K. Chipsona I and II- Suitable for processing (Chips), Also tolerant to frost and late blight.

- Late blight (*Phytophthora infestans*) is the most devastating disease occurring regularly in epiphytic form every year in hills and very frequently in plains.
- Potato virus – Aphid are vectors, important constraint in potato productivity.
- Bacterial Wilt – Common in Mid hills and Assam, Meghalaya, Maharashtra and Orissa.
- Wart Disease – (*Synchytrium endobioticum)* common in hilly regions of Darjeeling (W.B.).
- Cyst forming nematode – Common in Southern Hills.
- Potato Tuber Moth – More attack in warm plateau areas of Maharashtra, Karnataka and Madhya Pradesh.
- Varieties containing low sugar and high dry matter are suitable for preparation of special value-added products like chips, French fries, cubes and other dehydrated products.
- Potato is a self-pollinated crop but it is vegetatively propagated.
- The cultivated tetraploid varieties are highly heterozygous.
- Selfing or inbreeding in potato leads to loss of vigour of the progeny and non-flowering.
- Hybridization –Intervarietal or interspecific is used for potato.
- The parents for hybridization are generally selected on the basis of their combining ability.
- Resistant genes are found scattered in wild and semi cultivated species.
- Available in center of origin and diversity in S. America.
- Potato flowers are hermaphrodite (bisexual), emasculation is done in selected female parents in evening.
- Pollination is done in morning in emasculated flower.

- The bacterial "Plasmid" which is free floating rings of DNA is used in carrying and transferring the derived gene.
- Two commonly used bacteria in this technique are
  1. *Agrobacterium tumefaciens* – Causing crown gall disease
  2. *Agrabacterium rhizogenus* causing hairy roots in plant.
- "Restriction endonucleases" and "Ligase" are special enzyme which heip in cutting and sealing the DNA molecules at specific sites.
- In Potato the genetic transformation technique have been successfully used in developing varieties resistant /tolerent against viruses, insect herbicide, frost and salinity.
- Potato naturally flowers under cool climate and long day condition of more than 15 hours light (Hills).
- Hills are therefore ideal for hybridization work.
- Carbohydrates - Major constituents of potato.
- Potato is a short duration crop.
- Maximum area under Potato is in alluvial soils.
- Black Soils – Prone to cracking on drying and exposer of tuber to sun and tuber moth infestation.
- Excellent potato crop is grown under river bed system of cultivation in Deesa,, Banaskantha in Gujarat.
- Ph 5.5-7.5 (Acidic to neutral) is suitable for potato.
- In Potato crop temperate region crops response to thermoperiod.
- Favourable temperature for tuber formation – Below 21°C.
- Temperature above 21°C causes sharp fall in tuberization.
- Summer crop is taken in Hills.
- Winter crop – Tropical and Subtropical areas.
- For best yield potato need long day condition during growth and short day condition during tuberization.
- Good crop growth occurs when days are sunny and nights are cool.
- Most potato genotypes do not tuberize when the night temperature is more than 23°C (areas not suitable for cultivation).
- Maximum day temperature must be below 35°C.

Propagation – Traditionally through seed tuber.

- Each tuber has several eyes which (3-5 buds) develops into sprout.
- Potato tuber has a dormancy of 8-10 weeks.
- The multiple sprouting stage is the best stage for planting because it generate emergence of several strong stem.
- Storage under diffused light – To obtain seed tuber with numerous vigorous green sprouts.
- In India, potato is raised when day temperature is below 35°C and night temperature is 20°C or less.

- Sunny days and cool nights favour better growth and higher yield.
- Most of the potato genotypes do not tuberize when night temperature is more than 23°C. Hence, potato is grown as a summer crop in the temperate regions and as a winter crop in the tropical and the sub-tropical regions.
- It can be grown upto an altitude of 3000 meters above sea level. For growing potato meant for processing, the ideal temperature during tuberization is 14-20°C.
- At this temperature, the sugars formed are converted into starch by the enzyme called 'starch synthetase'. This is also the ideal temperature for translocation of photosynthates from source (leaves) to sink (tubers) resulting in higher dry matter accumulation in tubers.
- At low temperature, the enzyme 'invertase' is active which is responsible for accumulation of sugars in the tubers, resulting in low dry matter content.
- At higher pH, incidence of scab, a serious disease of potato, is more.
- Tuber are taken out 10-15 days before planting from cold storage and kept for 24 hours in pre cooling chambers.
- A stem number of 30 main stem have been found to be optimum.
- Potato is primarily propagated through tubers and appropriate tuber size for seed is 35-40 g.
- It can be obtained by using 40 g tuber at 60 cm distance between row and 20 cm between tubers.
- Tuber cutting is not recommended specially when producing a seed crop because it increases the incidence of damages of transmitting viruses and bacteria.
- Seed tubers accounts for nearly 40-50% cost of potato cultivation and being vegetatively propagated its quality are most important in production.
- In India prior to 1960, potato seed used to be imported to replace the local seed stocks of various regions.
- N. Hills used to be the major source of healthy seed production for planting potato crops in plains.
- Seed plot technique – For producing healthy seed in Northern India plains Punjab, Haryana, and Uttar Pradesh, important potato seed producing state in plains.
- The high hills were the traditional source of healthy seed as population of aphid remains low due to low temperature therefore incidence of virurses was quit low in hill grown seed.
- Hills account for only 5% of the area under potato.
- Seed Plot technique – Consist of raising the healthy seed crop during low aphid period, available in Northern plains during October - January and cutting of hulms and harvesting the crop before crossing damage level. (20 aphids/100 compound leaves).
- Tubers dip treatment with 3% boric acid for 30 minutes after harvest and before storage of produce is done in seed production technique (SPT).
- CPRI produces breeder seed (nearly 2600 tonnes) at four regional stations.
    1. Jalandhar (Punjab).

2. Modipuram (UP).
3. Patna (Bihar).
3. Gwalior (MP).

- Seeds are produced in North-Western plains and kept in cold storage during April September.
- Post cold storage is necessary to get rid of the sweating problem and brings abou respiratory equilibrium of tubers.
- N. Hills crop harvested in September – October.
- For breaking tuber dormancy tubers are soaked in 1% thiourea for 1 hour.
- Tuber size –25-30 mm or 25-75g- Better size.
  35-40 mm or 45-50 g. – Ideal.
- Plant population
  35-45 mm 60 cm × 20cm – 23,000 plants/ha.
  45-55 mm 60 mx 25cm – 67,000 plants/ha.
  25-35 mm 60 cm × 15cm – 1,11,000 plants/ha.
- **Seed rate** – 20-25 q/ha (15g)
  25-30 q/ha (30g)
  30-35 q/ha (45g)
- Dehaulming – Removal of the top portion of potato in the seed crop to avoid th infestation of virus carrying insect vectors is called as dehaulming. It should b done either by cutting the haulms or killed by spraying of total killer herbicides eg Peraquat.
- When aphid population reaches to critical number (20 aphids/100 leaves) chance for viral infection becomes very high.

Planting method and time

- Ridge and furrow method is the most popular method.
- In the plains the best time of planting is when the maximum temperature is betwee 30-32°C and 10-20°C.
- In hills the best time of planting is when maximum temperature is about 20-22°C an minimum temperature is about 12-15°C.

Planting time

Autumn / winter / spring – In plains
Summer / autumn – In hills

- Main crop in Gangatic plains is planted in October-November.
- Dhaincha and Sanai – Used for green manuring, reduces N, P and K requirement o crop by 20-30%, improves yield of potato 3 tonnes / ha.
- "N" is the most important nutrient for potato (120-150kg). N application increase the plant growth, leaf area, tuber number and tuber size.
- Max "N" response is recorded in alluvial soils.

- "Phosphorus" deficiency is more pronounced in acidic hill soil.
- "Phosphorus" deficiency leads to dull dark green leaves without lusture.
- "K" deficiency causes shortening of internodes, leaves become dark-bluish green. In mild "K" deficiency, terminal leaves show-bronzing accompanied with necrotic spots.
- Response to "K" is highest in alluvial soil.
- Earthing up is done – 40 days after sowing (DAS).
- Both Calcium ammonium nitrate (CAN) and ammonium sulphate are good source of "N" in potato crop.
- Irrigation is withdrawn 10 days before harvesting.
- Critical stage – germination, tuber formation, earthing up and bulking.
- Late blight attack on stem, leaves and tubers, appears in crop in the hills every year but in the plains it occurs occasionally.
- Brown rot cause wilting of plants hence it is also known as bacterial wilt.
- **Aphid** – Transmit viral disease.
- **Leaf hoppers** – Transmit mycoplasma diseases.
- Cyst nematode is present in the Southern hills only.
- True potato seed (TPS) is the natural botanical seed produced by the potato plant.
- About 100-120 g true potato seed is enough to raise a seedling crop or if the commercial crop is to be produced using seedling tubers.
- The produce of 40-45 g True Potato Seed (TPS) is enough to plant one-hectare crop.
- Hundreds of these tiny seeds are contained in the berries produced in dozens by each plant.
- TPS can be used as an alternative to seed potato tubers.
- TPS is usually sown in seedbeds 3-4 weeks prior to the potato-planting season.
- The plants in the beds produce small tubers, also called tuberlets, which are planted in the field the same way as the conventional seed tubers.
- The TPS has many advantages that include freedom from diseases, easy transportation, storage and low cost.
- The cost of 125 g TPS for one hectare is just 10 per cent of 20-30 quintals of tuber seed.
- Next year TPS provides better disease resistance due to higher heterogenity in population
- Water requirement – 350-550 MM.
- Water forms nearly 80% of the potato tubers.
- First irrigation is given one week after planting and subsequently as and when required.
- Critical stages of the crop development include stolon formation; and tuber initiation and development.
- The late sown crop in plains should be harvested latest by April end to avoid high temperature and charcoal rot infection in tuber.

- Harvesting is done before the temperature rise above 30°C.
- Harvesting is done by potato digger.
- Nearly one fifth of the total potato production in the country is used as plantin material.
- Drying and curing of skin – 10-15 days in heaps in cool place.
- Tubers are then graded normally in four sizes i.e. small (< 25 g), medium (25-50 g large (51-75 g) and extra large (>75 g).
- After grading potatoes meant for use as seeds next year are treated with 3% bori acid solution for 30 minutes for protecting against soil borne pathogens like blac scurf, common scab etc before storing in bags.
- Traditionally, disease (virus) free potato seed is produced at higher hills where aphi population remains low.
- **Storage** – 2-4°C with high relative humidity.
- The low temperature checks sprouting and rottage, whereas high R.H. reduce weight loss in tubers.
- The total cold storage capacity in the country is adequate only to store nearly 35-40% of total potato production.
- Value Added products – Chips, flakes, French fries, finger chips, granules, disc cubes and flour etc.
- For best yield, potato needs long day condition during growth and short day conditio during tuberization.
- Solanin – 5 mg/100g.
- If solanin is greater than 20 mg/100g, it is unfit for consumption.
- Protein – 1.6%
- Vitamin C– 17 mg/100g.
- Potassium – 568 mg/100g.

Nutrient content in Potato

Physiological disorders –

1. **Internal Brown Spot:** Appears particularly in light sandy soils which are nc irrigated regularly (moisture deficiency) .Variety Craig's defiance is prone to this.
2. **Black Heart:** When tubers are stored at 35-40° C inner tissue break down. Due t poor oxygen availability, tubers develop an internal brown discolouration, whicl eventually becomes black. The discolouration occurs in an irregular pattern, usuall with a distinct line between healthy and affected tissues.
   (a) Lack of O_2 is the probable cause (Poor ventilation).
   (b) At high storage temperature accumulation of CO_2 at the center of tuber migh contribute to damage by decreasing respiration.
   (c) High soil temperature and waterlogged soils contribute to black heart developmen in the field.
3. **Hollow Heart:** (Excess N_2 leads this disorder) –Irregular cavity in the center o tuber might contribute to damage, decrease respiration is probable cause. Excessiv

use of nitrogen and water application increases incidence of hollow heart whereas, application of potassium containing fertilizers, especially Potassium Chloride, reduces the incidence of hollow heart. It appears in variety which produces over size tubers e.g. – K. Jyoti and K. Sindhuri.

4. **Greening:** Exposure of tubers to bright light during post harvest handling or longer periods of low light intensity results in the development of greening in potatoes. It is associated with the formation of alkaloids such as 'solanine'. Solanine is also formed in response to bruising, wounding, sprouting and improper earthing up. Such potatoes have bitter taste and may result in temporary digestive discomfort. Peeling off green skin with a layer of the flesh removes most of the solanine content.
5. **Chilling injury:** Caused by low temperature of about 0°C.
6. **Freezing injury:** Caused by low temperature (Ice formation inside tuber when exposed to temperature between 1°C to 2°C). When cut in half a blue black discontinuous ring is observed in the vascular region.

"SWEET POTATO"

- Latin name: *Ipomoea batatas* L. (syn. *Batatas edulis*).
- Family: Convolvulaceae.
- Cultivars differ for ploidy level: 2n = 30 (diploid); 2n = 40 (tetraploid); 2n = 90, (hexaploid)).
- Dicotyledon, perennial with swollen, tuber-like roots.
- Sweet Potato occupies a prime place in terms of calories production / Unit area and time.
- It is grown as a starchy crop throughout the tropical, sub-tropical and frost free temperate climate zone in World.
- In world scenario more than 80% of the area under sweet potato cultivation and production lies in Asia.
- India hold 6th rank in total area of cultivation.
- Among Asian countries China ranks 1st in area and production.
- India with 1, 30,000 ha (0.11% of the gross cropped area in India) is the largest sweet potato producer in South Asia and 6th largest producer in the World.
- Productivity of India is very low (8.2 tonnes/ha) and is little more than half of the world average productivity.
- Orissa (40410 ha) is having maximum area under sweet potato cultivation followed by Kerala (20870 ha).
- In sweet potato production, Orissa (381000 MT) is the leader followed by Kerala (341000 MT).
- Orissa, UP and Bihar – Account for 89% area and 88% of production of sweet potato.
- It is used as vegetable and also in making chips.
- In USA about 60-70% of the sweet potato is utilized for human food.
- It is a source of starch and used in textile, paper, cosmetics and glucose.

- *Ipomoea trifida* – Probable ancestor of cultivated sweet potato.
- The skin colour ranges between red, purple, brown and white.
- The flesh colour ranges between white, yellow, orange and purple.
- Sweet potato was introduced in Europe, probably in Spain at least 60 years before Irish Potato.
- A tuber similar to sweet potato is cited in Vedic literature, generally it is considered to have been introduced in India by Portuguese in early 16th century.
- It contains 10% starch and 3-6% sugar.
- Sugars, primarily sucrose, glucose, fructose and maltose constitute 6-14 per cent of the dry matter content.
- Starch content, of which 20 – 28 per cent is amylose and the rest amylopectin, constitutes 50-80 per cent of the dry matter content.
- β-carotene content of tubers varies from 0.013 mg in cream fleshed (Indian types) tubers to as high as 8.8 mg per 100 g in orange-fleshed (American types) tubers.
- The mineral composition of root tuber is calcium 30 mg, magnesium 24 mg, potassium 373 mg, sodium 13 mg, phosphorus 49 mg, chlorine 85 mg, sulphur 26 mg and iron 0.8 mg per 100 g.
- Tubers are good source of vitamin A, B and C.
- Yellow and orange fleshed variety contains more carotene.
- Sweet potato reached Japan from Okinanea.
- Sweet potato is an annual crop. The plant is perennial vine.
- Flowers are solitary or cymose (White to purple).
- Fruits are capsule (5-8 mm diameter), having brownish black seed with a very hard Testa.
- Anthesis occurs at 4-5 AM.
- Rich in vitamin A. varieties are distinguished by leaf shapes, skin colour and flesh colour.
- Most sweet potatoes are propagated either by planting sprouts that rise from the roots or by rooting vine cuttings.
- The sprouts, slips or draws, are obtained by planting stored tuberous roots in a hotbed and are cut for transplanting to the production field when they have six to eight well-developed leaves and a good root system.
- Sweetpotato is propagated through stem or root cuttings or by adventitious roots called slips that grow out from tubers during storage.
- 90-100 kg tubers are used in 700-800 square meter area in January-February for raising vines in nursery beds for planting one hectare.
- Sixty thousand to seventy thousand vine cuttings, 15-25 cm long, are sufficient for one hectare.
- At planting, half of the length of the cutting is buried in the soil.
- The optimum time of planting is from April to July in north Indian plains and September-October in east Indian plains

- Sweet potato requires a long warm and humid growing season for optimum growth and tuber development.
- Plenty of sunshine and moderate rainfall with warm nights and days are the best suited for its cultivation.
- Optimum growth temperature: 21°to 29°C; chilling injury occurs when soil temperatures drop below 10°C.
- Harvested 130 to 150 days after transplanting.
- Mechanically harvested like Irish potatoes, but are more susceptible to bruising.
- Harvested tubers are often ‹cured› for several days at 29°C and 85 to 90% relative humidity to promote suberization and healing of bruises and/or cuts.
- The cured tubers are stored at 13° to 15°C and 85 to 90% relative humidity.
- If properly cured and stored, disease-free roots can remain edible for 6 to 7 months.

Varities	**characteristics**
H 42 (CTCRI, Thiruvanathapuram)	This is a hybrid between Vella Damph and Triumph. The plants are semi-spreading with unifoliate leaves and serrated leaf margins. Tubers are fusiform, medium sized with pink skin and white flesh. The yield varies from 200-250 q/ha. It is tolerant to potato weevil.
H 268 (CTCRI, Thiruvanathapuram)	This is a double cross hybrid. The plants are semi-spreading with 3-5 lobed dark green leaves. The tubers are big, fusiform with pink skin and light yellow flesh. Tuber yield varies from 200-250 q/ha.
Sree Nandini (CTCRI, Thiruvanathapuram)	It is an early maturing selection from 'S 32', an introduction from Sudan. It takes 100-105 days to reach marketable maturity. The tubers are medium sized with white skin and white flesh. Tuber yield varies from 200-250 q/ha.
Sree Vardhini (CTCRI, Thiruvanathapuram)	It is a selection from an open-pollinated progeny of 'S 13'. It is also an early maturing variety and takes 100-105 days to maturity. The plants are semi-spreading and leaves are simple with 3-5 shallow lobes. Tubers are medium sized with pink skin and light yellow flesh having carotene content of 1200 IU. Tuber yield varies from 200-250 q/ha.
Punjab Sweetpotato-21 (PAU, Ludhiana)	It is selection from a local material. The vines are medium long with thick stem and 4.5 cm long internodes. Leaves are dark green with purple tinge, broad and lobed. Tubers are about 20 cm long, 4 cm thick having deep red skin

Crop Improvement –

Clonal Selection (Which are self-incompatible)

1. Pusa Saffaid
2. Pusa Lal
3. Pusa Sundhuri – Rich in carotene.

4. Jawahar Sakarkand 114- Extra early variety.
5. Shree Nandini – Superior cooking quality.
6. Shree Vardhini – Rich in carotene (1200 IU/100g.).
7. Kalmegh – Assam, very early variety.
8. Samrat
9. Kiran – AP, spreading vine.
10. Gouri – Orissa
11. Sankar – Orissa
12. Varsha
13. Bhuban – Orissa
14. Shree Bhadra – Excellent trap crop for root knot nematode.
15. Rajendra Sakarkand 35-
16. Rajendra Sakarkand 43 - Moderately resistant to cercospora leaf spot and tolerant to weevil.
17. Rajendra Sakarkand 47- Tolerant to sweet potato weevil and cercospora leaf blight.
18. Co1, 2, 3 - From Tamil Nadu.
19. Konkon Ashwini

- Sweetpotato mosaic is a viral disease.
- The primary infection is caused by the infected propagation material where as aphids (Myzus persicae) cause secondary infection.
- Feathery mottle disease is a viral disease characterized by ring spot and leaf spot on foliage and russet crack and internal cork on the tubers.
- The primary infection is through the infected planting material where as aphids (Myzus persicae) cause the secondary infection.
- Sweetpotato weevil (Cylas formicarius) is the most serious pest of sweetpotato.
- Adult weevils are antlike and feed on all plant parts but prefer roots

82. "CASSAVA OR TAPIOCA"

- The word tapioca was derived from tipioca, the 'tupi' Indian name for the meal.
- Major starch crop of the tropics.
- Perennial and dicot shrub.
- Monoecious and cross pollinated crop.
- Female flowers open about 10 days before the male phase.
- Cassava grown from seed and develops a tap root system.
- Main constituent of tuber – Starch.
- Bitter principle – Cyanogenic glucoside (HCN).
- Yellow colour of flesh is due to Carotene.
- Cassava is a photo insensitive crop.

- Introduced into India by Portuguese during 17th century.
- Shri Visakam Thirunal encouraged cultivation of popular variety to overcome food shortage.
- Nigeria – Major cassava growing country (contribute 20% of total area and 20% of total production of world).
- In India Tamil Nadu (89610 ha) is having maximum area under tapioca cultivation followed by Kerala (54730 ha).
- In tapioca production, Tamil Nadu (2862000 MT) is the leader followed by Kerala (1725000 MT).
- In India cassava was first introduced in Kerala.
- Tamil Nadu has highest productivity of tapioca i.e. 38 tonnes/ha.
- Cassava is grown as rainfed crop in Kerala and Andhra Pradesh, but in Tamil Nadu it is grown both as irrigated and rainfed crop.
- Cassava is highly heterozygous in nature.
- Vascular blue lining or "Vascular streaking" is present in Cassava.
- Important food product from cassava is Sago.
- Gari, Farsirla, Fulgar and Macroni – Proceed product of Cassava.
- About 4.16% of total area and 7.79% of total production under vegetable is shared by cassava.
- Improvement research work on cassava was started since the establishment of Central Tuber Crop Research Institute (CTCRI), Thriuvanthpuram (1963).

Varieties –

1. Sree Harsha (1996) – Triploid clone (Diploid × Tetraploid Shree Shaya) maximum starch content of 39.05%, ideal for industrial purpose.
2. Sree Prakash (1987) – Resistant to cercospora leaf spot, yield 35-40t/ha, drought resistant, starch content – 30%.
3. Sree Jaya (1998) – Good cooking quality, suitable for low land cultivation.
4. Sree Vijaya (1998) - Short duration (6 months), suitable for low land cultivation.
5. Sree Sahya (H2304) (1977) – Multiple Cross Hybrid, nonbranching, involving five parent, yield – 35-40 t/ha.
6. Nidhi – Suitable for sandy loam soils.
7. Sree Vishakam – Hybrid variety.
8. H-165 (1971) – (Chadayamangalam Vella × Kalikalan), non branching type.
9. H-97 (1971) – (Manjavella × Brazilian Seedling).
10. H-226 – Highly susceptible to CMD, became dominant variety in cassava growing areas.
11. Co1 (1977) - Starch 35%.
12. Co2 – Starch 36.4%.
13. Co3 – Yield 42.6 tonnes/ha, starch 35.5%.

- Cassava is the drought tolerant crop but cannot withstand frost and growth ceases at temperature below 10-12°C.
- The roots, branches and leaves of cassava contain a cyanogenetic glucoside called phaseolunatin, which breaks down upon harvest into toxic hydrocyanic acid, acetone and glucose by the enzyme linase.

Propagation –

- Cassava is a typical tropical plant. It grows better between 20°N and 20°S latitudes.
- In general, the crop requires a warm and humid climate.
- The highest root production is expected when the average temperature is 25-27°C.
- The growth ceases completely when temperature falls to 10°C. The plant makes good growth when the rainfall is abundant.
- But it can be grown in areas receiving annual rainfall as low as 500 mm and as high as 5000 mm.
- Cassava is a short day plant and the optimum light period is about 12 hours.
- Longer light hours inhibit translocation of photosynthates to the roots.
- Cassava is propagated through stem cuttings that are 2-3 cm thick and 15-20 cm long.
- Cuttings are prepared by leaving about 10 cm from the base and 20-30 cm from the top.
- The cuttings are planted by hand or by planting machines.
- Hand planting is done by one of the three ways: vertical, flat below the soil surface or tilted to make an angle of 45° with the ground.
- Under low rainfall conditions, vertical-planting results in desiccation of the cuttings, while in areas of higher rainfall, flat planted cuttings may rot.
- In general, flat planting 5-10 cm below the soil surface is recommended in dry climates.
- Normally, cassava is cultivated as a rainy season crop. The best time of planting is the beginning of the monsoon season, especially for plantings meant for starch production.
- Under irrigated conditions, it can be planted any time of the year.
- In light soils, planting is done on flat beds so that it does not suffer from water logging.
- In heavy and poorly drained soils, planting is recommended on ridges.
- Planting the stakes deeper results in swelling of the mother stem and reduction in yield of root and size.
- Vertical planting is preferred because more uniform callus tissue formation takes place.
- Spacing – 75 × 75 cm – Non branching type.
  90 × 90 cm – Branching type.
- Harvesting is done after 5-6 months after planting in North India and 4 months after planting in South India.

- Harvesting is done after 3-4 months for short duration variety.
- Delay harvest results in cooking quality deterioration.
- Depending upon the variety, sweetpotato can be harvested three and a half to five and a half months after planting.
- Most varieties of cassava take 10-11 months from planting to marketable maturity. However, some early maturing varieties take only 7-8 months.
- Delay in harvesting of tubers leads to deterioration in quality.
- Most cassava is harvested by hand, lifting the lower part of the stem and pulling roots out of the ground and then removing them from base of the plant by hand.
- The upper part of the stem with leaves is removed before harvest.
- Yellowing of the leaves indicate maturity of the crop.
- Tubers from early harvest are comparatively soft and suitable for table purposes.
- Yield – 35-40 t/ha. - High yielding variety.
  28-30 t/ha. - Short duration variety.
- About 70% of total cassava production is used as food either directly or in processed form.
- Parboiled chips are a method of processing cassava tuber for storage (these chips are hard due to the partial gelatization).
- *Rhizophus orysal*, *Bacillus* sp., *Penicillium* sp. - These organism are responsible for rotting of cassava.
- Drying is the main mode of processing of cassava tuber to enhance its post-harvest storage.
- Curing in cassava is done for 3-4 days at 25-40°C and 80-85% R H.
- Cassava tubers are rich in carbohydrates (80-90% on a dry weight basis) and 20-30% starch on fresh weight basis.
- Cassava contains 70% ethanol soluble sugar and 1-6% free sugar.
- Cassava cultivars are divided into three categories based on cynide content in the edible portion of the root.
  1. Sweet or Nontoxic (<50 mg/kg. fresh weight)
  2. Intermediate (50-100mg/kg fresh weight)
  3. Bitter or toxic (>100mg/kg fresh weight)
- Cassava plant accumulates cynogenic glucosides – liamarin and losaustralin which are synthesized from amino acid valine and iso-leucine, respectively.
- These glucosides release HCN on hydrolysis.
- Enzymes responsible for hydrolysis are lina marinase, b-glucoside and hydroxynitrile lyose.
- The cynogenic potential of cassava leaves is 5-20 times higher than that of edible parts.
- Depending upon climate, soil type, variety and other production factors, a yield of 50-200 quintals per hectare is obtained.

- However, when the crop is given attention and recommended cultural practices are adopted, it is possible to obtain tuber yield upto 400 quintals per hectare.
- Cassava mosaic is a viral disease and is most serious of all cassava diseases.
- It spreads mainly through the infected planting material. The secondary infection is transmitted through whitefly (Bemsia tabaci).
- Variegated grasshopper (Zonocerus variegates) feeds on cassava leaves, petioles and tender stems
- The pest damage is more common on older plants than on younger plants, and is more severe in the dry then in the wet season.
- The insect is also controlled by parasitic fungi called Metarhizium anisopliae var. acridum. One of the commercial formulations of the fungi is known as 'Green Muscle'.

83. "COLOCASIA OR TARO"

- The root tubers are 2.5-10 cm in diameter, 10-40 cm long and are borne in clusters of 4-8 at the stem base.
- Two type of colocasia are there -
  1. Eddoe – *Colocasia esculenta var – antiquorum* – Arvi.
  2. Dasheen – *Colocasia esculenta var- esculenta* – Bunda.
- The eddoe type is commonly called arvi and dasheen as bunda.
- Eddoe type is most prevalent as vegetable.
- In 'dasheen' types, the corms are cylindrical and large.
- In 'eddoe' types, the corms are small, globoid and are surrounded by several cormels (stem tubers) and daughter corms.
- Africa rank 1st in both area and production followed by Asia.
- Colocasia tubers are rich in starch.
- Leaves and petioles are also used as vegetable.
- Fresh corms of taro have about 65 per cent water, 13-29 per cent carbohydrates and 1.4-3.9 per cent protein
- Colocasia tubers contain more protein, minerals, phosphorus and iron compared to other tuber crops.
- Tubers of cassava are rich source of carbohydrates (starch), minerals and vitamins.
- About 90 per cent of the dry matter content is starch.
- The starch has 20 per cent amylose and 80 per cent amylopectin.
- Digestibility of starch in raw state is 48 per cent and after cooking 78 per cent.
- Eddoe type have bigger cormels that the mother corms whereas Dasheen types have smaller cormels than the mother corms.
- Taro is recommended for gastric patient.
- Poi, a dish made by mashing bodied corms and by making a viscous fluid is used to settle the stomach.
- Colocasia responds to K fertilizer better than nitrogen and phosphorus.

Varieties –

1. Sree Rashmi (C-149)
2. Sree Pallavi (C-266) suitable for southern zone.
3. C - 16
4. Satamukhi
5. Satamukhi suitable for western zone.
6. Sahar Shamukhi
7. White Gauriya
8. Kakakachu
9. Panchamukhi
10. NDC1, NDC2, NDC3.
11. Saharshamukhi suitable for eastern zone.
12. Kadma
13. Muktakeshi
14. Nadia local
15. Ahina Local telia
16. Jhankhri
17. ML 1 (ICAR NEH Region, Umiam) it is a selection from local germplasm of Meghalaya. Average yield is 250-270 q/ha. It is moderately resistant to leaf blight.
18. ML 9 (ICAR NEH Region, Umiam) it is a selection from local germplasm of Meghalaya. Average yield is 200-220 q/ha. It is tolerant to leaf blight.

- Propagation - By small cormels (20-25g weight), 1 tonne planting material is enough for 1 ha.
- Plant spacing is maintained at 70-100 cm between rows and 50-90 cm between plants.
- Ideal time for planting – Rainy season.
- It is a warm season crop and flourishes in areas receiving plenty of rainfall.
- It requires average daily temperature above 21°C for better yields.
- Taro leaf hopper (*Tarophagus proserpina*) causes great crop losses in its cultivation.
- The irritant properties of taro are due to the calcium oxalate crystals in corms and leaves of *C. escultenta.*
- Calcium oxalate content is 0.008%.
- Major crystal types are raphides and druses.
- Tuber becomes very hard to cook due to water stagnation in field in both edode and dasheen types.
- Harvesting 120-150 days after planting.
- In height of plants and yellowing of leaves are indications of maturity.
- Yield – 30-40 tonnes/ha.
- Dasheen mosaic virus (DMV) is spread by Aphids is characterized by chlorotic and feathery mosaic pattern on leaves.

YAMS"

- Three species of yams are important and cultivated widely.
  1. Greater yam/Ratalu (*Dioscorea alata*)
  2. Lesser yam/Chinese Yam/Asiatic Yam (*Dioscorea esculenta*) Both are major yams of India.
  3. White yam/water Yam (*Dioscorea rotundata*) is extensively cultivated in Africa.
- Africa alone prod· ce 90% tuber and cover 95% of the area.
- **Greater yam:** Popularly known as ratalu, it possesses four or more wings on the stem which twines to the right, tuber skin is black or brown.
- **Lesser yam:** Popularly known as Suthani or Kangar, is usually thin, often pubescent and spiny which twines to the left. Tuber skin yellow-brownish with white flesh. After cooking sweeter than other yam.
- **White yam:** It has wingless stem which is twining to the right.
- *D. composita, D. floribunda, D. spiculiflora, D. friedrichsthalli and D. deltoida* – Exploited for pharmaceutical purpose and are non-edible.
- Yams contain 18-20% starch (contain mainly anylopectin).
- The edible part is underground tuber that varies in size and shape, averaging 1.25-3.5 kg.
- The major edible species of African origin are white Guinea (*D. rotundata* Poir.), yellow Guinea yam (*D. cayenensis* Lam.), and trifoliate or bitter yam (*D. dumetorum Kunth*).
- Yams are under-utilized annual or perennial climbers with heart shaped leaves.
- The edible part is underground tuber that varies in size and shape, averaging 1.25-3.5 kg.
- They are one of the important staple food crops of Africa and are cultivated by small farm holders to secure their own subsistence.
- Fugu is an important product made from yams.
- Yam flour is also used for human consumption as "Kokonte" in northern zones of W. Africa.
- Fresh yam tubers are also peeled, chipped, dried and milled into flour that is used to prepare dough called amala or telibowo
- *D. alata is* used for making wafers, cripies and biscuits.
- Diosgenin present in many wild species is used in drugs for oral contraceptives.
- Yams are rich sources of carbohydrates (contains 15 per cent starch) and are valuable sources of vitamins C, B, folate and minerals like potassium, iron and magnesium.
- They also contain 1.9-3.9 per cent protein on fresh weight basis.
- Yams are cooked to remove bitter substances including dioscorine, an alkaloid and diosgenin, one of the triterpenes.
- Dioscorea genus is dioecious in nature.
- Fruit type – Trilocular capsule.

- In *D. alata* no seed dormancy exists.
- 2x=20 in most primitive section but in *D. alata* majority of male clones are reported to be tetraploid (2x=40).
- Lesser yam (*D. esculenta*) is sweet and matures early than other yams.
- In *D. alata* bulb is a minor storage organ while in *D. bulbifera* the bulbiles is the major storage organ.
- Amylose content ranges from 10-25%.
- Diosgenin, a precursor of progesterone and cortisone is produced commercially from yams and used to treat arthritis, asthma, eczema, regulates metabolism and controls fertility.

Improved Varieties –

A. Greater yam/Ratalu (*Dioscorea alata*)

- **Sree shilpa:** First hybrid variety released in *D. alata*, released from CTCRI.
- **Sree Roopa:** Cooking quality is excellent, for Kerala. It is developed by clonal selection. Tuber yield varies from 250-300 q/ha.
- **Sree Keerthi:** Excellent cooking quality and taste, for Kerala. It is developed by clonal selection. Yield potential is 300 q/ha. It is developed by clonal selection. The starch content varies from 20–22 per cent.

B. Lesser yam/Chinese Yam/Asiatic Yam (*Dioscorea esculenta*)

1. **Sree Latha (CTCRI, Thiruvanathapuraam):** Suitable for AP, Assam, Kerala, Maharastra, WB and Bihar. It is developed by clonal selection. Average tuber yield is 250 q/ha
2. **Sree Kala (CTCRI, Thiruvanathapuraam):** Excellent culinary quality. It is a selection from the material introduced from Puerto Rico. It possesses excellent cooking quality. Average tuber yield is 200 q/ha.
3. Konkan Kanchan – Suitable for Maharashtra.

C. White yam (*Dioscorea rotundata*): Winged Yam

Winged yam, also known as water yam, is a vigorous twining herbaceous vine. It is the most commonly cultivated of all the edible yams. Its stem is square shaped with the corners compressed into wings and hence the name winged yam. It is more tolerant to long dry spells than any other yam.

1. **Sree Subhra:** It is developed by clonal selection from 'Iwo', an introduction from Africa. Excellent cooking quality, yield 35-40 t/ha.
2. **Sree Priya:** It is developed by clonal selection from 'Umidika', an introduction from Africa. Yield 35-40 tonnes/ha., for Kerala.
3. **Sree Dhanya:** Dwarf bush variety, do not require stacking. The tuber yield varies from 200-250 q/ha.

Propagation: By tuber pieces/ small whole tuber.

- Lesser yam – Tuber weighing 125-150 g.
- Greater yam and white yam – Tuber weighing 200-250 g.

- Propagation by vine cuttings is not advised because of slow tuber production.
- Yam tuber have long dormancy period (2-3 months) depending on species which is major problem in yam cultivation.
- The period of dormancy varies from 4-8 weeks in lesser yam, 12-14 weeks in greater yam and 14-16 weeks in white yam.
- Dormancy is broken by dipping tubers quickly in 4-8% solution of ethylene chlorohydrin followed by dry storage.
- To delay sprouting in storage, soaking of tuber for 10 hrs in 0.1% solution of malic hydrazide is done.
- Spacing – 75 × 75 cm – For lesser yam.

  90 × 90 cm – For Greater and white yam.
- Anthracnose of Yam (*Colletotrichum gloeospoioides*) was first found in Rajasthan, responsible for 70-80% loss in yield, severe form in *D. alata*.
- Harvesting: The lesser yam varieties take lesser time and reach marketable maturity in 7-8 months after planting where as the greater yam and white yam varieties take longer and reach marketable maturity in 9-10 months after planting.
- **Yield**- 25-30 tonnes/ha – Lesser yam.

  33-40 tonnes/ha – Greater yam.

  18-25 tonnes/ha – White yam.
- Tuber top portion is known as proximal.
- Tuber middle and bottom portion is known as distal.

84. "ELEPHANT FOOT YAM"

- It is also known as Amorphophallus or Suran or Jimmikand.
- Inflorescence – Spadix.
- In Japan it is used as food crop.
- Recommended as blood purifier.
- Paste of tuber is used to reduce pain in arthritis.
- Acridity causes irritation in mouth and throat which is due to the presence of calcium oxalate.
- A minimum of 22°C temperature during the daytime and 19°C during the night is required for proper growth and development.
- It prefers partial shading and is therefore commonly grown as an intercrop in coconut grooves and orchards.
- A well-distributed rainfall of 1000-1500 mm is helpful for optimum crop growth and tuber yield.

Varieties –

1. Gajendra – Cooking quality is very good, free from acridity, free from daughter corms, Yield 400-600 q/ha, suitable for East and North India.

2. Santragachi – Having several rough daughter corms (East India), slight acridity, suitable for East India.
3. Sree Padma – Improved variety.
4. Kovvur

ropagation: Cut or small tuber (high yielding and non acrid varieties)

- 50-100 g weight of tuber.
- Whole tuber should be preferred over cut tuber to minimize rotting.
- Tuber dormancy can be broken by treating them with thiourea (0.1%).
- GA_3 and Ethrel are also effective in breaking dormancy.
- Elephant foot yam has dormancy period of 3-7 months.
- Best planting time – Feb-March.
- It is a long duration crop.
- Harvesting – 7-8 month after planting (MAP) (November).
- Yield – 50-100 tonnes/ha.
- Water stagnation during crop growth results in tuber that is very hard to cook.
- Mealy bug attack corms of elephant foot yam, both in the field and in stores.

35. "YAM BEAN"

- B. N. – *Pachyrhizus* spp.
- Family – Fabaceae.
- Origin - South America.
- Edible part – Root.
- Populartly known as Potato bean and Mishrikand.
- High sugar content in tubers imparts sweet taste when eaten as raw.
- Its starchy conical (as turnip) shaped fleshy tubers are eaten.

  Varieties: Two types of cultivars are available in market.

  1. Mexican Type – Tuber is large 500-700g., less sweet compared to local, they have tendency to develop cracking on their tubers, less sweet preferred in Indian market.
  2. Local Type: Tuber is small 200-300g., sweeter, white flesh, less fiber, no cracking.

- Rajedra Mishrikand 1– Improved variety, popular in North Bihar.
- Yield 40 55 t/ha, matures in 110 140 days.
- Propagation – By seed (20-60 kg/ha).
- Yam beans bear pods 7-15 cm long with 8-10 brownish yellow to red seed.
- Mature pods containing seeds are source of a toxic substance "rotenone" which is harmful to grazing cattle.
- Spray of 2, 4-D (50 ppm) at flower initation stage induces flower thinning and increases tuber yield.

- Harvesting – Traditionally the trend is to harvest (110-140 DAS) the crop on the occasion of "Saraswati Puja" with the start of spring season because of market demand.
- **Yield:** 18-20 tonnes/ha - Local cultivars.
  36-40 tonnes/ha - Rajendra Mishri 1.
- Seed should be harvested after 240 days during March-April.
- Tuber cracking is the main problem due to delay harvest.
- Application of "K" reduces cracking of tuber.

86. "TANNIA"

- B. N. – *Xanthosoma sagittifolius.*
- Family – Araceae.
- Origin – South America.
- Edible part – Corms, leaves and petioles.
- Tannia or Xanthosoma is an herbaceous edible species.
- Suitable for intercropping in Banana.
- Popular in Maharashtra and Gujarat for making roll of leaves.
- Propagation – Corms and cormels (20-25 cm long).
- In Maharashtra and Gujarat where only leaves are harvested the main corms and carmels are left in the field.
- They act as new planting material.

87. "QUEENSLAND ARROWROOT / PURPLE ARROWROOT"

- B. N. – *Canna edulis.*
- Family – Cannaceae.
- Origin – Tropical America .
- Edible part – Tuber.
- Perennial herbaceous monocotylendon.
- Propagation – By rhizomes.
- Yield – 25-30 tonnes/ha.

88. "JERUSALEM ARTICHOKE"

- B. N. – *Helianthus tuberosus.*
- Family – Compositae (Asteraceae).
- Origin – Eastern United State.
- Edible part – Tuber.
- Cold-hardy, frost tolerant.
- Autumn-blooming sunflower.
- Harvested organ: Tuberous, potato-like roots (rhizomes).

- Native of N. America.
- Cultivated by the American Indians before the 16th century.
- Introduced to Europe in the 17th century.
- Two important sp. *H. annus* and *H. tuberosus* are economically important.
- Carbohydrate content – 8-19% (Inulin-Polymer of fructose).
- Jerusalem artichoke is commercial source of levulose used as sweetening agent by diabetic patients.
- Glucofructosans – From sucrose, are principle carbohydrate in tubers.
- Tuber also used for alcohal and beer beverage.
- Propagation – By small sets or tubers (50g. weight) (2-3 eye buds)
- Mammoth white French – Widely grown variety.
- Sutton's white – Widely grown variety.

89. "COLEUS"

- B. N. – *Solenostemon ratundifolius* (syn. *Coleus parviflorus*).
- Family – Lamiaceae.
- Origin – Tropical Africa, Asia and Australia.
- Edible part - Tuber
- Popularly known as Chinese potato or Koorka.
- Annual herbacoleus bushy plant.
- Sree Dhara – Improved variety, released by CTCRI.
- Propagation – By vine cuttings (10-15cm length).
- Yield 20-25 tonnes/ha.

90. "ARROWROOT/WEST INDIAN ARROWROOT"

- B. N. – *Maranta arundinacea.*
- Family – Marantaceae.
- Origin - Tropical America.
- Edible part – Rhizome.
- High quality starch (27.8%) is used as food for infants.
- Biscuits are known in every corner of India made by it.
- Its starch is also used as special glue and paste as a base for face powder and ice-cream stabilizer.
- Recently it is also used in production of carbonless paper for computer printout.
- Propagation – By small pieces of rhizomes (known as bits) (4-7 cm long, 2-4 nodes).
- 3 tonnes/ha planting material is enough.
- Inflorescence -Teminal bearing.
- Harvesting – 10-11 Months after planting.
- Yield – 4-7 tonnes/ha.

91. "GIANT TARO"

- B. N. – *Alocasia macrorrhiza* (Syn. *A. indica*).
- Family – Araceae.
- Origin – Asia (Srilanka).
- Edible part – Rhizomes and shoots (stem tuber).
- The fleshy aerial stem of giant taro is the primary product.
- The edible corm of tropical taro has higher starch content than either potatoes sweet potatoes.
- The species cultivated in Assam and West Bengal as food crops are – *A. indica*, *macrorhiza*, *A. cucullata*.
- Propagation – By suckers, shoot tip and cormels.

92. "CHINESE WATER CHESTNUT"

- B. N. – *Eleocharis dulcis*.
- Family – Cyperaceae.
- Origin – China.
- Edible part – Tuber.
- Perennial leafless sedge commonly found in marshes and swams through the worl
- Edible tuber rich in starch (7.3%) and sucrose (6.4%).
- Starch grains are very large, upto 27 μ in size.
- In India it is sold in Culcutta under the name Singapuri Keysur.
- First cultivated in South East China.

93. IMPORTANT PEST AND DISEASE OF POTATO AND TUBER CROPS

Sr. No.	Pest and Disease	Symptoms	Management
1.	Greasy Cutworm/ Black cut worm (*Agrotis ipsilon*)	Young larvae feed on the epidermis of the leaves and older larvae come out at night and feed young plants by cutting their stems.	Set up light trap @1/ha c pheromone traps @12/ha t attract male moths. Spray insecticide lik chloropyriphos 20 EC @ lit/ha or neem oil @ 3%.
2.	Potato tuber moth (*Phthorimaea operculella*)	It is a pest of field and storage, larva tunnels into foliage, stem and tubers	Spray NSKE (Neem see kernel extract) @5% c quinalphos 25 EC @2ml lit of water to manage folia damage.

3.	White grubs (*Holotrichia sp)*	Grubs feed on roots and tubers voraciously during night time.	Dust Quinalphos 5% @25 kg/ha at 10 days after first summer rain.
4.	Tobacco Caterpillar (*Spodoptera litura)*	Young larvae firstly scrape the leaves and older larvae spread out and devour the leaves resulting in poor growth of plants.	Spray chlorpyriphos 20 EC 2 lit/ha or dichlorovos 76 WSC 1 lit/ha.
5.	Green Leaf Hopper (*Empoasca kerri)*	Tips of affected leaves become brown turn upwards and get dried up.	Spray dimethoate 30 EC or phosphomidon (Dimecron) 2ml/lit.
6.	Aphid (*Myzus persicae)*	Aphids suck sap of plants, as a result of which leaves turn pale and dry up.	Spray dimethoate 0.3%.
7.	Whitefly (*Bemisia tabaci)*	Yellowing and curling of leaf.	Avoid common solanaceous crops in the endemic areas and spray dimethoate 0.3%.
8.	Sweet potato weevil (*Cylas formicarius)*	Discoloration, cracking, or wilting of damaged wines. An infested tuber is often riddled with cavities or tunnels.	Follow crop rotation and use insect-free vines as planting material.
9.	Tortoise beetles (*Aspidiomorpha miliaris)*	Young grub scrape on the upper surface of the leaves and older grub and adults bite large round holes in the leaves.	Spray Carbryl @1.25 kg/ha at 21 days interval at the time of harvesting sweet potato.
10.	Cassava scale (*Aonidomytilus albus)*	Infest stem portion and leaves of affected plants discoloured and dry up.	Spray methyl demeton 0.25%.
11.	Cassava mealy bug (*Paracoccus marginatus)*	Causing deformation of terminal shoots, which become stunted "bunchy tops".	Spray malathion or fenitrothion @1ml/ha.
12.	Late blight of potato (*Phytopthora infestans)*	Water soaked spots appear on leaves, increase in size, turn purple brown& finally black colour.	Use resistant verities and protective spraying with mancozeb or zineb 0.2 % should be done to prevent infection of tubers.

13.	Early blight (*Alternaria solani*)	Brown-black necrotic spot-angular, oval shape characterized by concentric rings.	Use disease free seed tubers for planting and very early spraying with Zineb or captan 0.2% and repeating it for every 15 – 20 days gives effective control.
14.	Post-harvest tuber rots (*Sclerotium rolfsii*)	Yellowish brown/milky white Sclerotia appeared on the infected tuber.	Treating seeds with mercury compounds after harvest reduces tuber rot.
15.	Black scurf (*Rhizoctonia solani*)	Black speck, black speck scab, russet scab on tubers.	Treating seed tubers with mercuric chloride solution.
16.	Brown rot or Bangle blight (*Ralstonia solanacearum*)	Bacteria ooze coming on infected tuber surface and emits a foul odour.	Use resistant varieties and follow crop rotation.
17.	Bacterial Soft rot (*Erwinia carotovora*)	Black lesion appears on the base of the plant and browning of infected tubers.	Treatment of the seed tubers with TBZ + acetic acid + Carbendazim 1% for 15 minutes effectively controls the disease.
18.	Charcoal Rot (*Macrophomina phaseolina*)	In the field, brown to black, water-soaked lesions appear on stems and petioles. In storage two distinct zones become apparent within the infected tissue.	Properly cure sweet potatoes immediately after harvest to reduce the incidence of charcoal rot.
19.	Cassava Mosaic Disease (*Indian cassava mosaic virus*)	The virus spreads by infected setts and whitefly *Bemisia tabaci* and infected plants stunted, tuber splitting and yield reduction.	Management of whitefly by installing yellow sticky traps, removal of weed hosts, spray neem oil 3% or Methyl dematon @ 2ml/lit.
20.	Black Rot (*Ceratocystis fimbriata*)	Small, circular, slightly sunken, dark brown spots are the initial symptoms of black rot.	Apply a postharvest fungicide.

Spices and Condiments

C.N.	B.N.	Family	2x=	Origin Place	Chemical Content	Part Used	Major Growing (Country / States)	Propagation/ Seed rate	Yield
Black Pepper	*Piper nigrum*	Piperacea	128	Tropical Forest of W. Ghat of S.India	Piperine	Berry	Vietnam Kerala (94%)	Runner Shoot Cutting	800-100kg/ha.
Cardamon (Small)	*Elettaria cardamom*	Zingiberaceae	24	Tropical Forest of W. Ghat of S.India	α-Terpinyla-cetate and 1,8 Cineol	Capsule	Kerala (60%	Suckers	500 kg/ha.
Cardamon (Large)	*Amomum subulatum*	ZIngiberaceae	7	E. Himalayan	-	Capsule & Seed	Nepal & Sikkim	Suckers	450 kg/ha.
Ginger	*ZIngiber Officinale*	Zingiberaceae	22	S.E. Asia	Pungency-Oxymethylo phenol (Gingerol)	Rhizome	Kerala (25%)	Seed Rhizomes	15-25 Tonnes/ ha.
Turmeric	*Curcumal Longa*	ZIngiberaceae		S.E. Asia	Curcumin Yellow	Rhizomes	A.P.	Seed Rhizomes	25-30 Tonnes/ ha.
Coriander	*Coriandrum sativum*	Apiaceae	22	Mediterranean region	Linalool (Essential Oil)	Fruit	Rajasthan	Seed 10-15 kg Irrigated Condition 25-30 kg Unirrigated Condition 40-42 Kg For Veg	400-500 kg (Rainfall) 600-1200 kg (Irrigated)
Cumin	*Cuminum cyminum*	Apiaceae	14	Mediterranean region	Thymol., cuminol	Fruit	Guj.(I) Raj. (II)	12-15 kg/ha.	7-8 q./ha.

Fenugreek	*Trigonella foenum graecum*	Leguminacean	16	S. E. Europe	Diosgenin	Seed	Rajasthan (80%) A.P.	20-25 kg/ha.	9-10 q./ ha.
Fennel	*Foeniculum Vulgare*	Apiaceae	22	Mediterranean region	Anethole (70%) (Sweetness)	Fruit	Gujarat, Rajasthan	10-12 kg./ha.	9-10 q./ ha.
Dill	*Anethum graveolens*	Apiaceae		Eurasia	Carvone	Fruit	Gujarat, Maharashtra	5-6 kg/ha for line sowing 10 kg/ha for Broadcast	8-9 q./ha.
Celery	*Apium graveolens*	Umbelliferae	22	Abyssinia and Asia	d-limonene d-selinene	Leaves & Seed	Punjab	375-500 g./ha	330-560 q./ ha green leaves 175-275 Kg Seed/ ha.
Clove	*Syzygium aromaticum*	Myrataceae	44	Indonesia	Eugenol (70-90%) Eugenylacetate (5-12%)	Unopened Flower bud	T.N., Kerala	By Seed (Mother Clove)	2 kg/tree
Nutmeg	*Myristica fragrans*	Myristicaceae		Indonesia	Terbein (7-16%)	Dried Kernel Mace-Aril	Kerala, T. N.	By Seed Epicotyl Grafting	800 kg Nutmeg / ha. 100 kg Mace /ha.
Cinamom	*Cinnamomum verum*	Lauraceae	24	Sri Lanka	Eugenol (70-80%) Cinnamic aldehyde	Bark	Kerala, T.N.	By Seed	200-300 kg Dried Bark
All Spice	*Pimenta dioica*	Myrtaceae		West Indies	-	Fruit	Jamaica		

Vanilla	*Vanilla planifolia*	Orchidaceae		Mexico-Brazil	Vanillin	Fruit (Beans)	Malagasy Republic Kerala, T.N.	By Stem Cutting	300-400 kg cured bean /ha.
Saffron	*Crocus stativus*	Iridaceae		India	Cicrocrocin	Stigma	Only in Kashmir	By Corms	5 kg dried saffron / ha.
Curry Leaf	*Murrya koenigii*	Rutaceae	18	India	Ksenigin Muragin	Leaves Flower	Andhra Pradesh	seeds	1000 kg / ha. (From 4th Year)
Ajowan	*Trachyspermum ammi*	Apiaceae	18	Eqypt	Thymol	Seed	Rajasthan, Gujarat	3-4 kg/ha. Seed	70-80 g/ ha
Thyme	*Thymus vulgaris*	Labiatae	28/30	Egypt	Thymol	Leaf & Flower	USA	By Seed & Cutting	1100-2200 kg
Rosemarry	*Rosemarinus afficivalis*	Labiatae	-	Med. Region (S. Europe)	1,B- Cineol	Leaves & Flower	-	-	-
Kokum	*Garcinia indica*	Guttiferae	24	India	B-hydroxy acetic acid	Fruit	-	-	-
Arafoetida	*Ferulla foetida*	Apiaceae	-	-	Ferumine	Oleoresin gum	-	-	-
Juniper	*Juniperus communis*	-	22	Arctic South of Tropical America	a-pinene, Campene b-pinene, Sabinene, Myrcene a-phellandrene	Berry	Australia	Seedling and cutting	80 g/ha

- Spices are those plants, the product of which are made use as food adjuncts to add aroma and flavour.
- Condiments are also spices, products of which are used as food adjuncts to add taste only.
- Both spices and condiments contain essential oil which provides the flavour and taste.
- They are used whole, ground, paste or liquid farm mainly for flavouring and seasoning.
- They are about 63 spices grown in India.

Classification of Spices

A. Classification based on life cycle: –

(a) Annual – Coriander, Fennel, Fenugreek.

(b) Biennial – Onion, Garlic.

(c) Perennial – Ginger, Turmeric, Black pepper, Clove, Cardamom.

B. Classification based on growth habit: –

(a) Herb – Ginger, Turmeric.

(b) Shrub – Caper, Juniper.

(c) Climber – Black Pepper.

(d) Trees – Clove, Cassia, Nutmeg, Cinnamom, All spices.

C. Based on Economic importance.

(a) Major Spices: Black pepper, Ginger, Cardamon and Turmeric.

(b) Minor Spices: Cumin, Coriander, Fennel, Fenugreek.

(c) Seed Spices: Dil, Aniseed, Celery, Bishopweed

D. Based on Plant part used –

(a) Leaves: Cassia, Curryleaf, Chive.

(b) Bark: Cinnamon.

(c) Unopened flower buds: Clove.

(d) Rhizome: Ginger, Turmeric, Asafoetida (exudate).

(e) Seeds: Coriander, Cumin, Fennel, Fenugreek.

(f) Fruits : Cardamon, Chilli, Dill, Celery, Cumin, Black Pepper, Allspice, Vanilla (pod), Aniseed, Ajowan

(g) Stigma: Saffron.

(h) Buds: Onian, Garlic, Shallot, Leek.

(i) Kernel: Nutmeg.

(j) Root: Horse radish.

(k) Herbal Spices: Rosemary, Thyme, Parsely, Horse radish.

Botanical Classification:-

1. Monocotyledonous

(1) Family Liliaceae

1. Meadow Saffron – *Colchicum autumnale*

(2) Family – Amarylidaceae

1. Onion – *Allium cepa*
2. Garlic – *Allium sativum*

(3) Family — Iridaceae

Saffron – *Crocus sativus*

(4) Family – Zingiberaceae

1. Ginger – *Gingeber officinale*
2. Turmeric – *Curcuma longa, C. domestica*
3. Small Cardamom – *Electtaria cardamom*
4. Nepal Cardamom – *Amomum sublatum*
5. Bengal Cadamom – *A. aronaticum*

(5) Family – Orchidaceae

1. Vanilla – *Vanilla planifolia, V. fragrans*

2. Dicotyledenous –

(1) Family Myrtaceae

1. Allspice – *Pimenta dioicea*
2. Clove – *Syzygium aromaticum*
   Eugenia caryophyllus

(2) Family Labiateae

1. Basil – *Ocimum basilicum*
2. Mint – *Mentha arvensis*
3. Peppermint – *Menthe piperita*
   Spearmint – *M. spicata*
   M. viridis
4. Sage - *Salvia officinalis*

(3) Family – Lauraceae

1. Cassia – *Cinnamomum inners*
2. Cinnamon – *C. verum*
3. Tejpat – *C. tamala*

(4) Family – Piperceae

1. Black pepper – *Piper nigrum*
2. Long pepper – *P. longum*

(5) Family – Leguminoseae

1. Fenugreek – *Trigonella foenum graceum*
2. Karturi methi – *Trigonella corniculata*

(6) Family – Myristicaceae

1. Nutmeg – *Myristica fragrans* / mace

(7) Family – Guttiferae

1. Kokam – *Garcinia Indica* / Garcinia

(8) Family – Umbelliferae (Apiaceae)

1. Coriander – *Coriandrum sativum*
2. Cumin/White cumin – *Cuminum cyaminum*
3. Cumin Black – *Nigella sativa*
4. Fennel – *Foeniculum vulgare*
5. Celery – *Apium graveolens*
6. Dill/Sowa – *Anethum graveolans*
7. Ajowan / Bishop's Weed – *Carcum capticum, Trachyspermum ammi*
8. Parsely – *Petroselinum crispum*
9. Asafoetida – *Ferula asafoetida*
10. Aniseed – *Pimpinella anisum, Anisum vulgare, Anisum officinalis*
11. Caraway – *Apium carvi, Carum carvi, Seseli carvi*

(9) Family – Rutaceae

1. Curry Leaf – *Murraya koenigi*

(10) Family – Capparidaceae

1. Caper – *Capparis Spinosa* –unopened flower bud is used.

General about spices and Condiments –

- Indian share in world trade of spices is around 18%.
- Nearly 90-95% of the total production is consumed locally and the rest is exported.
- India produces spices on 2.0 million hectares with an annual production of about 2.3 million tonnes valued at about 4500 crores, contributing nearly 20% of world's production, 30% of the trade in terms of quantity but only 10% in terms of value.
- India is the biggest exporter of spices, annually exporting about 1.9 lakh tonnes of different spices and spice products, earning a foreign exchange of about 700 crores.
- India is known as "Home of Spices" because most of the spices are native to our country.
- Spice oil is obtained by steam distillation method.
- Largest area under spices – Rajasthan.
- Largest production of spices – Andhra Pradesh.
- India is the largest producer, consumer and exporter of spices in the world.
- Southern States produce 30% of Cardamon and 25-30% of pepper.
- Dry and arid areas of Gujarat and Rajasthan produce 80% of Coriander and 90% of Cumin produced in the world.
- India – 90% turmeric production of world.
- Jammu and Kashmir – Produces world best saffron.
- Punjab – 90% celery production of total production of the world.

- Black pepper and Turmeric constitute more than 50% of total export earnings among spices.

INDIA – Production facts

- Black pepper – 54 %
- Black pepper – 26.3% area under cultivation of world.
- Ginger – 70%
- Turmeric – 76%

KERALA alone produces

- Black pepper – 94%
- Cardamom – 60%
- Ginger – 25%

RAJASTHAN

- Fenugreek – 80% area
- Fennel
- Coriander

GUJARAT

Cumin is major produced spice.

A.P.: Turmeric, Chilli

- Major Importer and consumer of oleoresin – USA (50%).
- Foot rot, slow decline, Pollu beetle – Important disease and pest of Black pepper.
- Capsule rot, ''Katte'', Kokke Kandu, Thrips – Important disease and pest of Cardamom.
- Rhizome rot – Important disease of Ginger and Turmeric.
- King of Spices – Black pepper.
- Queen of Spices – Small Cardamom.
- Cochin ginger / Indian ginger – Best in the world.
- Clove 2nd most important spice next to black pepper in the world.
- Nutmeg, All spice, Juniper- dioecious in nature.
- Turmeric is antioxidant due to phenolic character of curcumin.
- Large cardamom – Shade loving plant (Seophyte).
- India alone contributes about 10-12% of the total world trade in spices.
- Aril – Commercially important part of Mace.
- Pistil – Commercially important part of Saffron.
- International Organization for Standardization (ISO) approved Indian Vernacular names, common english names as well as scientific names to help identify correct spices because spices in known by different vernacular names in different parts .

- "Spices and Herbs in food industry" – Book written by Lewis (1984).
- Oil of black pepper consists mainly of terpenes.
- Per capita consumption of spice in USA is ½ times of Indian consumption.
- "Ajowan ka Phool" is crystal form of thymol.
- Essential oil in dill is 1.5-4%.
- Male musk deer smell resembles to Ambrette seed.
- ISI - Now BIS (Bureau of Indian Standards).
- ISO - International Organization for Standardization is situated at Budapest, Hungary.
- "Vasco De Gama" was the first to reach Malabar Coast, Kerala (India) and took with him a cargo of spices and other Indian products.
- In the 16th century "Cloves" for instance were among the spices used to preserve food without refrigeration.
- Cloves contain a chemical called eugenol that inhibits the growth of bacteria.
- Clove is still used to preserve "Virginia Ham" food. Likewise later mustered and ground mustared were also found to have preservative qualities.
- Ministry of Food and Agriculture established in 1951.
- Govterment of India constituted a high level of committee -Spices Enquiry Committee (1951-53) which submitted report to the Govt. of India in October 1953.
- Indian Central Spices and Cashewnut committee – 1961.
- CPCRI, Kasarged (Kerala) established in 1970.
- All India co-ordinated Improvement projects for spices was established in December 1970.
- Actually started functioning – 1971
- Spices Development Council, Division of Horticulture and Department of Agriculture & Co-operation – 1966.
- All spices development Council – 1966
- Under the aegis of the Directorate of Marketing and Inspection (DMI), Ministry of Agricullture Govt. of India was set up in 1937 under the APGM.
- The first "Agmark" specification was formulated for Curry powder in 1956, for Cardamom and Chilli in 1962 and for Turmeric. Ginger, Coriander, Chilli powder, Onion, Garlic, Ambadi, Cumin, Poppy seed in 1964.
- Central Agmark Laboratory (CAL) – Nagpur (1967).
- CAL is tempo of quality control to assist growers, merchants and exporters of spices in quick compulsory pre shipment inspection and quality control.
- Whole spice grading units – Bombay, Cochin and Virudhunagar.
- The Central Agmark Laboratory served both as an Apex and National Appellate Laboratory.
- APGM – Agricultural Produce Grading and Marketing Act-1937. Indian Spice quality control.
- Out of 26 spices covered under the Agmark Grading Scheme the 13 covered under

the "Compulsory Grading Scheme" under the Agmark for export purpose are Black pepper, Cardamom, Turmeric, Ginger, Onion, Celery Seed, Chillies, Coriander, Cumin, Fennel, Fenugreek and Curry powder.

- Agmark Organization (DMI) called mother of standardization in county, rendered pioneering services in the quality control of spices for over 50 years.
- CFTRI – Established in 1950 Mysore (Krt).
- First National Symposium on spices – "Spice Role in National economy" in collaboration with the spice export promotion council.
- Spices Export Promotion Council – Cochin (Kerala) was set up under the Ministry of Commerce, Govt. of India.
- Cardamom Board – 1966 (Under ministry of commerce) on 14th April 1966.
- Spices Board – 1986 (under ministry of commerce).
- Spices act – 1986.
- National Research Centre for spices, Peruvannamuzhi, Calicut was established in 1986 but later the name NRC for spices was changed to IISR (Indian Institute of Spices Research) in the year 1st July 1995.
- NRC on seed spices – Tabiji, Ajmer (Rajasthan).
- International Pepper Community (IPC) – Jakarta (Indonesia).
- At the International level, the three leading producer India, Malaysia and Indonesia formed IPC in March 1972 but started permanent functioning on August 1976 at Jakarta.
- Indian Standards Institution (ISI) is now renamed as "Bureau of Indian Standards (BIS), New Delhi.
- ISG – International Spice Group.
- About 95% pepper is grown in Kerala, 3-4% pepper is grown in Karnatka and rest pepper is grown in Tamil Nadu.
- India is the largest producer and supplier of Ginger and Turmeric.
- India cut throat competition from Guatemala for Cardamom and from Malaysia, Indonesia and Brazil for pepper.
- 275 kg/ha yield of pepper in India is lowest in world as compared to 4,500 kg/ha. in Malaysia.
- Large Cardamom is mainly confined to Sikkim.
- Area under Spices in India – 25 Lakh ha.
- Production of Spices in India – 29.07 Lakh tonnes.
- Chilli share 32% of total spice production.
- Andhra Pradesh – Leading state in chilli production.
- Chilli has oleoresin and natural colouring material.
- Andhra Pradesh 49.1%, Karnataka 12.9% of Total Chilli production.
- Turmeric share 19% of total production of spices mainly grown in Andhra Pradesh, Tamil Nadu and Orissa.
- Contribute of total spice production -

Garlic – 18%

Coriander – 12%

Ginger – 9%

Cumin – 4 %

Chilli -32%

Turmeric – 19%

- More than 90% of spices produced in the country are used for domestic consumption and rest exported as raw as well as value added products.
- Per capita consumption for spices in India – 2.639 Kg (1999-2000).
- Per capita consumption for spices in USA – 1.448 Kg (1999-2000).
- Major Seed Spices–
  1. Coriander
  2. Cumin
  3. Fenugreek

94. "BLACK PEPPER" (KALI MIRCH)

- Family: *Piperacea*
- Botanical name *Piper nigrum*
- Native to West coast (Malabar) of India
- King of Spices.
- Branching vine or climbing, perennial evergreen shrub.
- Black Pepper is the dried, mature but unripe berry (fruit) of *piper nigrum.*
- Area under cultivation – 1.36 lakhs ha.
- Production/Annum – 32 Thousand tonnes.
- India accounts 54% of the total area under pepper in the world but share of production is only 26.2%.
- India exports about 41000 tonnes annually.
- India earns a foreign exchange of Rs. 240 crores annually.
- Area of black pepper - Kerala (94%), Karnataka (5%) rest Tamil Nadu, Andhra Pradesh and Assam.
- Fruiting branches: Plagiotropes.
- Topshoot: Orthotropes.
- Hanging Shoot: Geotropes.
- Leaves: Lanceolate.
- Inflorescence: Catkin produced at nodes opposite to upper leaves.
- Flowers: Monoecious or dioecious or hermaphrodite forms occur in different varieties.
- High yielding – More percentage of bisexual flowers (80%), male flower 1 to 19%.
- Fruit - Single seeded berry.
- Fruit takes approximately 6 months to mature after flowering.

- Fruit setting – Normally 50% in cultivated spices.
- Perennial woody climber, growing upto 15 meters.
- First harvest is taken after 2 years of planting whereas commercial bearing starts from 6th year onwards.
- Spike shedding occurs to the extent of 14-65%, causes considerable loss.
- Spraying of IAA 50 ppm or planofix 50 ppm and or Zinc 0.5% at the time of berry setting stage reduces spike shedding.
- Indian pepper reaches to 75 countries, North America region being the major importer of Indian peppers.
- NAFED – National Agricultural Co-Operative Marketing Federation of India established – 2nd Oct 1958.
- Pepper Research Station – Penniyur.
- Largest producer of Black pepper – Vietnam.
- Malaysia leads in yield and give about 16 times higher yield than India.
- India average yield – 273 Kg/ha.
- Black pepper – Gynomonoecious.
- First hybridization in black pepper was done in 1954.
- First hybrid of Black pepper released in India – Penniyur – 1.
- Pungent taste is due to – Piperine (Found only in berries).
- Piperine is toxic to houseflies than pyrethrum and a mixture of 0.05% piperine and 0.01% pyrethrum is more toxic than 0.10% pyrethrum alone so it can used as an insecticide.
- Important breeding method – Clonal selection.
- *Piper colubrinum* – Resistant to foot rot, introduced in India from Brazil.
- Main buyers of Indian pepper – USSR, USA, Canada, Italy.
- USSR and USA take about 55-60% of total pepper export.
- Consumption period of pepper – 6-8 months.

Varieties:–

- Majority of cultivated types are monoecious.
- In India over 75 cultivars are cultivated.
- Black pepper starts yielding from 3-4 years.
- Penniyur-1 and Penniyur-3 are hybrids of black pepper.
- Required quality characteristics for high yielding varieties are
    - Be rich in bite factor (Piperine and Pipretine)
    - Have high density berries with more starch contents.
    - High yielding having long spikes with compact set berries.
    - Be rich in essential oil and oleoresin content
    - Resistant to disease and pest.
    - Low in fibre content.

- Karimunda is the most popular cultivar of pepper among growers of Kerala.
  1. Panniyur-1 – F1 hybrid between Uthirankottah × Cheriyakania Kadan, (Released in 1966). Oleoresins – 11.8%, Oil 3.5%, dry recovery -35.3%, piperine – 5.3%. Sensitive to excess shade.
  2. Panniyur– 2 (Krishna) – Open pollinated seeding of Balankottah 2159 Kg/ha, Oleoresin – 10.6%, Piperine – 6.2%, Released in 1989.
  3. Penniyur– 3 (Shima) – F1 hybrid between Uthirankottah × Cheriyakania Kadan), Released in 1989, Oleoresin- 12.6%, Piperine – 5.2%.
  4. Panniyur-4 – Selection from Kuthiravally type II, performs well even under adverse condition, Released in 1989.
  5. Panniyur-5 – Open pollinated progeny of Perumkodi, suitable for intercropping in arecanut, oleoresin – 12.33%, piperine- 5.3%.
  6. Sreekara – Selection from Karimunda, essential oil 7%, tolerant to drought, Oleoresin 13%, released in 1989.
  7. Subhakara – Selection from Karimunda (KS 27), released in 1989.
  8. Panchami – Selection from Aimpiriyan Coll-856, 8320 Kg green pepper/ha. Late maturing, released in 1991.
  9. Pournami – Oleoresin – 13.87% tolerant to root knot nematode (*Meloidogyne incognita*), released in 1991.
  10. PLD-2 – Suited to Thiruvananthapuram and Quilon district of Kerala.

Other important varieties –

- Tellichery, Malabar – Large, attractive, dark reddish, brown to black, very aromatic
- Malabar Garbled – Alone contribute nearly 90% of the total pepper export from India.
- Lampong and Singapore – Smaller, more shrivelled.
- Arakkulamunda – Regular bearer, early maturity, berries bold.
- Balancotta – Largest leaves among Kerala cultivars, it's oil content is one of the highest so very high flavour quality, tolerant to moisture stress and phytophthora wilt.
- Cheriyakaniyakkadan – Small leaves, regular bearer, wilt resistant.
- Cheriakodi – Fruits smallest among all types, bearing in alternate years.
- Daddagya – Fruit largest among mysore type. Esteemed for making white pepper.
- Kalluvally – regular bearer, tolerant to water stress and disease Kaniakkadam.
- Karimunda – Most popular cultivars grown throught Kerala, bisexual, Flowering May June.
- Kottanadan- Most popular in South Kerala, bisexual, highest oleoresin content 17.8%. Pierine – 6.6% Volatile oil – 2.5-4.2%
- Kuthiravally – Alternate bearer, spike very long and slender (Like horse's tail hence the name Kuthiravally) flowers bisexual, oleoresin 15%, piperine 6% essential oil 4.5%. Produces very high quality pepper.

- Thomanankodi, Aranavalan – Other prominent varieties.
- New released variety

  Thevam

  Shakti – Tolerent to fruit rot.

  Girimunda – Hybrid, for higher elevation

  Malabar excel- Tolerent to foot rot, hybrid, for higher evaluations.
- Pepper is a crop of humid tropic (75-95% R.H.).
- It grows successfully between 20^0 North and South latitude and from sea level upto 1500 m above Mean Sea Level (MSL).
- Crop tolerates temperature between 10^0-40^0C Ideal rainfall – 125-200 cm.
- Western Ghats are ideal place for cultivation.
- Best thrives on virgin soil (Ph 4.5-6.0).
- Propagation by – Runner shoot cutting (2-3 nodes).
- Prunned shoots are separated from vine during – Feb.-March.
- Cuttings from middle 1/3 of the shoot are desirable as they are high yielding.
- Cutting becomes ready for planting in May - June.
- A rapid multiplication technique has been developed by National Research Centre for spices.
- As a mixed crop grown with arecanut, coconut, mango and Jack fruit.
- Pepper – Planting is done with the onset of south west mansoon.
- Erytherina, *Grevelia robusta*, Ailanthus & *Garuga pinnata* are the tree used for climbing of pepper.
- Excessive shading during flowering and fruiting encourages pest infestation.
- Prunning –When the top reach 6 m (March – April).
- Panniyur 1 and Karimunda – Suitable variety for raising bush pepper by planting horizontal side branches or fruiting branches (plagiotropes), cutting with 2-3 nodes are suitable.
- Pepper vine start yielding usually from 3-4th year.
- The orthotropic branched are vegetative and have monopodial growing habit.
- The plegiotropic branches produce flowers and fruits.
- Flowering – May June (It takes 6 to 8 months from flowering to ripening stage).
- Harvesting – In plains – November to February.
- In Hills – January to March.
- Harvesting is done when one or two berries on the spike turn bright or red, the whole spike is plucked.
- White pepper is prepared by removing the outer skin and pulp below it before drying the berries.
- Tripod-stand is used for harvesting.
- Recovery of white pepper is about 25% of ripe berries.
- Recovery of Black pepper is about 33% of ripe berries.

- Yield – 800-1000 Kg. Black Pepper from 1 hectare of 7-8 years old.
- Full bearing start 7-8th years after planting.
- Decline start after 20-25 year of planting.
- In bush pepper cultivation all the leaves except the flag leaf are removed.
- Cuttings dipped in 1000 ppm IBA for 45 seconds are used for propagation in bush pepper cultivation.
- White pepper commends a higher market price for use in products as such mayonnaise where the black speeks of black pepper is not liked.
- For preparation of black pepper the berries are harvested slightly early while for white pepper it is harvested at slightly advanced stage of ripening when berry turns red.
- Plucking of black pepper is done when one or two berries turn red or orange.
- Blackening of pepper is due to enzymatic oxidation of phenols which is present in epicarp and mesocarp.
- Average pepper yield –
  In India – 273 Kg/ha.
  In Indonesia – 425 Kg/ha.
  In Malaysia – 2000 Kg/ha.
  In Sri Lanka – 431 Kg/ha.
- Piperine usually ranges from 4-10%.
- The alkaloid piperine (Melting point 129-130°C) is responsible for bitting taste of pepper.

95. "CARDAMOM" (SMALL)

- Indigenous to South India and Srilanka.
- Member of group of plants known as Scitamineae.
- Found a position on Queen Shiba gift box to King Solomon.
- Known as Queen of spices.
- It is 2nd most important spice of India after Black pepper.
- Cardamom is dried fruit of a tall perennial herbaceous plant.
- Area under cultivation – 100,000 ha.
- Annual production – 4000 metric tonnes, 40% of which is exported to more than 60 countries.
- Kerala 60%, Karnatka 31%, Tamil Nadu 9% contribution of total area in India.
- Recently, Guatemala has emerged as a main competitor.
- Inflorescence – Panicle.
- Flower – Bisexual and fragrant.
- Fruit – Capsule.
- Pollination – By honey bees.

- Cardamom flower remains open for 15-18 hours and stigma receptivity and pollen viability are maximum during morning hours.
- Four to fives bee hives/hactare should be maintained for pollination.
- Major producer country – India, Srilanka and Guatemalan.
- Guatemalan has overtaken India from second position to first rank; now supplying 62% of the world's requirement of Cardamom with Indian share reduced to 28% only.
- Productivity of India – 60 Kg/ha.
- Guatemalan – 300 Kg/ha. (5 times more).
- Aroma is due to α-terpinylacetate and 1, 8 – Cineol.
- Flower initiation – March to April.
- From initiation to full bloom it takes nearly 30 days and from bloom to maturing it takes 5 to 6 months.
- Cardamom found a place on "Queen Sheba" gift box presented to King Solomon.
- Guatemala has become the world's leading cardamom supplier accounting for three quarter of world trade in this commodity.
- In Arabian country beverages, cardamom flavoured coffee called Gawa is generally offered to guests at social and religions functions.
- Varieties are based on fruit size (two varieties).
- *Elettaria cardamomum* var. major (Wild).
- *Elettaria cardamomum* var. minor (Cultivated) is devided into three-

  (Maysore Type 900-1200 m. MSL)

  (Malabar Type 600-900 m. MSL)

  (Vazukha-Hybrid (Mysore × Malabar)

  New Varieties.

  1. Coorg Cardamom Malabar Sel. 1 – Malabar type released by NRCS.
  2. ICRI-1 – Malabar type, by ICRI, Mayladumpra.
  3. ICRI-2 – Mysore type by ICRI, Myladumpra.
  4. Mudigree 1 – Malabar type by RRS Mudigree.
  5. PV-1 – Malabar type by CRS Pumpadumpra.
  6. SKP-14 – Malabar type, By ICRI, Salekpur (Karnataka) Prostrate type of panicle, suited to Karnataka. yield – 260-409 Kg/ha.
- Primary and Secondary nursery is required.
- Malabar cardamom indicates ripening by paling of fruits.

  Propagation – Through suckers.
- 1 Kg of seed capsule may produce 5000 seedlings.
- Earthing up is done when seedling attain 5-6 leaf.
- In Kerala and Tamil Nadu 18 month's old seedling are used while in Karnataka 10 months old seedlings are used.
- Trashing – Removal of old drying shoots of the plant once in a year with the onset

of mansoon under rainfed condition and 2-3 times in high density plantation (HDP) provided with irrigation facility.

- Cardamom is a pseophyte plant, very sensitive to moisture stress.
- Shade regulation is done to –
  1. Regulate soil moisture and temperature.
  2. Provides congenial micro climate.
- Palangi (*Artocarpus fraxinifolius*), Jack, Red Cedar (*Cedirella toona*), Karimaram (*Diospyres ebenum*) and Elangi (*Mimusropes elangi*) are most suitable shade tree.
- Temporary shade trees like *Erythrina lithosperma* and *E. India* are most unsuitable as they compete for nutrients and soil moisture.
- Regular irrigation (25mm-37.5mm/Irriagation) helps in initiation of panicales, flowering and fruit set.
- Malabar Cardamom – Drought resistance, mostly suitable for lower hills.
- Mudigree-1 – Tolerant to thrips infestation.
- MCC – Tolerant to azhukal disease.
- Cardamom normally starts bearing after 2 years of planting
- Harvesting is done in October –November, average yield – 500 Kg/Ha.
- Ripe capsules are harvested in order to get maximum green colour during curing.
- Cardamom grading and marketing rules were formed in 1962.
- Malabar type is superior in flavour quality.
- Soaking the freshly harvested green cardamom capsule in 2% washing soda solution for 10 minute prior to drying helps to retain green colour during drying Temperature in Drier – 45° – 50°C for 14-18 hours.

  Temperature in Kiln – 50° – 60°C for overnight.
- Summer showers are essential for flowering.
- ECCARD drier (Economical Cardamom Drier) is based on the principle of economically drying using firewood for drying cardamom. It can dry 80-100 kg of Cardamom in 26 hours.
- MELCARD – (Modern Economical Low-Cost Cardamom Drier).
- Cardamom grades/Agmark/ISI specification.
  1. Alleppey Green, coorg green, "Bleached" half bleached, bleached white and mixed Cardamom and Cardamom seed.
  2. Alleppy Green is by far the most important type of Cardamom exported from India, accounting for about 85% of the total cardamom export.
  3. Volatile Oil – Average 8.3%.
- Malabar type – Most pleasant and mellow aroma and flavour character, no well suited for green cardamom, cured fruits are round in shape, 18.5 mm length, and lower cineol content.
- Mysore Type - Harsher aroma and flavour well suited for green cardamom, 21 mm length, and higher cineol content sp.

- Cardamom oil is produced commercially by steam distillation of crushed fruits.
- Best oil is obtained with alleppy green.
- Pyrosis – Excess water in mouth.
- Indian cardamom is low in fat and high is protein, iron, Vitamin – B and C.
- Eating a cardamom once daily with a tablespoon of honey improves eye-sight, strengthens nervous system and keeps one healthy.
- Excessive use of cardamom causes impotency.
- Varities Avinash, Vijetha and PV2 are recommended for release.
- Malabar tupe superior in flavour quality.

"CARDAMOM" (Large)

- Large cardamom also known as Nepal cardamom or greater Indian Cardamom is the dried fruit of *Amomum subulatum* (Seed contain 2-3% essential oil).
- It is shade loving plant (seophyte).
- It is cultivated in Sikkim, Darjeeling and Assam hills.
- India is main producer of large cardamom.
- Sikkim and Darjeeling District of West Bengal produces about 4200 tones of cured cardamom annually.
- About 150 tones are annually exported earning 70 lakhs/annum.
- Large cardamom is a perennial crop.

  Propagation – Rhizomes / Suckers.
- seed propagation requires 10g. seed for 1 M area.
- Fruit 2.5 cm long ovoid containing 40-50 seeds.
- Average yield – 300 - 1000 Kg/ha from 4-5th year.
- It starts bearing after 3 to 5 year.
- Primary and secondary nursery is required.
- Ideal time for gap filling – May - June.
- Curing is done to retain only about 10-13% moisture.

Varieties –

1. Bedo – Deep root makes it drought tolerant, tolerant to foorkey and viral infection, 2% essential oil.
2. Bharlangey – 500kg cured capsule/ha. 15-20 capsules at each spike, 2.5 essential oil.
3. Golsey – Self shaded, require less shade, suited to close spacing, immune to chirkey but not to foorkey disease, susceptible to fungal leaf-spot disease, 2-3% essential Oil.
4. Ramla – Moderately tolerant to chirkey but susceptible to foorkey disease, 2-3% volatile oil.
5. Ramsey – Major area under cultivation in Sikkim and Darjeeling, susceptible to Chirkey and foorkey especially at lower altitudes.

6. Sawney – Susceptible to both viral and leaf spot and spike rot disease, Heavy yield under congenial condition.
7. Two viral disease occurs in Large Cardamom.
   1. Chirkey – Transmitted by Corn aphid.
   2. Forkey – Transmitted by Banana aphid.

96. "GINGER"

- Flower – Bisexual.
- Inflorescence – Spike.
- Fruit – Capsule.
- Tropical crop which originated in South East Asia (India and China).
- Ginger is one of the oldest spices with a distinct flavour and pungency.
- India is the largest grower of ginger and also largest producer of dry ginger in the world.
- Major ginger growing states are Kerala, Orissa, Andhra Pradesh, Himachal Pradesh, Meghalaya and West Bengal.
- Kerala – 60% area and 25 % production of our country.
- India account 70% of total world production and 50% in world export.
- India annually exports 5000 tonnes earing Rs. 20 Crores.
- China is our major competitor in the world market.
- India – 53300 ha. area under cultivation with 130000 tonnes of dry ginger production.
- Ginger oleoresin is commercially known as gingerin.
- Ginger of commerce or Adrak is the dried underground stem or rhizomes of Zingiberous, herbaceous plant.
- It stands 3rd to 4th in pungency when competing with chillies.
- Ginger oleoresin (gingerin) contains 18-20 % gingerol which is the main pungent principal.
- The sundried rhizomes in the trade are known as "Hands" or "races".
- Jamaica and India produce the best quality ginger.
- Ginger in Saudi Arabia is predominantly used for flavouring coffee.
- Major Importing countries from India –

  Saudi Arabia – For fresh and bleached ginger.

  USA, UK, Canada and Singapore – For unbleached dry ginger.
- Indian ginger or Cochin ginger is considered one of the best in the world.
- Ginger rank 3rd in export from India in spices after Black pepper and Cardamom.
- The major bottle necks in boosting exports of Indian ginger are its higher fiber content than that of it's main competitor, the Jamaican ginger in world market and it's higher cost of production.

- Classification
  1. Z. officinale – main source of ginger for commerce, Rhizomes are thick and lobed.
  2. *Z. cassumunar* – widely cultivated in tropical Asia for medicines, also used for flavouring food and have stimulant and carminative properties.
  3. *Z. zerumbet* - a herb of 1 m height widely cultivated for its rhizome which are used for cough, stomachache, asthma, worms and in leprosy and skin diseases.

Varieties –

1. Rio-de-Janerio – 25-30 tonnes/ha., dry ginger recovery 16-18%, fiber content 5.9%, best for green ginger and oleoresin content, Introduced from Brazil in Kerala, high moisture requirement.
2. Wynad Manantody – Widely grown in Kerala, best for dry ginger, contributes 50% of total production.
3. China – Suitable for green ginger.
4. Maran – Suitable for dry ginger, most popular cultivar of Assam.
5. Suprabha – Oleoresin 8.9%.
6. Suruchi – Oleoresin 10%.
7. Surari – An X-ray inducted mutant, oleoresin 10.2%
8. Jamaica –
9. Bangkok –
10. Ernad –
11. Chernad –
12. Kuruppampadi –
13. Tafengiya –
14. Himgiri – Tolerant to rhizome rot.

- Pungecy is due to gingerol and shogaol.
- Volatile oil – Gingeberene.
- Ginger is always propagated by portions of the rhizomes known as seed rhizomes (Bits).
- Rhizomes length – 2.5-5.0 cm, weight 20-25g. Each having one or two good buds is suitable for sowing. Seed rate (rhizomes) – 1500-1800Kg/Ha.
- Rhizomes sets should be treated with 0.3% dithane M–45 solution for 30 minutes to control fungal diseases.
- It takes 20-25 days from bud initiation to full bloom and 23-28 days to complete flowering.
- Spacing – 30x30cm under bed system of planting.
- Sowing time – April-May.
- Harvesting – For green ginger 5-6 months (Dec.-Jan.).
- For dry ginger – Harvesting is done between 245 to 260 days when leaves turn yellow and gradually start drying up.

- Average yield – 15-25 tonnes/ha.
- For preparing dry ginger the outer skin is removed with split bamboo having pointed ends.
- **Unbleached ginger:** Washing over night – rubbed well to clean – outer skin is removed-dried in sun (for a week) – Rhizomes rubbed together for a week.
- **Bleached ginger:** To get good appearance, Rhizome are peeled and soaked in 2% lime water for 6 hours and dried.
- The yield of dry ginger is 16-25% of the fresh ginger.
- For processing into it's products, ginger rhizomes are harvested 5 months after planting.
- The major constaints are non-availability of better genotypes, higher cost of production and high fiber content in varieties.
- There is no flowering and seed set because of chromosomal sterility.
- Polyploidy breeding therefore offered some hope to explore the possibilities of obtaining flowering and seed set.
- Varieties of Ginger suitable for different end uses

End use	Varieties
For dry ginger	Karakkal, Nadia, Maran, Wynad, Manantody, Valluvanad, Emad and Kuruppampalli.
For green ginger	Rio-De-Janerio (preferably for extraction of oleoresin), China, Wynad Local and Tabengiya.
For high volatile oil	Sleeva Local, Narsapattam, Emad, Chemad and Himachal Pradesh.
For high oleoresin	Emad, Chemad, China, Kuruppampalli and Rio-De-Janerio.
Low fibre content	Jamaica, Bangkok Vareda and Nadia are also good for green ginger.

- Some or the are wise commonly cultivated varieties for Ginger in India

Area of cultivation	Cultivars
Assam	Thingpui, Jorhat, Nadia, Thinladium and Maran
West Bengal	Burdwan
Kerala	Wynad Local, Wynad Manantody, Emad and Thodupuzha
Karnataka	Kurappampadi and Karakkal
Andhra Pradesh	Narasapattam

Physical forms of ginger –

- **Hands and fingers:** As dug up from the ground.
- **Black ginger:** Whole rhizomes which have been scaled for 10-15 minutes in boiling water before being scraped and dried.
- The two types of Indian ginger entering international market are Cochin and Calicut ginger.

- Dry ginger is packed in gunny bags. Size of gunny bag most commonly used is 122cm × 67cm which is known as Liverpool bag.
- The pungent principle of ginger is oxymethyl phenols.
- Commercial preparation of oleoresin contains gingerol, shogaol and zingerone.
- Freshly prepared oleoresin contains gingerol as the main constituent (approximately 1/3 of weight of oleoresin).
- The mode of ginger consumption are 42-43% dry, 36% green ginger, 17-18% seed and 3-4% spoilage.

Physical farms of ginger –

1. **"Peeled", Scraped or uncoated:** Whole rhizomes from which the cork skin has been cleanly removed without damaging the underlying tissue.
2. **"Rough scraped":** Whole rhizome from which the skin has been partially removed usually on flat sides.
3. **"Unpeeled" or "Coated":** Whole rhizome which have been dried with the skin intact.
4. **"Bleached" or "Limed":** Clean peeled-whole rhizome which have been treated with lime or sulphurous acid to achieve a lighter colour.
5. **Splits:** Unpeeled rhizome which have been split longitudinally to accelerate drying.
6. **Slices:** Unpeeled rhizomes which have been sliced to accelerate drying.
7. **Ratoons:** Second growth rhizomes from plant.

97. "TURMERIC" (HALDI)

- Center of origin is South East India and Java.
- The spice "turmeric" constitutes boiled, dried, cleaned and polished rhizomes (the underground swollen stem of plant) of *Curcuma longa*.
- The plant is herbaceous, perennial and tropical crop.
- India is by far the largest producer and exporter of turmeric in the world.
- India accounts for 76% of total world production.
- Turmeric occupies about 6% of the total area under spices and condiments in India.
- Andhra Pradesh – Leading turmeric producer state.
- Yellow colour is due to curcumin.
- Rhizome has 1.8 to 5.4% curcumin.
- Rhizome has 2.5 to 7.2% essential oil.
- Rhizomes are called ginger.
- There are four important species of curcuma.
- *C. longa* – widely cultivated, used as colouring matter and condiment.
- *C. aromatica* – Cochin turmeric / Kasturi Manjal.
- *C. angustifolia* – East Indian Arrow root having plenty of starch in it's rhizome.

- *C. amada* – Mango ginger (Which has the taste and flavour of raw mango.) – used for pickle.
- In International trade turmeric is graded into "finger – wholes" and splits.
- Best quality – Fingers.
- Alleppey (fingers) – Well known in trade for it's colour.
- Turmeric is antioxidant due to phenolic character of curcumin.
- Flowering occurs in July to October.

Varieties –

- CO1: X-ray induced mutant from Erode local, suitable for drought and saline soil, rhizome big size, yield 35t/ha, duration 270 days, curcumin 32%.
- BSR 1: X-ray induced mutant from Erode local, suitable for waterlogged condition, 31t/ha., curcumin 4.2%.

1. Suguna: Clonal Selection, short duration cultivar (190 days), 603 qtl/ha yield.
2. Suvarna: Clonal Selection. Released by CPCRI, Calicut, curcumin content (8.7%).
3. Sudharshana: Clonal Selection. High curcumin content (7.9%)
4. Sugandham: Clonal Selection. Curcumin content (3.6%)
5. Krishna: Clonal Selection moderately resistant to rhizome fly and leaf spot disease.
6. Roma: Less susceptible to diseases, curcumin content (9.3%).
7. Suroma: Curcumin content (9.3%).
8. Rajendra Sonia.
9. Ranga.
10. Rashmi.
11. Sobha: Oleoresin 15%.
12. IISR Prabha: Oleoresin 15%.
13. IISR Prathiba: Oleoresin 16.2%.
14. Kanti.
15. Allepy turmeric: Best in world.
16. Amba: A variety of Mango ginger.
17. Pant Peetabh: Latest variety of turmeric.

- Highest amount of curcumin has been found in variety Edapalayam (10.9%) and oleoresin in "Konni" (19.2%).
- Propagation – Seed rhizomes (35-44g.) 2500 kg of rhizomes are required for 1 ha., spacing – 30x15 cm, 30x20cm in beds, 40-60cm × 25 cm (in ridge or furrow), 50x15 cm in flat system.
- Turmeric is a Shade loving plant (Seophyte).
- Planting time – April – June.
- Harvesting – 7-9 months after planting.

- Harvesting time – November to December.
- Early variety – Matures in 7-8 months.
- Medium variety – Matures in 8-9 months.
- Late variety – Matures after 9 months.
- Suitable temperature for different stages of turmeric is given below

Suitable Temperature	Stage
30-35°C	Germination
25-30°C	Tillering
20-25°C	Rhizome initiation
18-20°C	Bulbing

- Average yield – 20-25 tonnes of green turmeric. 25-30 tonnes/ha. (HDP)
- Turmeric processing involves three steps curing, polishing and colouring.
- Curing involves boiling of fresh rhizomes in water.
- Colouring is done to give good appearance and better finish to the product.
- Composition of the emulsion required for coating 100kg of half boiled turmeric is Alum 0.4kg. Turmeric powder – 2kg, Castor Seed oil-0.14kg, Sodium bisulphate – 30 g., Concentrated HCl-30 ml.
- Dried rhizomes – 5-6% oil.
- About 58% of the oil is composed of turmerones (sesquiterpene ketons) and 9% tertiary alcohol.
- Acetone, alcohol and ethylene dichloride were found suitable for extracting oleoresin from turmeric.
- Oleoresin contains 30-35% curcumin and 15-20% volatile oil.
- Dry recovery of turmeric is 15-30%.
- Trade name – Alleppy, Erode, Duggirala, Nizamabad, Rajapuri and Chudlapah.
- Dead heart is caused by shoot borer of turmeric.

98. "SEED SPICES" (CORIANDER)

- Term coriander is derived from Greek word "Karis" meaning "bedbug" because of pleasant fetid, bug like odour of green and unriped fruits.
- It is an annual herb, cultivated for its green leaves as well as for its fruits
- Highly cross pollinated crop.
- It is one of the first seed spice to be used by mankind as early as 5000 BC.
- Flowers are umbel, hermaphrodite, sometimes unisexual or polygamous white or rose coloured.
- Fruits – Schizocarp.
- Essential oil – Linalool (80%), fatty oil (19-21%).
- The main importing countries are those in the gulf region.
- Pakistan, Egypt and Iran are the main competitors of India in international trade.

- Coriander is employed for flavouring liquor particularly gin.
- The odour and taste are due to the essential oil content which varies from 0.1 to 1.0% in dry seed.
- Essential oils are used for flavouring liquors and cocoa preparation.
- Dried fruit – Major ingredient of curry powder,
- Good quality oleoresin is extracted from coriander seed.
- Decydenyde (0.1% volatile oil) is also used in perfumery.
- It is a tropical crop and can be grown throught the year (except very hot season i.e. March-May) for leaf purpose.
- Black cotton soils with high retentivity of moisture are suitable for its cultivation.
- Coriander is the leading seed spice crop grown in India.
- Coriander is the most heavly consumed herb plant.

Varieties –

1. RCr 41 – Highly resistant to stem gall and wilt but moderately tolerant to powdery mildew.
2. RCr 20 – Moderately tolerant to powdery mildew, wilt as well as stem gall.
3. RCr 435, 436
4. RCr 446 – Avg. yield 12q./ha.
5. GC 1- Yellow coloured grain, moderately tolerant to wilt and powdery mildew.
6. GC 2 - Yellow coloured grain, moderately tolerant to wilt and powdery mildew.
7. Sindhu – Resistant to aphid and moderately tolerant to wilt and powdery mildew.
8. Sadhna – Tolerant to white fly and mites.
9. Swathi – Tolerant to white fly, grain mould and wilt.
10. RD 44 (Rajendra Swathi) – Resistant to stem gall and moderately resistant to wilt aphid and weevil suitable for Bihar.
11. Co 1, 2, 3 – Dual purpose.
12. DH-5, Hissar Anand.
13. Pant Haritama – For Punjab.
14. NRCSS 1, 2 – New released varieties.
15. Karan – Small seeded variety suitable for Rajasthan.
16. Lam sel. – For Andhra Pradesh.
17. Merrocon

- Coriander is propagated by seed.
- Seed rate

  10-15 kg – In irrigated condition.

  25-30 kg – In rainfed Condition.

  40-42 kg – For leafy vegetable.
- Sowing Time

Kharif – June – July.

Rabi – Oct.-Nov.

- A spray of GA 100 ppm at five leaf stage increases the male flower while Ethrell and CCC reduce the ratio of bisexual to male flowers.
- GA @ 5-50 ppm hastens and improves flowering when sprayed at 5 leaf stage.
- Harvesting – 90-140 days after sowing (DAS)
- Yield –

  400-500kg (rainfed crop)/ha.

  600-1200 kg (Irrigated crop)/ha.
- Drying is done to maintain a moisture level below 13% in seeds.
- Spanish people eat fresh coriander leaves "*cilantro*" and it is also called Chinese parsley.
- Hissar Sugandh – Latest variety.
- IARI Sheetal – 36-3, CIMPO Sel. S-28 and S-52 are recommended for green leaf crop.
- The plants are andromonoecious i.e. both staminate and the perfect flowers are produced on the same plant and even in the same umbellet.

99. "CUMIN"

- It is an annual herb.
- Native of Egypt and Syria, Turkistan and Eastern Mediterranean region.
- Flower colour – White or pink.
- Inflorescence – Umbel.
- Volatile oil – 2.5-3.55 produce by steam distillation.
- The seed also contain 17.7% protein, 23.8% fat, 35.5% carbohydrate and 7.7% minerals.
- In India mainly cultivated in Rajasthan and Gujarat.
- It is commercially cultivated in Iran, India, Turkey, China, Morocco, Indonesia and Japan.
- Iran – Major cumin exporter country in the world.
- Fruits are rich in thymol.
- Cumin produces strong and heavy flavour due to an alcohol, cuminol and is slightly bitter in taste.
- It is used in preparation of curry powder.
- The main ingredient of volatile oil is aldehyde.
- It also contains 10% fixed (non-volatile) oil.
- Moderately cool and dry climate is best for cumin cultivation.
- High humidity and frequent rains particularly after flowering period increase incidence of disease like blight and powdery mildew.

- Planned work on collection and improvement of cumin was started by ICAR in 1957 at IARI.
- Aroma is due to cuminol.

Varieties –

1. RZ 19 – Tolerant to wilt and blight, released in 1988 developed at RAU, Jobner.
2. RZ 209 – Highly resistant to blight, selection from local collection of Ahore (Jalore).
3. GC 1 – Tolerant to wilt and blight, 3.6% volatile oil released in 1988.
4. GC 2 – Developed by GAU through pure line selection released in 1991.
5. Vijapur 5 – Popular variety of Gujarat, susceptible to diseases.
6. MC 43 –
7. S – 404
8. RS 1
9. UC 19
10. UC 194
11. Topolka – Introduce from Bulgaria.

- Seed rate – 12-15kg/ha.
- Best sowing time – Middle of November.
- Jeeri (*Plantago pumila*) a serious weed of cumin, can be effectively controlled by fluchloralin.
- Harvesting – 120-150 DAS (Days After Sowing).
- Yield – 7-8q/h.
- Cumin is one of the earliest known minor spices used by mankind.
- A drink prepared by cumin in Germany is known as Kummel.
- Black Cumin (Kalongi) – *Nigella sativus* – Rabi season crop var- kala jeera.
- Kalongi has a carbonyl compound "nigellone"".
- It prefers dry climate during February-March.
- Cumin wilt is the most serious disease.

100. "FENUGREEK"

- Annual herb (both tropical and temperate crop).
- Flower – Papilionaceous (white or yellow).
- Fruits – Legume.
- The species is a typically self- pollinated and destogamous.
- Pollen fertility ranges from 95-98% in the unopened flower bud and 67-80% in open flowers.
- The cultivated forms are diploids 2x=16.
- A sterile hexaploid 2x=44 has been recorded.

- Anthesis – 9AM and 6 PM, 11.30AM (peak period).
- Anther dehiscence – 10.30 am – 5.30 pm, 11.30-12.30 (peak period).
- Stigma become receptive 12 hours before flower opening and remains receptive for about 10 hours after opening of flower.
- Duration of flowering phase – 7-18 days.
- Common Methi – T. *foenum graecum.*
- Kasthuri Methi – T. *corniculata.*
- Kasthuri methi is slow growing type and remains in rosette condition during vegetative phase of growth period.
- Common methi take 5-6 days to germinate while kasthuri takes 7-8 days.
- Fenugreek is third largest seed spice in India after coriander and cumin.
- Fenugreek – Used for curry powder.
- Fenugreek is mainly cultivated in Rajasthan, Gujarat, Madhya Pradesh and Tamil Nadu.
- More than 80% area and production of the country is contributed by Rajasthan alone.
- India is the major producer and exporter of fenugreek.
- Fenugreek seeds contain "diosgenin" which is used as a starting material in the synthesis of sex harmones or preparation of contraceptive pills.
- Diosgenin in seed – 0.40-1.26%.
- Fenugreek is rich in amino acid and trigonelline 150mg/100g.
- The seed also good for elimination of bad breath and body odour and also help to combat dandruff and is a cure for baldness in man.
- Indian women use the seeds power of methi to promote lactation.

Recommended Varieties of fenugreek

S.No.	Variety	Features
1	Pusa Early Bunching	It is quick growing with upright shoots, good yielder, suitable for 2-3 cuttings. Developed at Indian Agricultural Research Institute, New Delhi.
2	Pusa Kesuri	A late flowering variety, rosette type leaves with special fragrance, heavy yielder and gives 5-6 cuttings. Developed at Indian Agricultural Research Institute, New Delhi.
3	Methi No. 47	Leaves are broad. Succulent and rich in Vitamin C. Released by Maharashtra state.
4	Co-1	A high yielding variety suitable for both greens as well as for seed. Developed at TNAU, Coimbatore.
5	Lam Selection	Plants of medium height, bushy, more branched. Developed by APAU.
6	Hissar Sonali	It is a variety for green leaves. It is developed by HAU, Hissar.

Varieties –

1. RMt 1 – Pure line selection from local cultivar. Suitable for Rajasthan 1500kg grain/ha, moderately resistant to root rot and tolerant to powdery mildew.
2. RMt 143 – Pure line selection from local cultivar from Jodhpur. Moderately resistant to powdery mildew. It is especially recommended for heavier soils of Chittor, Bhilwara, Jhalawar and Jodhpur of Rajasthan.
3. CO 1 – Dual purpose variety suitable for Tamil Nadu.
4. Rajendra Kanti – Moderately resistant to powdery mildew suitable for Bihar.
5. Lam Sel. 1- Tolerant to root rot, powdery mildew caterpillar and aphid suitable for Andhra Pradesh.
6. Hisar Sonali – Moderately resistant to leaf spot and root rot complex diseases.
7. Pusa early Bunching – Champa Fenugreek.
8. Kasuri selection – Champa Fenugreek.
9. HM 1 – Moderately resistant to leaf spot disease. 100kg/ha. grain suitable for Haryana.
10. Prabha (NLM).
11. Plume – 20.

- Last week of Oct. – first week of Nov. is best time for sowing
- Seed rate – 20-25 kg/Ha.
- Seed rate – 15-20 kg/Ha. (Kasuri methi)
- 40kg – broadcasting method.
- Spacing – 30 × 10 cm.
- Seed should be treated with *Rhizobium* culture.
- Harvesting – 70-165 Days after sowing (DAS), depends on variety.
- Average seed yield –

  10-11q./ha.

  8-10 tonnes of leaves/ha.
- Storage – 7-8% seed moisture having 40% R.H.
- For good yield field inoculation with *Rhizobium meliloti* is useful,
- A spray of ascorbic acid at 250-450 ppm improves plant growth and enhances number and size of pod.
- Rajendra Khushba – Latest released variety of fenugreek.
- Pant Ragini- Latest released variety of fenugreek.
- Powdery mildew is the most serious disease of fenugreek.
- Methi mosaic wilt is caused by virus.
- Isolation distance for seed production is 10 m. for foundation seed and 5 m. for certified seed.

101. "FENNEL"

- Annual herb
- *Foeniculum vulgare* sub species *Piperitum* is wild fennel.
- Fennel is native of South Europe and Mediterranean region.
- The crop is protandrous and cross pollinated.
- Fennel is mainly cultivated in Romania and India.
- In India – Gujarat is major producer followed by Rajasthan and Uttar Pradesh.
- The fruit of fennel yield a volatile oil, the main constituent is anethole.
- Bitterness is due to – Fenchone.
- Sweetness is due to – Anethole (70%).
- In India seeds are used for mastication and chewing alone or with betel leaves and scented soaps.
- Seeds also contain limone.

Varieties

1. RF 35 – Moderately tolerant to sugary disease, leaf spot and leaf blight, green seeds.
2. RF 101- Bear large umbel, developed by RAU, Bikaner through recurrent half sib selection
3. RF 125 – Yield 17.5 q/ha.
4. Gujarat Fennel 1 – Moderately tolerant to sugary disease and leaf spot, suitable for early sowing, tolerant to drought. Selection from Vijapur local (1985)
5. CO 1- Developed by TNAU, Coimbatore through recurrent selection form F 35. Suitable for drought prone water logged saline and alkaline condition, suitable for hilly areas. Volatile oil ranges from 1-1.5%.
6. PF 35 – Developed by GAU, Gujarat, highest volatile oil (1.5%)
7. S-7-9 – Dwarf variety
8. Mutant 1,2,3 –

- Fennel require cool and dry climate for its cultivation.
- Seed Rate – 10-12 kg/ha. (direct sowing)
- Seed Rate – 3-4 kg/ha. (transplanting crop)
- Sowing time – Mid September to mid October (Best time).
- The seed are sown in nursery beds during May-June and seedlings are transplanted in August-September (60-80x40-80 cm).
- Chewing type "Lucknawi" Fennel is produced by plucking the umbel 30-40 days after pollination when the size of seed is just half of the full development seed and then dried in shade.
- Fennel matures in 170-180 days, Yield 9-10 q./ha.
- Fennel is most vulnerable to frost damage at the flowering and seed formation stage.
- Control – Spray 0.1% H2So4 and irrigate the crop.
- Harvesting of umbels is done 4-5 times at 5-10 days interval.

102. "DILL" (SOWA)

- *Anethum graveolens* is the true Dill.
- Dill's most famous culinary used is for Dill pickle.
- Dill leaves are known as Dill weed.
- It is a winter annual herb.
- Flower – Pale yellow.
- Sowa is a long day plant (LDP).
- On distillation an essential oil containing high carvone content is obtained.
- The seed oil of sowa has superior aroma but the oil contains an oxygenated compoun "Dill apiole" which is toxic in large doses.
- The herb oil harvested at vegetative stage is rich in phellandrene where as seed o has high carvone content.
- It is grown in Europe, USA and India.
- In India – Gujarat, Maharastra, Rajasthan and Madhya Pradesh are main producers
- *Anethum sowa* has longer fruits than the European sp.
- *Anethum panmorium* has strong resemblance to the *A. faniculum*.
- India dill or sowa – *A. graveolens var sowa* (native to north India)
- Seed Rate –

  5-6 Kg. Line sowing.

  10 kg – Broadcasting.

  Sowing Time – Mid October (Best Time).

  Yield – 8-9 q./ha.

  Varieties – NRCSS-AD 1 (Contains 3.5% essential oil).

  NRCSS-AD 2 (Contains 3.5% essential oil).

  Haldwani Selection.

  Sel-163, NP-179.
- Aroma in herb is due to hexahydrobenzafuran.
- Crop is harvested when first flower head turns brown.
- Dill water is very common use of Dill for flatulence, colic and hiccups of infants ar children.

103. "CELERY"

- Latin name: *Apium graveolens* L. var *dulce*.
- "*Graveolens*" means strong scented; "*dulce*" means sweet.
- Family name: Umbelliferae (Apiaceae).
- Diploid (2n = 22) and Biennial.
- Related species used as vegetables: carrot, parsnip and celeriac.

- Origin -Mediterranean and Eurasia.
- It is a winter annual herb.
- Introduced from France.
- Commercially grown in Amritsar, Gurdaspur and Jalandhar district in Punjab, Uttar Pradesh, Himachal Pradesh and Tamil Nadu.
- Harvested organ: petiole, eaten raw or cooked.
- Used as a medicinal herb by the Egyptians and Romans.
- Older varieties required blanching.
- Modern cultivars are grown unblanched.
- Principle constituents of celery seed oil are-
  1. d-limonence (60%)
  2. d-Selinence (10%)

  Sedanonic acid anhydride (0.5%)
  Sedanolide (2.5-3.0%) → Both are responsible for the aroma of oil.
- Celery has 16-20% fatty oil.
- It is sensitive to very low day temperature (5-10°C) during vegetive growth causing early bolting.
- Mainly grown for salad purpose.
- Days to harvest maturity: 90 to 140 days after planting.

Varieties – Two type

A. Self blanched or yellow – (a) Golden (b) Florida Golden.

B. Green Leaved –

(a) Fork Hook Emperor – Late variety.

(b) Standard Beared – Early Variety.

(c) Wright Grove Giant – High yielding.

RR 185-1 – Improved variety.

Seed rate – 375-500g./ha.

Sowing Time – Aug-Sept.

Yield – 300-560q./ha, green leaves.

175-275 kg/ha. Seed yield.

Physiological disorder –

1. Pencil Strip – Excess of phosphorus.
2. Black Heart – 'Calcium' deficiency.
3. Cracked Stem – 'Boron' deficiency.

- Major insect pests are aster leafhopper, tarnished plant bug, aphids, carrot weevil, variegated cutworms, loopers, celery leaf tier, leafminer and mites.

104. "CLOVE"

- Clove is derived from French word *Clov* and English word *Clove* both meaning nails.
- It is an evergreen tree spice.
- Clove is the dried unopened flower bud of *syzygium aromaticum* (syn. *Eugeneia caryophyllus*).
- In India introduced around 1800 AD by the East India Company in their garden in Courtallam (Tamil Nadu).
- Indonesia, Zanzibar and Modagascar – Major producer.
- Indonesia alone contributes for 66% of world production.
- 2nd most important spice next to black pepper in the world.
- India – Nilgiris, Tirnelveli and Kanyakumar district of Tamil Nadu, Calicut, Kottayam districts of Kerala are major producer.
- Clove is also known as **chicken tongue spice**.
- Clove has alternate bearing habit.
- In Indonesia major part is absorbed for making KRETEK cigarette industry.
- Flower – Hermaphrodrite.
- Fruit – One seeded drupe (SSD).
- Clove is a strictly a tropical plant.
- Propagation – Seed (Known as mother clove).
- Seeds are sown on raised nursery beds at spacing of 2-3 cm and depth of 2 cm.
- Protection from coconut leaves is necessary in first year and in the second year to prevent sun burn.
- Seeds germinate within 45 days starting from third to fifth week.
- Safe conditions for storage – moisture 6.5%; storage temperature 25-28^{0}C and RH 30-40%.
- Seed become available from June to October and lose their viability within one week.
- Approach grafting of clove on it's own root stock is successful.
- The 18-24 month old seedlings are ready for transplanting in the field (June-July).
- Intercropping with Banana is very good.
- Bearing start after 7-8 years of planting and full bearing start after 15 to 20 years.
- Flowering season –

  Sept. – Oct. (In plains).

  Dec. – Feb. (At high altitude).
- Flowers are produced on young flush.
- Optimum stage of clove picking is when bud changes colour from green to slighty pinkish tinge.
- It takes 4-5 days for drying; well dried cloves will be only about 1/3rd of the weight of the original.

- About 11,000 to 15,000 dried cloves make 1 kg.
- Yield : 4-8 kg/tree from well maintained full grown tree.
- 2 kg/tree average annual yield after 15 years.
- Clove Oil – 16-21% in the buds.
- Oil contains – Eugenol (70-90%) (JRF 2008).
- Eugenyl acetate -5 to 12%.
- Aroma is due to – Methyl amylketone.
- Varieties – Pevang, Lezipore, Amulya.

105. "NUTMEG"

- It is an evergreen, conical tree.
- Nutmeg is the dried kernel of seed.
- Mace is dried aril surrounding the seed.
- Nutmeg was introduced in India towards the end of the 18th century.
- Over 15% of the world exports originate from Indonesia.
- Greneda is the second largest exporter of Nutmeg and Mace.
- In India – Kerala, Tamil Nadu and Karnataka are major producers.
- Nutmeg is dioecious tree (JRF 2008).
- Fruit – Drupe.
- Pollination – Wind and Insect.
- Propogation by seed.
- About 18-24 months old seedlings are used for transplanting in field.
- Epicotyl grafting technique developed by NRCS, Calicut is found to be promising now.
- Sraight growing shoots are known as orthotropic or chupan.
- Spacing – 8 × 8 M.
- Shade plant – *Erythrina indica*, *E. lithosperma* are planted for Nutmeg.
- Bearing start in 4 to 5 years when grafted.
- Bearing starts in 7 to 8 years when seedling tree is grown.
- Takes 15 to 20 years for full bearing.
- Harvesting – June-August.
- Fruit take 9 month from flowering to harvesting.
- Freshly pericarp is used for making pickles, jams and jellies.
- Nutmeg fruit weight 60g. of which seed weight 6 to 7g., mace 3 to 4g. and rest pericarp.
- Yield –

  2000-3000 fruits/year.

  800 kg nutmeg/ha.

  100 kg mace/ha.

- 10: 1 (female: male) plants should be planted.
- In established plantation the unproductive male plants can be converted into productive female tree by top working.
- Volatile oil terbein have weedicidal property.
- Pericarp contains 12-14% pectin.

106. "CINNAMOM"

- Family: Lauraceae (2n=14)
- Srilanka and Malabar are the center of origin
- True cinnamom also called as Ceylon cinnamom or Sweet Wood, one of the oldest known spices is the dried bark of *C. verum* (syn. C. *zeylanicum*).
- Cassia is popularly known as "Chinese Cinnamom" false cinnamon or bastard cinnamom also one of the oldest known spice.
- Indian cassia (Tejpet) – *Cinnamomom tamala.*
- Cinnamom bark has 0.5-2% essential oil, eugenol is the main constituent in the leaf oil.
- Cassia bark has 1 to 4.5% essential oil, cinnmaldehyde is the main constituent in the leaf oil.
- Cinnamom leave oil is equal to clove oil in eugenol content (70-95%)
- Seeds are sown in beds (10x20 cm apart) in June-July in nursery beds.
- Seedlings are transplanted in April-May before monsoon.
- First cutting is made after 2-3 years of transplanting when plant reaches 150-200 cm height.
- Quills of 60-125 kg/ha can be obtained after 3 years.
- From 10-11 year old tree 225-300 kg quills per hectare can be obtained.
- The National conservatory of cassia germplasm consists of 30 acessions.
- The oldest cinnamom plantation is the Anjarakandy Estate (250 ha.) in Cannore district of Kerala.
- Srilanka – Largest producer of cinnamom bark with best quality.
- In India grown in Kerala, Tamil Nadu and Karnataka.
- Indonesia, China and Vietnam are the largest producer of Cassia and it is not grown in India.
- Commercial method of propagation – By seed.
- Flowering - January.
- Fruit – June-Aug.
- Shoot harvest – Sep.–Nov. (Ideal for bark extraction).
- Cut 1.0-1.25 m length and 1.25cm thickness when they are about 2 years old.
- During drying the bark contracts and assumes the shape of quill.
- The quills are graded from 00000 being the finest quality to 0 the coarsest quality.

- The small pieces of bark, left after preparing the quills are graded as quillings.
- The very thin inner pieces of bark are dried as featherings.
- Other grade – Scrapped chips (Bark is scrapped instead of peeling).
- Yield – 200-300 kg dried bark/ha.
- Leaf and bark oil contains eugenol (70-80%) cinnamic aldehyde.

Varieties –

1. Navashree: 6-7 shoot/year, high yield 56kg/ha. in the first 4 year, high bark recovery (40.6%), bark 2.7% oil, Cinnamaldehyde content 73%, in oil bark oleoresin 8%, leaf oil 2.8%, recommended for all cinnamom growing areas.
2. Nithyashree – Released from IISR, Calicut. Selection, high yielding (44 kg/ha in first 4 years)
3. Konkan Tej –
4. Yercaud 13 –

107. "ALLSPICE"

- Its flavour is said to resemble a blend of cinnamom, clove and nutmeg.
- The allspice of commerce is the dried immature fruits of the tree *Pimenta dioica* (small evergreen tree).
- Jamaica: Main producer of allspice.
- Flower structurally hermaphrodite but functionally dioecious.
- Fruit – Two seeded berry.
- Flowering – March-June (white colour).
- Edible part – Dried immature fruits (Berries).
- Berries are rich source of vitamin A.
- Male tree – Flower earlier.
- Propagation – By seed.
- After 3 weeks of seed extractions they start reduce viability and loose completely viability after 9 week.
- Yield – 50-60 kg dry berries/tree.
- The pimento oil is distilled commercially in Jamaica.
- Oil in berries – 3.3-4.5% (Eugenol 65-66%).
- Pimento leaf oil is also distilled only in Jamaica (0.7-2.7% in dried leaf), (0.35-1.25 % in fresh leaves)
- Cultar show positive response to flowering and fruiting in all spices.

108. "VANILLA"

- Malagasy Republic grows 70-80% of the world's crop of Vanilla bean followed by Reunion.

- USA is the largest importer.
- Introduced in India as early as 1835.
- In India grown in Wynad (Kerala) and Nilgiris (Tamil Nadu).
- Vanilla is an orchid.
- *V. planifolia* – Produce short thick pods.
- *V. pompana* – Produce largest pods.
- Natural pollination is impossible due to presence of rostellum which prevents contact of anther cap and stigma.
- In it's native home, melapone bees and humming birds aid in pollination.
- Propagation – By stem cutting (60-120cm long vine) having 5-6 internodes.
- Due to peculiar structure of the flower artificial polliniation by hand is the rule for fruit setting.
- 85-100% success is obtained by hand pollination.
- Ideal time for pollination – 6 am to 1 pm.
- Pods take 6 weeks to attain full size from fertilization but takes 4 to 10 months to reach full maturity.
- Harvesting – When immature, bean is dark green in colour but when ripe yellowing start from it's distal end.
- Oil is called balsom of vanilla (vanillin).
- Vanillin – 2.3-2.9% in beans.
- Vanillin is developed as a result of the enzyme (B-glycosidaes) action on the glucoside contained in the beans during the process of curing.
- The maxican process yield, 4.15 to 4.40% of vanillin content.
- Vanillin is extracted from the cured beans by hydroalcholic extraction.
- Vanillin content will not be less than 2.5% if properly cured beans are used.
- Caumarin – Cheap substitute for vanilla.
- Yield – 300-400 kg cured beans/ha.

109. "SAFFRON"

- Saffron is as "precious as gold" not only because of its low production and high demand but also because of its wide range usage in medicine and religious rituals.
- It is used to colour food even at 1 ppm concentration which gives distinct yellow tinge.
- Saffron contains 0.5 to 1.0% essential oil.
- Main principle is cicrocrocin – Responsible for colour.
- Stigma is funnel shaped, used a spice.
- Aroma is due to Saffranol.
- In India it is cultivated only in Kashmir.
- Bitterness is due to – Pecrocrosin.

- Propagation – By bulbs /corms best method (8 tonnes/ha.) in the month of August.
- Flowering – End of October (At this time picking of flower should be done early in morning).
- Shahi Saffron – The finest, purest and most expensive saffron in trade.
- Mogra Saffron and lache saffron – Inferiour grade.
- Stigma is used to prepare shahi saffron and mogra saffron and lache saffron are prepared by rest flower parts.
- About 1.5 million flowers on drying give 1 kg of saffron.
- Yield –

  160 kg fresh flower /hectare.

  5 kg dried saffron /ha.

110. "CURRY LEAF"

- It plays an important role as a condiment in the culinary preparation of South Indian dishes.
- Self pollinated crop.
- Richest source of calcium.
- Crystaline glucoside ksenigin - Obtained from leaves.
- Muragin – Obtained from flower.

Varieties –

- DWD 1 is sensitive to winter, during winter growth is poor, whereas DWD 2 is winter insensitive (gives high yield).
- Suwasani.
- Senkaampu.
- Mainly propagated through seed.
- Yield : 20-25 tonnes of leaves.
- DWD -1 having 5.22% oil where as DWD contains 4.09% oil, both have good aroma.

111. "PAPRIKA CHILLI"

- B.N. – *Capsicum annum.*
- F – Solanceae.
- Bright red, sweet to mild pungent, rich in vitamin C.
- Mainly used for adding colour to finished product.
- Colouring matter – Capsanthin (Outer fleshy pericarp).
- Pungent due to – Capsaicin (Inner tissue and seed).
- Spain, Hungary, Morocco, Yugoslavia and USA are the chief producer of paprika.
- Seed rate – 1 kg/ha.
- Yield – 200 kg/ha.

- Bird chilli – *Capsicum fruitiscens*.
- India is largest producer of chilli in World, accounting 45% of total area under cultivation.
- Variety suitable for pickle making – Sindhur.
- Variety suitable for capsicin extraction – (1) Aparna, (2) Pachas Yellow.
- About 90% capsicin in chilli has been noticed in placenta conecting seed with pericarp.
- Green chillies retard cancer due to natural enzyme asperginase.

112. "BISHOP'S WEED/AJOWAN"

- Alkaloid – Thymol.
- Thymene (dethymolized oil).
- Xanthotoxin (0.4%).
- Uses – Treatment of vitilago or leucoderma and in making suntan lotin.
- Seed rate – 3-4 kg/ha.
- Variety – Lam sel. 1,2.
- Essential Oil – 3-4%.
- Fatty Oil – 26 %.
- Sowing Time – Oct.-Nov.
- Harvesting – May-June.
- Yield – 8-10q/ha.

113. "THYME"

- Volatile oil 1.08-2.0%.
- Thymol (45%) – It is responsible for aroma and has germicidal property.
- Propagation – By seed and cutting.

114. "ROSEMARRY"

- Propagation is done by cuttings (15cm) long.
- Flowering – May-June.
- Harvesting is done after full flowering stage.
- Essential oil – Pinene, Comphene, Cineol, Camphore, Borneol and Bormylacetate.

115. "KOKUM"

- Dioecious in nature.
- Konkan Amrit is an important variety.

116. "ASAFOETIDA"

- Hadda is most priced and strongest trade.
- Hingashakta is prepared from this, used in ayurvedic medicines.
- Tapping is done foliage turn green to yellow.

117 "JUNIPER"

- Dioecious in nature.
- Juniperin – Mixture of tannins and sugar is obtained from juniper.

118. IMPORTANT PEST AND DISEASE OF SPICES AND CONDIMENTS CROPS

Sr. No.	Pest and Disease	Symptoms	Management
1.	Pollu beetle *(Longitarsus nigripennis)*	Grubs bore into the berries and feed on entire internal contents. Such hollow berries are called **'Pollu'** berries.	Spraying quinalphos (0.05% each) during July and Neemgold (0.6%) (neem-based insecticide) during August, September.
2.	Pepper mussel scale *(Lepidosaphes piperis)*	Infected parts become yellowish and later dry up.	Spray dimethoate 0.1 % followed by one more spray on 21st day of first spray.
3.	Top shoot borer *(Laspeyresia hemidoxa)*	Larvae bore into tender terminal shoots and feed on internal tissues resulting in blackening and decaying of shoots.	Spray quinalphos (0.05% each) on tender terminal shoots and repeat spraying at monthly intervals.
4.	Thrips (Sciothrips cardamom)	Formation of corky encrustations on pods resulting malformed and shrivelled.	Spray insecticides like quinalphos 0.025% or phosalone 0.07%.
5.	Shoot and capsule borer *(Dichocrocis (Conogethes) punctiferalis)*	Larva bores the unopened leaf buds and feeds the leaf tissue. Resulting drying of leaf and produce **'dead heart'** symptom.	In early stage it controlled effectively with insecticide but in later stage of higher dose of insecticide becomes ineffective.
6.	Beetle borer *(Onthophgus spp.)*	Adults beetle bores the flowers and feed on tissues.	Spraying of contact insecticide viz., quinalphos @2ml/lit.
7.	White flies *(Dialeurodes cardamom)*	Nymphs and adults remain in colonies on lower leaf surface and suck the sap.	Spray methyl demeton or dimethoate (0.05% each) on the foliage.

8.	Hairy caterpillars (*Eupterote cardamom)*	They fed voraciously on leaves and defoliating the whole cardamom plants.	Spray monocrotophos or quinalphos 3ml/lit.
9.	Cardamom aphid (*Pentalonia nigronervosa)*	The aphids act as a vector of the mosaic or **'Katte'** virus of cardamom.	Foliar spraying with dimethoate 2 ml/lit at regular interval.
10.	Root grubs (*Holotrichia serrata)*	Entire root system is found damaged resulting in drying and rotting.	Applying chlorpyriphos 0.075% or phorate 2-4 g a.i./plant in May and Sept-October.
11.	Shoot borer (*Conogethes punctiferalis)*	The presence of a bore-hole on the pseudo stem.	Spraying malathion 0.1% at 30 day intervals during July to October
12.	Leaf roller (*Udaspes folus)*	Leaves become folded or rolled longitudinally.	A spray with carbaryl (0.1%) or dimethoate (0.05%).
13.	Cinnamon butterfly (*Chilasa clytia)*	Larvae feed on tender and slightly mature leaves.	Spraying quinalphos 0.05% on tender and partly mature leaves.
14.	Scale insects (*Ceroplastes floridensis*)	Feed on plant sap and cause yellow spots on leaves, defoliation and wilting of shoots.	Spraying of dimethoate (0.05%) is recommended.
15.	Root-knot nematodes (*Meloidogyne incognita*)	Small leaves and bear fewer flowers. "Giant cells" within the root.	Fumigate or drench the nursery beds with 2 % Formalin.
16.	Wilt disease (*Fusarium Oxysporum* f.sp. *coriandèrii)*	The roots of affected plants shows vascular discolouration.	Soil drenching with Carbendazim 0.1% or copper oxy chloride 0.25%.
17.	Powdery Mildew (*Erysiphe polygoni)*	Small, white spots appear on leaf surface and later cover the entire surface.	Spray with wettable sulphur 0.2 per cent.
18.	Stem gall (*Protomyces macrospores)*	Galls appear on the leaves and stems of the plants affected by this disease.	Seed treatment and spray 0.1% solution of Carbendazim.
19.	Soft rot or rhizome rot (*Pythium aphanidermatum)*	Water soaked and the rotting spreads to the rhizome resulting in soft rot.	Seed tretment with mancozeb 0.3% for 30 minutes before storage and planting.
20.	Bacterial wilt (*Ralstonia solanacearum)*	Water soaked spots appear at collar region and vascular tissues show dark streaks.	Drenched with Bordeaux mixture 1% or copper oxychloride 0.2%.

21.	Leaf spot (*Phyllosticta zingiberi)*	Water soaked spot and later turns as white spot surrounded by dark brown margin.	Controlled by spraying Copper oxychloride 0.25 % or Mancozeb 0.2%.
22.	Slow decline/slow wilt (F*usarium, Rhizoctonia, Diplodia* and nematode)	Tip burn of leaves, die-back of twigs, root knot and vascular browning appear. The affected vines gradually die.	Soil drenching with Copper oxychloride in May– June and September– October.
23.	Azhukal disease/ capsule rot/fruit rot (*Phytophthora sp.)*	Large water soaked spots with black colour appear on leaves and the underground plants and rhizomes become rots.	Soil drench with Bordeaux mixture 1 % (or) Copper oxychloride 0.25%.
24.	Cardamom mosaic virus (CDMV)	General chlorosis of young leaves. Vector: Aphid (*Pentalonia nigronervosa).*	Spray with dimethoate (or) Phosphomidon to kill the vector.
25.	Anthracnose / Pollu disease (*Gloeosporium Spp.)*	Circular or irregular grey spots on the leaves and berry become chaffy (Pollu means hallow thing).	Three rounds of spray with Bordeaux mixture 1%.
26.	Phytophthora foot rot / quick wilt (*Phytopthora capsici)*	Die back, foot rot or collar rot and root rot	Spraying with 1% Bordeaux mixture (or) COC 0.25% (or) Alitte 0.3%.

MULTIPLE CHOICE QUESTIONS

ANNEXURE – I

1. Which vegetable crop is day neutral (JRF, 2004)
   (a) Tomato (b) Potato (c) Sweet potato (d) Spinach
2. In freezing which principle is followed (JRF, 2001)
   (a) Condensation (b) Crystallization (c) Drying (d) All
3. The long day type variety of onion is __________
   (a) Early Grano (b) Pusa White flat
   (c) Brown Spanish (d) Arka pitamber
4. AGMARK standard: (JRF, 2014)
   (a) Is related with agricultural products
   (b) approved by the Directorate of marketing and Inspection
   (c) Both (a) and (b)
   (d) None of the above
5. Seed Control Order was passed in the year: (JRF, 2016)
   (a) 1963 (b) 1966 (c) 1968 (d) 1983
6. Great lakes is a variety of – (ARS, 2018)
   (a) Amaranthus (b) Lettuce (c) Spinach (d) Portulaca
7. The economical part in broccoli is termed as _________
   (a) Curd (b) Flower bud (c) Flower stalk (d) Head
8. Ripening fruits exit which type of gas (JRF, 1999)
   (a) Ammonia (b) Ethylene (c) Methane (d) All
9. The vegetable having 70% exchange earnings in export of fresh vegetable is
   (a) Onion (b) Okra (c) Cucumber (d) Tomato
10. First Agricultural University established in India was: (JRF, 2014)
    (a) IIHR, Bangalore (b) PAU, Ludhiana
    (c) IARI, New Delhi (d) GBPU&T, Pantnagar
11. Which of the following is single seeded vegetable (JRF, 2001)
    (a) Cucumber (b) Leek (c) Chow-chow (d) Tomato
12. Onion variety suited for export purpose - (ARS, 2018)
    (a) Agrifound Dark Red (b) Agrifound light Red
    (c) Arka Niketan (d) Arka bindhu

13. Potato contains how much % of protein (JRF, 1999

(a) 1.8 (b) 2.2 (c) 1.4 (d) 3.2

14. Softness of Pea is measured by (JRF, 2002

(a) Derometer (b) Refractometer (c) Alanometer (d) Tendrometer

15. The native place of brinjal is: (JRF, 2014

(a) China (b) India (c) USA (d) Mexico

16. Seed rate of garlic for 1ha- (ARS, 2018

(a) 300-500 kg/ha (b) 500-600 kg/ha (c) 700-900 kg/ha (d) 900-1200 kg/ha

17. Indian institute of Vegetable Research is situated at: (JRF, 2014

(a) Lucknow (b) Delhi (c) Kanpur (d) Varanasi

18. Cryo - preservation is associated with (JRF, 2001

(a) Liquid nitrogen (b) Liquid oxygen

(c) Liquid potassium (d) Liquid carbon di-oxide

19. The process of bringing a wild species under human management is referred to as

(a) Acclimatization (b) Introduction (c) Selection (d) Domestication

20. Which vegetable is rich Calcium (JRF, 2004

(a) Amaranthus (b) Cowpea (c) Potato (d) Brinjal

21. Pusa Pragati is a variety of: (ARS, 2016

(a) Okra (b) Cucumber (c) Garden pea (d) Carrot

22. Study of wine making is known as: (JRF, 2014

(a) Enology (b) Fermentation (c) Enzymology (d) None of these

23. pH of tomato for processing should be less than: (ARS, 2016

(a) 2.5 (b) 4.5 (c) 6.5 (d) 7.5

24. For inactivation of microbial growth which method is followed (JRF, 2001

(a) Freezing (b) Sulphuring (c) Blanching (d) None

25. In tomato red colour is due to (JRF,1998

(a) Anthocynin (b) Lycopene (c) Xanthophylls (d) None

26. How many types of gametes (2n) will be produced by F2 in a dihybrid cross?

(a) 4 (b) 2 (c) 9 (d) 16

27. 'Blossom end rot' of tomato is due to deficiency of (JRF, 2002

(a) Mg (b) Bo (c) Mo (d) Ca

28. Which phenotypic ratio is expected in F_2 in case of complementary gene action?

(a) 12: 3: 1 (b) 13: 3 (c) 15 : 1 (d) 9 : 7

29. Onion bulb dipping is done for good quality, flowering using chemical growth regulator- (ARS, 2015

(a) NAA (b) 2,4 D (c) GA (d) IBA

30. Pungency of Garlic is due to (JRF, 2002

(a) Allyl propyl disulphide (b) Diallyl disulphide

(c) Isothiduinate (d) No

31. Saur- rubben is processed product of - (ARS, 2015

(a) Turnip (b) Beet root (c) Knol-Khol (d) Cabbage

32. Inbreeding depression is found in: (JRF, 2014)
(a) Radish (b) Cabbage (c) Onion (d) None of these
33. Pusa deepti is a variety of – (ARS, 2018)
(a) Cucumber (b) Tomato (c) Capsicum (d) Chilli
34. The crop which can be grown in acidic soils (JRF,1997)
(a) Spinach (b) Cauliflower (c) Watermelon (d) French bean
35. Arka Navneet is the variety of (JRF, 2001)
(a) Brinjal (b) Tomato (c) Okra (d) None
36. TSS of watermelon is (JRF, 2002)
(a) 10-12^0 Brix (b) 14-16^0 Brix
(c) 18-22^0 Brix (d) 20-22^0 Brix
37. Which of following is associated with vegetable research (JRF, 2001)
(a) Swaminathan (b) G. Kallo (c) T K Bose (d) Randhawa
38. The first step in genetic improvement of plant species is
(a) Domestication (b) Germplasm collection
(c) Selection (d) Hybridization
39. First India who won World Food Prize was awarded to: (JRF, 2014)
(a) Vergese Kurien (b) M.S. Swaminathan
(c) K.L. Chadha (d) S.K. Roy
40. Seed rate of cucumber per hectare is (JRF, 1999)
(a) 2.5-3.5 kg (b) 5-6kg (c) 1-2 kg (d) 6-8kg
41. Daria cultivation is followed in which crop (JRF, 2001)
(a) Tomato (b) Root crops (c) Cucurbits (d) Bulb crops
42. Bottleneck is a disorder of (JRF, 2002)
(a) Cucumber (b) Bottle gourd (c) Carrot (d) Rose
43. *Zanthosoma* is: (JRF, 2016)
(a) Tuber crop (b) Leafy crop (c) Fruit crop (d) Edible flower
44. Tomato leaf curl virus is transmitted by: (JRF, 2014)
(a) Thrips (b) White fly (c) Mites (d) None of the above
45. Johansen (1903) proposed the pure line theory on the basis of seed size in crop
(a) Potato (b) Sweet potato (c) French bean (d) Dolichos bean
46. In preservation of food how much quantity of colour is permitted (JRF, 2001)
(a) 250 ppm (b) 300 ppm (c) 200 pmm (d) 500 pmm
47. Among which variety of cucumber pure line method used - (ARS, 2018)
(a) Swarna Ageti (b) Pusa Sanyog (c) Himangi (d) Phule Shubha
48. Late blight of potato and tomato is caused by: (JRF, 2014)
(a) Phylophthora infestans (b) Alternaria solani
(c) Eruvinia crotovora (d) Pythium aphanidermatum
49. Edible part of Jerusalem artichoke is: (ARS, 2016)
(a) Stem (b) Leaf (c) Flower (d) Tuberous root
50. _______ state has monopoly in Black pepper production
(a) Karnataka (b) Kerala (c) Tamilnadu (d) AP

51. Leafy vegetables are stored through: (JRF, 2014)
    (a) Air blasting (b) Vacuum cooling (c) Freezing (d) Room cooling
52. Sree Dhara is a var. of ________
    (a) Sweet Potato (b) Cassava (c) Coleus (d) None
53. Which is prerequisite for genetic improvement of a vegetable crop?
    (a) Male sterility (b) Genetic variation
    (c) Sexual reproduction (d) Apomixis
54. In cabbage shot hole symptoms are due to: (ARS, 2016)
    (a) Fruit borer (b) White flies
    (c) Diamond Back Moth(d) Aphids
55. Bt brinjal is developed against the insect: (JRF, 2014)
    (a) Fruit and shoot borer(b) Whitefly
    (c) Epilachna beetle (d) MLO
56. Mudigree is a var. of ________
    (a) B. Pepper (b) Cardamom (c) Cinnamon (d) Clove
57. For measurement of TSS which instrument is used (JRF, 2001)
    (a) Dendrometer (b) Thermometer (c) Refractrometer (d) None
58. Arka Jyoti is an improved variety of
    (a) Muskmelon (b) Bottle gourd (c) Watermelon (d) Snap melon
59. State accounting maximum area & production of cardamom is- (JRF, 2003)
    (a) Tamilnadu (b) Kerala (c) Karnataka (d) Meghyala
60. Among the following Plant Growth Regulators, which one prevents abscission of young fruits? (JRF, 2014)
    (a) NAA (b) ABA (c) Ethrel (d) Mepiquat chloride
61. Which variety of sweet potato is rich in β-carotene content? (JRF, 2015)
    (a) Gauri (b) Rajendra shakarkand-35
    (c) Co-3 (d) Rajendra shakarkand-43
62. Tomato is a: (JRF, 2014)
    (a) Long-day plant (b) Short-day plant
    (c) Day neutral plant (d) None of the above
63. CIAH is located at: (JRF, 2015)
    (a) Jodhpur (b) Bikaner (c) Lucknow (d) Anand
64. The crop showing greatest inbreeding depression (JRF,1997)
    (a) Onion (b) Lettuce (c) Tomato (d) Lima bean
65. Size of cabbage butterfly larva is: (ARS, 2016)
    (a) 2 cm (b) 3 cm (c) 5 cm (d) 7 cm
66. Triploid watermelon is: (JRF, 2015)
    (a) Tetra-2 (b) Tetra-4
    (c) Arka Madhura (d) Pusa Rasal

67. For processing tomatoes are harvested at: (ARS, 2016)
(a) Red ripe stage (b) Green stage
(c) Yellow green stage (d) Pink stage
68. In mustard which type of incompatibility exist (JRF, 1999).
(a) Sporophytic (b) Gametophytic
(c) Both (d) None
69. ________ has antioxidant property
(a) Black pepper (b) Cardamom (c) Kokum (d) Turmeric
70. Hybrid RS-841 is a rootstock for __________
(a) Pointed gourd (b) Watermelon (c) Muskmelon (d) Cucumber
71. Betalains is a pigment in__________
(a) Beet (b) Bougainvillea (c) Cactus (d) All
72. Which growth regulator is substitute of low chilling treatment (JRF, 2004)
(a) Auxins (b) Cytokinin (c) Gibrillin (d) ABA
73. Why does blanching is practiced in vegetable (JRF, 2000)
(a) Micro organism removal (b) In-activate enzyme
(c) For softening the vegetable (d) All
74. Which one is a seed spice: (JRF, 2015)
(a) Saffron (b) Cumin (c) Clove (d) Kokum
75. Selection provides opportunity to isolate the promising genotypes from a
(a) Homogenous Population (b) Homozygous Population
(c) Heterogeneous Population (d) Heterozygous Population
76. The breeding line used in maintained A line is – (ARS, 2018)
(a) B line (b) C line (c) Both (d) None of these
77. Crossing over takes place in which stage of meiosis: (JRF, 2015)
(a) Zugotene (b) Doplotene (c) Pachytene (d) Liptotene
78. The hybrids of crop, mostly preferred in India are (JRF,1997)
(a) Brinjal (b) Cucumber (c) Onion (d) Tomato
79. Following seed rate is recommended for sowing one hectare watermelon (JRF, 2002)
(a) 1-1.5 kg (b) 2-2.5 kg (c) 4.5-5.0 kg (d) 5.5 – 6.5 kg
80. Mass selection is practiced on the basis of
(a) Genotype (b) Phenotype (c) Environment (d) All of Above
81. Homozyocity and homogeneity is maximum in (JRF, 1999)
(a) Pure line (b) Inbred line (c) Land races (d) Synthetic varieties
82. Which one is a shallow rooted vegetable (JRF, 2015)
(a) Tomato (b) Pea (c) Carrot (d) Leek
83. Seed Plot Technique is related to: (JRF, 2014)
(a) Potato (b) Tomato (c) Sweet Potato (d) Radish
84. Which of the following is used to check sprouting of onion under storage?
(a) NAA (b) GA (c) MH d. PCPA

85. Which of the following seed rate of TPS is recommended for planting one ha area?
(a) 150 g (b) 250 g (c) 350 g (d) 450 g
86. Which one of the following is a major disease of onion in India
(a) Leaf blight (b) Purple blotch (c) Smut (d) White rot
87. Which one of the following soil types is best for vegetable cultivation?
(a) Sandy (b) Clay (c) Sandy loam (d) Clay loam
88. The main function of cytokinin is (JRF, 2004)
(a) Cell enlargement (b) Cell division (c) Cell elongation (d) All
89. Lincoln is an exotic variety of: (JRF, 2015)
(a) Cowpea (b) Pea (c) French bean (d) Broad bean
90. The varieties prefixed with Arka are released from
(a) IIHR (b) IARI (c) CPRI (d) CTCRI
91. The concept of Centre of origin of crop plants was proposed by
(a) JR Harlan (b) NI Vavilov (c) JG Mendal (d) C. Linnaeus
92. Which is the anti gibberellin factor (JRF, 2004)
(a) Cycocel (b) BA (c) Cultar (d) IBA
93. Bordeaux mixture was given by: (JRF, 2015)
(a) Millardet (b) Mendel (c) Robert Hook (d) None of the above
94. Double bean is a common name of: (ARS, 2016)
(a) Broad bean (b) Indian bean (c) French bean (d) Lima bean
95. A vegetable crop showing less than 5 per cent cross-pollination is considered as
(a) Cross-pollinated (b) Self-pollinated
(c) Often Self-pollinated (d) Often cross-pollinated
96. Which variety of carrot is released from IARI Katrain Station: (JRF, 2015)
(a) Pusa Kesar (b) Pusa Meghali (c) Pusa Yamdagni (d) Zeno
97. Cleistogamy conditions of flower promotes
(a) Self-pollination (b) Cross-pollination (c) Apomixis (d) Male sterility
98. According to boswell, cole crops are originated from: (ARS, 2016)
(a) Italy (b) Cyperus (c) Japan (d) China
99. Vegetative reproduction in vegetable crops may occur through
(a) Rhizome, corm, stem cuttings and grafting
(b) Tuber, bulb and rhizome
(c) Runner, stolon, sucker and meristem culture
(d) All of the above
100. Family of sweet potato is (JRF, 2002)
(a) Euphorbiacae (b) Aizoiaceae (c) Agavaceae (d) Convolvulaceae
101. Rio-De-Janerio is a variety of: (JRF, 2015)
(a) Black pepper (b) Fenugreek (c) Ginger (d) Turmeric
102. Which is correct about the cells present in embryosac?
(a) Egg cell, two antipodal cells, a secondary nucleus and two synergids
(b) Three antipodal cells, two synergids, one egg cell and one 2n secondary nucleus

(c) One egg cell, two synergids and two secondary nucleus

(d) A 2n secondary nucleus, one egg cell and three antipodal cells

103. Which cucurbit have highest vitamin A (JRF, 2001)

(a) Bitter gourd (b) Cucumber (c) Pumpkin (d) All

104. Bolting of onion takes place if temperature goes below (JRF, 2004)

(a) 10°C (b) 15°C (c) 20°C (d) 25°C

105. Most pungent part of chilli is: (JRF, 2015)

(a) Seeds (b) Pericarp (c) Placenta (d) Tendril

106. Plant introduction consist of

(a) Procurement and quarantine (b) Multiplication and distribution

(c) Cataloguing and evaluation (d) All of above

107. In which of the following underground root is edible (JRF, 2002)

(a) Potato (b) Ginger (c) Onion (d) Radish

108. Greening in potato is due to (JRF,1998)

(a) High temperature (b) High light intensity

(c) Low temperature (d) All

109. Bitterness in bitter gourd is due to: (JRF, 2015)

(a) Momordicidin (b) Cucurbitacin

(c) Allyl propyl di sulphide (d) Cheratin

110. Cross-pollination in plants is encouraged by

(a) Monoecy (b) Dicliny (c) Dichogamy (d) All above

111. Storing of seed at freezing temperature (JRF, 2004)

(a) Orthodox (b) Recalcitrant

(c) Ex-situ conservation (d)In situ conservation

112. Anther mutant cultivar of Tomato is (JRF, 2002)

(a) Pusa Divya (b) Pusa Gaurav (c) Pusa Ruby (d) None

113. How much isolation distance is maintained for foundation seed production in onion? (JRF, 2015)

(a) 400 m (b) 200 m (c) 1000 m (d) 100 m

114. The varieties prefixed with Pusa are released from

(a) IIHR (b) IARI (c) CPRI (d) CTCRI

115. Oxalates are present more in (JRF, 2001)

(a) Leafy vegetables (b) Root vegetable

(c) Bulb crops (d) None

116. Which phenotypic ratio is expected in F_2 in case of masking gene action

(a) 12 : 3 : 1 (b) 13 : 3 (c) 15 : 1 (d) 9 : 7

117. Potato verity more suitable for chips making (JRF, 2002)

(a) Kufri Jyoti (b) Kufri Naveen (c) Kufri Sheetman (d) Kufri Chipsona

118. The ability of cell to generate into a whole plant is known as (JRF, 2001)

(a) Regenaration capacity (b) Totipotancy

(c) Autolysis (d) None

119. Temporarily self-incompatibility may be over come through
(a) Irradiation and bud pollination
(b) Bud pollination and heat shock
(c) Grafting and by removal of stigmatic surface
(d) All of the above

120. Which operation for temporary suppression of self-incompatibility reaction is most successful with both sporophytic and gametophytic systems.
(a) Bud pollination (b) Removal of stigma
(c) Grafting (d) Irradiation

121. White rust of crucifer is caused by (JRF,1998)
(a) Crostopus candida (b) Albugo cadida
(c) Peranosporra (d) Cleviceps

122. Complete genetic heteterozygosity & heterogenity is found in (JRF, 2002)
(a) F_1 hybrids (b) Pureline (c) Multilines (d) Synthetics

123. Which of the following is /are wind pollinated? (JRF, 2013)
(a) Radish (b) Beet (c) Amaranthus (d) Both(b) and (c)

124. In a typical monohybrid cross, genotypic ratio in F_2 will be
(a) 1 : 2 : 1 (b) 1 : 1 : 1 (c) 3 : 1 : 1 (d) 2 : 1 : 1

125. Which of the following vegetable belong to family- composite (JRF, 2002)
(a) Spinach (b) Sorrel (c) Endive (d) Chive

126. Phytoton-30 is used for inducing flowering in ---- (ARS, 2018)
(a) Long day (b) Short day (c) Day neutral (d) All of these

127. Sabudana/sago is prepared from – (JRF, 2015)
(a) Tapioca (b) Cassava (c) Both a and b (d) Potato

128. Which phenotypic ratio is expected in F_2 in a monohybrid cross?
(a) 1 : 2 : 1 (b) 1 : 1 : 1 (c) 3 : 1 (d) 2 : 1 : 1

129. Safed musli belongs to genus: (JRF, 2015)
(a) Chlorophytum (b) Anethum (c) Casia (d) Cephaelis

130. Photophosphorylation is (JRF, 1999)
(a) Hill reaction
(b) Formation of ATP in presence of light
(c) Formation of NADP in presence of light
(d) None

131. Which of the following is four angled Bean (JRF, 2002)
(a) Winged Bean (b) Cluster Bean (c) Lima Bean (d) Broad Bean

132. Which phenotypic ratio is expected in F_2 in case of polymeric gene action?
(a) 12 : 3 : 1 (b) 9 : 6 : 1 (c) 15 : 1 (d) 9 : 7

133. SPT in potato was given by- (1965) (ARS, 2018)
(a) Rattan lal (b) Pushkarnath (c) V.L Chopra (d) M.S. Swaminathan

134. _________ is barrier of self pollination in vanilla
(a) Labellum (b) Rostellum (c) Anther cap (d) Column

135. Fruit of brassica is called as (JRF, 1999)
(a) Pome (b) Berry (c) Siliqua (d) Drupe

136. Catface is an important disorder of: (JRF, 2014)
(a) Tomato (b) Chilli (c) Brinjal (d) Apple

137. Calcium deficiency causes blossom end rot in (JRF-2008)
(a) Potato (b) Tomato (c) Cauliflower (d) Cabbage

138. Pungency in radish is due to: (JRF, 2015)
(a) Isothiocyanate (b) Allyl propyl di sulphide
(c) Di allyl di sulphide (d) Allyl isothiocynate

139. Which of the following cucurbits exhibits vivipary
(a) Cucumber (b) Chow Chow (c) Pumpkin (d) Muskmelon

140. In 1948, which of the following terms was coined by Thimann: (JRF, 2016)
(a) Ethylene (b) Hormone (c) Phytohormone (d) Animal hormone

141. Application of which one can reduce enzymatic oxidation of patato: (JRF, 2015)
(a) N fertilizer (b) SSP (c) DAP (d) Kcl

142. French bean variety Pusa parvati is developed by X-rays irradiation by the utilizing parents- (ARS, 2018)
(a) Contender (b) Wax pod (c) Tender Green (d) Kentucky Wonder

143. The dioecious crop among the following is
(a) Brinjal (b) Sweet potato (c) Asparagus (d) Ridge gourd

144. Delayed harvesting in radish causes: (JRF, 2015)
(a) Pithiness (b) Forking (c) Splitting (d) Hollow root wart

145. The no of cultivated yam spp in India- (ARS, 2018)
(a) 2 (b) 4 (c) 3 (d) 5

146. Which of the following cucurbit is dioecious
(a) Pointed gourd (b) Chayote (c) Little gourd (d) All of these

147. Absence of axillary bud is the typical feature of
(a) Cauliflower (b) Cabbage (c) Brussels sprout (d) Sprout broccoli

148. Important insect vector of plant viruses are: (JRF, 2016)
(a) Aphids (b) Thrips (c) Bugs (d) Bacteria

149. How many genotypes will be possible in F_2 (3^n) in case of a trihybrid cross
(a) 16 (b) 9 (c) 27 (d) 81

150. Most suitable radiation for disinfection of food is (JRF, 2001)
(a) UV rays (b) Gamma rays (c) Alpha rays (d) All

151. Which acid is present in turmeric? (JRF, 2015)
(a) Citric acid (b) Tartaric acid (c) Iso citric acid (d) Mallic acid

152. The main constituent of cinnamon bark is ________
(a) Eugenol (b) Cinnamaldehyde (c) Eugeny l acetate (d) Limone

153. ________ var. of tomato sets fruits at low temp.
(a) Pusa Ruby (b) Pant bahar (c) Hissar Arun (d) Pusa sheetal

154. Seed rate of cauliflower is (JRF, 1997)
(a) 100 g/ha (b) 200 g/ha (c) 350-600 g/ha (d) 1000 g/ha

155. Rossette formation is been due to deficiency of (JRF, 1999)
(a) Mn (b) Zn (c) Cu (d) Mo

156. How many types of gametes will be expected in F_1 (2^n) in a trihybrid cross
(a) 4 (b) 8 (c) 27 (d) 16

157. Highest content of polyphenols is present in: (JRF, 2016)
(a) Apple (b) Beet root (c) Pepper (d) Blue berries

158. _________ is a self blanched var. of cabbage
(a) Pusa shubhra (b) Pusa sharad (c) Pusa Him jyoti (d) Pusa snowball

159. Male sterile line is maintained by (JRF, 1999)
(a) Crossing with heterozygous (b) Male fertile line
(c) Both (d) None

160. The smallest perfect F_2 population (4^n) will be in case of a dihybrid cross
(a) 4 (b) 8 (c) 27 (d) 16

161. Sweet potato is: (JRF, 2016)
(a) Diploid (b) Triploid (c) Hexaploid (d) Tetraploid

162. In plant spacing 2x2, how much plant can be accommodated (JRF, 2001)
(a) 2000 (b) 2500 (c) 3000 (d) 3500

163. In a monohybrid cross, in case of incomplete dominance, phenotypic ration in F_2
(a) 1 : 2 : 1 (b) 1 : 1 : 1 (c) 3 : 1 (d) 2 : 1 : 1

164. In tomato blossom end rot is due to (JRF, 1999)
(a) Ca deficiency (b) Low Temp (c) High temp (d) Very high humidity

165. _______ element is necessary for root growth
(a) N (b) P (c) K (d) Zn

166. The Indian institute of Vegetable Research is located at (JRF, 2001)
(a) Ranchi (b) Delhi (c) Varanasi (d) Bangalore

167. Which of the following is a perennial in nature? (JRF, 2014)
(a) Bitter gourd (b) Cucumber (c) Chow Chow (d) Ridge gourd

168. The active principle of clove oil is (JRF-2008)
(a) Macine (b) Oleum (c) Aldehyde (d) Eugenol

169. Synthetic variety of cabbage is: (JRF, 2016)
(a) Pusa Mukta (b) Pusa Sambandh (c) Pusa Ageti (d) Pusa giant

170. Bareja system is the term associated with
(a) Nutmeg (b) All spice (c) Betelvine (d) Black pepper

171. Vegetable included in solanaceous fruit group (JRF-2005)
(a) Potato (b) Tomato (c) Sweet Potato (d) Okra

172. Leading producer of large cardamom is
(a) Indonesia (b) China (c) Sri Lanka (d) India

173. Crossing over occurs during
(a) Leptotene (b) Pachytene (c) Zygotene (d) Diplotene

174. _______ acid present in Kokum
(a) Citric (b) Malic (c) Acetic (d) Hydroxy citric

175. Which vegetable crop is propagated by stem cutting (JRF, 1999)
(a) Raddish (b) Carrot (c) Cucumber (d) Sweet potato

176. Andromonoecious sex for is generally found in: (JRF, 2016)
(a) Cucumber (b) Musk melon (c) Pumpkin (d) Pointed gourd

177. Saffron belongs to family
(a) Fabaceae (b) Lauraceae (c) Iridaceae (d) Myrtaceae

178. Number of all possible crosses in a diallel cross (excluding reciprocals) are
(a) n (n-1) (b) n (n-1)/2 (c) n (n-1)/3 (d) $n^2/2$

179. Damping off is caused by (JRF, 1999)
(a) Pythium spp (b) Fusarium (c) Alternaria (d) All

180. Arka Ajit is a improved variety – (ARS, 2018)
(a) Tomato (b) Pea (c) Okra (d) Muskmelon

181. Which is the following is used for production of Virus free plants? (JRF-2006)
(a) Nodes (b) Meristem (c) Anther (d) Ovule

182. Sporophytic system of incompatibility is found in: (JRF, 2018)
(a) Cabbage (b) Cucumber (c) Tomato (d) Onion

183. Which type of broccoli is most nutritive
(a) Purple coloured (b) White coloured (c) Green coloured (d) None of the above

184. Aijowan ka phool is a
(a) Crystal form of thymol (b) Liquid form of Thymol
(c) Crystal form of cineol (d) Crystal form of Linalool

185. Quarantine inspection of Agricultural and Horticultural crops is carried out by (JRF-2008)
(a) Botanical syrvey of India-kolkatta
(b) Forest Research Institute-Dehradun
(c) Indian Agicultural Research Institute-New Delhi
(d) National Bureau of Plant Genetic Resources- New Delhi

186. Pusa nasdar is a variety of – (ARS, 2018)
(a) Ridge gourd (b) Sponge gourd (c) Bottle gourd (d) Pointed gourd

187. Which of the following is the oldest known spice
(a) All spice (b) Nutmeg (c) Black pepper (d) Cassia

188. _______ is called as chinease potato
(a) Coleus (b) Canna indica (c) Giant taro (d) WI Arrowroot

189. Gerkhins belong to which genus (JRF, 2001)
(a) Cucurbit (b) Cucumis (c) Citrulus (d) None

190. Number of single crosses in full diallel are
(a) n (n-1) (b) n (n-1)/2 (c) 2n (n-1)/2 (d) n^2

191. Soft rot of vegetables is caused by (JRF, 1999)
(a) Verticillium (b) Erwinia carotovora
(c) Fusarium (d) None

192. Heterosis is the result of which type of gene effects
(a) Over-dominance (b) Dominance (c) Additive (d) Epistatic

193. Cheapest method of preservation is - (ARS, 2018)
(a) Washing (b) Drying (c) Freezing (d) Sugaring

194. Seeds of fenugreek contain ___________
(a) Eugenol (b) Cheratin (c) Trigonelin (d) Diosgenin

195. Spinach belongs to family _________
(a) Compositae (b) Spinaceae (c) Umbeliferae (d) Chenopodiaceae

196. Which refrigerant is commonly used in cold storage? (JRF, 2016)
(a) Ethylene (b) Ammonia (c) Carbide (d) Sodium benzoate

197. All India Co-ordinated Potato Improvement Project started in________
(a) 1960 (b) 1982 (c) 1975 (d) 1971

198. Jade cross is a F1 hybrid of ________.
(a) Brussels sprout (b) Spouting brocoli (c) Knol-khol (d) Cabbage

199. The variety resistant to black leg in cabbage is ________
(a) Pusa Mukta (b) Golden
(c) Pusa Drum head (d) Pusa sambandh

200. Temporary method of preservation is- (ARS, 2018)
(a) Chilling (b) Dehydration (c) Sugaring (d) Freezing

201. How much area is suitable for kitchen gardening for a family of five (JRF, 2004)
(a) 50 m^2 (b) 100 m^2 (c) 200 m^2 (d) 250 m^2

202. Akashin a disorder in Radish is due to deficiency of __________
(a) Manganese (b) Boron (c) Magnesium (d) Galcium

203. The single seeded cucurbit is _______
(a) L. acutangula (b) L. culindrica (c) C. moschata (d) S. edule

204. Which can is used in canning of pea- (ARS, 2018)
(a) Coloured can (b) Plain can (c) Both (d) None of these

205. The main site for dark reaction of photosynthesis is: (JRF, 2016)
(a) Stroma (b) Grana (c) Chloroplast (d) Chromoplast

206. Ideal storage temperature for brinjal is (JRF, 2002)
(a) 14-15°C (b) 10°C (c) 0°C (d) 20°C

207. Sweet potato is an autopolyploids and has the ploidy level of
(a) 3x (b) 4x (c) 5x (d) 6x

208. Nitrogen deficiency in plants leads to- (JRF, 2003)
(a) Excessive (b) Profuse flowering
(c) Lush green colour (d) Chlrosis

209. Number of all possible three way crosses among n lines should be
   (a) n (n – 1) (n – 2)/2 (b) n (n – 1) (n – 2)/3
   (c) n (n – 1) (n – 2)/4 (d) n (n – 1) (n – 2) (n – 3)/3
210. Self-pollination increases
   (a) Homozygosity (b) Heterozygosity (c) Homogeneity (d) Heterogeneity
211. Which of following is hybrid variety of black pepper (JRF, 2001)
   (a) Penyur-3 (b) Penyur-2 (c) Penyur-4 (d) None
212. In purelines variety genetic variation may arise over a period of time due to
   (a) Mechanical mixture (b) Out-crossing
   (c) Mutation (d) All above
213. Clove belong to family (JRF,1997)
   (a) Leguminoceae (b) Myrtaceae (c) Stirculiaceae (d) Rosaceae
214. Name the non-parasitic disorder due to Ca deficiency in tomato? (JRF, 2016)
   (a) Blossom end rot (b) Radial cracking
   (c) Blotchy ripening (d) Catface
215. Ethylene absorbent- (ARS, 2018)
   (a) $KMNO_4$ (b) $MgSo_4$ (c) $NaNo_3$ (d) K_2So_4
216. Purity of pureline varieties can be maintained through regular
   (a) Pedigree selection (b) Mass selection
   (c) Recurrent selection (d) Pure line selection
217. Bolting problem in onion occur if temperature goes below (JRF, 1999)
   (a) 25°C (b) 20°C (c) 15°C (d) 10°C
218. Swarna rekha is the var of (JRF, 2001)
   (a) Pointed gourd (b) Bitter gourd (c) Bottle gourd (d) None
219. First auxin developed by T.W. Went in the year 1928 was: (JRF, 2014)
   (a) IAA (b) IBA (c) PAA (d) None of the above
220. Which crop is highly tolerant to soil acidity? (JRF, 2016)
   (a) Onion (b) Potato (c) Sweet potato (d) Okra
221. Pureline selection can be applied for the improvement of
   (a) Local varieties (b) Introduced varieties
   (c) Old pureline varieties (d) All the above in self-pollinated crops
222. Preservation of cells, tissues and organ in liquid nitrogen is known as: (JRF, 2014)
   (a) Cryopreservation (b) In-situ preservation
   (c) Cold preservation (d) None of these
223. First commercial product of C3 plant in photosynthesis – (ARS, 2018)
   (a) CO_2 (b) Oxalic acid (c) Acetic Acid (d) Phosphoric Acid
224. In Brinjal only those flowers set fruits which have (JRF,1997)
   (a) Short (b) Long style (c) Medium (d) Both B & C
225. The process of synthesis of mRNA from a DNA template is known as: (JRF, 2016)
   (a) Translation (b) Transcription (c) Transduction (d) Transformation

226. In a random mating population , gene and genotypic frequency remains same generation after generation provided that there is no
(a) Selection (b) Migration (c) Mutation (d) Random drift
(e) All of above

227. Browning in cauliflower is due to (JRF, 1997)
(a) Cu (b) Bo (c) Ca (d) Mo

228. Fruit and shoot borer is a serious pest of which vegetable crop? (JRF, 2015)
(a) Tomato (b) Brinjal (c) Onion (d) Chilli

229. The gain under selection for quantitative traits is impeded by
(a) Non-additive gene action (b) Low selection intensity
(c) Low heritability of the trait (d) All of the above

230. In which crop the adventitious root is present (JRF, 1999)
(a) Sweet potato (b) Turnip (c) Radish (d) Carrot

231. Effect of self-fertilization on the frequency of one of the two homozygote (AA/aa) at a single locus (A/a) after n generation of selfing will be calculated by formulae
(a) $2^{n-1/2n+1}$ (b) $2^{n/2n+1}$ (c) $2^{n-1}/2^{n}$ (d) $2^{n+1}/2^{n}$

232. Value added and fermentation product from white cabbage is: (JRF, 2014)
(a) Sauerkraut (b) Pickle (c) Silage (d) All of the above

233. Tag colour of certified seed is: (JRF, 2015)
(a) Golden yellow (b) White (c) Purple (d) Blue

234. Vanilla grows best at a height of (JRF,1997)
(a) 1 m (b) 2 m (c) 3 m (d) 0.5 m

235. The micro-organism causing rotting in vegetable crop is – (ARS, 2018)
(a) Alternaria (b) Phomopsis (c) Diplodia (d) All of these

236. Television, radio, fairs and exhibitions are the media of: (JRF, 2016)
(a) Group communication (b) Individual communication
(c) Community communication (d) Mass communication

237. Multiple diseased resistance variety for DM, PM, Anthracnose, Angular leaf spot in cucumber- (ARS, 2018)
(a) Poinsette (b) Japanese long green
(c) Pusa Sanyog (d) Phule Shubhangi

238. Which of following is responsible for sourness of radish (JRF, 2000)
(a) Isothiosinate (b) Anthocynins (c) Auxins (d) All

239. For seed production of cucumber, maintain following m. isolation between two cultivars
(a) 800 m (b) 1000 m (c) 1200 m (d) 1600 m

240. What is the threshold level (μL/L) of ethylene in fruits and vegetables? (JRF, 2016)
(a) 0.03 (b) 0.02 (c) 0.01 (d) 0.04

241. Heterosis in bottle gourd was reported by
(a) I.D. Tyagi (b) G.S. Gill (c) V.S. Seshadri (d) U.K. Verma

242. Self incompatibility results due to (JRF-2008)
(a) Genetic causes (b) Morphological causes
(c) Physiological and Biochemical causes (d) All of the above

243. How can one protect the stem-end of radish root from becoming green
(a) By thinning (b) By earthing (c) By staking (d) All of the above

244. The staple vegetable in our food is
(a) Potato (b) Brinjal (c) Chillies (d) Tomato

245. India is second largest producer of vegetables after (JRF, 1999)
(a) Brazil (b) China (c) Africa (d) Japan

246. Part of onion we eat is called
(a) Corm (b) Bulb (c) Rhizome (d) Tuber

247. The pigment present in watermelon is (JRF, 2002)
(a) Carotenoid (b) Anthocynin (c) Lycopene (d) Both b & c

248. Food storage organs of ginger and turmeric are called
(a) Rhizome (b) Tuber (c) Corm (d) Root

249. Inbreeding is commercially used in improvement of - (ARS, 2018)
(a) Soleneceous (b) Cucurbits (c) Crucifers (d) Compositae

250. Name the chemical which is used to preserve coloured juices: (JRF, 2016)
(a) Sodium chloride (b) Sodium benzoate (c) Tartaric acid (d) Sugar

251. First time cytoplasmic genetic male sterility is used in (JRF, 2000)
(a) Onion (b) Pea (c) Ground nut (d) Carrot

252. Population of cross-pollinated vegetable crops (composites and synthetics) are
(a) Homozygous and homogenous (b) Homozygous and heterogeneous
(c) Heterozygous and heterogeneous (d) heterozygous and homogenous

253. Mass selection is based on the
(a) Progeny test (b) Maternal parent only
(c) Paternal parent only (d) Both parents

254. Proportion of completely homozygous plants after m generation of selfing when n number of genes are segregating
(a) $[2^m - 1/2^m]^n$ (b) $[2^n - 1/2^n]^m$ (c) $[2^m - 1/2^{m+1}]^n$ (d) $[2^m - 1/2^{m+1}]^n$

255. The colour of tag used for breeder seed is: (JRF, 2016)
(a) Golden Yellow (b) White (c) Blue (d) Breen

256. IIHR headquarter is located at (JRF,1997)
(a) Kasargod (b) Calicut (c) Kottayan (d) Bangalore

257. VK-1/ Priya is a cultivar of ________
(a) Snake gourd (b) Bitter gourd (c) Ridge gourd (d) None

258. The ETL of aphid population in case of Potato is ________
(a) 20/100 leaves (b) 30/100 leaves (c) 10/100 leaves (d) 40/100 leaves

259. Anther culture is usually used for the production of: (JRF, 2016)
(a) Diploid plants (b) Haploid plants (c) Triploids (d) Tetraploids

260. Propagation of curry leaves is generally done by ________
(a) Suckers (b) Seed (c) Both i and ii (d) Stem cutting

261. Which is non-pungent vegetable when uncooked (JRF, 2004)
(a) Lettuce (b) Radish (c) Onion (d) Chilli

262. Whiptail of cauliflower is due to deficiency of (JRF, 1999
(a) Mo (b) Zn (c) Co (d) Cu
263. Simple recurrent selection is most suitable for characters with
(a) High heritability (b) Low heritability
(c) Moderate heritability (d) All of the above
264. Potato is a (JRF, 2002
(a) Allohexaploid (b) Autohexaploid (c) Auto tetraploid (d) Allotetraploid
265. Which micronutrient is mostly deficient in Indian soils? (JRF, 2016
(a) Mn (b) Zn (c) Fe (d) B
266. The difficulties encountered in the production of interspecific hybrids may be broadl classified into
(a) Failure of zygote formation and hybrid sterility
(b) Failure of zygote development and F1 seedling development
(c) Failure of zygote formation and development
(d) All of the above
267. Bacterial milt resistant variety of Brinjal (JRF, 2004
(a) Arka Abha (b) Arka Vardan (c) Arka Gaurav (d) All
268. Most efficient species of earthworm for vermicomposting in tropical regions is: (JRF, 201
(a) Eisenia foetida (b) Eudrilus eugeniae
(c) Perionyx excavates (d) Bimastos parvus
269. Bush pepper is produced from- (JRF, 200
(a) Runners (b) Laterals (c) Top shoots (d) Hanging shoots
270. Efficiency of mass selection principally depends on the
(a) Heritability of trait (b) Gene frequency
(c) Number of genes controlling the trait (d) All of the above
271. Food Corporation of India (FCI) was established in year (JRF, 200
(a) 1950 (b) 1955 (c) 1960 (d) 1965
272. Arka Abir is the variety of (JRF, 200
(a) Brinjal (b) Cucurbit (c) Tomato (d) Chilli
273. Sterility in distant hybrids may be due to the
(a) Cryptic structural hybridity (b) Cytoplasm of alien species
(c) Lack of chromosome pairing (d) All of the above
274. Which vegetable is dioecious in nature and propagated by stem cutting (JRF, 2002
(a) Spinach (b) Sweet potato (c) Tobacco (d) Pointed gourd
275. Which of following is responsible for cracking in tomato (JRF, 200
(a) B deficiency (b) Ca deficiency (c) S deficiency (d) Mn deficiency.
276. Name the Onion variety which is pickling type and suitable for export: (JRF, 2016
(a) Agrifound Dark Red (b) Agrifound Light Red
(c) Agrifound Red (d) Agrifound Rose

277. The FPO licence is given by (JRF, 2001)
(a) Govt of India (b) State Govt (c) Agri University (d) None
278. Development of a variety usually takes
(a) 4 – 6 years (b) 7 – 8 years (c) 10 – 12 years (d) 15 – 20 years
279. Yellow colour of onion is due to (JRF, 2002)
(a) Qucertin (b) Myrcetin (c) Allicin (d) Malvelin
280. High CO2 and low O2 storage refers to: (ARS, 2016)
(a) Zero Energy Cool Chamber (b) Controlled Atmospheric Storage
(c) Both (d) None of these
281. For simply inherited characters, selection may be started from
(a) F_4 (b) F_3 (c) F_2 (d) F_1
282. PK-ft-16 is the variety of (JRF, 2001)
(a) Chilli (b) Tomato (b) Brinjal (d) Cucurbirts
283. Segregation for different characteristics does not fit Mendelian ration in F_2 from wide crosses due to
(a) Irregular meiosis
(b) Impaired recombination
(c) Reduced survival of recombinant gametes and zygotes
(d) All of the above
284. How many agro-cliamtic zones (ACZ) are here in India: (JRF, 2016)
(a) 5 (b) 15 (c) 25 (d) 30
285. Recommended seed rate (g/ha) of cabbage
(a) 300-400 g (b) 400-500 g (c) 100-200 g (d) 500-600 g
286. Tospo virus infects which crop: (JRF, 2016)
(a) Tomato (b) Cabbage (c) Brinjal (d) Palak
287. In India who has introduced climate controlled green house cultivation (JRF-2006)
(a) ICAR (b) Mahyco seed Co.
(c) Indo-American Hybrid Seed Co. (d) Suton & Sons
288. Seed rate per hectare of cluster bean is
(a) 10-20 g (b) 5-10 kg (c) 40-50 kg (d) 30-40 kg
289. Vegetables are rich source of
(a) Carbohydrate (b) Vitamins (c) Proteins (d) None of the above
290. Eugenol is the main constituent in the leaf oil of (JRF-2006)
(a) Cassia (b) Cinnamon (c) Bal leaf (d) Curry leaf
291. The most favourable range of soil pH for tomato cultivation is: (JRF, 2016)
(a) 4.5-5.5 (b) 7.5-8.5 (c) 6.0-7.0 (d) 7.2-8.0
292. Chemical constituent present in pepper
(a) Piperine (b) Curcumine (c) Cineol (d) Eugenol

293. Vegetables for processing are subjected to drying after
(a) Blanching (b) Without blanching
(c) Sulphuring (d) Any other means

294. Which of the following is the early variety of cabbage
(a) Pride of India (b) Early Drum Head (c) Chieftain (d) None of the above

295. Most important factor in a hybridization programme is
(a) Choice of parents for hybridization (b) Emasculation
(c) Handling of segregating generations (d) Pollination

296. Browning due to boron deficiency is common in- (JRF, 2003
(a) Radish (b) Cauliflower (c) Onion (d) Tubers

297. Which vegetable crop are rich source of niacin (JRF, 2004
(a) Spinach (b) Cowpea (c) Raddish (d) Sweet potato

298. Heterostyle is found in: (JRF, 2018
(a) French bean (b) Brinjal (c) Chilli (d) Tomato

299. ________ is called as stress hormone
(a) Auxine (b) Gibberellins (c) Cutotokinins (d) ABA

300. ________ crop/crops is /are suitable for greenhouse cultivation
(a) Sweet pepper (b) Cucumber (c) Tomato (d) All

301. 'Hortscience' is published by
(a) ICAR (b) Hortil. Soc. of India
(c) ASHS (d) Govt. of India

302. The ________ is done to obtain virus free plants
(a) Pollen culture (b) Cell culture (c) Embryo culture (d) Meristem culture

303. ________ is the site of B-oxidation in cell.
(a) Chloroplast (b) Vacuole (c) Mitochondia (d) Ribosomes

304. Edible part in Jerusalem artichoke is- (JRF, 2003
(a) Seed (b) Cutting (c) Layers (d) Tubers

305. Which vegetable has high calorie value (JRF, 2004
(a) Fenugreek (b) Amaranthus (c) Potato (d) Tomato

306. ________ is 1st tropical variety of cabbage developed for cultivation under high temperatur conditions at IARI, New Delhi.
(a) Pusa Sambandh (b) Pusa Mukta
(c) Pusa Drum Head (d) Pusa Ageti

307. Yellow anther type in case of chilli is ________
(a) Bhagy lakshmi (b) Jwalamukhi (c) Andhra jyoti (d) Bhaskar

308. Vegetable of 20th century is- (JRF, 2003
(a) Cluster bean (b) Amaranthus (c) Winged bean (d) Chikurmains

309. Variety resistant to cyst nematodes in Potato is ________
(a) Kufri swarna (b) Kufri sutlej (c) kufri Giriraj (d) Kufie Asoka

310. The mechanism of protogyny is observed in________
(a) Onion (b) Cabbage (c) Carrot (d) Radish

311. Which is called as a protective food (JRF, 2004)
(a) Vegetable (b) Fruits
(c) Fruits & vegetable (d) Cereals

312. In combination breeding, which is most important consideration for choice of parents
(a) Per se performance (b) Combining ability
(c) Genetic variability (d) Heritability

313. Intercropping of tomato with cabbage can reduce the damage done by the: (JRF, 2016)
(a) Diamond-back moth(b) Aphids
(c) Butterflies (d) Leafhoppers

314. Sulphur containing food is stored in (JRF, 2001)
(a) AR cans (b) SR cans (c) Both (d) None

315. Main objective of emasculation in breeding programme is
(a) Prevention of selfing (b)Encourage crossing
(c) Prevention of out crossing (d) Hybrid development

316. Large cardamom is propagated through- (JRF, 2003)
(a) Seed (b) Cutting (c) Layer (d) Sucker

317. Which vegetable is rich source of Vit. C (>100 mg/100g) (JRF, 2004)
(a) Tomato (b) Cabbage (c) Turnip (d) Radish

318. Most widely accepted method of emasculation in vegetable crops is
(a) Hand emasculation (b) Alcohol treatment
(c) Suction method (d) Hot water treatment

319. Better principle in better gourd is (JRF, 2003)
(a) Cucurbitacine (b) Momordicin (c) Solanin (d) None

320. Cassava belongs to family: (JRF, 2016)
(a) Rhaminaceae (b) Solanaceae (c) Euphorbiaceae (d) Myrtaceae

321. Edible portion of Turnip is: (JRF, 2018)
(a) Hypocotyl (b) Root (c) Tuber (d) Stem

322. Arka Pragati is a variety of (JRF, 2002)
(a) Okra (b) Onion (c) Palak (d) Lettuce

323. Cleistogamy favours (JRF, 1999)
(a) Self pollination (b) Cross pollination (c) Both (d) None

324. 'R' lacquered cans are used for canning of: (JRF, 2016)
(a) High acid foods (b) Non-acid foods (c) Low acid foods (d) Neutral foods

325. Incompatibility is common is family (JRF,1998)
(a) Malvaceae (b) Rosaceae (c) Liliaceae (d) Poaceae

326. National Horticulture Board (NHB) was established in: (JRF, 2018)
(a) 1984 (b) 1990 (c) 1965 (d) 1978

327. According to ISI tomato has been specialized into (JRF, 2002)
(a) 4 grades (b) 3 grades (c) 2 grades (d) Only one grade

328. Spice crop largely exported from India (JRF, 1997)
(a) Black pepper (b) Cardamom (c) Cumin (d) Turmeric

329. Highest hybrid varieties in vegetable are grown in (JRF, 2002)
(a) Tomato (b) Cabbage (c) Brinjal (d) Carrot

330. Good storage require (JRF,1998)
(a) High temp & high humidity (b) Low temp & high humidity
(c) High temp & low humidity (d) None

331. The minimum total soluble solids content in jam should be (as per FPO specification): (JRF, 2018)
(a) 62°B (b) 65°B (c) 78°B (d) 68°B

332. Which of following elements is common to all proteins and nucleic acid? (JRF, 2018)
(a) Suphur (b) Magnesium (c) Nitrogen (d) Phosphorus

333. Totipotency In plant is present in: (JRF, 2018)
(a) Phloem (b) Cork (c) Cambium (d) Meristem

334. For handling wide crosses which is most suitable breeding scheme
(a) Backcross method (b) Pedigree
(c) Diallel selective mating (d) Biparental mating scheme

335. Common blight of bean is caused by (JRF, 2002)
(a) Virus (b) Physiological (c) Bacteria (d) Mycoplansma

336. Resistance to which disease was transferred from Solanum demissum into potato
(a) Late blight (b) Early blight (c) Virus Y (d) Virus X

337. Little leaf of brinjal is caused due to (JRF,1998)
(a) Virus (b) Mycoplasma (c) Bacteria (d) Fungi

338. Which group of crops show severe inbreeding depression
(a) Cross-pollinated crops (b) Vegetative propagated crops
(c) Self-pollinated crops (d) Seed propagated crops

339. Clones generally degenerates due to
(a) Viral infection (b) Bacterial infection
(c) Mutations (d) All above

340. Heterobeltiosis is estimated over the (JRF, 2002)
(a) Wild parent (b) Better parent (c) Popular variety (d) Popular hybrid

341. Corm is an example of a: (JRF, 2018)
(a) Modified stem (b) Modified root (c) Modified leaf (d) Modified scales

342. Genetic variation within a clone may arise due to the
(a) Somatic mutations (b) Mechanical variation
(c) Segregation and recombination (d) All of the above

343. Daria method of cultivation is followed in which vegetable crop (JRF,1998)
(a) Cucurbits (b) Cole crops (c) Tubers (d) Root crops

344. Which amino acid is precursor of ethylene (JRF, 1999)
(a) Cystine (b) Purine (c) Methionine (d) Proline

345. In radish which substance is present (JRF, 2001)
(a) Allyl propyl disulphide (b) Dially1 disulphide
(c) Isothiocynate (d) None

346. Diploid apogamy is found in (JRF, 2002)
(a) Allium spp (b) Perithelium spp (c) Tobacco (d) Rice
347. Gynoecious lines are common in (JRF,1997)
(a) Water melon (b) Musk melon (c) Long melon (d) Cucumber
348. Leguminous leafy vegetable is (JRF,1998)
(a) Cowpea (b) Fenugreek (c) Spinach (d) Palak
349. Which of the following is protoandraus (JRF, 2002)
(a) Onion (b) Tomato (c) Cauliflower (d) French bean
350. Origin of Brinjal is (JRF, 1997)
(a) Africa (b) America (c) India (d) China
351. Triploid variety is commercially used in (JRF, 2004)
(a) Muskmelon (b) Cucumber (c) Watermelon (d) All
352. The family of Sauropus androgynus is – (ARS, 2018)
(a) Chenopodiaceae (b) Umbelliferae (c) Euphorbiaceae (d) Cruciferae
353. Genetic analysis in clonally propagated crops is very difficult due to the following
(a) Reduced flowering (b) Perennial life cycle
(c) Sterility (d) All of the above
354. Which dioecious vegetable is propagated by stem cutting (JRF, 1999)
(a) Yams (b) RUBP (c) Pointed gourd (d) All
355. Large cardamom is commercially propagated through.
(a) Rhizomes (b) Tubers (c) Suckers (d) Cormels
356. The triploid variety water melon suited for polyhouse cultivation- (ARS, 2018)
(a) Sugar Baby (b) Pusa Bedana (c) Arka Madhura (d) Arka Manik
357. Black pepper starts yielding from________ year onwards
(a) 2nd year (b) 3rd year (c) 2½ year (d) 4th year
358. The F_1 hybrid in Brinjal is ________
(a) Pusa Uttam (b) Pant Rituraj
(c) Arka Kusumakar (d) Arka Navneet
359. Pusa rasraj is improved variety of – (ARS, 2018)
(a) Muskmelon (b) Watermelon (c) Long melon (d) None of these
360. Per capita consumption of potato in India/year is ________
(a) 15 kg (b) 20 kg (c) 30 kg (d) 50 kg
361. Daily requirement of vegetable per capita/day is (JRF, 2004)
(a) 150 gram (b) 200 gram (c) 285 gram (d) 400 gram
362. Rhubarb is propagated through________
(a) Stem cutting (b) Tubers
(c) Division of crowns (d) Root suckers
363. Higher temp and high light intensity induces in cucurbits- (ARS, 2018)
(a) Female Flower (b) Male flower (c) Both (d) None of these

364. Arka Mamik and Arka Jyoti are the sister seedlings of a cross between
(a) Tetra-2 × Pusa Rassal (b) IIHR-21 × Crimson Sweet
(c) IIHR-20 × Crimzon Sweet (d) None of the above

365. "Sylvia" is edible podded variety of
(a) Pea (b) Broad bean (c) Lima Bean (d) French bean

366. Breeder seed production in onion is carried out by – (ARS, 2018)
(a) Seed-Bulb method (b) Bulb-seed method
(c) Both (d) None of these

367. Pusa Sanyog, is a F-1 hybrid of
(a) Cucumber (b) Muskmelon (c) Watermelong (d) Longmelon

368. Onion variety suitable for kharif season is
(a) N-53 (b) Pusa Red (c) Early Graino (d) Pusa Madhvi

369. The processing attributes determining quality of potato- (ARS, 2018)
(a) High dry matter and low reducing sugar
(b) Low dry matter and high reducing sugar
(c) High dry matter and reducing sugar
(d) Low dry matter and reducing sugar

370. Which of the following is precursor for auxin synthesis (JRF, 2002)
(a) Methionine (b) Tryptophon (c) Adenine (d) None

371. Botanical name of bitter gourd is (JRF,1998)
(a) Cucumis melo (b) Momordica (c) citrullus (d) None

372. The quality of tomato for processing of purees and paste- (ARS, 2018)
(a) High consistency and low viscosity (b) Low consistency and high viscosity
(c) High consistency and viscosity (d) Low consistency and viscosity

373. 'Pusa Parvati' is a variety of French bean developed through.
(a) Clonar selection (b) Mass selection
(c) Mutation breeding (d) None

374. Indian institute of Vegetable Research is located at (JRF, 2004)
(a) New Delhi (b) Allahabad (c) Varanasi (d) Mumbai

375. First improved variety of EFY- (ARS, 2018)
(a) Sree Jaya (b Sree Prakash (c) Sree Arun (d) Sree Padma

376. Pusa Komal is the variety of (JRF, 2001)
(a) French bean (b) Cluster bean (c) Cow pea (d) Goa bean

377. Hetrobeltosis is estimated over the: (JRF, 2018)
(a) Popular variety (b) Mid parent (c) Better parent (d) Popular hybrid

378. Adhunik and Panchali are varieties of ________
(a) Chilli (b) Okra (c) Greater yam (d) Better gourd

379. Which of following originated in India.
(a) Pea (b) French bean (c) Cow-pea (d) Cluster bean

380. Main constraint in commercial cultivation of spices is (JRF,1997)
(a) Lack of planting material (b) Lack of market
(c) Lack of knowledge (d) All
381. A chimera can be maintained through
(a) Cuttings, budding and graftings (b) Somatic embryogenesis
(c) Apomixis (d) Adventious shoot regeneration
382. In plant spacing 2x2, how much plant can be accommodated (JRF, 2001)
(a) 2000 (b) 2500 (c) 3000 (d) 3500
383. Sweet potato belongs to the family: (JRF, 2018)
(a) Chenopodiaceae (b) Convolvulaceae (c) Fabaceae (d) Compositae
384. Kufri Safed is a selection from
(a) Phulwa (b) Kufri Sindhuri
(c) Darjeeling Red Round (d) Uptodate
385. Which type of heterosis is of practical importance in breeding
(a) Average heterosis (b) Heterobeltiosis
(c) Economic heterosis (d) Luxuriance
386. The highest Broccoli producing country in world- (ARS, 2018)
(a) USA (b) China (c) Japan (d) India
387. The state leads in vegetable production (JRF, 1997)
(a) Bihar (b) Punjab (c) U.P. (d) W.B.
388. Watermelon bud necrosis is caused by: (JRF, 2016)
(a) Fungus (b) Bacteria (c) Virus (d) Mycoplasma
389. Heterosis can be fixed by
(a) Vegetative propagation (b) Apomixis
(c) Balanced lethal system (d) All of the above
390. Which variety of tomato can be sown throughout year (JRF, 2000)
(a) Pusa ruby (b) Pusa sadabahar (c) Sioux (d) None
391. The term 'climacteric' was first used by: (JRF, 2016)
(a) Gane (b) Kidd and West (c) Cruess (d) Bleaker
392. Heterosis can be fully exploited in the form of (JRF, 2002)
(a) Hybrids (b) Composites (c) Synthetics (d) Multilines
393. CIMAP is located at: (JRF, 2018)
(a) Lucknow, U.P (b) Pantnagar, U.K
(c) IARI, New Delhi (d) Anand, Gujarat
394. Capacity of soil that resists sudden change in pH of soil is called: (JRF, 2018)
(a) Buffer capacity (b) Cation exchange capacity
(c) Anion exchange capacity (d) Biological Oxygen
395. Marsh spot in peas is due to the deficiency of: (JRF, 2016)
(a) Cu (b) Ge (c) Cl (d) Mn

396. General combining ability can be estimated from performance of
(a) Parents (b) F_1s (c) Backcross (d) F_2s

397. Maturity in pea is judged by – (ARS, 2018)
(a) Refractometer (b) Tendrometer (c) Tensiometer (d) Hygrometer

398. Potato is a .... plant for its tuber production
(a) Short-day (b) Day-neutral (c) Long-day (d) None

399. Potato tuber production is maximum at following temperature
(a) 15°C (b) 20°C (c) 25°C (d) 30°C

400. Arkel is a variety of pea introduced from: (ARS, 2016)
(a) USA (b) England (c) UAE (d) Germany

401. Optimum temperature for curd initiation and development in Snowball cauliflower is
(a) 20-27°C (b) 12-16°C (c) 20-25°C (d) 10-16°C

402. Puffiness in tomato is caused by
(a) Viral attack (b) Bacterial attack
(c) Physiological disorder (d) None of the above

403. The condition where a flower cannot be fertilized by the pollen of same is called: (JRF, 2018)
(a) Dichogamy (b) Homogamy
(c) Unisexual (d) Self incompatibility

404. Pusa Deepti is the variety of (JRF, 2001)
(a) Brinjal (b) Pumpkin (c) Melon (d) None

405. Most commonly used male sterility system for hybrid seed production in vegetables is
(a) Cytoplasmic genetic male sterility
(b) Genetic male sterility
(c) Photoperiod sensitive genetic male sterility
(d) Cytoplasmic male sterility

406. Pusa snow-ball is variety of – (ARS, 2018)
(a) Cauliflower (b) Cabbage (c) Turnip (d) Broccoli

407. Which of following bears yellow colour flower (JRF, 2001)
(a) Bottle gourd (b) Luffa (c) Cucumber (d) All

408. Heterosis may be fixable if it is mainly due to
(a) Additive × dominance effects (b) Dominance × dominance effects
(c) Additive and additive × additive effects (d) All of the above

409. Intervarietal hybridization is also called
(a) Interspecific hybridization (b) Intergeneric hybridization
(c) Intraspecific hybridization (d) Distant hybridization

410. Metaxenia is observed in (JRF, 2001)
(a) Root crops (b) Tuber crops (c) Bulb crops (d) Cucurbits

411. Economic or useful heterosis can be estimated by

(a) $\frac{\text{Mean value of } F_1 - \text{Mean value of commercial cultivar}}{\text{Mean value of commercial cultivar}} \times 100$

(b) $\frac{\text{Mean value of } F_1 - \text{Mean value of min-parent}}{\text{Mean value of mid-parent}} \times 100$

(c) $\frac{\text{Mean value of } F_1 - \text{Mean value of best parent}}{\text{Mean value of best parent}} \times 100$

(d) $\frac{\text{Mean value of } F_1 - \text{Mean value of } F_2}{\text{Mean value of } F_1} \times 100$

412. Commercial part of mace is
(a) Dried kernel (b) Seed (c) Flower (d) Dried aril

413. The most effective method for the transfer of oligogenic character is (JRF, 2002)
(a) Pedigree breeding (b) Bulk breeding
(c) Back cross breeding (d) None

414. In Bitter gourd, seed extraction, fruit harvesting is done at - (ARS, 2018)
(a) Light yellow stage (b) Dark yellow stage
(c) Dark green stage (d) Light red stage

415. The F1 hybrid is a cross between- (ARS, 2018)
(a) Two inbred lines (b) Two outbreed lines
(c) Both (d) None of these

416. The large cardamom is propagated by (JRF, 2001)
(a) Seeds (b) Suckers (c) Cutting (d) All

417. Which type of male sterility system is most convenient to use
(a) Genetic (b) Cytoplasmic
(c) Cytoplasmic-genetic (d) Temperature sensitive genetic

418. Term heterosis was given by
(a) East (1908) (b) Shull (1914) (c) Powers (1908) (d) Davenport (1910)

419. Modern technique of inbred evaluation is based on
(a) Testing of GCA and SCA
(b) Phenotypic evaluation
(c) Prediction of double cross performance from single cross data
(d) All of the above

420. Main objective of vegetable blanching….. (JRF, 2019)
(a) Stop Microbial activity (b) Stop enzymatic activity
(c) Stop Physiological Disorder (d) None of these

421. India's share in world trade of spice is
(a) 8% (b) 18% (c) 28% (d) 38%

422. Seed stock degeneration in potato mainly due to- (ARS, 2018)
(a) High temperature (b) Low temperature
(c) Mild temperature (d) High humidity

423. The tuberization in potato stop at temperature above- (ARS, 2018)
(a) 15°C (b) 25°C (c) 20°C (d) 5°C

424. Term mutation was first introduced in 1900 by
(a) Muller (b) Bateson (c) Hugo de Vries (d) Gaul

425. For population improvement in any crop, best method is (JRF, 2002)
(a) Pureline selection (b) Pedigree selection
(c) Mass selection (d) Recurrent selection

426. Generally, the frequency of spontaneous mutations is
(a) 10^{-6} (b) 10^{-4} (c) 10^{-5} (d) 10^{-7}

427. Pusa shandar is a variety of : (ARS, 2016)
(a) Watermelon (b) Snapmelon (c) Muskmelon (d) Longmelon

428. CFTRI was established in the year: (ARS, 2016)
(a) 1925 (b) 1935 (c) 1950 (d) 1965

429. Pusa Jyoti is a variety of
(a) Amaranth (b) Bottle gourd (c) Spinach (d) Brinjal

430. Golden Acre is a variety of
(a) Cauliflower (b) Broccoli (c) Cabbage (d) Brussels sprout

431. Diosgenin is found in amongst which crop- (ARS, 2018)
(a) Mustard (b) Oat (c) Fenugreek (d) Cumin

432. Jade Cross is a variety of
(a) Cauliflower (b) Broccoli (c) Cabbage (d) Brussels sprout

433. Riceyness in cauliflower is the malformation of: (ARS, 2016)
(a) Petiole (b) pedicel (c) Stem (d) Leaf

434. The pungency of chilli is due to (JRF, 2001)
(a) Capsanthin (b) Capsacin (c) Isothiocynate (d) None

435. Mutagenic changes are the results of following
(a) Deletion (b) Addition (c) Substitution (d) All of the above

436. Pusa Mukta is a variety of
(a) Cauliflower (b) Broccoli (c) Cabbage (d) Brussels sprout

437. Chillies are good source of (JRF,1997)
(a) Protein (b) Fat (c) Carbohydrates (d) Vitamin

438. Which vegetable is underground stem (JRF, 1999)
(a) Potato (b) Radish (c) Carrot (d) Okra

439. Which type of mutations is used in crop improvement
(a) Lethal (b) Sublethal (c) Vital (d) Subvital

440. Somatic chromosome number of okra: (ARS, 2016)
(a) 2n = 100 (b) 2n = 130 (c) 2n = 140 (d) 2n = 160

441. Early cauliflower seed production is not possible in plains due to: (ARS, 2016)
(a) Low temperature & RH (b) Mild temperature & RH
(c) High temperature & RH (d) All of these

442. Most effective wavelength of ultraviolet rays is
(a) 254 nm (b) 280 nm (c) 325 nm (d) 354 nm
443. Largest producer of cauliflower is (JRF, 1997)
(a) China (b) Kenya (c) India (d) Australia
444. For Pea seed storage temperature is: (ARS, 2016)
(a) 4°C (b) 6°C (c) 8°C (d) 10°C
445. Mutagen treatment reduces
(a) Seed germination and fertility (b) Growth rate
(c) Vigour (d) All of the above
446. In radish which substance is present (JRF, 2001)
(a) Allyl propyl disulphide (b) Diallyl disulphide
(c) Isothiocynate (d) None
447. A male sterile line controlled by single recessive gene in muskmelon: (ARS, 2016)
(a) ms-5 (b) ms-6 (c) ms-3 (d) ms-7
448. Ratio of beneficial to useless mutations is one in
(a) 500 (b) 800 (c) 1200 (d) 1500
449. Which is the drought resident crop (JRF, 1999)
(a) Sweet potato (b) Okra (c) Cauliflower (d) Potato
450. Dominant mutations can be selected in
(a) M_1 generation (b) M_2 generation
(c) M_3 generation (d) M_2 and M_3 generation
451. NASA planted which vegetable in space station is: (ARS, 2016)
(a) Broccoli and Lettuce (b) Palak and Chicory
(c) Lettuce and Chinese cabbage (d) Chinese cabbage and Chicory
452. A cross between an inbreed line and an open pollinated variety is called as (JRF, 2002)
(a) Test cross (b) Back cross (c) Poly cross (d) Top cross
453. Mutation breeding can be used to
(a) Improve specific character (b) Induce desirable mutant alleles
(c) Improve polygenic traits (d) All of the above
454. Which plant part can be used for mutagenic treatments
(a) Seeds (b) Pollen grains
(c) Cuttings(d) All of the above
455. Root–seed method is used planting root stock called- (ARS, 2018)
(a) Head (b) Steckling (c) Knobling (d) None of these
456. Brinjal variety Arka Anand is resistant to: (JRF, 2018)
(a) Fusarium wilt (b) Bacterial wilt (c) Damping off (d) Phomopsis blight
457. Leek is botanically known as: (JRF, 2018)
(a) Allium fistulosum (b) Allium porrum
(c) Allium schoenoprasum (d) Allium aggregatum
458. Generally what concentration of Cochicine is used for seed treatment?
(a) 0.05 % (b) 0.20 % (c) 0.50 % (d) 1.0 %

459. De-topping is mechanical harvesting followed in – (ARS, 2018)
(a) Cauliflower (b) Brussel sprout (c) Cabbage (d) Broccoli

460. For hybrid seed production in tomato pollen parent ratio is: (ARS, 2016)
(a) 6:1 (female:male) (b) 12:1 (female:male)
(c) 18:1 (female:male) (d) 24:1 (female:male)

461. Which planting material is most conveniently and effectively used for mutagenic treatment
(a) Seeds (b) Pollen grains (c) Cuttings (d) All of the above

462. Whiptail of cauliflower is caused due to deficiency of (JRF, 2004)
(a) Mo (b) Cu (c) Zn (d) B

463. Which growth regulator is commonly used as weedicide (JRF, 1997)
(a) ABA (b) IBA (c) NAA (d) 2, 4-D

464. Night temperature requirement for tomato is (JRF, 1999)
(a) 5-10°C (b) 15-20°C (c) 25-30°C (d) 10-15°C

465. Pithiness is a disorder of: (ARS, 2016)
(a) Beetroot (b) Radish (c) Carrot (d) Turnip

466. Mutagenesis serves which function
(a) Parents for hybridization (b) Direct release as a variety
(c) Material for genetic studies (d) All of the above

467. Wintering is done to (JRF, 1999)
(a) Expose plants to cold temperature (b) It is alternative to root pruning in rose
(c) It protects plant against cold (d) All

468. Broccoli contain Anti-cancer chemical compound- (ARS, 2018)
(a) Saponin (b) 3-4 thiocarbamate
(c) Sulphoraphane (d) Dimethyl pyrazine

469. Which operation is essential during M1 generation
(a) Prevention of out crossing (b) Prevention of mechanical mixture
(c) Selection of superior types (d) Both (a) and (b)

470. Spontaneous mutation used mostly for (JRF, 1999)
(a) Ornamentals (b) Fruits (c) Vegetable (d) Spice

471. Gajendra is a variety of: (ARS, 2016)
(a) Elephant Foot Yam (b) Colocasia
(c) Yam (d) Cassava

472. Most commonly used agent for chromosome doubling is
(a) Nitrous oxide (b) Colchicine (c) Gibrellic acid (d) Alcohol

473. SSCA refers to: (ARS, 2016)
(a) State Seed Certification Agency (b) Seed State Certification Agency
(c) Supplier Seed Certification Agency (d) Super Seed Certification Agency

474. Purple blotch is problem in (JRF, 2004)
(a) Okra (b) Tomato (c) Onion (d) Turmeric

475. Clubrot of cauliflower is caused by (JRF, 2002)
(a) Xanthomonas compestris (b) Plasmodiophora brassicae
(c) Fruining crotovera (d) None
476. The highest/maximum productivity of potato is in _______
(a) U.P. (b) Bihar (c) Punjab (d) W.B.
477. Salt tolerant species of tomato…. (JRF, 2019)
(a) Solanum peruvianum (b) Solanum chilense
(c) Solanum arcanum (d) Solanum pennellii
478. Seed less watermelon developed by utilized breeding – (ARS, 2018)
(a) Amphidiploid (b) Autopolyploid (c) Aneuploid (d) None of these
479. Pungency in onion is due to (JRF, 1998)
(a) Allyl propyl disulphide (b) Di-allyl disulphide
(c) Capsacin (d) None
480. Generally all cole crops ancestors are found in: (ARS, 2016)
(a) Indo-Burma region (b) Chinese region
(c) Mediterranean region (d) Persian centre
481. Sreekara is variety of : (JRF, 2018)
(a) Ginger (b) Turmeric (c) Black pepper (d) Cardamom
482. Isolation distance for foundation seed production in Cole crop- (ARS, 2018)
(a) 800m (b) 1000m (c) 1600m (d) 1200m
483. Vegetable crop mostly affected by thrips (JRF, 2004)
(a) Tomato (b) Okra (c) Chilli (d) Onion
484. ________ ranks 1st in total vegetable production in India.
(a) U.P. (b) Bihar (c) Maharashtra (d) W.B.
485. ________ 1st in both area and production of Coriander in India.
(a) Gujarat (b) U.P. (c) M.P. (d) Rajasthan
486. A disease having regular occurrence at a particular area is called as (JRF, 1999)
(a) Epidemic (b) Endemic (c) Sporadic (d) None
487. India is the major importer of
(a) Ginger (b) Garlic (c) Black Pepper (d) Coriander
488. Following variety of black pepper possess resistance to nematodes and is high yielding released in the year 1991 by NRCS
(a) Sreekara (b) Subhakara (c) Pournami (d) Panchami
489. Chemical formula of colchine is
(a) $C_{22}H_{25}O_6N$ (b) $C_{25}H_{22}O_6N$ (c) $C_{22}H_{27}O_8N$ (d) $C_{22}H_{25}O_6N_2$
490. Highest respiration is found in which of the following vegetable: (ARS, 2016)
(a) Watermelon, Muskmelon and Roundmelon
(b) Potato, Onion and chilli
(c) Cabbage, Pea and Beans
(d) Chilli, Carrot and Tomato

491. Indian cardamom export is set back by_______country.
(a) Malaysia (b) Indonesia (c) Guatemala (d) Chana

492. The productivity of vegetable in India is_______
(a) 11.2% (b) 14.4% (c) 12.8% (d) 15%

493. Nepal cardamom is botanically known as
(a) Amomum aromaticum (b) Elettaria cardamom
(c) Amomum sublatum (d) None of these

494. Ricyness in cauliflower is due to (JRF,1997)
(a) Growing in rice field (b) Unfavourable temperature
(c) Mo deficiency (d) None

495. Scented inflorescence Allium species is: (ARS, 2016)
(a) Allium schoenoprasum (b) Allium fistulosum
(c) Allium cepa (d) Allium tuberosum

496. Pre-Cooling method suited in leafy vegetable crops – (ARS, 2018)
(a) Room cooling (b) Forced air cooling
(c) Vacuum cooling (d) None of these

497. Nullisomics produce which type of gametes?
(a) n +1 (b) n – 1 (c) n (d) n + 1 –1

498. Meristem culture can be used for
(a) Clonal propagation (b) Recovery of virus free stocks
(c) Germplasm exchange and conservation (d) All of the above

499. The quick freezing is prefect over slow freezing, due to (JRF, 2001)
(a) Formation of large crystal (b) Formation of small crystal
(c) To save time (d) None

500. Mint is botanically known as
(a) Mentha piperita (b) M. arvensis (c) M. spicata (d) None of the above

501. NBRI is situated at _______
(a) New Delhi (b) Culcutta (c) Lucknow (d) Panpur

502. Bitterness in cucumber is due to_______
(a) Glucosides (b) Acids (c) Alkaloides (d) Metaxenia

503. Which one of the following is highly cross-pollinated (JRF,1997)
(a) Okra (b) Bitter gourd (c) Chilli (d) Cowpea

504. Effect of selfing on the frequency of both the homozygotes (AA + aa) with reference to a single locus (A/a) after n generation of selfing will be calculated by formulae
(a) $2^{n-1/2n+1}$ (b) $2^{n/2n+1}$ (c) $2^{n-1}/2^{n}$ (d) $2^{n+1}/2^{n}$

505. Multistyled flowers are common in (JRF,1998)
(a) Brinjal (b) Okra (c) Chilli (d) Limabean

506. Which of the following crop show protandry.
(a) Carrot (b) Tomato (c) Chilli (d) Brinjal

507. Tag colour of certified seed- (ARS, 2018)
(a) Yellow (b) Purple (c) White (d) blue

508. Pusa Ruby is a variety of (JRF, 2002)
(a) Potato (b) Brinjal (c) Tomato (d) Chilli

509. ________ coloured onion var. are suitable for dehydration
(a) Red (b) Yellow (c) White (d) All

510. ________ is the seed rate for pole var. of French bean
(a) 60 kg/ha (b) 80 kg/ha (c) 40-45 kg/ha (d) 25-30 kg/ha

511. PGR is sprayed in cucurbits crops at – (ARS, 2018)
(a) One leaf stage (b) 1-2 leaf stage (c) 2-4 leaf stage (d) 4-6 leaf stage

512. Variety of chilli resistant to leaf curl virus (JRF, 1997)
(a) Bhaskar (b) Pusa Deepti (c) Jwalamukhi (d) Pusa Jawala

513. Rajendra Swathi is variety of : (JRF, 2018)
(a) Coriander (b) Fenugreek (c) Black pepper (d) Cardamom

514. The metal nutrient present in the structure of chlorophyll is (JRF, 2002)
(a) Fe (b) Zn (c) Ca (d) Mg

515. What is the optimum dose of a mutagen for effective treatment
(a) LD_{50} (b) Close to LD_{50}
(c) Lower than LD_{50} (d) Higher than LD_{50}

516. Major constraint in vegetative propagation of cardamom is (JRF, 1997)
(a) Katte disease (b) Thrips (c) Capsule Borer (d) None

517. Effect of selfing on the frequency of heterozygote (Aa) at a single locus (A/a) after n generation of selfing will be calculated by formulae
(a) $2^{n-1/2n+1}$ (b) $2/^{2n+1}$ (c) $2^n/2^{n-1}$ (d) $2/2^n$

518. Vegetable crop which can tolerate saline soil condition- (ARS, 2018)
(a) Sugarbeet (b) Pea (c) Cucumber (d) Pumpkin

519. Curing of harvested sweet potato tubers takes place if the tubers are kept for 10 days at
(a) 20°°C and 90% RH (b) 40°C and 70% RH
(c) 80°C and 30% RH (d) None of the above

520. Delayed harvesting in radish may cause
(a) Forking (b) Deformed roots (c) Pithiness (d) All of the above

521. Multiple disease resistant variety of watermelon............. (JRF, 2018)
(a) Sugar Baby (b) Arka Manik (c) Pusa Bedana (d) Arka Jyoti

522. In case of tomato blotchy ripening is due to deficiency of
(a) N (b) P (c) K (d) Ca

523. Early cauliflower is ready for harvesting in the period of
(a) Mid-September to mid-November (b) Mid-January to April
(c) Mid-November to mid January (d) None of the above

524. Earthing of potato crop is done
(a) 20 days after sowing (b) 40 days after sowing
(c) 50 days after sowing (d) None of the above

525. Most serious disease of vegetable seedlings (JRF, 1998)
(a) Root rot (b) Damping off (c) Wilt (d) None

526. The vegetable which is richest source of Vit. A is ________
(a) Turnip leaves (b) Carrot (c) Pumpkin yellow (d) Drumstick leaves

527. The per capita consumption of vegetable/availability of vegetable in India is,
(a) 300 g (b) 285 g (c) 175 g (d) 225 g

528. All the species of capsicum are diploid with chromosome number: (JRF, 2018)
(a) 2n = 2x = 24 (b) 2n = 2x = 48 (c) 2n = 2x = 14 (d) 2n = 2x = 34

529. C 4 vegetable (ARS 2014)
(a) Amaranthus (b) Pea (c) Bitter gourd (d) Lettuce

530. The bean mainly cross pollination in nature- (ARS, 2018)
(a) Broad bean (b) Pea (c) Cluster bean (d) French bean

531. Arka Suman is an variety of ____________
(a) French bean (b) Cowpea (c) Garden pea (d) Dolichus bean

532. The diploid chromosome number of cluster bean is ________
(a) 2n = 22 (b) 2n = 24 (c) 2n = 14 (d) 2n = 1

533. The crop which is not transplanted (JRF,1997)
(a) Tomato (b) Brinjal (c) Chilli (d) None of the above

534. Standard unit used in expressing refrigeration capacity is (JRF, 2002)
(a) Calorie (b) Ampere (c) Ton (d) Joule

535. Carrot is a rich source (JRF, 1998).
(a) Vit A (b) Vit B (c) Vit C (d) Vit D

536. The optimum size of potato tuber for planting is ________
(a) < 25g (b) 30-40 g (c) > 50g (d) 100 g

537. The edible portion of Jerusalem artichoke is (JRF, 2001)
(a) Leaf stalk (b) Tuber (c) Fruits (d) Bulb

538. __________ is called as multivitamin plant
(a) Amaranthus (b) Spinach (c) Chekurmanis (d) Asparagus

539. Inbreeding is deleterious in (JRF, 1999)
(a) Tomato (b) Maize (c) Onion (d) Cucumber

540. The growth regulator related with abscission of leaves is (JRF, 2002)
(a) ABA (b) Ethylene (c) GA (d) Cytokinin

541. Family of sweet potato is (JRF,1998)
(a) Solanaceae (b) Convolvulaceae (c) Mavaceae (d) Compositae

542. In garlic, bulbils develop naturally from modified
(a) Flower (b) Root (c) Stem (d) Fruit

543. For packing of spinach, cauliflower and sweet potato which type of can is used (JRF,1998)
(a) AR cans (b) SR cans (c) Plain (d) All

544. In flowering plants each anther has
(a) One pollen sac (b) Two pollen sacs
(c) Three pollen sacs (d) Four pollen sacs

545. Which of the following insect in non polyphagous (JRF, 2000)
(a) Lady bird beetle (b) White grub (c) Termite (d) Grass hopper

546. Asexually propagated crops are
(a) Highly heterozygous (b) Generally polyploids
(c) Mostly perennials (d) Shows all characteristics of a, b & c

547. Which one of the following is used as growth retardant (JRF,1997)
(a) GAB (b) NAA (c) Ethophon (d) MH

548. In which variety of cauliflower seed production is not possible in north Indian plains (JRF, 1999)
(a) Pusa deepali (b) Pusa subhra (c) Pusa synthetic (d) Pusa snowball -1

549. Cross pollination in radish due to (ARS, 2018)
(a) Protoandry (b) Male sterility
(c) Protogyny (d) Self-incompatibility

550. Storage temperature of cucumber is (JRF, 1999)
(a) 0°C (b) 4-5°C (c) 6-7°C (d) 10°C

551. The term epistasis is given by (JRF, 2001)
(a) Mendel (b) Halman (c) Bateson (d) Hull

552. Lady bird beetle attack on family (JRF, 2000)
(a) Cucurbitaceae (b) Solanaceae (c) Lilliaceae (d) None

553. Which of the following hormone is most effective in induction of bolting in rosette plant (JRF, 2002)
(a) Auxin (b) GA_3 (c) Cytokinin (d) Ethylene

554. Genetic assortative mating is closest form of
(a) Outcrossing (b) Inbreeding (c) Panmixis (d) Open pollination

555. Irish famine was caused by (JRF, 1999)
(a) Phytopthora (b) Alternaria
(c) Fusarium oxysporum (d) none
(c) Gibbrilin (d) Auxin

556. Carrot belongs to family (JRF, 2001)
(a) Compositae (b) Umbelliferae (c) Leguminoceae (d) None

557. Phytic acid a toxic substance present in________
(a) Leafy vegetable (b) Peas and beans (c) Cucurbits (d) Root vegetable

558. Best quality of oleoresin is extracted from which crop- (ARS, 2018)
(a) Capsicum (b) Chilli (c) Black Pepper (d) Tomato

559. The vegetable crop that tops in the share of total vegetable production
(a) Brinjal (b) Tomato (c) Potato (d) Cauliflower

560. Late blight resistant variety of potato is (JRF, 1999)
(a) Kufri badshah (b) Kufri sindhuri (c) Kufri megha (d) None

561. In India Saffron (Cross sativus) cultivated only in_________
(a) Punjab (b) Kashmir (c) Assam (d) TamilNadu

562. The spacing for celery is ________
(a) 15 × 20 (b) 15 × 40 (c) 20 × 40 (d) 25 × 40

563. Allspice (Pimenta dioca) belongs to family ________
(a) Umbelliferae (b) Apiaceae (c) Myrtaceae (d) None

564. Which is non bulb forming member of onion family (JRF, 2001)
(a) Leek (b) Garlic (c) Onion (d) None

565. East India Company introduced cabbage, cauliflower etc. from
(a) Mediterranean (b) China (c) Afghanistan (d) Central Asia

566. Daily requirement of vitamin –D for a healthy adult is (JRF, 2002)
(a) 200 I.U. (b) 400 I.U. (c) 600 I.U. (d) 800 I.U.

567. In 1883 late blight of potato was inadvertently introduced in India from
(a) Europe (b) Mexico (c) USA (d) Peru

568. The world collection of sweet potato germplasm is maintained at
(a) Peru (b) China (c) New Zealand (d) Austalia

569. The fruits & vegetables are known as (JRF, 2001
(a) Nutritive food (b) Energy food (c) Protective food (d) None

570. To retain green colour of capsule of cardamom what is used
(a) Washing soda 2% (b) Calcium carbide
(c) Ethylene (d) All

571. Fruit of Portulaca (a summer leafy vegetable) is (JRF, 2002
(a) Capsule (b) Berry (c) Siliqua (d) None

572. Family of clove is
(a) Myrtaceae (b) Apiaceae (c) Alliaceae (d) None

573. Anther culture is ordinarily used for production of
(a) Diploid plants (b) Haploid plants (c) Polyploid plants (d) Triploid plants

574. In brinjal seed and fruit set can be improved by spray of - (ARS, 2018
(a) NAA @200ppm (b) 2-4D @2ppm (c) IBA @20ppm (d) Boron @0.2%

575. The 2n number of cucumber is (JRF, 2001)
(a) 14 (b) 24 (c) 38 (d) 18

576. Tetraploid variety of Palak………….. (JRF, 2018
(a) Pusa Bharti (b) Pusa Jyoti (c) Pusa Harit (d) All green

577. Precursor of sulforaphane in broccoli: (ARS, 2016
(a) Glucoraphanin (b) Ethylene (c) Oxalic Acid (d) All of these

578. The browning occur in (JRF, 2001
(a) Cauliflower (b) Brinjal (c) Okra (d) None

579. Failure of embryo development in interspecific crosses may be due to
(a) Chromosome elimination (b) Genotypic disharmony
(c) Cytoplasmic incompatibility (d) All of the above

580. Most expensive spice crop of the world is
(a) Saffron (b) Coriander (c) Turmeric (d) Celery

581. Arka mangala is a variety- (ARS, 2018)
(a) Tomato (b) Cowpea (c) French Bean (d) Yard long bean

582. Part used for consumption of clove is
(a) Seed (b) Bark (c) Flower (d) Unopened bud

583. In cole crops which mechanism is commonly prevailing for hybrid seed production
(a) Male sterility (b) Self-incompatibility
(c) Apomixis (d) All above

584. In carrot cross-pollination is encouraged by which mechanism?
(a) Protogyny (b) Protandry
(c) Self-incompatibility (d) Male sterility

585. Phomopsis blight in brinjal is 1st reported in : (ARS, 2016)
(a) 1910 (b) 1912 (c) 1914 (d) 1916

586. Optimum night temperature for fruit set in tomato (JRF, 2002)
(a) 8°C (b) 13°C (c) 17°C (d) 21°C

587. In 1900, potato moth came in to India from
(a) Italy (b) Ireland (c) USA (d) New Zealand

588. The term of epistatis was coined by - (ARS, 2018)
(a) Marks (b) J. Roberts (c) Mendal (d) Bateson

589. Plant exploration would be most successful when it is targeted at
(a) Farmer's field (b) Primary centre of origin
(c) A microcenter (d) Secondary centre of origin

590. Recessive bud mutation is most likely to expressed in
(a) An inbred (b) A clone (c) A pureline (d) A hybrid

591. Kufri Red is a selection from
(a) Phulwa (b) Kufri Sindhuri
(c) Darjeeling Red Round (d) Up to date

592. Rupali variety of tomato is developed by (JRF, 2001)
(a) Indo America Hybrid seed company (b) Ankur seeds
(c) Mhycho seeds (d) None

593. Selfing reduces heterozygosity in each generation by the factor
(a) 1/4 (b) 1/3 (c) 1/2 (d) 1/8

594. Potato contains following amount of vitamin-C (mg/100g of edible portion) (JRF, 2002)
(a) 16 (b) 17 (c) 100 (d) 25

595. Which of following hormone is used as a herbicide (JRF, 2001)
(a) 2,4-D (b) NAA (c) ABA (d) IAA

596. Temperature required for dehydration of vegetables: (JRF, 2015)
(a) 60-66°C (b) 66-71°C (c) 70-75°C (d) 50-55°C

597. The present DG of ICAR is ________
(a) Dr. R.S. Paroda (b) Dr. T Mahapatra
(c) Dr. Mangala Rai (d) Dr. G. Kaloo

598. Which of the following is class one preservative (JRF, 2001)
(a) Silver nitrate (b) KMS (c) Sugar (d) Vinagar

599. India stands 1st in the production of _______
(a) Tomato (b) Cabbage (c) Onion (d) Okra

600. _________ is the sexually sterile diploid crop
(a) Chekurmanis (b) Rhubarb (c) Garlic (d) Beet

601. Which of the following hormone increase senescence (JRF, 2001)
(a) Ethylene (b) NAA (c) IAA (d) ABA

602. ________ is variety of cabbage, which is resident to black leg
(a) Pusa Mukta (b) Pusa drum head (c) Golden acre (d) None

603. Arka Abhay is a variety of
(a) Brinjal (b) Chilli (c) Tomato (d) Bhindi

604. Male sterile line in onion is isolated from________
(a) Pusa Red (b) Pusa Ratnar (c) Pusa Madhavi (d) Arka Bindu

605. Which one is cylindrical crook neck free variety of bottle gourd: (JRF, 2015)
(a) Pusa Naveen (b) Pusa Meghdut (c) Pusa Manjari (d) PSPL

606. Fufu is important product made from_________
(a) Bottlegourd (b) Muskmelon (c) Calocasia (d) Yam

607. Sree Harsha is a var. of _______
(a) Yam (b) Cassava (c) Calocasia (d) Sweet Potato

608. NARS refers to which- (JRF, 2003)
(a) National Ayurvedic Research System (b) National Agriculture Research System
(c) National Agriculture Research Scheme (d) National Agriculture Review System

609. Arka Jeet is a popular cultivar of
(a) Muskmelon (b) Watermelon (c) Watermelon (d) Bitter gourd

610. Late blight resistant variety of Potato (JRF, 2004)
(a) Kufri jyoti (b) Kufri giriraj (c) Kufri chipsona (d) All

611. The cultivated turmeric is (JRF, 2001)
(a) Tetraploid (b) Triploid (c) Diploid (d) None

612. ________ is also called as 'four angled bean'
(a) broad bean (b) French bean (c) Winged bean (d) Indian bean

613. Pusa Sheetal is tomato while Arka Sheetal is
(a) Long melon (b) Water melon (c) Musk melon (d) Snap melon

614. When peeling is done with the help of NaOH, it is known as: (JRF, 2015)
(a) Lye peeling (b) Flame peeling
(c) Steam peeling (d) Mechanical peeling

615. Isolation distance of Brinjal for certified seed production is _______
(a) 200 m (b) 5 c (c) 100 m (d) 50 m

616. __________ is a thermosensitive vitamin.
(a) Vit. A (b) Vit. E (c) Vit. C (d) Vit. D

617. FPO was promulgated in the year_______
(a) 1949 (b) 1955 (c) 1961 (d) 1963

618. Which variety of cauliflower is suitable for South Indian condition (JRF, 2004)
(a) Pusa subhra (b) Pusa early synthetic
(c) K-3 (d) All

619. CIPHET is situated in (JRF, 2000)
(a) Amritsar (b) Nagpur (c) Hyderabad (d) Ludhiana

620. _________ is a veg. of 20th century
(a) Clusterbean (b) Broad bean (c) Winged bean (d) Chekurmanis

621. ACC synthetase inhibitor in Flavr –savr Tomato is...... (JRF, 2019)
(a) Lycopene production (b) Ethylene production
(c) Both (d) None of these

622. Which type of cucurbits (Cucumber) is planted in polyhouse............... (JRF, 2018)
(a) Monoecious (b) Cross pollinated (c) Dioecious (d) Parthenocarpy

623. Mushrooms are an excellent source of ________ a mineral that the body needs to produce red blood cells
(a) Mn (b) Cu (c) Fe (d) Cl

624. The fruit in mustard is ________
(a) Siliqua (b) Hespericium (c) Pepo (d) Lomeninum

625. Pellagra is due to deficiency of _________
(a) Niacin (b) Folic acid (c) Biotin (d) Pyridoxine

626. Brinjal fruit and shoot borer resistant species... (JRF, 2019)
(a) S. xanthocarpum (b) S. sisymbrifolium (c) S. nigrum (d) All of these

627. A cross in which order of male & female parent is reversed is called as (JRF, 2002)
(a) Reciprocal cross (b) Back cross (c) Test cross (d) Top cross

628. _________ instrument is used to measure growth of plant.
(a) Fathometer (b) Galvanometer
(c) Crescograph (d) Anemometer

629. Sulphur fungicide can be freely used on all vegetable except one- (JRF, 2003)
(a) Peas & beans (b) Root vegetable (c) Cucurbits (d) Okra

630. _________ vitamin is also called as Tocopherol.
(a) K (b) D (c) B_6 (d) E

631. The biofortified variety of cauliflower (yellow colour) is developed at .... (ARS, 2018)
(a) IIVR (b) AVRDC (c) IIHR (d) IARI

632. Horticultural Society of India (HST) was established in ________
(a) 1942 (b) 1947 (c) 1954 (d) 1950

633. Which variety of Onion is suitable for kharif season (JRF, 2004)
(a) Pusa red (b) Pusa madhri (c) N-53 (d) Early Grano

634. Nutmeg and clove are native to
(a) India (b) Africa (c) Indonesia (d) America

635. In our diet staple vegetable is (JRF, 2000)
(a) Brinjal (b) Potato (c) Tomato (d) Chilli

636. ______ been is also known as Goa bean.
(a) Hycinth bean (b) French bean (c) Winged bean (d) Dolichus bean

637. ________ is yellow coloured variety of Onion.
(a) Hisar-11 (b) Arka kalyan (c) Punjab-48 (d) Early Grano

638. Origin place of coriander is.... (JRF, 2019)
(a) Mediterranean region (b) China
(c) Indo-Burma region (d) Africa

639. Time & temperature for vegetable processing in canning depend upon (JRF, 2001)
(a) Amount of microbes present in food (b) Texture of product
(c) Nutrients (d) None

640. In onion cross-pollination is encouraged by which mechanism?
(a) Protogyny (b) Protoandry (c) Cleistogamy (d) Heterostyly

641. Genetic variability can be created through
(a) Somatic hybridization (b) Hybridization
(c) Mutation (d) All above

642. Nantes is the variety of (JRF, 2004)
(a) Radish (b) Carrot (c) Onion (d) Turnip

643. NPK requirement in legumes is- (JRF, 2003)
(a) 3: 1: 0 (b) 4:2:1 (c) 3: 1:1 (d) 1:2:2

644. The amount of solanin/tuber of potato considered as unfit for consumption is.
(a) 5 mg (b) 10 mg (c) 20 mg (d) 25 mg

645. The origin of sweet potato is________
(a) Asia (b) Mexico (c) S. America (d) Africa

646. Object of precooling is (JRF, 2000)
(a) To fresh the product (b) To improve the quality
(c) Both a and b (d) To remove field heat

647. Sree Shilpa is first hybrid of ________
(a) Greater yam (b) Elephant foot yam
(c) Lesser yam (d) White yam

648. The main aim of hybridization in plants is
(a) Improvement in one or more quantitative trait
(b) Use of F1 as a hybrid
(c) Transfer of one or more quantitative trait
(d) All of the above

649. In vegetable improvement programme which is the most commonly used
(a) Interspecific hybridization (b) Intergeneric hybridization
(c) Intervarietal hybridization (d) Distant hybridization

650. Contribution of vegetable in Total horticulture production is.... (JRF, 2019)
(a) 40% (b) 50% (c) 60% (d) 70%

651. 'Pusa Kanchan' is variety of ________
(a) Tapioca (b) Colocasia (c) Turnip (d) Onion

652. Type of inflorescence in Beet root is ________
(a) Raceme (b) Cyme (c) Spike (d) Spadix

653. Which is not a legume crop- (JRF, 2003)
(a) Pea (b) beans (c) Wheat (d) Fenugreek

654. White rust resistant variety of radish- (ARS, 2018)
(a) Pusa Desi (b) Arka Nishant (c) Pusa Chetki (d) Arka Samrat

655. Which of the following is responsible for controlling of fruit drops (JRF, 2000)
(a) Auxin (b) Gibberellins (c) Cytokinin (d) Ethylene

656. Major bottlenecks in breeding clonally propagated crops are
(a) Perennial life cycle, reduced flowering and fertility
(b) Long life cycle and difficulty in genetic analysis
(c) Severe inbreeding depression
(d) All of the above

657. Which is not the variety of Brinjal (JRF, 2004)
(a) Pant samrat (b) Pant bahar (c) Pant rituraj (d) Pusa ankur

658. Vitamin-C is commonly known as (JRF, 2002)
(a) Citric acid (b) Acetic acid (c) Lactic acid (d) Ascorbic acid

659. ________ is leading supplier of chilli during off-season.
(a) A.P. (b) Maharashtra (c) M.P. (d) H.P.

660. The variety suitable for both cold and hot sector in Tomato is
(a) Pusa sheetal (b) Pusa sadabahar (c) Pusa Hybrid-1 (d) Arka sambandh

661. Onion is stored at a temperature of (JRF, 2004)
(a) 4-5°C (b) 25°C (c) 6-8°C (d) 10°C

662. Completely tuberization in potato is stopped at temperature of (JRF, 2004)
(a) 20°C (b) 25°C (c) 30°C (d) 23°C

663. Which is not a cytokinin (JRF, 2004)
(a) BA (b) Zeatin (c) Etheral (d) Kinatin

664. The chilli hybrid CH-1 is bred using the mechanism of ________
(a) Self-incompatibility (b) GMS
(c) CGMS (d) Exerted stigma.

665. Seed rate of ginger is: (JRF, 2018)
(a) 15-18 q/ha (b) 9-10q/ha (c) 18-22 q/ha (d) 35-40 q/ha

666. Maximum seed productivity per unit area among solanaceous crops is of.... (JRF, 2019)
(a) Tomato (b) Chilli (c) Capsicum (d) Brinjal

667. Drying recovery in Turmeric..... (JRF, 2019)
(a) 10-15% (b) 30-40% (c) 15-30% (d) 5-10%

668. The variety of Amaranthus which is resistant to white rust- (IIHR) (ARS, 2018)
(a) Arka saguna (b) Arka Arunima (c) Both (d) None of these

669. Konkan Bushan is a variety of....
(a) Cowpea (b) Indian bean (c) Pea (d) Jack bean

670. Which vegetable crop has leading area in India- (ARS, 2018)
(a) Tomato (b) Potato (c) Cauliflower (d) Cabbage

671. The optimum temperature for seed germination in case of pea is_______
(a) 5°C (b) 10°C (c) 20°C (d) 22°C

672. Florida Butter is an important variety of ________
(a) Broad bean (b) Lima bean (c) Indian bean (d) Winged bean

673. Which organization is related to agricultural marketing (JRF, 2004)
(a) NAPHED (b) APHEDA (c) NABARD (d) All

674. Indian bean is ready for harvesting after-------months of sowing.
(a) 2, (b) 3, (c) 4, (d) None of these.

675. The F_1 hybrid in case of Muskmelon is ________
(a) Pusa Madhuras (b) Pusa rasraj (c) Arka rajhans (d) Arka jeet

676. Largest importer of cauliflowers in the world is ________
(a) Netherlands (b) Germany (c) England (d) Switzerland

677. The origin of Dolichus bean is _________
(a) China (b) India
(c) Mexico (d) Mediterranean region

678. The book HORTI. SCIENCE is written by (JRF, 2004)
(a) Willium gadd (b) Swaminathan (c) Jennick (d) Randhawa

679. _________ is the selective herbicide for broad leaved plants
(a) Paraquat (b) Diquat (c) 2, 4-D (d) None

680. The vegetable in which all parts of plant can be used as edible part.
(a) Lima bean (b) Winged bean (c) Radish (d) Broad bean

681. Which of following hormone is used in cucubits for inducing flowering (JRF, 2000)
(a) IBA (b) NAA (c) Gibberlin (d) Auxin

682. ---------- variety of dolichos is a cross of Hebbal Avarae and Pusa Early Prolific.
(a) Arka Vijay (b) Deepaliwal (c) Dasarawal (d) Pusa sem-2.

683. Maximum life of Vanilla is (JRF,1997)
(a) 10 years (b) 15 years (c) 30 years (d) 50 years

684. Suitable type of cauliflower... (JRF, 2019)
(a) European types
(b) Cornish types
(c) Inter crossing between European and Cornish types
(d) None of these

685. Khihara techniques related vegetable crop- (ARS, 2018)
(a) Musk melon (b) Water melon (c) Okra (d) Pumpkin

686. A pureline may be used
(a) As a parent for hybridization (b) For mutation studies
(c) As a variety (d) All the above

687. The ornamental type in case of cucurbits is ___________
(a) Summer squash (b) White squash (c) Pumpkin (d) White squash

688. Herkogamy favours (JRF, 2002)
(a) Self pollination (b) Cross pollination (c) Mutation (d) Wide hybridization

689. Hybrid Onion programme originated in 1925 by (JRF-2006)
(a) T.W. Whitaker (b) B. D. Dowker (c) V. Swarup (d) H.A. Jones

690. In _________passed the fruit product order (FPO)
(a) 1950 (b) 1955 (c) 1960 (d) 1965

691. A phenomenon is which a gene has more than one phenotypic effect (JRF, 2002)
(a) Polymorphism (b) Penetrance (c) Expressively (d) Pleiotropism

692. Murriya koenigin is indigenous to.
(a) India (b) Myanmar (c) Bangladesh (d) Srilanka

693. Basic chromosome number in snake gourd is
(a) 14 (b) 20 (c) 24 (d) 34

694. Petha Kaddu is known as
(a) Pumpkin (b) Ridge gourd (c) Smooth gourd (d) Ash gourd

695. Test cross is crossing F1 with (JRF-2006)
(a) Dominant parent (b) Recessive parent
(c) Either of the parents (d) None of these

696. Self-incompatibility can be effectively utilized in ....crops (JRF-2006)
(a) Tomato (b) Cole crops (c) Bhendi (d) Peas

697. Triploid variety of Cassava… (JRF, 2019)
(a) Sree Harsha (b) Sree Jaya (c) Sree Vijaya (d) Sree Prakash

698. Acid tolerant vegetable crop- (ARS, 2018)
(a) Cole crops (b) Watermelon (c) Okra (d) Muskmelon

699. In cucumber production of male flowers can be induced by the application of (JRF-2006)
(a) Boron (b) Ethephon 600 ppm
(c) GA 2000 ppm (d) Auxin

700. Kashi sweta is a variety – (ARS, 2018)
(a) Brinjal (b) Tomato (c) Radish (d) Carrot

701. Single gene male sterility is transferred in .... years
(a) 3-4 (b) 5-6 (c) 7-8 (d) 4-5

702. .... is the progenitor form of the cultivated cucumber (JRF-2006)
(a) Cucumis sativus var. hardwickii (b) Cucumis melo
(c) Cucumis anguria (d) Citrus lanatus

703. CAM mechanism is observed in plants having: (JRF, 2016)
(a) Oxidative stress (b) Salt stress (c) Water stress (d) Cold stress

704. Murriya paniculata, Murriya exotica and Murriya Koenigin are realities of
(a) Drumstick (b) Curry leaf (c) Tree tomato (d) None of above

705. If potato contain ...., it is not suitable for processing: (JRF, 2015)
(a) High dry matter (b) High phenol and sugar
(c) Low phenol (d) Low sugar

706. Pollen sterility in tomato can be identified by using (JRF-2006)
(a) Acetocarmine (b) $KMnO_4$ (c) Methylene blue (d) Phenolphthalein

707. PPVFRA act-2001; PPVFRA Rules 2003; PPVFRA came to force in: (ARS, 2016)
(a) 2001 (b) 2002 (c) 2003 (d) 2005

708. The proportion of additive genetic variance to phenotypic variance is called (JRF-2006)
(a) Heterobeltiosis (b) Heritability
(c) Dominance relationship (d) Linkage relationship

709. Swarna Rekha is a variety of - (ARS, 2018)
(a) Ivy gourd (b) Pointed gourd (c) Bottle gourd (d) Snake gourd

710. The average daily requirement of vegetables for human adult is (JRF-2006)
(a) 130 g (b) 180 g (c) 285 g (d) 510 g

711. Pre harvest spray of which of the following is done to reduce storage loss of onion (JRF-20006)
(a) Maleic hydrazide (b) Gibberellic acid
(c) Indole butyric acid (d) Ethephon

712. Red colure in chilli is basically due to presence of- (ARS, 2018)
(a) Lycopene (b) Capsanthin (c) Carotene (d) All of these

713. Diploid apagomy is found in (JRF-2009)
(a) Allium spp. (b) Knol-khol (c) Okra (d) Cluster bean

714. The percentage of recovery of black pepper is about.........of the ripe berries (JRF-2006)
(a) 20 (b) 23 (c) 48 (d) 33

715. The raciness in cauliflower is due to- (ARS, 2018)
(a) Unfavourable Temperature (b) Unfavourable Humidity
(c) Unfavourable Light (d) Unfavourable Irrigation

716. India Cassia is (JRF-2006)
(a) CInnamomum cassia (b) C. burmannii
(c) C. loureirii (d) C. tamala

717. The hottest chilli bhoot jolokia is a variety belongs to: (ARS, 2016)
(a) C. annum (b) C. baccatum (c) C. frutescence (d) C. chinensis

718. Which method is costlier in storing of product? (ARS, 2018)
(a) Cold storage (b) C A storage
(c) Hypobaric storage (d) M A storage

719. ....is used as a donor for brinjal fruit and shoot borer (JRF-2006)
(a) Solanum nigrum (b) S. melongena
(c) S. Gilo (d) S. muricatum

720. The optimum temperature for Potato tuberization is (JRF-2006)
(a) 7°C (b) 17°C (c) 14°C (d) 24°C

721. The whole edible pod variety of pea – (ARS, 2018)
(a) Arkel (b) UN-53 (PAU) (c) Arka Ajit (d) Bonneville

722. ZECC was developed by – (ARS, 2018)
(a) P. Hazra (b) S. K. Roy and Khurdiya
(c) B. Singh (d) M. K. Roy

723. Highest potato producing state in India is: (JRF-2009)
(a) Uttar Pradesh (b) Madhya Pradesh (c) Bihar (d) West Bengal

724. Rajendra Swati is a variety of (JRF-2006)
(a) Cumin (b) Fenugreek (c) Fennel (d) Coriander

725. For Cole crops certified seed, isolation distance is : (ARS, 2016)
(a) 500m (b) 1000m (c) 800m (d) 1200m

726. Common blight of bean is caused by (JRF-2009)
(a) Vurus (b) Phytoplasma (c) Bacteria (d) Fungus

727. The most useful insect in seed production of vegetable is: (JRF-2009)
(a) Bumble bees (b) Honey bees (c) Wasps (d) Thrips

728. The inbreeding depression is found in – (ARS, 2018)
(a) Cabbage (b) Cauliflower (c) Knol- khol (d) All of these

729. The popular variety of radish for summer and rainy season in north Indian plains is (JRF-2009)
(a) Pusa Himani (b) Pusa Chetki (c) Japanese white (d) White Icicle

730. A serious disease of onion commonly found in seed crop is: (JRF-2009)
(a) Late blight (b) Purple blotch (c) Red rot (d) Leaf curl virus

731. Which of the following is a mutant ginger variety
(a) Surari (b) Surbhi (c) Rio de Janerio (d) Suruchi

732. Which of the following sugar is sweetest in fruits : (JRF, 2018)
(a) Glucose (b) fructose (c) Galactose (d) Sucrose

733. Cluster bean belongs to the genus: (JRF, 2018)
(a) Phaseolus (b) Dolichos (c) Cyamopsis (d) Pisum

734. Cochin/Kasturi Turmeric…. (JRF, 2019)
(a) Curcuma longa (b) Curcuma caesia
(c) Curcuma aromatica (d) Curcuma angustifolia

735. Head office of NBPGR is located at (JRF-2007)
(a) Raipur (b) Hyderabad (c) New Delhi (d) Bhubaneshwar

736. Pusa Bedana is a seedless variety of: (JRF, 2018)
(a) Musk melon (b) Water melon (c) Long melon (d) Round melon

737. The chemical used in synthetic seed production- (ARS, 2018)
(a) Sodium alginate (b) Sodium nitrate
(c) Sodium hydroxide (d) None of these

738. In cardamom, mainly pollination occurs due to
(a) Honey bee (b) House fly (c) Air (d) Water

739. The cleistogamous vegetable is: (JRF-2009)
(a) Garden pea (b) Lettuce (c) Capsicum (d) Tomato

740. Which vegetable crop is commercially used for processing in India? (JRF-2007)
(a) Tomato (b) Ashgourd (c) A & B both (d) Tapioca

741. Sex form present in betelvine is
(a) Monocious (b) Dioecious
(c) Andromonoecious (d) None

742. Riceyness is a physiological disorder associated with: (JRF-2009)
(a) Cabbage (b) Cauliflower (c) Broccoli (d) Chinese cabbage

743. Which one of the following is a yellow vein mosaic resistant variety of okra. (JRF-2009)
(a) Pusa Sawani (b) Harbhajan (c) Arka Anamika (d) Pusa Makhmali

744. Which spp of okra is used in back-crossing for developing of YVMV resistant Arka Anamika - (ARS, 2018)
(a) A. manihot (b) A. tetraphyllus (c) A. tuberculentus (d) A. moschatus

745. Alternate form of a single gene at same loci is known as (JRF-2009)
(a) Diallel (b) Chromosome (c) Allel (d) Plastid

746. Katte disease mainly problematic in
(a) Cinnamon (b) Cumin (c) Cardamom (d) Pepper

747. Gametes are usually...........cells. (JRF-2007)
(a) Diploid (b) Haploid (c) Germ (d) Polyploid

748. Hardening is a process in the tissue culture for (JRF-2007)
(a) Cellus formation
(b) Morphogenesis
(c) Making the plants fit for growing under field conditions after tissue culture
(d) Plantlet regeneration

749. D- leaf associated with which crop............ (JRF, 2018)
(a) Tomato (b) Potato (c) Muskmelon (d) Lettuce

750. While applying basal dose of DAP, it should never mixed with (JRF-2007)
(a) Zn (b) Cu (c) Fe (d) Mg

751. Interspecific hybrid variety of Tomato.................. (JRF, 2018)
(a) Arka Rakshak (b) Pusa 120 (c) Pusa Red Plum (d) Pusa Gaurav

752. Non growth retardant for onion sprouting is: (ARS, 2016)
(a) Maleic Hydrazide (MH) (b) Ethephon
(c) Both (d) None of these

753. Which crop gives higher protein recovery per unit area? (JRF-2007)
(a) Pea (b) Gherkin (c) Carrot (d) Potato

754. TPS seed rate of potato- (ARS, 2018)
(a) 50-100 gm (b) 100-120 gm (c) 120-150 gm (d) 150-200 gm

755. Carrot roots become longer
(a) Above 10°C (b) Below 10°C (c) Above 25°C (d) Below 25°C

756. India Institute of Spices Research is situated in
(a) Cochin (b) Trivandrum (c) Calicut (d) Dharwad

757. The seed rate for one ha planting of palak is
(a) 2-5 kg (b) 5-9 kg (c) 15-20 kg (d) 25-30 kg

758. Rio-de-Janerio is an improved variety of
(a) Turmeric (b) Garlic (c) Ginger (d) Large Cardamom

759. The red colour of chilli fruit at ripening stage is due to the pigment (JRF-2007)
(a) Capsaicin (b) Capsicutin (c) Capsanthin (d) Lycopene

760. Recombinant inbred lines exhibits monogenic trait in: (ARS, 2016)
(a) 2:1 segregation ratio (b) 3:1 segregation ratio
(c) 1:1 segregation ratio (d) None of these

761. Which of the following spices is known as 'King of Spices'
(a) Ginger (b) Cardamom (c) Turmeric (d) Black Pepper

762. Which of the following spices is known as 'Queen of Spices'
(a) Ginger (b) Cardamom (c) Turmeric (d) Balck Pepper

763. Which of the followings has the highest cation exchange capacity? (JRF-2007)
(a) Sandy loam soil (b) Loam soil (c) Loamy sand soil (d) Clay soil

764. The chromosome number of Coriandrum sativum is
(a) 2n = 11 (b) 2n = 33 (c) 2n = 22 (d) None of the above

765. Gene for Powdery mildew resistance in Pea..... (JRF, 2019)
(a) Or (b) ar2 (c) pm1 (d) er1

766. The capacity of a cell to develop into a complete plant is called............ (JRF-2007)
(a) Tissue culture (b) Fecundity (c) Totipotency (d) Haploidy

767. ELISA is ....linked test (JRF-2007)
(a) Bacteria (b) Virus (c) Enzyme (d) Protozoa

768. Temperature for heading in cabbage..... (JRF, 2019)
(a) 10-15°C (b) 25-30°C (c) 15-20°C (d) 30-40°C

769. The onion variety suited for dehydration – (ARS, 2018)
(a) Pusa Red (b) Pusa Ratnar
(c) Pusa white round (d) Pusa Madhavi

770. Sweet potato is a (JRF-2007)
(a) Short day plant (b) Long day plant (c) Day neutral (d) Intermediate

771. The anti-cancer properties in crucifers is due to the presence of: (JRF, 2016)
(a) Indole-3-carbinol (b) Anthocyanin (c) Diadzein (d) Carotenoids

772. Self pollination in tomato is due to (JRF-2009)
(a) Allogamy (b) Cleistogamy (c) Chasmogamy (d) Autogamy

773. Delphinidin a plant pigment is responsible for development of .....colour (JRF-2009)
(a) Red (b) Blue (c) Orange (d) Pink

774. Man made vegetable is: (JRF-2009)
(a) Rutabaga (b) Cauliflower (c) Kale (d) Cabbage

775. Oleoresin in extracted from (JRF-2007)
(a) colocasoa (b) chilli (c) Brinjal (d) Onion

776. Which one of the following states has maximum coefficient of variation in rainfall? (JRF-2005)
(a) Assam (b) Maharashtra (c) Rajasthan (d) West Bengal

777. Pre Cooling is meant to (JRF-2005)
(a) Improve fruit quality (b) Remove field heat
(c) Fresh the object (d) None of these

778. Round shape variety of Radish (JRF-2009)
(a) Japanese White (b) Rapid Red White Tipped
(c) Pusa Himani (d) Chinese Pink

779. Early blight of tomato is caused by (JRF-2005)
(a) Fungus (b) Virus (c) Bacterium (d) None of these

780. In a DNA molecule, thymine always pairs with (JRF-2009)
(a) Adenine (b) Guanine (c) Cystocine (d) Urecil

781. In radish the system of pollination is (JRF-2005
(a) Self pollination (b) Cross pollination
(c) Often Cross-pollination (d) None of these

782. The class of seeds which must maintain relatively high moisture content in order to remain viable are known as: (JRF, 2016
(a) Orthodox seeds (b) Exaibuminous seeds
(c) Recalcitrant seeds (d) Labyrinth seeds

783. Muriate of Potase is called (JRF-2005
(a) K_2SO_4 (b) KCI (c) K_2O (d) $KMNO_4$

784. TSS for tomato processing… (JRF, 2019
(a) 1.5-3.5°Brix (b) 3.5-5.5°Brix (c) 5.5-6.5°Brix (d) 6.5-8.5°Brix

785. Sodium rich vegetable crop – (ARS, 2018
(a) Lettuce (b) Brinjal (c) Okra (d) Watermelon

786. Chromosome number of cucumber is: (JRF-2005
(a) 2n = 24 (b) 2n = 14 (c) 2n = 12 (d) 2n = 20

787. Chromosome number of okra is (JRF-2005
(a) 2n = 22 (b) 2n = 130 (c) 2n = 88 (d) 2n = 20

788. Triploid water melon was released in the year…. (JRF, 2019
(a) 1950 (b) 1948 (c) 1939 (d) 1956

789. Apical meristem culture is used to produce plants free from: (JRF, 2016
(a) Viruses (b) Bacteria (c) Fungus (d) Nematode

790. Indian share in world vegetable production (ARS, 2018)
(a) 13.5% (b) 13.7% (c) 13% (d) 14.3%

791. Little leaf of brinjal is caused by (JRF-2005)
(a) Bacteria (b) Fungi (c) Mycoplasma (d) Zn deficiency
792. Tripping mechanism in dancing stigmatic lobe comes out of anther the characteristics features of which crop (Imp.)
(a) Pea (b) Cowpea (c) Okra (d) Limabean
793. Varsha Uphar and Hissar Unnat are two important cultivar of: (Imp.)
(a) Tomato (b) Okra (c) Capsicum (d) Brinjal
794. Which of the following crop has low inbreeding depression [loss in vigour due to inbred] (Imp.)
(a) Onion (b) Cabbage
(c) Carrot (d) Snowball cauliflower
795. STG (Shoot Tip Grafting) is done for.... (JRF, 2019)
(a) Mango (b) Banana (c) Citrus (d) Litchi
796. Which one is the correct scientific name of cauliflower (JRF-2005)
(a) Brassica oleracea var capitata (b) Brassica oleracea var gummifera
(c) Brassica oleracea var botrytes (d) Brassica oleracea var italica
797. Vegetable forcing means – (ARS, 2018)
(a) Growing of vegetable in main season (b) Growing of vegetable in off season
(c) Both (d) None of these
798. The bitterness is colocasia corms is felt due to the presence of (JRF-2005)
(a) Calcium carbonate (b) Capscicin
(c) Calcium oxalate (d) Sodium chloride
799. Which compound imparts colour and flavour to Chilli? (JRF, 2016)
(a) Oleoresings (b) Capsanthin (c) Solanin (d) Lycopene
800. Celery edible part is – (ARS, 2018)
(a) Fleshy leaf stalks (b) Roots (c) Fleshy Stem (d) Fleshy tubers
801. Contender is an important variety of: (JRF, 2014)
(a) Cowpea (b) French bean (c) Sweet pea (d) None of these
802. Heretosis can be fully exploited in the form of (JRF-2005)
(a) Composites (b) Hybrids (c) Multilines (d) Synthetics
803. Ea-1 gene is related to: (ARS, 2016)
(a) Potato (b) Brinjal (c) Tomato (d) Maize
804. Which Gene Action determines color development in cauliflower ... (JRF, 2019)
(a) Single recessive gene (b)Overdominance gene
(c) Single dominance gene (d) Predominance of non-additive gene
805. Mosaic virus in cucumber spreads by (JRF-2005)
(a) Seeds (b) Aphids (c) Jassids (d) White fly
806. Indira kakoda -1 is a variety – (ARS, 2018)
(a) Ivy gourd (b) Pointed gourd (c) Spine gourd (d) Ridge gourd
807. In self-pollinated crops, cross pollination varies from (Imp.)
(a) 5-10% (b) 20-30% (c) 30-50% (d) 50-70%

808. Okra hybrids are very popular as F1 , all are resistant to YVMV governed by (Imp.)
(a) Singh dominant gene (b) Single recessive gene
(c) Two recessive gene (d) All

809. Which crops have staminal sterility (Imp.)
(a) Tomato (b) Brinjal
(c) Cabbage (d) Cucumber (gynoecious line)

810. Brinjal belongs to the species (JRF-2005
(a) Melongena (b) Khasianum (c) Sisymbrifolium (d) None of these

811. In tomato which part is used for induction for mutation (Imp.)
(a) Seeds (b) Roots (c) Shoot apex (d) All

812. Leafleass mutants are produce in which crop (Imp.)
(a) Brinjal (b) Onion (c) Okra (d) Pea

813. Breeder seed is the progeny of (JRF-200
(a) Registered seed (b) Nucleus seed (c) Foundation seed (d) Certified seed

814. Which cucumber spps. have most resistant against nematodes.
(a) Cucumis anguria (b) Cucumis ficifolius
(c) Cucumis metuliferus (d) None of these

815. Little leaf resistant spps. of Brinjal (Imp.)
(a) S. torum (b) S. nigrum (c) S. gilo (d) S. khasianum

816. Characteristics flavor of cole crops is due to volatile compounds
(a) Diallyl-disulphide (b) Dimethyl-trisulphide
(c) Isothiocynate (d) Diphenylamine

817. Pusa Purple Round is a variety of. (JRF, 201
(a) Tomato (b) Brinjal (c) Chilli (d) Turnip

818. In male sterility, the pollen is (JRF-200
(a) Abortive (b) Shrivelled (c) Non functional (d) All of these

819. French bean seed storage at vapour proof container having optimum moister conte ranged from ….. (ARS, 201
(a) 7-10% (b) 10-12% (c) 12-15% (d) 15-18%

820. International Seed Testing Association [ISTA] is situated at
(a) Colambo (Srilanka) (b) New Delhi (India)
(c) Norway (Switzerland) (d) Rome (Italy)

821. New Zealand spinach (Tetragonia expansa) belongs to family
(a) Chenopodiaceae (b) Solanaceae (c) Apiaceae (d) Aizoaceae

822. Crop having max/highest productivity in India (Imp.)
(a) Cabbage (b) Cassava (c) Potato (d) Okra

823. Botanical name of lima bean is – (ARS, 201
(a) P. Vulgaris (b) P. Lunatus
(c) P. Tetragonolobus (d) P. Sativum

824. Lima bean(Butter bean) is (Imp.)
(a) Self-pollinated (b) Cross-pollinated
(c) Often-cross-pollinated (d) Highly cross-pollinated

825. In Sweet potato root system is (JRF-2005)
(a) Deep (b) Medium (c) Medium deep (d) Shallow

826. Gametophytic self-incompatibility is found in
(a) Beet (b) Cabbage (c) Turnip (d) Tomato

827. Bacterial blight resistant variety of cowpea.
(a) Pusa Rituraj (b) Pusa Barsati (c) Pusa Komal (d) Arka Suman

828. Seed ball (2-6) method is used in propagating of – (ARS, 2018)
(a) Beet root (b) Carrot (c) Radish (d) Sweet Potato

829. Pusa Red Plum variety of tomato has highest vitamin-C content and it is developed through (Imp.)
(a) Intervarietal crosses (b) Interspecific crosses
(c) Intergeneric (d) Intraspecific crosses

830. Sree Dhara is a variety of (Imp.)
(a) Elephant foot yam (b) Sweet potato
(c) Topioca (d) Coleus

831. Self- core Asiatic Carrot variety is...... (JRF, 2019)
(a) Pusa Yamdagni (b) Early Nantes (c) Zeno (d) Pusa Rudhira

832. Which method of seed production is used in onion for both kharif and rabi- (ARS, 2018)
(a) seed-seed (b) bulb-seed (c) Both (d) None of these

833. Pusa snowball -1 variety of cauliflower of which seed production is only possible in
(a) Plains (b) Hills (c) Both (d) None

834. Which type of flower in brinjal have maximum fruit cutting.
(a) Long styled (b) Medium styled
(c) Pseudo short-styled (d) Short styled

835. In private sector first F1 hybrid of tomato "Karnatka Hybrid" and of sweet pepper "Bharat" was released for commercial cultivation in 1973 by a private seed company.
(a) IAHS- Banglore (b) Nath (c) MAHYCO (d) Beejosheetal

836. Multistyle flowers are common in (JRF-2005)
(a) Okra (b) Brinjal (c) Chillies (d) Lima Bean

837. Which was the first example of successful breeding for disease or pest resistance?
(a) Pusa Sawani – YVMV (b) Pusa Komal – bacterial blight
(c) Pusa Shubra – black rot (d) Pusa Drum Head – Black leg

838. Little leaf of Brinjal is transmitted by (Imp.)
(a) Hopper (b) Aphids (c) Jassids (d) White fly

839. Sex form of cultivated Black pepper... (JRF, 2019)
(a) Monoecious (b) Gynomonoecious (c) Dioecious (d) Andromonoecious

840. Share of vegetable is about............% of the total cultivable land in India is
(a) 1.2 (b) 2.8 (c) 3.6 (d) 4.8

841. Which of the following is leading vegetable crop of India in production
(a) Cabbage (b) Potato (c) Tomato (d) Brinjal

842. Drumstick leaves are rich in source of: (ARS, 2016)
(a) Vitamin B and D (b) Vitamin A and C
(c) Vitamin B and E (d) All of these

843. Which of the following is a long day type onion variety
(a) Nasik Red (b) Brown Spanish
(c) Udaipur102 (d) Pusa Red

844. Karpenchenko (1997) made first intergeneric hybrid between
(a) Radish & Cabbage (b) Cabbage & cauliflower
(c) Radish & cauliflower (d) Radish & Turnip

845. Pusa Parvati an improved cultivar of French has resistant to mosaic and Powdery Mildew disease and suitable for hills and it was developed through
(a) Selection (b) Hybridization (c) Mutation (d) Polyploidy

846. Pusa Kirti & Pusa Kiran are the variety of
(a) Spinach (b) Palak (c) Lettuce (d) Amaranthus

847. Asparagus is a (Imp.)
(a) Monoecious (b) Dioecious
(c) Andromonocious (d) Hermis

848. "Akashin" a physiological disorder observed in radish is due to deficiency of
(a) N (b) Ca (c) K (d) B

849. Which one of the following is used to enhance more fruit set (JRF,1997)
(a) NAA (b) 2,4-D (c) 2,4,5-T (d) Ethylean

850. Following variety of Black Pepper is one of the open pollinated progeny of Balankotta
(a) Panniyur -1 (b) Panniyur – 2 (c) Panniyur -3 (d) None of the above

851. The bitterness / acridity in colocasia corm is due to
(a) Calcium oxalate (b) Calcium carbonate
(c) Calcium chloride (d) Potassium oxalate

852. The country where maximum genetic diversity in brinjal occurs
(a) India (b) Peru and Bolivia (c) Iran (d) China

853. International Institute of Horticulture is situated in (JRF-2006)
(a) Brazil (b) Italy (c) USA (d) None of these

854. Long term storage, which is most commonly used
(a) Liquid nitrogen (b) Liquid oxygen
(c) Carbon dioxide (d) All these

855. The fruit of tomato is (JRF, 2002)
(a) Pone (b) Stone (c) Berry (d) Drupe

856. ------------ is also known as chinchida or Pallakaya
(a) Bitter gourd (b) Snake gourd (c) Ash gourd (d) Pumpkin

857. Capsicum is generally ____________ pollinated crop
(a) Self (b) Cross (c) often cross (d) none of the above
858. Cultivated okra is --------- in Nature
(a) Polyploid (b) Tetraploidy (c) Aneuploidy (d) All of the above
859. Wax gourd is mainly grown in
(a) Tamil Nadu (b) Kerala (c) North India (d) None of the above
860. Lutein alkaloid is found in
(a) Cucurbita muschata (b) Cucurbita maxima
(c) Pointed gourd (d) None of the above
861. Colchicine is soluble in
(a) Alcohol (b) Cold water (c) Chloroform (d) All of the above
862. Micropropagation is most popularly followed in: (ARS, 2016)
(a) Tomato (b) Potato (c) Brinjal (d) Cauliflower
863. August & September are varieties of
(a) Knol-khol (b) Cabbage (c) Radish (d) Kale
864. Source of Vit B-12... (JRF, 2019)
(a) Tomato (b) Potato (c) Leafy vegetable (d) Radish
865. The productivity of vegetable in India is
(a) 11.2% (b)14.4% (c) 12.2% (d) 15.1%
866. According to Vavilov (1951) onion originated in ... (ARS, 2018)
(a) Iran (b) Iraq (c) Afghanistan (d) China
867. India contributes.......... % in world vegetable production
(a) 5.38 (b) 23.5 (c) 13.48 (d) 31.31
868. USA stands at which position in total vegetable production
(a) 1st (b) 3rd (c) 2nd (d) 4th
869. Kashi Sweta is a variety of............ (JRF, 2018)
(a) Brinjal (b) Bottle Gourd (c) Radish (d) Pumpkin
870. India ranks in vegetable production
(a) First (b) Second (c) Third (d) Fourth
871. Which of the following is leading vegetable crop of India in productivity
(a) Cabbage (b) Tapioca (c) Tomato (d) Brinjal
872. Which of the following is leading vegetable crop of India in area
(a) Cabbage (b) Potato (c) Tomato (d) Brinjal
873. Which of the following is the leading export commodity of India
(a) Fresh onion (b) Fresh cucumber (c) Fresh potato (d) Fresh tomato
874. Inbreeding depressions is much more among the following crops
(a) Winter cauliflower (b) Broccoli
(c) Summer cauliflower (d) Brussels sprout
875. Mitosis takes place in which of the following organs (JRF-2008)
(a) Reproductive organs (b) Somatic organs
(c) Anthers (d) Both (a) and (b)

876. Transgenics are produced in tomato for
(a) Modified ripening (b) Resistance to herbicides
(c) Resistance to virus (d) All the above

877. Introduction of genes coding for Bacillus thuringenesis (Bt.) insecticidal crystal protein confer resistance to
(a) Insects (b) Bacteria (d) Diseases (d) Viruses

878. Strains of Bt produce toxins, which confer resistance against
(a) Diptera (b) Coleopteran
(c) Lepidoptera larvae (d) All the above

879. Tomato variety which can with stand high as well as low temperature (JRF-2007)
(a) Pusa Naveen (b) Hissar Lalit (c) Pusa Sheetal (d) Pusa Sadabahar

880. "Newleaf" is transgenic of which crop commercialized by Monsanto since 1989
(a) Potato carrying 3A a gene (b) Cabbage carrying 3A a gene
(c) Tomato carrying 3A a gene (d) Brinjal carrying 3A a gene

881. Which of the following tomato varieties are good for processing
(a) Arka Shaurabh (b) Arkas Ashish (c) Arka Ahuti (d) All the above

882. Colour blindness genes are located on (JRF-2008)
(a) Y-chromosomes (b) X-chromosomes (c) Autosomes (d) All of the above

883. Colour contents of tomato are controlled by
(a) Two genes (b) Four genes (c) Three genes (d) Multiple gene

884. Global diversity of vegetables crops consists of about
(a) 200 spp. (b) 400 spp. (c) 300 spp. (d) 500 spp.

885. Arka Manorama is variety of…. (JRF, 2019)
(a) Marigold (b) Gladiolus (c) Carnation (d) Crossandra

886. High starch content Curcuma species is…. (JRF, 2019)
(a) Curcuma longa (b) Curcuma caesia
(c) Curcuma aromatica (d) Curcuma angustifolia

887. Progenitor of cucumber- (ARS, 2018)
(a) C. foetidissima (b) C. hardwickii (c) C. metuliferus (d) C. anguria

888. The origin/parents of Arka Anand (ARS, 2018)
(a) IIHR-3 X IIHR-322 (b) IIHR-2X SM -6
(c) A pure line selection from IIHR-194 (d) None of these

889. Which of the following centers is not primary center of diversity for any vegetable crops
(a) Chinese (b) South America (c) North America (d) Central America

890. Which of the following is primary center of diversity for cucumber
(a) Chinese-Japanese (b) Hindustani Center
(c) Indo-Chinese (d) Mediterranean region

891. Which of the following is primary center of diversity for bottle gourd
(a) Chinese-Japanese (b) Hindustani Center
(c) Indo-Chinese (d) Mediterranean region

892. Which of the following is a stem vegetable? (JRF-2007)
(a) Colocasia (b) Amaranth (c) Sweet potato (d) Knolkhol
893. Which of the following is primary center of diversity for yams
(a) Chinese-Japanese (b) Hindustani Center
(c) Indo-Chinese (d) Mediterranean region
894. The variety of EFY free from acridity- (ARS, 2018)
(a) Sree Varun (b) Gajendra (c) H-42 (d) Gouri
895. Pusa Swarnima is a variety of
(a) Summer Squash (b) Turnip (c) Winter Squash (d) Carrot
896. Which of the following is primary center of diversity for radish
(a) Chinese-Japanese (b) Hindustani Center
(c) Indo-Chinese (d) Mediterranean region
897. Kanji is made from: (ARS, 2016)
(a) Pusa asita (b) Pusa kesar (c) Pusa meghali (d) Pusa Vrishti
898. Which of the following is component of National Gene Bank
(a) National seed repository (b) Cryobank
(c) National tissue culture (d) All the above
899. Development of embryo from egg cell without fertilization is called as (JRF-2008)
(a) Apospory (b) Autogamy (c) Parthenogenesis (d) Apogamy
900. Pusa Nasdar is a variety of
(a) Amaranth (b) Bottle gourd (c) Ridge gourd (d) Brinjal
901. Cryopreservation (cryobank), can be adopted for preservation of which type of seed
(a) Orthodox (b) Both a & c (c) Recalcitrant (d) None a & c
902. Pusa Kesar is a variety of
(a) Summer Squash (b) Turnip (c) Winter Squash (d) Carrot
903. The cultivated potato is - (ARS, 2018)
(a) Autodiploid (2x) (b) Autotetraploid (4x)
(c) Autohexaploid (6x) (d) Autooctaploid (8x)
904. Arka Harit is a variety of
(a) Amaranth (b) Bottle gourd (c) Bitter guard (d) Tomato
905. Which of the following is primary center for spinach
(a) Central Asia (b) Africa (c) Near East (d) Central America
906. Capsaicin formula: (ARS, 2016)
(a) $C_6H_{12}O_8$ (b) $C_{12}H_{24}No_4$ (c) $C_{18}H_{27}No_3$ (d) $C_6H_{24}H_3$
907. Pusa Katki is a variety of
(a) Cauli flower (b) Bottle gourd (c) Bitter guard (d) Brinjal
908. N-53 is a variety of
(a) Amaranth (b) Bottle gourd (c) Onion (d) Brinjal
909. Which of the following is monocot vegetable crop- (ARS, 2018)
(a) Beet root (b) Colocasia (c) EFY (d) Potato

910. Arka Nishant is a variety of
(a) Amaranth (b) Bottle gourd (c) Radish (d) Carrot

911. Which of the following is primary center for cassava
(a) Central Asia (b) Africa (c) Near East (d) Central America

912. Pusa Alankar is a variety of
(a) Summer Squash (b) Turnip (c) Winter Squash (d) Carrot

913. In a bisexual flower, the phenomenon when gynoecium matures earlier than androecium is called: (JRF, 2016)
(a) Protandry (b) Autogamy (c) Heterogamy (d) Protogyny

914. Which of the following varieties are mutants of tomato
(a) CO1 (b) Pusa lal Meeruti (c) S-12 (d) All the above

915. Pusa Yamdagini is a variety of
(a) Carrot (b) Turnip (c) Beetroot (d) Radish

916. Which of the following is chilli variety is a result of hybridization?
(a) G1 (b) G2 (c) G3 (d) G5

917. Coated or pelleted seeds are used for: (JRF, 2015)
(a) Low germination rate seeds (b) Bold seeds
(c) To prevent odour (d) Dormant seeds

918. Blanched stem is the edible portion of: (ARS, 2016)
(a) Parsley (b) Celery (c) Portulaca (d) Lettuce

919. Which of the following is a small seeded variety of coriander
(a) Karan (b) Swathi (c) Sindhu (d) Sadhna

920. "Yercaud-1" is an improved variety of
(a) Cinnamon (b) Clove (c) Nutmeg (d) Kokum

921. Triple disease resistant variety of Tomato.............. (JRF, 2018)
(a) Pusa 120 (b) Arka Rakshak (c) Pusa Sadabahar (d) Arka Vikas

922. Which one of the following in not a cole crop: (JRF, 2015)
(a) Knol-khol (b) Cabbage (c) Turnip (d) Kale

923. Which of the following turnip varieties is not the result of hybridization
(a) Pusa Swarnima (b) Pusa Kanchan (c) Pusa Sweti (d) Pusa Chandrima

924. Little leaf of brinjal is transmitted by- (ARS, 2018)
(a) Cestius phystatic (b) Bemisia tabaci
(c) Dysdercus cingulatus (d) Pempherulus affinis

925. Pusa Kranti is a variety of
(a) Tomato (b) Okra (c) Brinjal (d) Sweet potato

926. Improved Japanese, cauliflower is introduced in India from
(a) USA (b) Philippines (c) Denmark (d) Israel

927. Golden Acre, cabbage is introduced from
(a) USA (b) Phillipines (c) Denmark (d) Israel

928. Pusa Barsati, cowpea is an improvement over introduction from
(a) USA (b) Phillipines (c) Denmark (d) Israel

929. Bonneville' pea is introduced from
(a) USA (b) Phillipines (c) Denmark (d) Israel

930. GMS in Chilli is maintained by.... (JRF, 2019)
(a) A line (b) C line (c) B line (d) None of these

931. Parbhani Kranti is variety of
(a) Tomato (b) Okra (c) Brinjal (d) Sweet potato

932. TSS of jam should be: (JRF, 2015)
(a) 65% Brix (b) 68% Brix (c) 70% Brix (d) 69% Brix

933. The critical day length requirement for bulbing in garlic is: (JRF, 2016)
(a) 10 hrs (b) 12 hrs (c) 14 hrs (d) 8 hrs

934. Sugar Baby, watermelon has been introduced in India from
(a) USA (b) Phillipines (c) Denmark (d) Israel

935. The highest export earning vegetable crop- (ARS, 2018)
(a) Chilli (b) Onion (c) Tomato (d) Pumpkin

936. Chemical present in turmeric is
(a) Cineol (b) Zingiberene (c) Cucurmine (d) Linalool

937. Which of the following spice is native to India
(a) Cardamom (b) Kokum (c) Saffron (d) All the above

938. Blanching is done to: (JRF, 2015)
(a) Inactivate enzymes (b) Inactivate microbes
(c) Remove moisture (d) Storage

939. Arkel, pea is introduced from
(a) France (b) Phillipines (c) Denmark (d) Israel

940. The first F1 hybrid to be released from IARI, New Delhi in 1973 was
(a) Pusa Meghdoot in bottle gourd (b) Pusa Sanyog in cucumber
(c) Pusa Alanka in summer squash (d) None of the above

941. Area (%) under hybrid is maximum in
(a) Tomato (b) Brinjal (c) Cabbage (d) Okra

942. Dill is a native of
(a) India (b) Eurasia (c) Mediterranean (d) South America

943. Which of the following crops is wind pollinated
(a) Spinach (b) Tomato (c) Cucumber (d) Watermelon

944. The often cross pollinated vegetable crop is (JRF-2008)
(a) Pea (b)Tomato (c) Amaranth (d) Limabean

945. Palam Samridhi is an improved variety of
(a) Cabbage (b) Broccoli (c) Cauliflower (d) Brussels sprout

946. Which of the following is hybrid variety of black pepper
(a) Penniyur-1 (b) Penniyur-2 (c) Penniyur-4 (d) None

947. Kt-1 (Pusa Deepti) developed at IARI, Regional station Katrain belongs to
(a) Tomato (b) Brinjal (c) Chilli (d) Capsicum

948. CH-1, hybrid developed at IARI, New Delhi belongs to
(a) Tomato (b) Brinjal (c) Chilli (d) Capsicum

949. Crossing over takes place during (JRF-2008)
(a) Leptotene (b) Zygotene (c) Pachytene (d) Diplotene

950. Leading cardamom growing state in India is
(a) Tamilnadu (b) Andhra Pradesh (c) Kerala (d) Karnataka

951. Arka Pitambar and Arka Kirtiman are hybrids of
(a) Okra (b) Capsicum (c) Onion (d) Chilli

952. Hybrid seed production in cucumber involves
(a) Use of gynoecious lines (b) Both a & c
(c) Use of monoecious lines (d) None of the above

953. Which of the following state is the largest producer of Black pepper
(a) Kerala (b) Andhra Pradesh (c) Tamil Nadu (d) Karnataka

954. Sex expression in cucurbits is manipulated by
(a) AgNo3 (b) GA3 (c) Etherl (d) All the above

955. Pinching of male flowers is effective in
(a) Bottle gourd (b) Cucumber (c) Water melon (d) Muskmelon

956. Application of ethrel @ 200-300 ppm at two true leaf stages and four leaf stages suppresses male flowers in
(a) Bottle gourd (b) Squashes (c) Pumpkins (d) All the above

957. Black pepper starts yielding from............years onwards
(a) 2nd (b) 3rd (c) 4th (d) 5th

958. In a DNA molecule , thymine always pair with (JRF-2008)
(a) Adenine (b) Guanine (c) Cytosine (d) Urecil

959. Which of the following monoecious line is used for production of Pusa Rasraj
(a) M1 (b) M3 (c) M2 (d) M4

960. Non-lobing leaf character is used as single recessive marker gene for hybrid seed production in
(a) Watermelon (b) Muskmelon (c) Cucumber (d) None of the above

961. Bitterness in cucumber is due to effect of external pollens and is called
(a) Alkaloids (b)Metaxenia (c) Acids (d) Glucosides

962. Quantity of vine used for transplanting sweet potato in 1 ha- (ARS, 2018)
(a) 25000 vine cutting/ha (b) 45000 vine cutting/ha
(c) 65000 vine cutting/ha (d) 35000 vine cutting/ha

963. Hayes and Jones (1916) first reported heterosis in
(a) Cucumber (b) Muskmelon (c) Muskmelon (d) Lone melon

964. Nutritional quality breeding deals with genetic improvement in (JRF-2008)
(a) Vitamins (b) Oil (c) Protein (d) All of the above

965. Hybrids in cabbage developed by using cytoplasmic male sterility are suitable for
(a) Hills (b) Plateau (c) Plains (d) Islands

966. Onion umbel dip treatment is given before emasculation at temperature of — (ARS, 2018)
(a) 11.5°C (b) 14.7°C (c) 13.5°C (d) 12.8°C

967. Hara Madhu, Pusa Madhuras and Arka Rajhans are varieties of ---------------
(a) Watermelon (b) Muskmelon (c) Snapmelon (d) Long melon

968. ---------- is also known as balsam pear
(a) Ridge gourd (b) Ash gourd (c) Snake gourd (d) Bitter gourd

969. -------variety of bottle gourd is a cross of PSPL X Selection -2
(a) Pusa Naveen (b) Pusa Manjari (c) Pusa Gomal (d) Pusa Meghdoot

970. Cucumber is --------------- in nature
(a) Monoecious (b) Dioecious
(c) Gynomonoecious (d) Andromonocious

971. Tender crop and cascade are cultivars of _______ suited for processing purpose.
(a) Indian bean (b) French bean (c) Cowpea (d) Winged bean

972. For best growth and yield of French bean , the optimum temperature should be
(a) 25-30°C (b) 10-15°C (c) 15 -25°C (d) 35-40°C

973. The number of chromosomes is reduced to half in plants obtained by (JRF-2008)
(a) Anther culture (b) Ovule culture (c) Shoot tip culture (d) Both (a) and (b)

974. Type of incompatibility in cabbage and cauliflower is
(a) Sporophytic (b) Neither a nor c (c) Gametophytic (d) Both a and c

975. Welsh and Grimbell in 1947 first reported male sterility in
(a) Radish (b) Onion (c) Cucumber (d) Carrot

976. The % of seed recovery of tomato fruit- (ARS, 2018)
(a) 10% (b) 20% (c) 30% (d) 35%

977. Regeneration of a plant from a single cell in culture medium is called as (JRF-2008)
(a) Protoplast culture (b) Meristem culture (c) Organ culture (d) Cell culture

978. CH-1, A hybrid in chilli is result of a cross between
(a) MS-12 X Punjab Lal(b) MS-13 X Punjab Lal
(c) MS-1 X Punjab Lal (d) MS-11 X Punjab Lal

979. Which of the following is used as androcides in okra
(a) GA3 (b) MH (0.4-0.5%) (c) NAA (d) None of the above

980. Which of the following is used as maintainer line in onion breeding
(a) Line A (b) Line D (c) Line B (d) Line C

981. Which of the following is used as maintainer line in onion breeding
(a) Line A (b) Line D (c) Line B (d) Line C

982. MCPA is a - (ARS, 2018)
(a) Natural retardant (b) Synthetic growth stimulant
(c) Synthetic retardant (d) All of these

983. The variety of sweet potato having purple skin and yellow flesh- (ARS, 2018)
(a) Sree Vardhini (b) Varsha (c) H-42 (d) H-41

984. In India, male sterile line have been isolated from which of the following varieties
(a) Niphad-2-4-1 (b) Pusa Red
(c) Nasik White Globe (d) All of the above

985. In onion the type of self sterility is
(a) Genic (b) Cytoplasmic-genic
(c) Cytoplasmic (d) None of the above

986. Crucifer vegetables are value for (JRF-2008)
(a) Magnesium containing compounds isothicynates
(b) Sulphar containing compounds glucosinolates
(c) Molybdenum containing compounds isoflavonones
(d) Boron containing compounds vanillin

987. Pinching of male flower is effective in hybrid seed production of
(a) Bottle gourd (b) Ash gourd (c) Bitter gourd (d) None of the above

988. Which of the following organisms are the most destructive in vegetable crops
(a) Fungus (b) Virus (c) Bacteria (d) None of the above

989. Chimera can be maintained by (JRF-2008)
(a) Somatic embryogenesis (b) Apomixis
(c) Adventitious shoot formation (d) Cutting and Grafting

990. Which of the following tomato varieties are resistant to root knot nematodes
(a) Hissar Lalit (b) Arka Vardhan (c) Sel. 120 (d) All of these

991. Which of the following tomato varieties are resistant to bacterial wilt
(a) Arka Abha (b) Arka Alok (c) Arka Shresha (d) All of the above

992. Which of the following tomato varieties are resistant to fruit borer
(a) Pusa Uphar (b) BT-10 (c) BT-1 (d) None of the above

993. 2,4-D is a (JRF-2008)
(a) Fungicide (b) Auxin
(c) Growth hormone (d) Insecticide

994. Which of the following species of Lycopersicon are resistant to leaf curl virus
(a) L. peruvianum (b) L. glandulosum
(c) L. pimpinellifolium (d) All of these

995. Arka Manik watermelon is resistant to
(a) Powdery mildew (b) Anthracnose
(c) Downey mildew (d) All of these

996. Pusa Mukta is black rot resistant variety of
(a) Knol Khol (b) Cabbage (c) Cauliflower (d) Tomato

997. Which of the following are tolerant to salt stress in tomato
(a) L. peruvianum (b) L. pennellii (c) L. cheesmani (d) All of these

998. Arka Sheel and Manjari Gota are resistant to
(a) Little leaf (b) Root knot rematodes
(c) Aphid (d) None of these

999. Long day type variety of garlic is
    (a) Agrifounf White (b) Agrifound Parvati
    (c) Yamuna Safed (d) G-282

1000. Which of the following brinjal varieties is resistant to aphid
    (a) Arka Sheel (b) Manjari Gota (c) Annamalai (d) Pusa Bhairav

1001. Which of the following chilli varieties is resistant to leaf curl
    (a) Pusa Jawala (b) Pant C-1
    (c) Pusa Sadhabahar (d) All of these

1002. Which of the following are resistant to powdery mildew in chilli
    (a) Arka Lohit (b) Pusa Jawala (c) Pant C-1 (d) Pusa Sadhabahar

1003. Poinsette cucumber is resistant to
    (a) Powdery mildew (b) Angular leaf spot (c) Anthracnose (d) All of these

1004. Black leg is an important disease of which of the following vegetable crops (JRF-2008)
    (a) Cauliflower (b) Tomato (c) Winged bean (d) Cabbage

1005. Allium fistulosum is resistant to
    (a) Neck rot (b) Onion smut (c) Pink rot (d) All of these

1006. Which of the following are phomopsis blight resistant varieties of brinjal
    (a) Pant Samrat (b) Pusa Purple Cluster
    (c) Pusa Bhiarav (d) All of these

1007. Pusa Sadhabahar, a chilli variety is resistant to
    (a) CMV (b) Leaf curl (c) TMV (d) All of these

1008. Pant C-1 is resistant to
    (a) Bacterial blight (b) TMV (c) CMV (d) All of these

1009. In green vegetables the limiting amino acid is (JRF-2008)
    (a) Tryptophan (b) Methionine
    (c) Lysine (d) Arginine

1010. Okra is a
    (a) Self-pollinated (b) Often cross-pollinated
    (c) Cross-pollinated (d) None of these

1011. Which of the following mechanisms in brinjal favours cross pollination
    (a) Heterostyly (b) Protoandry
    (c) Protogyny (d) None of these

1012. Seed rates of kharif okra is… (ARS, 2018)
    (a) 7.5 kg (b) 15 kg (c) 20-25 kg (d) 15-18 kg

1013. Large Cardamom is widely cultivated in________
    (a) Kerala (b) Karnataka
    (c) Tamilnadu (d) Sikkim

1014. In which of the following state potato is generally not grown
    (a) Kerala (b) Tamil nadu
    (c) Uttar Pradesh (d) Andhra Pradesh

1015. Which of the following is Summer season crop
(a) Broad bean (b) Lima bean (c) Pea (d) None of these

1016. Type of inflorescence in carrot is
(a) Spike (b) Panicle (c) Umbel (d) Catkin

1017. Self pollination is a form of (JRF-2008)
(a) Outbreeding (b) Inbreeding (c) Random mating (d) None of the above

1018. Which of the following is sexually sterile diploid among all
(a) Onion (b) Garlic (c) Leek (d) Beet

1019. Palak is a
(a) Self-pollinated (b) Often cross-pollinated
(c) Cross-pollinated (d) None of these

1020. Which of the following is wind pollinated
(a) Onion (b) Radish (c) Carrot (d) None of these

1021. Indole-3-Carbinol is present in: (ARS, 2016)
(a) Tomato (b) Cauliflower (c) Cabbage (d) Turnip

1022. Hormone used for drought resistance is: (JRF 2016)
(a) 2,4-D (b) Cytokinin (c) IBA (d) ABA

1023. Amaranthus seed rate is- (ARS, 2018)
(a) 0.5-0.8kg/ha (b) 0.8-1kg/ha (c) 1-1.5kg/ha (d) 1.5-2kg/ha

1024. Lima bean is a
(a) Self-pollinated (b) Often cross-pollinated
(c) Cross pollinated (d) None of these

1025. Which of the following mechanisms exists in cole crops
(a) Heterostyly (b) Protoandry (c) Protogyny (d) None of these

1026. Trypsin inhibitor is found in which tuber crop? (ARS, 2016)
(a) Beet root (b) Carrot (c) Sweet Potato (d) Radish

1027. Which of the following mechanisms exists in carrot
(a) Heterostyly (b) Protoandry (c) Protogyny (d) None of these

1028. Type of inflorescence in cassava is
(a) Spike (b) Panicle (c) Umble (d) Catkin

1029. --------------- variety of chilli is cross of NP 46A x Pure red
(a) Pant C-1 (b) Pusa Jwala (c) Pusa Sadabahar (d) None of the above

1030. Ash gourd is considered good for people suffering from
(a) Dysentery (b) Stomach ache
(c) Nervousness (d) High blood pressure

1031. Cinavakacherri Murunga, Puna murangai and PKM-1 are varieties of.
(a) Curry leaf (b) Basella (c) Broad bean (d) Drumstick.

1032. --------------- a grain type amaranths is widely grown in Gujarat and Maharashtra.
(a) Jobner Green (b) Rajgarh (c) Pusa Harit (d) Pant Haritima.

1033. ---------------- is popularly known as malbar night shade, poi or Indian spinach.
(a) Methi (b) Basella (c) Spinach (d) Spinach belt

1034. The edible part of rutabaga is
(a) Root (b) Spear (c) Stem (d) Corm

1035. In onion, pink colour is due to: (JRF, 2016)
(a) Anthocyanin (b) Quercetin (c) Znathophyll (d) Carotene

1036. Which of the following is winter crop
(a) Broad bean (b) Lima bean (c) French bean (d) Cluster bean

1037. Edible part of most of the summer season vegetable crops is
(a) Fruit (b) Stem (c) Leaf (d) Root

1038. Tetrad is seen in: (JRF, 2016)
(a) Diplotene (b) Pachytene (c) Diakinesis (d) Krebs cycle

1039. Sesbania grandiflora, agathi belongs to family
(a) Leguminosae (b) Labiateae (c) Portulacae (d) Moringaceae

1040. Cultivation of two crops together in the same field by mixing their seeds is called (JRF-2008)
(a) Mixed farming (b) Intercropping (c) Mixedcropping (d) Doublecropping

1041. In cabbage which method of seed production is good for seed yield with breeder seed production? (ARS, 2016)
(a) Head intact method (b) Stump method
(c) Stump with central core intact method (d) Non of these

1042. Which of the following vegetable belong to C4 plant
(a) Carrot (b) Potato (c) Amaranths (d) Bean

1043. "Potherbs" are generally
(a) Cool season crops (b) Rainy season crops
(c) Warm season crops (d) None of these

1044. Self pollination in leaf Amaranthus is mainly due to – (ARS, 2018)
(a) More no. of female flower (b) More no. of male flower
(c) More no. of female and male flower (d) None of these

1045. The non-preferance insect resistant mechanism is known as (JRF-2008)
(a) Antichlorosis (b) Antibiosis (c) Antixenosis (d) Antipheromonosis

1046. Sauropus androgynous, chekurmanis belongs to
(a) Euphorbiaceae (b) Solanaceae (c) Convolvulaceae (d) Polygonaceae

1047. Asparagus belongs to the family: (ARS, 2016)
(a) Cucurbitaceae (b) Araceae (c) Liliaceae (d) Compositae

1048. Which of the following is deep rooted crop
(a) Tomato(Wolf apple) (b) Onion
(c) Cabbage (d) Garlic

1049. The highest TSS containing species used in breeding is - (ARS, 2018)
(a) S. pseudocapsicum (b) S. chimelweskii
(c) S. nigrum (d) S. gracile

1050. Mamordica cochinchinensis is commonly known as
(a) Bitter gourd (b) Kakrol (c) Sweet gourd (d) Round melon

1051. C4 vegetable is: (ARS, 2016)
(a) Palak (b) Amaranthus (c) Parsley (d) Chicory

1052. Which of the following genus belong to cucurbitaceae
(a) Benincasa (b) Trichosanthes (c) Legenaria (d) All of these

1053. In which crop, Gynoecious sex is commercially exploited for hybrid seed production: (JRF, 2016)
(a) Water melon (b) Cucumber (c) Ridge gourd (d) Musk melon

1054. Cyamposis tetragonolobus is botanical name of
(a) Broad bean (b) Cluster bean (c) Lima bean (d) Scarlet bean

1055. Name the highly salt tolerant vegetable crop: (JRF, 2016)
(a) Summer squash (b) Bottle gourd (c) Ash gourd (d) Cucumber

1056. Xanthosoma sagittifolium is botanical name of
(a) Giant Taro (b) Tannia (c) Taro (d) Arrowroot

1057. Anthesis in which of the following cucurbits takes place in the evening
(a) Cucumber (b) Muskmelon (c) Sponge gourd (d) Bottle gourd

1058. Which crop is a good source of L-Dopa (dopamine) to cure Parkinson disease in humans? (JRF, 2016)
(a) Broccoli (b) Broad bean (c) Celery (d) Lettuce

1059. In which of the following crops, cleistogamy is observed
(a) Lettuce (b) Capsicum (c) Tomato (d) Onion

1060. Satputia variety of ridge gourd have sex form
(a) Monoecious (b) Dioecious (c) Hermaphrodite (d) Gynoecious

1061. Herkogamy is observed in
(a) Lima bean (b) Capsicum (c) Tomato (d) All of these

1062. In which of the following vegetable crops maximum varieties are released in India
(a) Tomato (b) Pepper (c) Brinjal (d) Potato

1063. Genetic male sterility is observed in
(a) Cole crops (b) Watermelon (c) Muskmelon (d) All of these

1064. Chromosome number of Brinjal….. (JRF, 2019)
(a) 36 (b) 48 (c) 24 (d) 12

1065. Cytoplasmic male sterility is observed in
(a) Carrot (b) Cucumber (c) Sweet pepper (d) All of these

1066. God send vegetable called- (ARS, 2018)
(a) Winged bean (b) Indian bean (c) French bean (d) Cluster bean

1067. "Solasodine" is present in
(a) Tomato (b) Onion (c) Brinjal (d) Garlic

1068. The seed of cabbage can only be produced commercially under: (JRF, 2018)
(a) Tropical climate (b) Temperate climate
(c) Central Indian plains (d) Sub tropical climate

069. Which of following variety of carrot is excellent for canning
(a) Chantaney (b) Pusa Kesar (c) Pusa Lal (d) Pusa Meghali

070. Dioscorine is present in
(a) Yam (b) Elephants foot yam
(c) Cassava (d) Colocassia

071. Yellow colour in Tomato is due to ..... (JRF, 2019)
(a) Lycopene (b) Prolycopene
(c) Anthocyanin (d) Carotenoids

072. Peas and beans contain which of the following toxic substances
(a) Phytic acid (b) Anti vitamin E factor
(c) Trypsin inhibitor (d) All of these

(JRF, 2018)

073. Water use efficiency is the ration of:
(a) Yield/ET (b) ET/ Yield (c) ET/ DM (d) PET/ Yield

(ARS, 2018)

074. Ca rich vegetable crop-
(a) Amaranthus (b) Cucumber (c) Tomato (d) Brinjal

075. Which of the following toxic substances is present in celery
(a) Apiin (b) Sinigrin (c) Cucurbitacin (d) Choline esterase

076. "Cheratin" is present in
(a) Bitter gourd (b) Watermelon (c) Bottle gourd (d) Muskmelon

077. Which of the following is a commercial source of Levulose
(a) Queensland arrowroot (b) Jerusalem Artichoke
(c) Colocasia (d) West Indian Arrowroot

078. Male sterility in Cole crops is due to.... (JRF, 2019)
(a) CMS (b) GMS (c) CMS and GMS (d) CGMS

079. Which of the following acts as an anti-oxidant
(a) Selenium (b) Iodine (c) Boron (d) Calcium

080. In IPM techniques following flowering plant are used in tomato- (ARS, 2018)
(a) Rose (b) Marigold (c) Tulip (d) Chrysanthemum

081. White brinjal is recommended for
(a) Diabetes (b) Cancer
(c) Heart disease (d) Skin problems

082. Chilli hybrid – 1 was developed at : (JRF, 2018)
(a) IIHR- Banglore (b) PAU-Ludhiana
(c) IARI – New Delhi (d) HAU-Hisar

083. Bitterness in carrot is due to
(a) Sulphoraphane (b) Isopentanol (c) Isocoumarin (d) Carotene

084. Ancestor of cole crop- (ARS, 2018)
(a) B.oleracea var. gongylodes (b) B. oleracea var. Sylvestris
(c) B.oleracea var. napobrassica (d) B.oleracea var. acephala

1085. Which is the rich source of vitamin C (JRF, 2001
(a) Cabbage (b) Tomato (c) Chilli (d) All

1086. Green manure should be incorporated at its (JRF, 2002)
(a) Early tender stage (b) Fully matured stage
(c) Flowering stage (d) Drying stage

1087. Tenderometer is used to measure the maturity of
(a) Cowpea (b) Cluster bean (c) Pea pods (d) French bean

1088. The per capita consumption of vegetable/availability of vegetable in India is
(a) 300g (b) 285g (c) 145-175g (d) 225g

1089. Which of the following countries has the highest per capita intake of vegetables
(a) Greece (b) Republic of Korea
(c) Turkey (d) India

1090. Which of the following are thermo-insensitive varieties of cauliflower
(a) Pusa Katki (b) Pusa Synthetic (c) Pusa Deepali (d) All of these

1091. Red flesh pumpkins are rich source of : (JRF, 2018
(a) Vitamin C (b) Protein (c) Vitamin A (d) Acids

1092. Which of the following is the first resistant variety released in vegetable in India
(a) "Pusa Sawani" in Okra (b) Pusa Jawala in chilli
(c) Sel. 120 in tomato (d) Annamalai in brinjal

1093. Which of the following is first hybrid released in vegetable crop in public sector
(a) Pusa Meghdoot in bottle gourd (b) Pusa Sanyog in cucumber
(c) Pusa Alankar in summer squash (d) Arka Navneet in brinjal

1094. Scooping is a practice that facilitates bolting in
(a) Cabbage (b) Brussels sprout (c) Cauliflower (d) Broccoli

1095. Durgapura Kesar is a variety of.... (JRF, 2019)
(a) Watermelon (b) Muskmelon (c) Round melon (d) None of these

1096. Curry leaf is a
(a) Self-pollinated crop (b) Cross-pollinated crop
(c) Often cross-pollinated (d) None

1097. Flower colour of Snake gourd: (ARS, 2016)
(a) Yellow (b) Red (c) Pink (d) White

1098. Pusa Manjari is an improved variety of
(a) Bittergourd (b) Ridgegourd (c) Bottlegourd (d) Cucumber

1099. Blanching is practiced in
(a) Okara (b) Asparagus (c) Tomato (d) Onion

1100. Powdery mildew resistant variety of watermelon – (ARS, 2018)
(a) Pusa Rasal (b) Arka Manik (c) Sugar Baby (d) None of these

1101. Which of the following is most commonly used rooting hormone (JRF-2006)
(a) IBA (b) NAA (c) IAA (d) GA

102. Which is the nodal agency for tissue culture propagule: (ARS, 2016)
(a) Department of Science Technology (DST)
(b) Department of Bio-Technology (DBT)
(c) CSRI, New Delhi
(d) None of these

103. Saffron belongs to the family (JRF-2006)
(a) orchidaceae (b) Iridaceae (c) Lauraceae (d) Apiaceae

104. Which of the following are self blanched varieties of cauliflower
(a) Pusa Snowball-K (b) Pusa Himjyoti (c) Pusa Sharadh (d) Pant Shubra

105. Cavity spot is due to deficiency of – (ARS, 2018)
(a) B (b) Mg (c) Ca (d) Zn

106. Which of the followings are self blanched varieties of cauliflower
(a) Pusa Deepali (b) Pusa Snowball – 16
(c) Pusa Snowball K-1 (d) All of these

107. In cryopreservation, seed are immersed in liquid nitrogen at............°C (JRF-2006)
(a) 100 (b) –196 (c) –100 (d) 0

108. Lincon is an exotic variety of
(a) Cowpea (b) Pea (c) French bean (d) Broad bean

109. Which of the following radish variety is resistant to rust
(a) Scarlet Globe (b) White Icicle
(c) French Breakfast (d) Arka Nishant

110. Which of the following vegetable require mulching
(a) Potato (b) Tomato (c) Lettuce (d) Brinjal

111. Which of the following is late maturing variety of cauliflower ? (JRF-2008)
(a) Pusa Him Jyoti (b) Pusa Synthetic
(c) Pusa Snowball – 16 (d) Pusa Deepali

112. Anthesis of sponge gourd is take place in : (ARS, 2016)
(a) Morning (b) Evening (c) Afternoon (d) Night

113. Which of the following palak variety is development through spontaneous mutation
(a) Pusa Palak (b) Jobner Green (c) Pusa Harit (d) Pusa Jyoti

114. Auxin is synthesized in the plant at (JRF-2006)
(a) Stem (b) Apex (c) Root hairs (d) Leaves

115. ....is commonly used for propagation of bush pepper (JRF-2006)
(a) Geotrophic shoot (b) Orthotropic shoot
(c) Plagiotropic shoot (d) Runner

116. Delayed harvesting in radish may cause
(a) Pithiness (b) Forking (c) Deformed roots (d) All

117. Which of the following varieties of Brussels sprout is dwarf one
(a) Long Island (b) Catskill (c) Amager Market (d) Danish Prize

118. "Rutin" and Vitmin-P are associated with which vegetable crop
(a) Chili (b) Tomato (c) Brinjal (d) Potato

1119. "Capsaicin" is chemically
(a) $C_{17}H_{27}NO_3$ (b) $C_{11}H_{22}NO_3$ (c) $C_{18}H_{28}NO_3$ (d) None of these

1120. Pillow disorder in cucumber is due to - (ARS, 201
(a) Zn (b) Ca (c) Mg (d) Na

1121. Botanical name of mango ginger is (JRF-200
(a) Curcuma longa (b) Curcuma amada
(c) Mangifera indica (d) Curcuma aromatic

1122. Pusa Bedana ,Durgapura Meetha, Durgapura Kesar are improved varieties of
(a) Watermelon (b)Muskmelon (c) Bottlegourd (d) Bittergourd

1123. Which of the following onion variety is suitable for dehydration
(a) Udaipur – 101 (b) Jobner Green (c) Udaipur – 102 (d) Pusa Jyoti

1124. Pusa Sumeet is an improved variety of
(a) Broad bean (b) Cluster bean (c) Lima bean (d) Peas

1125. All Green is a variety of (JRF-200
(a) Beet leaf (b) Spinach (c) Methi (d) Broccoli

1126. - maturity group in vegetable crop is found in- (ARS, 201
(a) Cauliflower (b) Potato (c) Okra (d) Brinjal

1127. Emasculation is done in: (JRF, 201
(a) Self pollinated crops (b)Cross pollinated crops
(c) Sterile crops (d) None of these

1128. Which of the following French Bean varieties is developed through mutation
(a) Pusa Parvathi (b) Arka Komal (c) Phule Surekha (d) Pant Anupma

1129. Sree Dhara is a variety of
(a) Sweet potato (b) Coleus(Chinese potato)
(c) Colocasia (d) Cassava

1130. Jobner Green is a high yielding variety of
(a) Spinach (b) Indian Spinach (c) Amaranthus (d) Onion

1131. Which of the following vegetable produces maximum amount of nutrient/unit area/ur time
(a) Potato (b) Carrot (c) Tomato (d) Onion

1132. Browning in cauliflower is due to deficiency of – (ARS, 201
(a) Ca (b) Na (c) Bo (d) Zn

1133. Little leaf of brinjal is transmitted by
(a) Fly (b) Leaf hopper (c) Whitefly (d) None of these

1134. Mosaic in chilli is transmitted by
(a) Hopper (b) White fly (c) Aphid (d) None of these

1135. Poisonous baiting is done to control (JRF-200
(a) Borer (b) Whitefly (c) Thrips (d) Fruitfly

1136. Which of the following is a recognized grading standard of tomato
(a) Super B (b) Super C (c) Super A (d) Super D

1137. Which of the following tomato varieties are resistant to nematodes
(a) Nemamukt (b) SI-120 (c) Nemared (d) All of these
1138. 'Arka Alok' is a tomato variety is resistant to (JRF-2008)
(a) Wilt (b) Nematode
(c) Fruit borer (d) High temperature stress
1139. "Muscilage substance" of okra used for clearing sugarcane juice is extracted from
(a) Flower (b) Leaf (c) Stem and root (d) Fruit
1140. Which of the botanical varieties of lettuce have excellent shipping and handling abilities? (JRF, 2016)
(a) Butter head type (b) Crisp head type (c) Cos type (d) Celery lettuce
1141. Lux is a unit of : (JRF, 2014)
(a) Light (b) Photosynthesis
(c) Transpiration (d) Respiration
1142. Which of the following are the methods of seed extraction in tomato.
(a) Fermentation (b) Alkali treatment
(c) Acid treatment (d) All of these
1143. Indeterminate habit of tomato is associated with (JRF-2008)
(a) Lower provitamin A (b) Lower total soluble solids
(c) Poor keeping quality (d) High total soluble solids
1144. In which of the following crop TIBA is exogenously applied
(a) Watermelon (b) Cucumber (c) Musk melon (d) Bottle gourd
1145. Which of the following is present in cluster bean
(a) Asparagine (b) Mannogalacton (c) Glutamine (d) All of these
1146. Which of the following is pole type variety of French bean
(a) Kentucky Wonder (b) Giant Stringless
(c) Contender (d) Pant Anupama
1147. Hottest chilli of India is: (JRF, 2015)
(a) Pusa Jwala (b) NP 46 A (c) Bhoot Jolokia (d) Bhaskar
1148. Favisim, an illness caused due to an allergy of which plant part of broad bean
(a) Green pod (b) Pollen (c) Root (d) Both a and b
1149. 'Scooping' is an important operation in raising seed crops in which of the following crops: (JRF, 2016)
(a) Broccoli (b) Brussel Sprouts (c) Cabbage (d) Cauliflower
1150. Brown Heart of turnip is cause by the deficiency of
(a) B (b) Water stress (c) Ca (d) High temperature
1151. Japanese White is a variety of (JRF-2005)
(a) Radish (b) Carrot (c) Turnip (d) None
1152. About 95 to 99.5 portion of the plant tissue are made of (JRF-2008)
(a) C, H and O (b) N, P and K (c) Ca, Mg and C (d) Cu, Zn and Fe

1153. Each beet root fruit (Capsule) contains 2-6 seeds hence requires
(a) Thinning (b) Weeding (c) Hoeing (d) None of these

1154. In Knol khol, fleshy edible part is and enlargement of (JRF-2005)
(a) Stem (b) Root (c) Leaf (d) bud

1155. "Cinkle of potato" is caused by
(a) Fungus (b) Virus (c) Bacteria (d) MLOs

1156. Which type of sex form is present in onion? (JRF, 2016)
(a) Protandry (b) Protogyny (c) Chasmogamy (d) Cleistogamy

1157. Golden nematode (Globodera Heterodera rostochinenesis) is serious pest of
(a) Potato (b) Chili (c) Tomato (d) MLOs

1158. Majority of the cucurbits are seed propagated except
(a) Pointed gourd (b) Kartoli and Kakrol
(c) Little gourd (d) All of these

1159. Balausta is a modified: (JRF, 2015)
(a) Berry (b) Drupe (c) Capsule (d) Pepe

1160. To breed for disease resistance which method is used (JRF, 1999)
(a) Back cross (b) Mass selection
(c) Pure line selection (d) All

1161. Arka Sheetal is an improved variety of
(a) Long melon (b) Snap melon (c) Water melon (d) Muskmelon

1162. Black leg in potato is caused by
(a) Fungus (b) Virus (c) bacteria (d) MLOs

1163. Macrophomina phaseoli, in potato causes
(a) Charcoal rot (b) Soft rot (c) Brown rot (d) Wart

1164. Virginia Savoy is a variety of
(a) Spinach (b) Beet (c) Palak (d) None of these

1165. Headquarter of 'AGMARK' is located at (JRF-2008)
(a) Jaipur (b) Nagpur (c) Lucknow (d) New Delhi

1166. Which of the following is a serious pest of sweet potato
(a) Hairy caterpillar (b) Horned caterpillar
(c) Sweet potato weevil (d) Leaf eating caterpillar

1167. Isolation distance in Brinjal for seed production is (JRF, 2002)
(a) 50-100 m (b) 10-20 m (c) 200-300 m (d) 500 m

1168. Arka Mohini is a variety of
(a) Tomato (b) Okra (c) Brinjal (d) Sweet Pepper

1169. The response of plants to relative day length of the day and night is called: (JRF, 2016)
(a) Photoperiod (b) Photoperiodism (c) Phytotron (d) Vernalization

1170. Male sterility is common in (JRF,1998)
(a) Carrot (b) Bean (c) Okra (d) None

1171. Gynoecious lines in cucumber is induced by: (ARS, 2016)
(a) 2-4D (b) Auxin (c) Ethrel (d) Gibberellin

1172. CH-1 variety of chilli is developed at (JRF, 2001)
(a) TNAU, Coimbatore (b) PAU, Ludhiana
(c) IIHR, Bangalore (d) IARI, New Delhi

1173. Saffron is native to
(a) India (b) America (c) Indonesia (d) Africa

1174. Quantitative characters are governed by (JRF-2007)
(a) Monogene (b) Bi-gene (c) Recessive gene (d) Polygene

1175. Dormancy of seed can be due to (JRF-2007)
(a) Impermeability of seed coat (b) Requirement of water
(c) Presence of ABA in the seed (d) All of these

1176. Ditylenchus dipsaci is a serious nematode of
(a) Onion (b) Brinjal (c) Tomato (d) Chili

1177. Majority of the cucurbits are
(a) Annual and monoecious (b) Parennial dioeclous
(c) Annual and dioecious (d) Perennial and monoecious

1178. "Potato Withes Broom" is caused by
(a) MLOs (b) Bacteria (c) Fungus (d) Virus

1179. Which of the following is a temperate turnip
(a) Pusa Kanchan (b) Pusa Chandrima (c) Pusa Swati (d) Pusa Swarnima

1180. Pionsette is the popular variety of (JRF-2005)
(a) Longmelon (b) Muskmelon (c) Watermelon (d) Cucumber

1181. Plant Quarantine Act came in to force in the year: (JRF, 2015)
(a) 1941 (b) 1914 (c) 1966 (d) 1973

1182. Splitting in carrot is basically due to
(a) B deficiency (b) Fluctuating moisture imbalance
(c) Ca deficiency (d) High temperature

1183. Wind breakers are planted in the direction: (JRF, 2014)
(a) South-East (b) North-East (c) South-West (d) North-West

1184. Which of the following is an early, dwarf but wrinkled seeded variety of pea
(a) Alaska (b) Meteor (c) Asauji (d) Arkel

1185. London Flag is improved variety of
(a) Leek (b) Garlic (c) Onion (d) Welsh Onion

1186. Photoblastism has been used in: (JRF, 2014)
(a) Tomato (b) Water melon (c) Potato (d) Lettuce

1187. In plant breeding nearly homozygous lines are produced by continued self-fertilization accomplish by selection for ………… generations (JRF-2006)
(a) 3-4 (b) 5-6 (c) 7-8 (d) 9-10

1188. Early Grano variety of onion was introduced from: (JRF, 2016)
(a) UK (b) USA (c) Canada (d) China

1189. Which of the following vegetable is tolerant to acidic soil? (JRF, 2014)
(a) Brinjal (b) Chilli (c) Okra (d) Potato

1190. Onion variety most suitable fpr export is
(a) N-53 (b) Arka Bindu (c) Arka Kalyan (d) Pusa Red

1191. Pusa Rohini is the variety of
(a) Tomato (b) Chilli (c) Cabbage (d) Brinjal

1192. Optimum temperature for tuber production
(a) 15-20°C (b) 20-25°C (c) 10-15°C (d) >30°C

1193. Which of the following enzymes converts starch to sugar in potato
(a) Starch phosphorylase (b) Both a and c
(c) Glucose-1-Phosphate (d) None of a and c

1194. Red pumpkin beetle (Aulacophora foveicollis) is less damaging in which of the following cucurbits
(a) Pumpkin (b) Bottle gourd (c) Muskmelon (d) Bitter gourd

1195. In onion.............is used as maintainer for the male sterile line (JRF-2006)
(a) Smsms (b) Nmsms (c) NMsMs (d) NMsms

1196. Which of the following cucurbits is less affected by powdery mildew
(a) Luffa (b) Cucumis (c) Lagenaria (d) Momordica

1197. Pointed gourd is a : (JRF, 2014)
(a) Self-pollinated crop (b) Evergreen vegetable
(c) Perennial vegetable (d) None of these

1198. Which of the following nutrient element effect sex expression in pumpkin
(a) B (b) Ca (c) Mo (d) Both a and c

1199. Common methi bears flowers of which colour
(a) White (b) Red (c) Pink (d) Yellow

1200. What is the full form of PCR? (JRF, 2015)
(a) Polymerase chain reaction (b) Polymer chain reaction
(c) Polymeric chain reaction (d) Polymerised chain reaction

1201. Which of the following is non-climacteric
(a) Cucumber (b) Tomato (c) Muskmelon (d) Watermelon

1202. Flat saur of canned products is caused by: (JRF, 2015)
(a) Bacillus (b) Clostridium (c) Salmonella (d) Staphylococcus

1203. Maleness in gynoecious line of cucumber is induced by
(a) Silver nitrate (b) ABA (c) NAA (d) IAA

1204. The genetics basis of pureline selection was first explained by (JRF-2006)
(a) C. A. Brim (b) J.B. Harrington (c) C.O. Gardner (d) W. Johannsen

1205. Self blanching is the characteristics of which group of cauliflower variety
(a) Late (b) Mid season (c) Early (d) None of these

1206. Which of the following is a physiological disorder of potato
(a) Phyllody (b) Leaf roll
(c) Purple top wilt (d) Hollow or black heart

1207. 'Brown anther' type male sterility is observed in which vegetable crop? (JRF, 2016)
(a) Radish (b) Turnip (c) Carrot (d) Beet

1208. Starch content % of sweet potato tuber is
(a) 16 (b) 6 (c) 26 (d) 0.6

1209. Pusa bedana is a variety of : (ARS, 2016)
(a) Muskmelon (b) Watermelon (c) Longmelon (d) Roundmelon

1210. Yerrusseri product is prepared from which vegetable
(a) Cucumber (b) Pumpkin (c) Carrot (d) Radish

1211. Which of the following variety of tomato have potato leaf marker.
(a) Pusa Ruby (b) Pusa Red Plum (c) Pusa Divya (d) Pusa Hybrid-1

1212. "Allinase" is an enzyme that converts alliin in garlic into
(a) Alicin (b) Allicine (c) Allinin (d) Allinine

1213. Optimum seed rate(g/ha) of cauliflower is
(a) 200-300 (b) 300-500 (c) 700-800 (d) 800-900

1214. Seed rate for raising one acre of radish is (JRF-2005)
(a) 1-5 kg (b) 2-3 kg (c) 4-5 kg (d) 7-8 kg

1215. The origin place of ridge gourd is
(a) Ethiopia (b) India (c) Brazil (d) China

1216. Basic chromosome number in bottle gourd is
(a) 11 (b) 12 (c) 9 (d) 10

1217. Arka Komal and pant Anupama are the varieties of
(a) Cowpea (b) French bean (c) Cluster bean (d) Pea

1218. Average daily temperature of -------------- is most favourable for cucumber growth
(a) 18-24 0C (b) 10-140 C (c) 28-300 C (d) None of the above

1219. Early December is the improved variety of (JRF-2007)
(a) Pea (b) Cauliflower (c) Cabbage (d) Carrot

1220. ISTA is established at: (ARS, 2016)
(a) Geneva, Switzerland(b) Italy, France
(c) Newyork, USA (d) New Delhi, India

1221. In sweet potato anti nutritional factor is
(a) Trypsin Inhibitor (b) Cynoglucoside
(c) Coumarin (d) Quercitin

1222. Flavour of cabbage leaves is due to the glucoside
(a) Cucurbitacin (b) Oleoresin (c) Sinigrin (d) Eruic acid

1223. Direct seeding in tomato is due to: (ARS, 2016)
(a) Late Yield (b) Early Yield (c) Mid Yield (d) None of these

1224. In cucumber common sex expression is (JRF-2005)
(a) Monoecism (b) Dioecism
(c) Andremonoecism (d) None of these

1225. Downy mildew of muskmelon controlled by systemic fungicide (JRF-2005)
(a) Ridomi MZ (b) Blitox (c) Indofil M45 (d) None of these

1226. Allin is an a odourless and colourless
(a) Acid (b) Pigment (c) Alcohol (d) Amino acid

1227. The predominant pigment, responsible for red colour of tomatoes is due to (JRF-2008)
(a) Beta-carotene (b) Anthocyanin (c) Lycopene (d) Xanthophyll

1228. The process of producing several identical copies of a gene sequence is known as: (JRF, 2016)
(a) Gene cloning (b) Gene sequencing (c) Gene splicing (d) Micro cloning

1229. "Pillow" is a physiological disorder of............ caused due to calcium deficiency
(a) Cucumber (b) Radish (c) Carrot (d) Turnip

1230. Tomato belongs to the species (JRF-2005)
(a) Peruvianum (b) Esculentum (c) Hirsutum (d) Pimpinellifolium

1231. Pusa Early Prolific and Blue Lake are varieties of.
(a) French bean (b) Indian bean (c) Cowpea (d) Jack bean

1232. Progeny of foundation seed is: (JRF, 2014)
(a) Certified seed (b) Registered seed (c) Nucleus seed (d) None of the above

1233. Frog eye rot is a physiological disorder of
(a) Chilli (b) Potato (c) Brinjal (d) Tomato

1234. At which temperature, potato should be stored for processing: (JRF, 2015)
(a) 10°C (b) 14°C (c) 12°C (d) 15°C

1235. "Blossom end rot" is a physiological disorder of
(a) Tomato (b) Watermelon (c) Chili (d) All of these

1236. Buttoning, Riceyness, Fuziness and Leafy stem are the physiological disorder of
(a) Onion (b) Garlic (c) Cabbage (d) Cauliflower

1237. "Metsubre" is a disorder of ----due to calcium deficiency
(a) Colocasia (b) Amorphophallus (c) Yam (d) Potato

1238. Which of the following is due Mg deficiency
(a) Browning (b) Hollow stem (c) Chlorosis (d) Whiptail

1239. Pungency of onion is due to the presence of (JRF-2005)
(a) isothiocynate (b) Allyl propyl disulphide
(c) Solanin (d) Betanin

1240. Basic chromosome number in dolichos is.........
(a) 14 (b) 18 (c) 16 (d) 20

1241. Which of the following mechanism exists in carrot? (JRF, 2016)
(a) Heterostyly (b) Protogyny (c) Protandry (d) Decliny

1242. Seed rate of cauliflower is: (JRF, 2015)
(a) 400-700 g/ha (b) 600-700 g/ha (c) 300-400 g/ha (d) 1-1.5 g/ha

1243. "Golden Flake" is due to excess Ca oxalate and low K: Ca ratio in
(a) Tomato (b) Carrot (c) Potato (d) Cabbage

1244. "Pencil strip" is physiological disorder of celery due to
(a) Phosphorus (b) N_2 deficiency (c) K calcium (d) Ca deficiency

1245. All cole crops are reportedly derived from
(a) Colewort (b) Cabbage (c) Polewort (d) Cauliflower

1246. Pumpkin grown throughout India is (JRF-2005)
(a) C. moshata (b) C. maxima (c) C. ficifolia (d) C. paradisi

1247. Which part of the world is considered as center of diversity for tomato
(a) Mexico (b) Peru (c) Chile (d) India

1248. Whiptail in Cauliflower is caused due to the deficiency of (JRF-2005)
(a) Potassium (b) Boron (c) Molybdenum (d) Copper

1249. Origin of tomato is from
(a) Peru Ecuador (b) Mexico (c) Chile (d) India

1250. Fruit cracking in tomato is due to: (JRF, 2015)
(a) B deficiency (b) Ca deficiency (c) K deficiency (d) Mg deficiency

1251. Which type of tomato are referred as "self topping"
(a) Determinate (b) Intermediate (c) Indeterminate (d) None of these

1252. Which growth regular is useful for fruit set at high temperature in tomato
(a) Auxin (b) Ethylene (c) Gibberllins (d) ABA

1253. Sweet potato is a (JRF-2007)
(a) Modified stem (b) Modified crow
(c) Modified rhizome (d) Modified root

1254. Which of the following carrot variety is introduced from Germany and suitable for Nilgiri hills
(a) Imperator (b) Zeno (c) Chanteny (d) Danvers

1255. Which of the following is a flower type observed in brinjal
(a) Long syled (b) Short styled
(c) Pseudo short styled (d) All of these

1256. Which organ of the chili fruit is richest in pungency
(a) Seed (b) Pericarp (c) Placental (d) Leaves

1257. Cat face is a disorder of (JRF-2007)
(a) Cho-cho (b) Pointed gourd (c) Tomato (d) Cauliflower

1258. Which of the following cauliflower group is highly self incompatible
(a) Early maturing Group -I (b) Mid season Group – II
(c)Late maturing Group – III (d) None of these

1259. Solanum acaule, a potato species is resistant to
(a) Frost (b) Heat (c) Nematode (d) Water stagnation

1260. Red colour development in tomato is favoured by
(a) Red light (b) Both (c) Far red light (d) None of these

1261. Genetically modified vegetable crop recommended for commercial cultivation in India (JRF-2007)
(a) Potato (b) Brinjal (c) Tomato (d) None of these

1262. "Thermoperiodicity" refers to response of the plant to high day temperature and low temperature rather then constant temperature. It is commonly observed in

(a) Tomato (b) Brinjal (c) Chilli (d) Okra

1263. Pungency in horse radish is due to

(a) Allyl iso thiocyanates (b) Butyle thiocyanates

(c) b and c (d) None of these

1264. Select the element constituent of amino acid which is essential for protein synthesis (JRF-2008)

(a) Calcium (b) Magnesium (c) Manganese (d) Sulphur

1265. Sponge gourd and ridge gourd are

(a) Intercrossable (b) Intersterile (c) Both a and b (d) None of these

1266. Which of the following source is is used for YVMV resistance transfer in Arka Anamika and Arka Abhay

(a) A. tetraphyllus (b) A.esculentus

(c) A.manihot (d) A.manihot spp.manihot

1267. Which of the following cucurbit groups have the highest chromosome number

(a) Pumpkin and squash(b) Cucumber

(c) Gourds (d) Melons

1268. Which one does not belong to family Zingiberaceae: (JRF, 2015)

(a) Ginger (b) Turmeric (c) Cardamom (d) Onion

1269. Ideal processing potato cultivator should have (JRF-2007)

(a) High TSS and low sugars (b) High dry matter & low sugars

(c) High sugars (d) High acidity

1270. Which of the following cucurbits have the lowest chromosome numbers

(a) Cucumber (b) Muskmelon (c) Watermelon (d) Longmelon

1271. A nullisomic individual is represented by(JRF-2008)

(a) 2n-1 (b) 2n+2 (c) 2n-2 (d) 2n+1

1272. The variety of okra resistant both YVMV and leaf curl virus- (ARS, 2018)

(a) Kashi lila (b) Kashi Kranti (c) Kashi Pragati (d) All of these

1273. The diploid chromosome number of cluster bean is

(a) 2n = 22 (b) 2n = 24 (c) 2n = 14 (d) 2n = 16

1274. The origin of Dolichus bean is

(a) China (b) India

(c) Mexico (d) Mediterranean region

1275. Major pollinating agent in cucurbits is (JRF-2005)

(a) Honey bee (b) House fly (c) Ants (d) Bumble bee

1276. Origin of water melon is

(a) Africa (b) India (c) Iran (d) None

1277. Which of the following cucurbits is also used as ornamental type

(a) Summer squash (b) White squash (c) Pumpkin (d) Winter squash

1278. Little leaf of brinjal is transmitted by (JRF-2008)
(a) Aphids (b) Jassids (c) Hopper (d) Whitefly

1279. Which of the following cucurbits is propagated by stem cuttings
(a) Kakrol (b) Kartoli (c) Little gourd (d) All of these

1280. Bad smell in leafy vegetables is due to… (JRF, 2019)
(a) Rotting (b) Over maturity (c) Discoloration (d) All of these

1281. Pusa rasdar is a variety of : (ARS, 2016)
(a) Bottle gourd (b) Sponge gourd (c) Bittergourd (d) Ridge gourd

1282. Tuberization in sweet potato is promoted under (JRF-2005)
(a) 7 hr. light (b) 9 hr. light (c) 11 hr. light (d) 13 hr. ligh

1283. Leafy vegetables are usually rich source of (JRF-2008)
(a) Iron (b) Protein (c) Carbohydrate (d) Carotene

1284. In which of the following vegetable all parts are edible
(a) Winged bean (b) Limabean (c) Broadbean (d) Carrot

1285. Which one is found in radish: (JRF, 2015)
(a) Non adienal (b) Dimethyl disulphide
(c) Dimethyl pyarzin (d) Isothiocynate

1286. Molecular markers are used for determining........variation in plants. (JRF-2007)
(a) Genetic (b) Phenotypic (c) Cytoplasmic (d) Endoplasmic

1287. Sree Shilpa is a first hybrid of
(a) Greater Yam (b) Elelhant foot yam (c) Lesser yam (d) White yam

1288. Shoot and fruit borer of okra belongs to
(a) Earias vitella (b) Amrasca bigttula bigttula
(c) Tetranychus spp (d) Heliothis armigera

289. Cucumis ficifolia is a type of species
(a) Perennial (b) Biennial (c) Annual (d) None of these

290. Scientific name of Wax Gourd: (ARS, 2016)
(a) Bennincasa hispida (b) Trichosanthus diocea
(c) Cucumis anguira (d) Praecitrullus fistulosus

291. Phytic acid a toxic substance is present in
(a) Leafy vegetables (b) Peas and beans (c) Cucurbits (d) Root vegetables

292. All spice belongs to family
(a) Umbelliferae (b) Apiaceae (c) Myrtaceae (d) None

293. Restriction fragments length polymorphism (RFLP) is a (JRF-2007)
(a) Marker technique (b) Selection method
(c) Mutation technique (d) Gene slicing technique

294. The amount of solanin/tuber is considered as unfit for consumption
(a) 5 mg (b) 10 mg (c) 20 mg (d) 25 mg

295. Linkage between gene affect
(a) Independent assortment (b) Fertilization
(c) Vernalization (d) Anaphase

1296. Organic soils contain: (JRF, 2016)
(a) >20% organic mater (b) >20% organic mater
(c) >30% organic mater (d) <30% organic mater

1297. Mass selection is essentially a method of
(a) Hybridization technique (b) Clonal selection
(c) Population improvement (d) Mutation breeding

1298. Highest uniformity is observed in: (ARS, 2016)
(a) Single cross hybrid (b) Double cross hybrid
(c) Triple cross hybrid (d) all of these

1299. Seed borne disease of cauliflower is managed by
(a) Formaldehyde (b) Hot water treatment
(c) Methyl bromide (d) Soil sterilization

1300. Arka Jyoti is watermelon while Pusa Jyoti is
(a) Palak (b) Sugarbeet (c) Amaranth (d) Beetroot

1301. Cavity spot disorder of carrot is caused due to deficiency of (JRF-2007)
(a) Ca (b) Fe (c) S (d) Cu

1302. Powdery mildew of pea is caused by
(a) Fungi (b) Bacteria (c) Virus (d) None

1303. Crossing over during meiosis results in
(a) Promoting linkage (b) Breaking linkage
(c) Help in mutation (d) None

1304. Time interval between irrigation is referred as: (JRF, 2016)
(a) Irrigation efficiency (b) Irrigation frequency
(c) Irrigation scheduling(d) Irrigation potential

1305. Pusa Harit is a palak while Arka Harit is a
(a) Beetroot (b) Bittergourd (c) Spinach (d) Lettuce

1306. Pusa Rasraj is a F1 hybrid of (JRF-2008)
(a) Tomato (b) Watermelon (c) Muskmelon (d) Cucumber

1307. Grain amaranthus species is: (ARS, 2016)
(a) Amaranthus spinosus (b) Amaranthus tricolor
(c) Amaranthus hypocondriacus (d) Amaranthus retroflexus

1308. Back cross method of breeding means
(a) Mutation breeding (b) Tissue culture
(c) Polyploidy breeding (d) Crossing F1 to desirable parents

1309. In heterosis breeding one should capitalize on
(a) Inbreeding depression (b) Homozygosity
(c) Over dominance (d) High seed cost

1310. Optimum night temperature range for fruit set in tomato is (JRF-2008)
(a) 10-20°C (b) 15-20°C (c) 15-25°C (d) 25-30°C

1311. Which vegetable has high calorific value
(a) Fenugreek (b) Potato (c) Amaranthus (d) Tomato

1312. Which of the following is used for controlling soil borne pathogens in organic agriculture? (JRF, 2016)
(a) Palsmodiophora brassicae (b) Scllerotiorum rolfsii
(c) Trichoderma virid (d) None of these

1313. The synthesis of auxin can be done by the nutrient: (ARS, 2016)
(a) Mg (b) Mn (c) Na (d) Zn

1314. -------------- is considered as King of Gourds
(a) Ridge gourd (b) Snake gourd (c) Little gourd (d) Pointed gourd

1315. Yard Long Ridge gourd variety ---------- in nature
(a) Monoecious (b) Dioecious
(d) Andromonoecious (d) Gynoecious

1316. Annual types of Moringa is propagated through.
(a) Seeds (b) Stem cuttings (d) Root cuttings (d) Limb cuttings

1317. Swarna is a hybrid of ---------------
(a) Pumpkin (b) Watermelon (c) Bottle gourd (d) Muskmelon

1318. The mechanism of SSI used for hybrid seed production is – (ARS, 2018)
(a) Tomato (b) Cole crop (c) Brinjal (d) Capsicum

1319. The relative percentage of sand, silt and clay in a given soil sample is known as: (JRF, 2016)
(a) Soil structure (b) Soil texture (c) Horizon (d) Epipedon

1320. Arka Vikas is a tomato where as Pusa Vikas is
(a) Cucumber (b) Pumpkin (c) Ridgegourd (d) Snakegourd

1321. Highest area of polyhouse cultivation in world – (ARS, 2018)
(a) India (b) USA (c) China (d) Germany

1322. Term 'golden revolution' is related to: (JRF, 2016)
(a) Mustard (b) Horticulture (c) Agroforestry (d) Sugarcane

1323. Transgenic plants can be produced under
(a) Open field condition (b) Controlled condition
(c) In a small plot (d) None

324. India is the largest producer, consumer as well as exporter of
(a) Spices (b) Fruits (c) Milk (d) Rice

325. Naveen is a tomato hybrid while Pusa Naveen is a
(a) Bottel gourd (b) Ridge gourd (c) Sponge gourd (d) Bitter gourd

326. Which of the following tomato variety has the indeterminate growth habit? (JRF-2008)
(a) Sioux (b) Roma (c) Pusa Gaurav (d) Pusa Early Dwarf

327. Cavity spot in carrot is due to deficiency of
(a) Calcium (b) Boron (c) Iron (d) Magnesium

328. The yellow colour in tomato is due to pigment
(a) Anthocyanin (b) Carotene (c) Lycopene (d) Pro-lycopene

329. Pusa Beedana, a triploid watermelon was released in
(a) 1970 (b) 1969 (c) 1989 (d) 1972

1330. Symptoms of 'Ca' deficiency in plants is first noticed in (JRF-2008)
(a) Middle leaves (b) Lower leaves
(c) Terminal leaves (d) All leaves simultaneously

1331. The matching variety of bitter gourd and special character is – (ARS, 2018)
(a) Arka Harit, long fruit
(b) Pusa vishesh, dwarf vine
(c) Priya, susceptible to green fruit
(d) Pant Karela -1, susceptible to red pumpkin beetle

1332. Which of the following vegetable is autogamous: (JRF, 2018)
(a) Palak (b) Onion (c) Radish (d) Tomato

1333. Which of the following serves as a vector for the spread of little leaf disease of brinjal: (JRF, 2016)
(a) Whitefly (b) Leafhopper (c) Jassids (d) Red Spider Mites

1334. Arka Suryamukhi is an improved variety of
(a) Summer squash (b) Pumpkin (c) Winter squash (d) Musk melon

1335. ....is the source of Bt gene (JRF-2007)
(a) Fungi (b) Bacteria (c) Virus (d) Mycoplasma

1336. 'Arkel' Garden pea is the most suitable for
(a) Early season (b) Late season (c) Mid season (d) None of the above

1337. Watermelon easily crosses with: (ARS, 2016)
(a) Musk melon (b) Long melon (c) Snape melon (d) Round melon

1338. Bitter taste in brinjal fruit is due to
(a) Anti vitamina E factor (b) Solasodine
(c) CN glycosides (d) Trypsineinhibitors

1339. Pugency in garlic is due to
(a) Diallyl disulphide (b) Isothiocynates
(c) Allyl propyl disulphide (d) Erucic acid

1340. Many fold effects of a gene refer to (JRF-2008)
(a) Penetrance (b) Epistasis (c) Pleiotrophy (d) None of these

1341. Garden of spices... (JRF, 2019)
(a) Karnataka (b) Tamil Nadu (c) Kerala (d) Puducherry

1342. Plug trays are related to (JRF-2007)
(a) Management nursery(b) Fruit harvesting
(c) Vegetable harvesting (d) All are true

1343. For the induction of female flower in watermelon TIBA is used at rate- (ARS, 2018)
(a) 10-50 ppm (b) 50-100 ppm (c) 100-150 ppm (d) 150-200 ppm

1344. Leaf edible Solanum Sp. is: (ARS, 2016)
(a) Solanum tuberosum (b) Solanum aethiopicum
(c) Solanum sysimbrifolium (d) Solanum melongena

1345. Black heart of potato develops upon storage of tubers at temperature (oC) of
(a) 5 (b) –5 (c) 0 (d) None of the above

1346. Black rot of cauliflower is
(a) Fungal disease (b) Viral disease
(c) Bacterial disease (d) None of the above
1347. In tomato Hybrid Hisar Anmol the donor for leaf curl virus is......... (JRF-2006)
(a) Lycopersicon peruvianum (b) L. pimpinellifolium
(c) L. esculentum (d) L. hirsutum L sp. glaberatum
1348. Pure line method largely used to improve- (ARS, 2018)
(a) Cross pollinated crops (b) Self-pollinated crops
(c) Both (d) None of these
1349. Pusa Early Bunching is a variety of: (JRF, 2016)
(a) Spinach (b) Radish (c) Fenugreek (d) Coriander
1350. Black rot of cabbage in transmitted by
(a) Mechanical means (b) Seed
(c) Vectors (d) None of the above
1351. While Pusa Gaurav is a tomato Arka Gaurav is
(a) Tomato (b) Capsicum (c) Brinjal (d) Chili
1352. 34, Use of rootstock is common in (JRF-2007)
(a) Cole crops (b) Cucurbits
(c) Tuber crops (d) Leafy vegetable crops
1353. Phyllody is a serious disease of radish transmitted by: (ARS, 2016)
(a) White flies (b) Leaf eating caterpillar
(c) Leaf hoppers (d) All of these
1354. Brown heart is a physiological disorder of
(a) Radish (b) Turnip (c) Carrot (d) Beet
1355. Which vegetable seed viability is less than 1 year – (ARS, 2018)
(a) Onion (b) Brinjal (c) Okra (d) Tomato
1356. About 95 to 99% portion of plant tissues are made of: (JRF-2009)
(a) C, H and O (b) N. P & K (c) Ca, Mg and C (d) Cu, Zn and Fe
1357. A short duration crop in between two main season crops is termed as: (JRF, 2016)
(a) Cash crop (b) Intercrop (c) Catch crop (d) Companion crop
1358. The part of Isabgol consumed as laxative is: (JRF, 2014)
(a) Husk (b) Roots (c) Stem (d) Leaves
1359. All spice, Pimenta dioca, belongs to family
(a) Myrtaceae (b) Rutaceae (c) Myrsticaeae (d) Solanaceae
1360. Predominant sex form of Cucumis melo is: (JRF-2009)
(a) Gynoecious (b) Monoecious (c) Bisexual (d) Andromonoecious
1361. Dormancy in potato can be broken by use of
(a) Thiourea (b) Ethylene chlorohydrins
(c) Potassium thiocynate (d) All of the above
1362. Which of the following is the leading producer of all spice?
(a) India (b) Sri Lanka (c) Jamaica (d) Indonesia

1363. Favism is a term associated with pollen allergy of
(a) Pea (b) Cow pea (c) Broad bean (d) French bean

1364. Self incompatibility is overcome by (JRF-2006)
(a) Bud pollination (b) Surgical techniques
(c) Irradiation (d) All of these

1365. Name the intensive type of vegetable garden? (JRF, 2016)
(a) Truck garden (b) Market garden
(c) Processing garden (d) Seed garden

1366. Protogyny is prevalent in: (JRF-2009)
(a) Cabbage (b) Onion (c) Cucumber (d) Spinach

1367. The method used to detect a viral disease is: (JRF, 2014)
(a) AFLP (b) RFLP (c) ELISA (d) None of these

1368. Gherkin, Cucumis anguria is a mutant of
(a) C. moschata (b) C. pepo (c) C. longipes (d) C. sativus

1369. Which of the following vegetable is known as 'God sent Vegetable'? (JRF-2009)
(a) Pea (b) Tomato (c) Winged bean (d) Brinjal

1370. Mostly vegetable dehydration is done at temp- (ARS, 2018)
(a) 40-54°C (b) 60-66°C (c) 70-86°C (d) 90-100°C

1371. Sauerkraut can be made from: (JRF-2009)
(a) Broccoli (b) Cabbage
(c) Chinese cabbage (d) Spinach

1372. Ploidy level of cultivated potato is
(a) Diploid (b) Tetraploid (c) Triploid (d) Hexaploid

1373. "Sui generis" is a Latin phrase which means (JRF-2008)
(a) A new system (b) A secure system
(c) A system of it's own (d)A conventional system

1374. Stamenless mutant is reported in: (JRF-2009)
(a) Cabbage (b) Brinjal (c) Tomato (d) Carrot

1375. Tuberization in potato is favoured by
(a) Short day low temperature (b) Short day high temperature
(c) Long day low temperature (d) Long day high temperature

1376. Cluster bean contains mucilaginous substance called
(a) Allantoin (b) Mannogalacton (c) Allantoin (d) Glycoprotein

1377. Pant Samrat is brinjal while Samrat is
(a) Potato (b) Sweet potato (c) Tapioca (d) Colocasia

1378. The type of germination in garden pea is: (JRF-2009)
(a) Epigeal (b) Hypogeal (c) Hypo-epigeal (d) Epi-hypogeal

1379. Colour of Pusa Swati variety of turnip is
(a) Red (b) White (c) Golden (d) None of the above

1380. Large sized potato tubers are more prone to
(a) Black heart (b) Hollow heart (c) Brown heart (d) All of the above

1381. Which of the following vegetable is known as 'God Sent vegetable' (JRF-2008)
(a) Pea (b) Tomato (c) Winged bean (d) Brinjal

1382. Based on the respiratory activity of vegetables highest respiration rate value is: (ARS, 2016)
(a) >20 mg of Co2/kg/hr. (b) >30 mg of Co2/kg/hr.
(c) >40 mg of Co2/kg/hr. (d) >60 mg of Co2/kg/hr.

1383. The recommended daily consumption of leafy vegetable is
(a) 50 g (b) 116 g (c) 250 g (d) None of the above

1384. Fruit setting in brinjal usually occurs in the flower having
(a) Long and medium style (b) Medium style
(c) Short and medium style (d) Shortstyle

1385. What is the optimum temperature for seed germination in case of pea
(a) 8°C (b) 10°C (c) 22°C (d) 30°C

1386. Potato tubers are borne on: (ARS, 2016)
(a) Stem ends (b) Stolen ends (c) Both (d) None of these

1387. While Pusa Lal is Sweet potato ;Pusa Red is a
(a) Tomato (b) Onion (c) Carrot (d) Tapioca

1388. Drying ratio of onion: (ARS, 2016)
(a) 8:1 (b) 12:1 (c) 16:1 (d) 24:1

1389. The chilli variety CH-1 developed by utilizing male sterility- (ARS, 2018)
(a) CMS types (b) GMS types (c) CGMS types (d) None of these

1390. Variety resistant to cyst nematode in potato is
(a) Kufri Swarna (b) Kufri Sutlej (c) Kufri Giriraj (d) Kufri Ashoka

1391. Storage temperature for processing of Potato.... (JRF, 2019)
(a) 0°C (b) 2-4°C (c) 1-2°C (d) 4-6°C

392. Triploid hybrid of tapioca is: (ARS, 2016)
(a) Sree Harsha (b) Sree Prakash (c) Sree Jaya (d) Sree Vijaya

393. Which variety of radish is developed by family/line breeding method- (ARS, 2018)
(a) Pusa reshmi (b) Kashi sweta (c) Pusa desi (d) Pusa chetaki

394. Garden pea belongs to the sub species
(a) arvense (b) elatius (c) hortense (d) None of the above

395. Golden nematodes is a serious pest of
(a) Tomato (b) Brinjal (c) Potato (d) Cabbage

396. Carrot type is used for hybrid seed production is: (ARS, 2016)
(a) Diploid type (b) Petaloid type (c) Triploid type (d) Hexaloid type

397. variety of cucumber suitable for summer season: (ARS, 2016)
(a) Pusa Uday (b) Pusa Sanyog (c) Pusa barkha (d) All of these

398. Browning of media causing explants drying is mainly due to (JRF-2006)
(a) Tannins (b) Abscissic acid (c) Benzoic acids (d) Gibberellins

1399. While Arka Komal is French bean, Pusa Komal is
(a) Pea (b) Cowpea (c) Cluster bean (d) Lima bean

1400. The growth regulator responsible for shoot elongation is (JRF-2006)
(a) BAP (b) GA_3 (c) 2, 4-D (d) IAA

1401. Yellow anther type variety in chilli is
(a) Bhagyalakshmi (b) Jawala mukhi (c) Andhra Jyoti (d) Bhaskar

1402. Pearson Head compactness formula is: (ARS, 2016)
(a) Z = C × 100/W3 (b) Z = F × 100/W3 (c) Z = B × 100/W3 (d) Z = D × 100/W3

1403. Haploid plants have been produced by anther culture in tomato by
(a) Greshof and Doy (b) Mehta and Upadhaya
(c) Tyagi and Prasad (d) H.B. Singh

1404. Triple disease resistance (tomato leaf curl virus, early blight, wilt) tomato- (ARS, 2018)
(a) Arka Rakshakh (b) Arka Alok (c) Arka Vardan (d) Arka Abha

1405. Name the country which ranks first in vegetable production (JRF-2007)
(a) India (b) USA (c) China (d) Japan

1406. Arka Rakshak is a variety of: (ARS, 2016)
(a) Brinjal (b) Chilli (c) Tomato (d) Capsicum

1407. Haulms in potato seed crop are cut during
(a) End of December or first week of January
(b) First fortnight of January
(c) Second fortnight of January
(d) None of the above

1408. The enhancing effect of coconut water in the promotion of growth and differentiation o excised tissue and organ has been attributed to the presence of (JRF-2006
(a) Cytokinin (b) Auxins (c) BA (d) ABA

1409. Hermophrodite cultivar of ridge gourd
(a) Pusa Nasdar (b) CO1 (c) Satputia (d) Jhinga Torai

1410. Pusa Bindu is brinjal while Arka Bindu is
(a) Tomato (b) Onion (c) Garlic (d) Cucumber

1411. Among earliest maturing variety of potato – (ARS, 2018
(a) Kufri Sindhuri (b) Kufri Jyoti
(c) Kufri Chipsona (d) Kufri Ashoka

1412. The variety souitable for both cold and hot areas of the country in tomato is
(a) Pusa Sheetal (b) Pusa Sadhabahar
(c) Pusa Hybrid-1 (d) Arka Alok

1413. Auxanometer is used to measure (JRF-2006
(a) TSS (b) Transpiration rate
(c) Acidity (d) Growth of the plant

1414. Type of self-incompatibility observed in cole crops is
(a) Sporophytic (b) None of theses (c) gametophytic (d) Both a and c

1415. Type of self-incompatibity observed in tomato is
(a) Sporophytic (b) None of theses (c) gametophytic (d) Both a and c

1416. Ancestor of cauliflower is: (ARS, 2016)
(a) Brassica cretica (b) Brassica rapa
(c) Brassica carinata (d) Brassica campestris

1417. Almost all the vegetable crops belong to the plant sub community
(a) Bryophyte (b) Spermatophyte (c) Thalophyte (d) None of these

1418. All the vegetable crops belongs to the Plant – Division
(a) Angiosperm (b) Both a and c (c) Gymnosperm (d) None of these

1419. Asiatic type turnip variety is: (ARS, 2016)
(a) Pusa Sweti (b) Pusa Swarnima
(c) Pusa Chandrima (d) Purple Top White Globe

1420. Which of the following crops is hardy
(a) Asparagus (b) Sweet potato (c) Potato (d) Tapioca

1421. The major garlic producing state in India- (ARS, 2018)
(a) Bihar > Karnataka > MP (b) Gujrat > Orissa > MP
(c) Orissa > Bihar > MP (d) Rajasthan > MP > Bihar

1422. Which of the following is shallow rooted crop
(a) Asparagus (b) Sweet potato (c) Artichoke (d) Onion

1423. Sauerkraut is fermented product prepared from the leaves of: (JRF, 2016)
(a) Cabbage (b) Palak (c) Lettuce (d) Amaranth

1424. Which of the following crops has the highest respiration rate
(a) Broccoli (b) Cabbage (c) Onion (d) Leek

1425. Which of the following has the lowest respiration rate
(a) Onion (b) Pea (c) Sweet corn (d) Parsley

1426. Which of the following is an hexaploid
(a) Sweet Potato (b) Water melon (c) Potato (d) Muskmelon

1427. Which of the following is an autotetraploid
(a) Potato (b) Leek (c) Okra (d) None of these

1428. Sponge gourd anthesis time is at : (ARS, 2016)
(a) Evening (b) Night (c) Afternoon (d) Morning

1429. Cross pollination by high temperature in tomato occurs due to which mechanism- (ARS, 2018)
(a) Exerted stigma condition (b) Non-exerted of stigma condition
(c) Both condition (d) None of these

1430. Seed rate of cloves(kg/ha.) as garlic
(a) 100-200 (b) 400-500 (c) 600-800 (d) 900-100

1431. Garlic contains amino acid mainly
(a) Allicin (b) Alliin (c) Allinage (d) Diallyl disulphide

1432. Vine crops referred to which group of crop

(a) Cucurbits (b) Bulb crop (c) Solanaceous (d) Cole crop

1433. Wind pollinated vegetable is: (ARS, 2016)

(a) Amaranthus (b) Tomato (c) Lettuce (d) Chilli

1434. Which of the following is also considered as segmental polyploidy? (JRF, 2016)

(a) Potato (b) Water melon (c) Tomato (d) Okra

1435. Cultivated single seeded cucurbits is

(a) Chayote (b) Kakrol (c) Kartoli (d) Pointed gourd

1436. Hybrid variety of brinjal most suitable for Bartha making is

(a) BH-2 (b) Pusa Anmol (c) BH-1 (d) None of the above

1437. Pinching of male buds before anthesis in the female parent in cucurbits was first suggested by

(a) Choudhary and Singh (b) Seshadhri and More

(c) Tyagi and Singh (d) V. Swaroop

1438. Little leaf of Brinjal is caused by (JRF, 2004)

(a) Bacteria (b) Fungus (c) Mycoplasma (d) Virus

1439. Mo deficiency causes whiptail in cauliflower in: (ARS, 2016)

(a) Saline Soil (b) acidic soil (c) Alkaline soil (d) None of these

1440. Most commonly used cryoprotectant is

(a) DMSO (b) Glycerol (c) Proline (d) Sorbitol

1441. ____________ is the major pest of Brinjal (JRF, 1997)

(a) Fruit fly (b) Thrips

(c) Fruit & shoot borer (d) Mealy bug

1442. Acridity in colocasia is due to (JRF, 1998)

(a) Calcium oxalate (b) Potassium (c) Sulphur (d) None

1443. Which spp. is used in YVMV resistance breeding – (ARS, 2018)

(a) A. esculantum – tuberculatus (b) A. esculantum – Manihot

(c) A. esculantum – tetraphyllus (d) All of these

1444. Heterosis can be estimated by

(a) $\dfrac{\text{Mean value of } F_1 - \text{Mean value of commercial cultivar}}{\text{Mean value of commercial cultivar}} \times 100$

(b) $\dfrac{\text{Mean value of } F_1 - \text{Mean value of mid-parent}}{\text{Mean value of mid-parent}} \times 100$

(c) $\dfrac{\text{Mean value of } F_1 - \text{Mean value of best parent}}{\text{Mean value of best parent}} \times 100$

(d) $\dfrac{\text{Mean value of } F_1 - \text{Mean value of } F_2}{\text{Mean value of } F_1} \times 100$

1445. A vegetable highly tolerant to soil acidity : (JRF, 2018)

(a) Okra (b) Cauliflower (c) Asparagus (d) Potato

1446. Fruit ripening hormone is (JRF, 2002)
(a) Auxin (b) GA (c) Cytokinin (d) Ethylene
1447. Stem cutting for Cassava for propagation are mainly taken from- (ARS, 2018)
(a) Upper portion (b) Middle portion (c) Lower portion (d) All parts
1448. Pusa hara baigan is a variety having: (ARS, 2016)
(a) Lycopene (b) Vitamin D
(c) Vitamin C (d) Anti-oxidant property
1449. In green vegetables, the limiting amino acids is
(a) Methionine (b) Lysine (c) Arginine (d) Tryptophan
1450. Seed rate (kg/ha) of fenugreek is
(a) 5-10 (b) 15-20 (c) 10-15 (d) 20-25
1451. Favism is seen in: (ARS, 2016)
(a) Broad bean (b) Lima bean (c) Indian bean (d) Cowpea
1452. Seed (tuber) rate per hectare of potato is
(a) 1 quintal (b) 5 quintal (c) 10 quintal (d) 15 quintal
1453. Netted varieties of muskmelon are known
(a) Honey Dews (b) Casabas (c) Cantaloupes (d) None of the above
1454. Grafting in monocots is difficult to perform because: (JRF, 2014)
(a) Absence of vascular cambium (b) Scattering of vascular bundles
(c) Absence of secondary growth (d) None of the above
1455. Application of chemical fertilizers using modern irrigation techniques is known as: (JRF, 2014)
(a) Chemigation (b) Fertigation (c) Foliar spray (d) None of these
1456. Commercially used growth retardant is (JRF, 2004)
(a) Paclobutrazol (b) BA (c) Both (d) GA3
1457. Generally, seedless water melon is: (JRF, 2014)
(a) Diploid (b) Triploid (c) Tetraploid (d) Hexaploid
1458. The chemical compound used for inducing male flower in gynoecium cucumber- (ARS, 2018)
(a) $AgNO_3$ (b) $NaNO_3$ (c) $FeSO_4$ (d) $CuSO_4$
1459. Louis de Vilmorin (1856) used the concept of progeny test for the improvement of
(a) Pea (b) Sugarbeet (c) Potato (d) Maize
1460. In garden pea,the flower colour is
(a) White (b) Purple (c) yellow (d) Pink
1461. Loss of vigour in productivity is due to: (JRF, 2014)
(a) Generic drag (b) Linkage
(c) Luxuriance (d) Inbreeding depression
1462. Capacity of refrigeration is measured by (JRF,1999)
(a) Gram (b) Ton (c) Pascal (d) Lux
1463. Repetition of treatments under investigation at the same time is known as: (JRF, 2014)
(a) Blocking (b) Pruning (c) Replication (d) Biasness

1464. While Arka Pragati is an Onion; Pusa Pragati is
(a) Cluster bean (b) Garden pea (c) Broad bean (d) Lima bean

1465. Tapioca belongs to the family (JRF-2005)
(a) Araceae (b) Euphorbiaceae (c) Rosaceae (d) Dioscoreaceae

1466. Anti-diabetic factor in bittergourd is (JRF-2006)
(a) Lipid (b) Protein (c) Pigments (d) None of these

1467. The variety of onion developed by utilizing male sterility- (ARS, 2018)
(a) Arka Kirtiman (b) Arka Niketan (c) Arka Kalyan (d) Arka Pragati

1468. ………. colour light favours stem elongation which unfavorably resemble etiolated plants (JRF-2006)
(a) Blue (b) Green (c) Ultra-violet (d) Red

1469. In tomato, which chemical is sprayed for fruit setting in adverse condition
(a) 2.4-D (b) PCPA (c) Etheral (d) Cycocel

1470. Isolation distance for certified seed production in radish is
(a) 1000m (b) 1600m (c) 800 m (d) 200 m

1471. Black rot is a most common disease of: (JRF-2009)
(a) Tomato (b) Potato (c) Celery (d) Cabbage

1472. Kasuri methi belongs to the following species
(a) Foenum-graecum (b) Corniculata
(c) oleracea (d) None of the above

1473. Phenolic factor present in onion which has antifungal properties is
(a) Sinigrin (b) Quercetin (c) Catechol (d) Allylpropyl

1474. In Nutmeg, male and female plants are planted in the ratio of (JRF-2006)
(a) 1:10 (b) 1:15 (c) 1:13 (d) 1:5

1475. Diara method of cultivation is common in (JRF-2005)
(a) Cucurbits (b) Cole crops (c) Root crops (d) Tuber crops

1476. In watermelon, the most important indicator of maturity is
(a) Days after pollination (b) Dull sound when fruit is thumped
(c) Metallic sound when fruits is thumped (d) None of the above

1477. Which one of the following vegetable crop can tolerate acid soils (JRF-2008)
(a) Cucumber (b) Bean (c) Okra (d) Watermelon

1478. The word auxin is derived from: (JRF, 2016)
(a) Latin (b) Greek (c) English (d) French

1479. King of vegetable in India is : (ARS, 2016)
(a) Tomato (b) Onion (c) Okra (d) Potato

1480. Commonly used carbon energy source in plant tissue culture media is (JRF-2006)
(a) Sucrose (b) Glucose (c) Lactose (d) Galactose

1481. Kufri Badshah variety of potato is resistant to
(a) Viral disease (b) Late blight (c) Wart (d) None of the above

1482. Leaf curl virus in nature is spread by
(a) Aphids (b) Leafhopper (c) Whitefly (d) None of the above

1483. Protein is found in abundance in (JRF-2005)
(a) Beans (b) Cauliflower (c) Cabbage (d) Bittergourd

1484. Nipple tipped variety of tomato is: (ARS, 2016)
(a) Arka saurabh (b) Arka Vikas (c) Arka Abhijit (d) Arka Abha

1485. Little leaf disease of brinjal is caused by
(a) MLOs (b) Fungi (c) Bacteria (d) None of the above

1486. Curd initiation and development in cauliflower is mainly influenced by (JRF-2009)
(a) Temperature (b) Photoperiod
(c) Relative humidity (d) None of these

1487. Little leaf disease of brinjal is spread by
(a) Aphid (b) Leaf hopper (c) Whitefly (d) None of the above

1488. The most commonly used solidifying agent in plant tissue culture media is (JRF-2006)
(a) Agar (b) Sucrose (c) Starch (d) Maltose

1489. Much hoeing is not advisable in carrot as the smell of the leaves attracts
(a) Wire worm (b) Eel worm (c) Carrot fly (d) Aphids

1490. Which of the following is black carrot variety? (JRF, 2016)
(a) Pusa Asita (b) Pusa rudhira (c) Pusa Shresti (d) Pusa Samrishti

1491. Cracking of knol-khol is due to: (ARS, 2016)
(a) Nutrition deficiency (b) Moisture deficiency
(c) High temperature (d) Heavy irrigation after prolonged drought

1492. Onion seeds remain viable for germination for following years
(a) 1-2 (b) 2-3 (c) 3-4 (d) 4-5

1493. Bt-Brinjal is commercially grown in: (ARS, 2016)
(a) Bangladesh (b) China (c) UK (d) India

1494. Pea edible podded cultivar is
(a) Arkel (b) Perfection New Line
(c) Bonneville (d) Sylvia

1495. ---------------------- is the richest source of vitamin A
(a) Carrot (b) Tomato (c) Potato (d) Sweet potato

1496. The AICVIP was started in
(a) 1968 (b) 1988 (c) 1986 (d) 1970-71

1497. Which of the following is not a dioecious vegetables: (JRF, 2018)
(a) Spine gourd (b) pointed gourd (c) Ash gourd (d) Asparagus

1498. Round head type of cabbage mature
(a) Early (b) Mid (c) Late (d) None

1499. Salt tolerant variety of onion
(a) Punjab Selection (b) Nasik Red (c) No. 404 (d) N-53

1500. The variety suited for rainfed cultivation of tomato- (ARS, 2018)
(a) Arka Ashish (b) Arka meghali (c) Arka Vishal (d) Arka Alok

1501. Optimum time for planting onion bulbs for seed production is
(a) Septermber (b) October
(c) December-January (d) None of the above

1502. In honey bee the pollen basket is present in: (JRF, 2018)
(a) Fore legs (b) Hind leg
(c) Both the leg (d) In between fore and hind leg

1503. Which type of cucurbits are planted in polyhome: (JRF, 2016)
(a) Andromonoecious (b) Monoecious
(c) Gynoecium with parthenocarpy (d) Gynoecious

1504. Polypoidy can be induced by: (JRF, 2018)
(a) Colchicine (b) Nitrous acid
(c) Methyl methane sulphonate (d) EMS

1505. Pusa Chikni is a variety of
(a) Sponge gourd (b) Bottle gourd (c) Ridge gourd (d) all season

1506. Removal of air from cane after filling, is known as: (JRF, 2015)
(a) Processing (b) Exhausting (c) Blanching (d) Brining

1507. Quality of pea is ascertained by
(a) Refractometer (b) Potentiometer (c) Tenderometer (d) None of the above

1508. Pusa Rasraj is
(a) A pure line variety of muskmelon (b) A hybrid variety of longmelon
(c) A hybrid variety of muskmelon (d) A seedless variety of watermelon

1509. Which type cauliflower is self blanching in nature? (ARS, 2016)
(a) Early type (b) Late type (c) Mid type (d) All of these

1510. Rat-tail radish is grown for its
(a) Long thin pod (b) Shoots (c) Tender roots (d) None of the above

1511. In turnip pithiness resistant variety is
(a) Pusa Kanchan (b) Pusa Swarnima (c) Pusa Swati (d) Pusa Chandrima

1512. The inheritance of Sclerotia stalk rot in cauliflower is controlled by: (ARS, 2016)
(a) Polygenic recessive gene (b) Polygenic dominant gene
(c) Both (d) None of these

1513. Active principal present in black pepper: (JRF, 2015)
(a) 1,8 cineol (b) Piperine (c) Anethol (d) Phenol

1514. Single fruit of palak contains seeds
(a) 2 – 3 (b) 6 – 7 (c) 4 – 5 (d) 8 – 9

1515. Multigerm seed is found in case of
(a) Onion (b) Carrot (c) Tomato (d) Beetroot

1516. Technique of combining 2 or more major genes is known as : (JRF, 2016)
(a) Genetic engineering (b) Gene pyramiding
(c) Genetic map (d) Combining ability

1517. The most suitable temperature for colour development in carrot is
(a) 10-15°C (b) 15-20°C (c) 20-25°C (d) None of the above

1518. The crop with which turnip crosses easily is
(a) Radish (b) Cabbage (c) Mustard (d) None of these

1519. The immediate precursor for ethylene synthesis is (JRF-2008)
(a) SAM (b) ACC (c) Polyamine (d) Tryptophan

1520. Clinching is related to: (JRF, 2018)
(a) Freezing (b) Drying (c) Canning (d) Fermentation

1521. The most ancient type of garden is
(a) Kitchen garden (b) Vegetable forcing (c) Market garden (d) None of the above

1522. The most common cucurbit grown under glasshouses
(a) Bottle gourd (b) Cucumber (c) Watermelon (d) None of the aboe

1523. Which of the following is ethylene absorbent? (JRF, 2014)
(a) $KMNO_4$ (b) K_2HPO_4 (c) KI (d) None of these

1524. The most commonly grown garden beet variety
(a) Detroit Dark Red (b) Pusa Chandrima
(c) Pusa Kanchan (d) None of the above

1525. The most serious disease of okra
(a) Yellow vein mosaic (b) Powdery mildew
(c) Root rot (d) None of the above

1526. Cassava is used for the manufacture of (JRF-2005)
(a) Pectin (b) Sugar (c) Sago (d) Latin

1527. Stomatal activity is controlled by which of the following: (JRF, 2014)
(a) Cl (b) K^+ (c) Na (d) Mg

1528. Onion variety suitable for kharif & Rabi season: (ARS, 2016)
(a) Arka Kalyan (b) Arka Pragati (c) Arka Niketan (d) N-53

1529. The pigment responsible for red colour in carrot is
(a) Carotene (b) Lycopene (c) Anthocyanin (d) Quercetin

1530. The first example for the use of bioagent for controlling postharvest diseases is: (JRF, 2018)
(a) Candida oleophila (b) Pseudomonas syringae
(c) Bacillus subtilis (d) Trichoderma harzianum

1531. The primary objective of greenhouse cultivation of vegetables is
(a) To get early crop (b) To protect the crop from disease
(c) To get off season crop (d) To protect the crop from excessive rains

1532. Tomato is rich source of: (JRF-2009)
(a) Vitamin a (b) iron (c) Phosphorus (d) Vitamin C

1533. Sweet potato belongs to the family (JRF-2005)
(a) Convolvulaceae (b) Solanaceae (c) Euphorbiaceae (d) Araceae

1534. Trigonella corniculata (Kasuri methi) produces flowers of following colour
(a) White (b) Purple (c) Red (d) Orange to yellow

1535. Triploid watermelon variety is
(a) New Hampshire Midget (b) Charleston Grey
(c) Pusa Bedana (d) None of the above

1536. Pro trays are related to (JRF-2006)
(a) Post harvest (b) Nursery (c) Transport (d) Processing

1537. Flavours include: (JRF, 2014)
(a) Taste (b) Aroma (c) Both (a) and (b) (d) Texture

1538. Which is the progenitor of watermelon: (JRF, 2018)
(a) C. edulis (b) C. vulgaris (c) C. colocynthis (d) C. citriodes

1539. Type of vegetable gardening followed on Dal lake of the Kashmir Valley is
(a) Floating garden (b) Kitchen garden (c) Hydroponics (d) None of the above

1540. Protandrous condition favours cross pollination in (JRF-2005)
(a) Cabbage (b) Onion (c) Spinach (d) Bottlegourd

1541. Variety of bottle gourd without 'crook neck'
(a) Pusa Summer Prolific Long (b) Pusa Summer Prolific Round
(c) Arka Bahar (d) Pusa Meghdoot

1542. Which one of the following vegetables produce maximum seeds per fruit?
(a) Tomato (b) Chilli (c) Brinjal (d) Potato

1543. For keeping the sugar content low, potato tubers should be stored at a temperature of
(a) 5°C (b) 10°C (c) 15°C (d) None of the above

1544. Fruits and shoot borer is a serious pest of: (JRF-2009)
(a) Cabbage (b) Brinjal (c) Bitter gourd (d) Tomato

1545. Which of the following potato varieties is resistant to late blight (JRF-2005)
(a) Kufri Badshah (b) Kufri Sinduri
(c) Kufri Bahar (d) Kufri Chandramukhi

1546. Akashin a physiological disorder in radish is due to deficiency of
(a) B (b) Ca (c) Mn (d) Cu

1547. In the plains of northern India, most of the varieties of pea are suitable for sowing from
(a) Mid-August to September (b) After mid-November
(c) October-November (d) None of the above

1548. A plant breeder wants to develop a disease resistant variety. What he should do first: (JRF, 2018)
(a) Hybridization (b) Mutation (c) Selection (d) Mass selection

1549. Kharif onion variety suited for commercially cultivation- (ARS, 2018)
(a) Arka Kalyan (b) Agrifound dark red
(c) N 53 (d) All of these

1550. Chemically pectin is a …………….. (JRF, 2018)
(a) Acidic (b) Saline (c) Alkaline (d) Neutral

1551. The safe seed storage moisture content should be between: (JRF-2009)
(a) 8-9% (b) 10-12% (c) 13-14% (d) 15-16%

1552. Although a deficiency of any one of the elements listed may result in chlorosis, only one of these elements is found in chlorophyll, which is it?
(a) Chlorine (b) Iron (c) Magnesium (d) Zinc

1553. Drought tolerant onion cultivar is
(a) Hissar-2 (b) Pusa Ratnar (c) Pusa Red (d) Baswant 780

1554. In case of chilli the perennial variety is
(a) Pusa Jawala (b) Arka Lohit (c) Pusa Sadabahar (d) Arka Abir

1555. Precursor of ethylene is: (JRF, 2015)
(a) Cyctein (b) Methionine (c) Leucine (d) Lysine

1556. In case of brinjal the phomopsis blight resistant variety is
(a) Pusa Ankur (b) Pusa Bhairav (c) Arka Sheel (d) Black Beauty

1557. Central Agricultural Research Institute is located at (JRF-2008)
(a) Kolkatta (b) Port Blair
(c) Visha Khapatnam (d) Delhi

1558. Optimum temperature and humidity for storage of sweet potato tubers are
(a) 5°C and 95-100% RH (b) 10°C and 90-95% RH
(c) 15°C and 85-90% RH (d) None of the above

1559. Precursor of auxin is: (JRF, 2015)
(a) Mn (b) Mg (c) Tryptophan (d) B

1560. In tissue culture............... is used to induce both callus growth and root initiation (JRF-2006)
(a) Auxin (b) GA (c) ABA (d) Cytokinin

1561. Pear shaped fruited variety of tomato is
(a) Punjab Kesri (b) Punjab Chhuhara
(c) PNR-7 (d) Punjab Tropic

1562. Seed potato takes about..........days to germination after sowing. (JRF-2007)
(a) 5-10 (b) 10-15 (c) 15-20 (d) 30

1563. Kihara's technique is related with which of the following crop (JRF-2006)
(a) Onion (b) Watermelon (c) Tomato (d) Muskmelon

1564. Ramsey and Golsey are important variety of
(a) Cinnomon (b) Small cardamom (c) Ginger (d) Large Cardamom

1565. Which of the following is a man made vegetable ? (JRF-2008,2009)
(a) Clusterbean (b) Artichoke (c) Rutabaga (d) Winged bean

1566. Plant regulator which improves fruit set in tomato is
(a) Gibberellic acid (b) Maleic hydrazide
(c) 2,4-Dichlorophenoxy acetic acid (d) Ethephon

1567. The carrot seed required for sowing one ha is
(a) 1-2 kg (b) 4-5 kg (c) 10-15 kg (d) 15-20 kg

1568. First intergeneric hybrid between radish and cabbage was made by (JRF-2008)
(a) Hull (1945) (b) Rimpu (1890)
(c) Andrew Knight (1800) (d) Karpenchenko (1927)

1569. Onion for breeder seed production is followed by: (ARS, 2016)
(a) Seed to seed method (b) Bulb to seed method
(c) Both (d) None of these

1570. 'Chirkey' and 'Foorkey' are major viral diseases of
(a) Small cardamom (b) Large Cardamom
(c) Nutmeg (d) Black Pepper

1571. The usual weight of seed piece of potato is
(a) 30-40 kg (b) 40-50 kg (c) 50-60 kg (d) None of the above

1572. River bed cultivation is suited to which group of vegetable - (ARS, 2018)
(a) Water and muskmelon (b) Potato and Tomato
(c) Onion and Radish (d) Cauliflower and cabbage

1573. ....cultivation is mostly concentrated in peri-urban areas (JRF-2007)
(a) Fruit (b) Vegetable (c) Medicinal crop (d) Plantation crop

1574. Sporophytic self incompatibility system is present in (JRF-2006)
(a) Cabbage (b) Tomato (c) Chilli (d) Okra

1575. Long duration variety of turmeric..... (JRF, 2019)
(a) Suguna (b) Rajendra Sonia (c) Suvarna (d) BSR.1

1576. Free living bacteria involved in nitrogen fixation: (JRF, 2018)
(a) Rhizobium (b) Azotobacter (c) Anabaena (d) Azospirillum

1577. The cultivation of crop from the re-growth of stubbles of crop after harvest is called (JRF, 2018)
(a) Arrowing (b) Reaping (c) Ratooning (d) Seasoning

1578. Blossom end rot is caused due to deficiency of (JRF-2007)
(a) Fe (b) Cu (c) Zn (d) Ca

1579. The somatic chromosome number of cumin are
(a) $2n = 11$ (b) $2n = 12$ (c) $2n = 14$ (d) None of the above

1580. Fenugreek is a diploid with chromosome number
(a) $2n = 14$ (b) $2n = 16$ (c) $2n = 12$ (d) $2n = 18$

1581. Which of the following state is the largest producer of Cardamom
(a) Kerala (b) Andhra Pradesh (c) Tamilnadu (d) Karnataka

1582. Isolation distance of pea for seed bed - (ARS, 2018)
(a) 20 m (b) 40 m (c) 50 m (d) 60 m

1583. Which one among the following is physiological disorder of tomato? (JRF, 2015)
(a) Puffiness (b) Hollow stem (c) Black heart (d) Splitting

1584. Major constraint in vegetative propagation of cardamom is
(a) Katte disease (b) Capsule Borer (c) Thrips (d) None

1585. Indian Farming is a publication of (JRF-2007)
(a) IARI, New Delhi (b) CAZRI, Jodhpur
(c) ICRISAT, Hyderabad (d) ICAR, New Delhi

1586. Konkan Tej is the variety of
(a) Cardamom (b) Clove (c) Cinnamon (d) Nutmeg

1587. Oil of black pepper consist mainly of
(a) Phenol (b) Alkaloid (c) Terpenes (d) Tetracycline

1588. Allepy and Duggirala are the trade name of
(a) Ginger (b) Garlic (c) Tamarind (d) Turmeric

1589. The removal of field heat from farm produce is called- (ARS, 2018)
(a) Washing (b) Freezing (c) Chilling (d) Pre-cooling

1590. Polyphenol oxydase enzyme inactivate at the temperature of (JRF-2007)
(a) 80-90°C (b) 100°C (c) 50°C (d) above 100°C

1591. Viral disease of potato are spread by (JRF-2005)
(a) Jassids (b) Aphids (c) White flies (d) Mites

1592. Indicator plant for B and Mo deficiency: (JRF, 2015)
(a) Cucurbits (b) Tomato (c) Okra (d) Cauliflower

1593. Trachyspermum ammi is the botanical name of
(a) Hing (b) Aijowan (c) Zeera (d) Parsley

1594. Cinnamon belongs to the family
(a) Lauraceae (b) Fabaceae (c) Iridaceae (d) Araceae

1595. In which type of soil, nematodes grow well: (JRF, 2015)
(a) Heavy soil (b) Soil rich in organic matter
(c) Sandy loam soils (d) No reaction with soil

1596. Botanical name of spine gourd- (ARS, 2018)
(a) M. angolensis (b) M. angustisepala (c) M. charantia (d) M. Dioca

1597. Important constituent in the Bishop's weed is
(a) Cineol (b) Limone (c) Thymol (d) Phenol

1598. Nutmeg plants are (JRF-2008)
(a) Monoecious (b) Andromonoecious
(c) Androdioecious (d) Dioecious

1599. Linalool, is important ingredient of
(a) Fennel (b) Coriander (c) Cumin (d) Aniseed

1600. Growth regulator which improves fruit set in tomato (JRF-2005)
(a) 2,4-D (b) GA (c) MH (d) Ethephon

1601. Origin of ginger is
(a) Mediterranean region (b) Coriander
(c) South East Asia (d) North West Asia

1602. Carrot is a good source of (JRF-2005)
(a) Vitamin A (b) Riboflavin (c) Vit-C (d) Protein

1603. Terbein is the important constituent of
(a) Ginger (b) Cinnamon (c) Nutmeg (d) Curry leaf

1604. Origin of Small cardamom is
(a) Sri Lanka (b) Indonesia (c) India (d) China

1605. Chromosome no. of cultivated okra spp. is- (ARS, 2018)
(a) 110 (b) 130 (c) 120 (d) 140

1606. Bitterness in brinjal is caused if glycol-alkaloid exceed- (ARS, 2018)
(a) > 10mg/100gm flesh (b) > 20mg/100gm flesh
(c) > 30mg/100gm flesh (d) > 40mg/100gm flesh

1607. Bush pepper is propagated from
(a) Runners (b) Laterals (c) Top shoots (d) Hanging shoots

1608. The biting taste of black pepper is because of
(a) Terpenine (b) Piperine (c) Pipermenthene (d) Pentanol

1609. Ber is a physiological disorder of: (JRF, 2015)
(a) Brinjal (b) Okra (c) Tomato (d) Chilli

1610. Feathering and scrapped chipps are important grades of
(a) Black pepper (b) Clove (c) Ginger (d) Cinnamon

1611. Pea type of germination: (ARS, 2016)
(a) Hypogeal (b) Epigeal (c) Both (d) None of these

1612. The leading state in production of Fenugreek is
(a) Rajasthan (b) Gujarat
(c) Madhya Pradesh (d) Uttar Pradesh

1613. The leading ginger producing country in the world is
(a) China (b) Srilanka (c) Italy (d) India

1614. Essential oil in Dill is
(a) 0.1-2.6% (b) 0.1-1.0% (c) 1.5-4% (d) None

1615. Highest production of spice is in which state
(a) Himachal (b) Andhra Pradesh
(c) Madhya Pradesh (d) Uttar Pradesh

1616. Pusa Chetki variety of radish is most suitable for sowing during
(a) March-August (b) October-November
(c) After November (d) None of the above

1617. Root system of sweet potato is basically
(a) Shallow (b) Deep (c) Medium (d) None of the above

1618. Nucleus was first discovered by (JRF-2008)
(a) Benda (1877) (b) Camillo Golgi
(c) Robert Brown (1882) (d) Flemming (1882)

1619. Seed production of which crop is not done on plains
(a) Early cauliflower (b) Cabbage
(c) Onion (d) Carrot

1620. Shape of head of cabbage variety Pride of India is
(a) Flat (b) Pointed (c) Round (d) None of the above

1621. In case of potato the Economic Threshold Level (ETL) of aphid population is
(a) 20/100 leaves (b) 30/100 leaves (c) 40/100 leaves (d) 50/100 leaves

1622. In Aspargus edible part used is - (ARS, 2018)
(a) Tuber (b) Leaf (c) Stem (d) Spear

523. Squash melon is believed to have originated in
(a) China (b) India (c) Malaysia (d) None of the above

524. Systemic fungicide used for the control of downy mildew of muskmelow is
(a) Indofil M-45 (b) Ridomil MZ (c) Blitox (d) None of the above

525. In cucumber gynoecious line is maintained by: (ARS, 2016)
(a) NAA (b) GA_3 (c) 2-4D (d) Cytokinin

526. Seed Plot Technique is used to produce virus free seed tuber in: (ARS, 2016)
(a) North Indian plains (b) South India plains
(c) Western India plains (d) Eastern India plains

527. The best time for transplanting cabbage seedling in the field is
(a) Morning (b) Afternoon (c) Evening (d) Night

528. Trichosanthes dioca is native of (JRF-2008)
(a) China (b) India (c) Iran (d) Peru

529. The stalk of the ovule is called: (JRF-2007)
(a) Pedicel (b) Petiole (c) Panicale (d) Micropyle

530. Pollination and fertilization before anthesis is called (JRF-2007)
(a) Cleistogamy (b) Chasmogamy (c) Homogamy (d) Dichogamy

531. Large cardamom is a
(a) Sun loving plant (b) Shade loving plant
(c) Both a and b (d) None

532. India accounts for _ _ _ _ _ % of total area under black pepper
(a) 56 (b) 55 (c) 54 (d) 50

533. Indian accounts for _ _ _ _ _ _ % of turmeric production in the word
(a) 80 (b) 70 (c) 76 (d) 90

534. Cinnamomum tamala is the botanical name of
(a) Cinnamon (b) Tejpat (c) All spice (d) None

535. Heterostyly is found in............ (JRF, 2018)
(a) Tomato (b) Brinjal (c) Longmelon (d) Watermelon

536. Highest respiration rate in sprouting broccoli is: (ARS, 2016)
(a) At the Flowering stage (b) At the Immature stage
(c) At the maturity stage(d) None of these

537. Major all spice producing country in the world
(a) India (b) Sri Lanka (c) Jamaica (d) America

538. Flowering stimulus is mostly perceived by (JRF-2007)
(a) Shoot apex (b) Leaves (c) Buds (d) Stems

539. Edible part of Asparagus is: (JRF, 2016)
(a) Root (b) Tender shoots (c) Fruits (d) Flowers

540. Plant part of saffron used as spice
(a) Seed (b) Bark (c) Flower stigma (d) Rhizome

541. Seed production is not possible under plains: (ARS, 2016)
(a) Palak (b) Tomato (c) Brinjal (d) Beetroot

1642. Commercial part of all spice is
(a) Whole tree (b) Leaves (c) Bark only (d) Fruits only

1643. Origin of clove is
(a) Indonesia (b) Africa (c) India (d) None

1644. Black heart in potato occurs at temperature of - (ARS, 2018
(a) 0°C (b) 10°C (c) 15°C (d) 20°C

1645. Navashree and Nithyashree are varieties of
(a) Cardamom (b) Cinnamom (c) Black Pepper (d) None

1646. Purelines are: (ARS, 201(
(a) Heterozygous and Heterogeneous (b) Homozygous and Homogeneous
(c) Both (d) None of these

1647. Indian Spinach/Poi/Basella belongs to the family
(a) Chenopodiaceae (b) Leguminosae (c) Basellaceae (d) None of the abov(

1648. Poinsette is the most popular variety of
(a) Muskmelon (b) Long melon (c) Cucumber (d) None of the abov(

1649. Alleles of F1 hybrids are in nature: (ARS, 201(
(a) Homozygous (b) Heterozygous (c) Both (d) None of these

1650. Core intact is one of the seed production methods used in which of the following crops (JRF, 201(
(a) Broccoli (b) Cabbage (c) Knol khol (d) Brussels' sprout

1651. Cells without cell well are known as (JRF-200
(a) Chloroplast (b) Protoplast (c) Proplastics (d) Chromoplast

1652. Black speck is a disorder of: (ARS, 201(
(a) Beetroot (b) Palak (c) Sweet potato (d) Cassava

1653. Pollination in beetroot is by
(a) Insect (b) Water (c) Wind (d) All of the above

1654. Phomopsis blight is serious problem in seed crop of – (ARS, 2018)
(a) Brinjal (b) Chilli (c) Tomato (d) Potato

1655. In muskmelon, the TSS content varies from ---------- per cent
(a) 8-10 (b) 5-6 (c) 8-17 (d) 3-4

1656. ---------------variety of muskmelon is cross Hara Madhu × Edisto
(a) Sugar baby (b) Arka Manik
(c) Punjab Sunhari (d) Hara Madhu

1657. Monohybrid cross is made for…. (JRF, 201
(a) F_2 Generation seed (b) F_1 Generation seed
(c) Both (d) None of these

1658. ---------------------- is the chief pollinating agent in cucumber
(a) Housefly (b) Bumble bee (c) Honey bee (d) None of the abov

1659. Straight eight is a variety of
(a) Pumpkin (b) cucumber (c) Tinda (d) Bottle gourd

660. Bottle gourd is a --------------- crop
(a) Self Pollinated (b) Cross Pollinated
(c) Often Cross pollinated (d) None of the above
661. The origin place of bitter gourd is
(a) Asia (b) Indo-Myanmar (c) China (d) Europe
662. Muskmelon seed does not germinate at temperature lower than
(a) 180 C (b) 240 C (c) 300 C (d) 250 C
663. French bean is sown during ________ month in south Indian plains.
(a) July (b) August (c) September (d) November
664. Insect vector or leaf curl of tomato is: (JRF, 2015)
(a) Aphid (b) Leaf hooper (c) White fly (d) Thrips
665. Pea is the richest source of
(a) Carbohydrate (b) Protein (c) Vitamin (d) Fat
666. Bloosm end rot of grape is due to deficiency of: (JRF, 2015)
(a) Ca (b) B (c) Mg (d) Mn
667. Fenugreek seed contain
(a) Allicin (b) Diosgenin (c) Ajmacline (d) All
668. For the transfer of which trait backcross breeding is followed? (ARS, 2016)
(a) Oligogenic (b) Polygenic (c) Both (d) None of these
669. In metaphase stage of cell division chromosomes are arranged on(JRF-2007)
(a) Equitorial plate (b) Polar plate (c) Centromere (d) None of these
670. .... is used as a source of YVM resistance in Okra breeding (JRF-2006)
(a) Abelmoschus angulosus (b) A. crinitus
(c) A. manihot sp. tetraphyllus (d) A. moschatus
671. Vanilla belongs to family
(a) Orchidaceae (b) Guttiferae (c) Fabaceae (d) Iridaceae
672. Ploidy level of Turmeric......... (JRF, 2018)
(a) Hexaploid (b) Tetraploid (c) Triploid (d) Diploid
673. Curry leaf belongs to family
(a) Rosaceae (b) Solanaceae (c) Rutaceae (d) Myrtaceae
674. Nutmeg belongs to family
(a) Myrtacea (b) Lauraceae (c) Myrsticaceae (d) Piperaceae
675. Multiplier onion is cultivated throughout the year in. (ARS, 2016)
(a) MP (b) Maharashtra (c) Tamil Nadu (d) Gujarat
676. Which is an antioxidant (JRF-2009)
(a) Vitamin-C (b) Vitamin-A (c) Sugar (d) Ethylene
677. White fly and mite resistant variety of Coriander..... (JRF, 2019)
(a) Gujarat Corinader-1 (b) Sadhana
(c) Rajendra Swati (d) CO-1
678. Origin of fenugreek is
(a) Europe (b) America (c) Asia (d) Australia

1679. ______________ is a good source of iodine in our diet: (JRF, 201

(a) Okra (b) Watermelon (c) Bitte gourd (d) Garlic

1680. Male musk deer smell resembles to

(a) Ambrette seed (b) Basil (c) Celery (d) Davana

1681. Phytochrome P730 stimulates flowering of (JRF-200

(a) Short day plant (b) Long day plants

(c) Day neutral plants (d) Dark

1682. Manno-galacton is obtained from vegetable crop- (ARS, 201

(a) Indian bean (b) Cluster bean (c) Winged bean (d) Lima bean

1683. Curcuma longa is botanical name of (JRF-200

(a) Zinger (b) Turmeric (c) Cardamom (d) Black pepper

1684. Pusa Early Bunching is a variety of

(a) Amaranth (b) Bottle gourd

(c) Fenugreek(methi) (d) Brinjal

1685. Mendelian population refers to .... sharing individuals (JRF-200

(a) Different gene pool (b) Mixed gene pool

(c) Same gene pool (d) Both A and B

1686. ....is used as donor parent for salt tolerance in Tomato breeding (JRF-200

(a) Lycpersicon (b) L. hirsutum

(c) L. cheesmanii (d) L. pimpinellifolium pennellii

1687. Monohybrids are derived by hybridizing two individuals which differ for (JRF-200

(a) Single-character (b) Double character

(c) Quadriple character (d) Multiple character

1688. Angur Lata is a variety of

(a) Amaranth (b) Bottle gourd (c) Tomato (d) Brinjal

1689. Massing and Selfing technique of breeding was first time suggested in (JRF-200

(a) Papaya (b) Turnip (c) Allium spp. (d) Mango

1690. Tomato was introduced in India by Royal-Agri Horticultural society in the year : (ARS, 201

(a) 1852 (b) 1924 (c) 1872 (d) 1828

1691. Which of the following is not the hybrid variety of tomato (JRF, 199

(a) Karnataka (b) Naveen (c) Rashmi (d) Pusa Gaurav

1692. Hill soils are deficient in (JRF-200

(a) Phosphorus (b) Potassium (c) Nitrogen (d) Calcium

1693. Pusa Chikni is a variety of

(a) Amaranth (b) Bottle gourd (c) Smooth gourd (d) Brinjal

1694. The common name of Ammomum subulatum is (JRF-2006)

(a) Large cardamom (b) Nepal cardamom

(c) Great Indian cardamom (d) All the above

1695. Potato is rich source of (JRF, 1999)
(a) Fat (b) Starch (c) Vitamin (d) Mineral
1696. Ipomea batatas requires photoperiod for flowering is (JRF-2009)
(a) 7 hrs. (b) 9.5 hrs. (c) 13 hrs. (d) 11 hrs.
1697. "Cicrocrocin" is present in
(a) Saffron (b) Sweet flag (c) Clove (d) Dill
1698. Cabbage seed production is carried out region- (ARS, 2018)
(a) Hills only (b) Plains only (c) Both areas (d) None of these
1699. Improved Japanese is a variety of
(a) Cauliflower (b) Broccoli (c) Cabbage (d) Brussels sprout
1700. Botanically pumpkin is (JRF-2007)
(a) Berry (b) Pepo (c) Catkin (d) Drupe
1701. Self incompatibility leads to cross pollination: (JRF, 2018)
(a) Cucumber (b) Carrot (c) Cauliflower (d) Okra
1702. The plant part used in case of Kokum
(a) Fruit (b) Leaves (c) Rhizome (d) Bark
1703. Brassica oleracia var-acephala is the botanical name of (JRF-2009)
(a) Cauliflower (b) Cabbage (c) Kale (d) Broccoli
1704. Kalmegh is a variety of: (ARS, 2016)
(a) Beet root (b) Sweet potato (c) Carrot (d) Amaranthus
1705. Saurkraut is a …. (JRF, 2019)
(a) Fermented cauliflower (b) Fermented cabbage
(c) Fermented Broccoli (d) None of these
1706. Organic group that are permanently bound to the enzyme is: (JRF, 2018)
(a) Prosthetic group (b) Activator
(c) Coenzyme (d) Co activator
1707. Inflorescence of small cardamom is
(a) Panicle (b) Spadix (c) Spike (d) Raceme
1708. Small cardamom is propagated by
(a) Suckers (b) Seed (c) Rhizomes (d) Root cuttings
1709. Sauerkraut is a product prepared by: (JRF, 2018)
(a) Freezing (b) Dehydration (c) Fermentation (d) Canning
1710. Which of the following varieties is an introduction of ginger
(a) Rio de Janerio (b) Suprabha (c) Surabhi (d) Suruchi
1711. Which spice is called the "queen of spices"
(a) Black pepper (b) Small cardamom
(c) Large cardamom (d) Clove
1712. Palak variety which is resistant to cercospora leaf spot… (JRF, 2019)
(a) Arka Anupama (b) Pant Composite (c) All green (d) Pusa Jyoti

1713. Which spice is referred as "King of spices"
(a) Black pepper (b) Small cardamom
(c) Large cardamom (d) Clove

1714. NSC was established in the year: (ARS, 2016)
(a) 1963 (b) 1982 (c) 1978 (d) 1968

1715. Self sterile variety of black pepper
(a) Balankotta (b) Both a and c (c) Kalluvalli (d) None of these

1716. The method of Lye peeling is through: (JRF, 2018)
(a) Sulphuric Acid (b) Nitric Acid (c) Caustic soda (d) Peelers

1717. Seedless watermelon is (JRF-2007)
(a) Triploid (b) Hexaploid (c) Tetraploid (d) Diploid

1718. Gold fleck is due to deposition of
(a) Calcium oxalate (b) MgSO4 (c) Boron (d) Zinc

1719. Delayed ripening of fruits is overcome by application of
(a) Borax (b) ZnSO4 (c) NAA (d) Ethrel

1720. Brinjal variety MDV 1 was evolved by -------------
(a) X-rays (b) Gamma rays (c) EMS (d) MMS

1721. In Carrot storage the bitterness is due to: (ARS, 2016)
(a) Acetic Acid (b) Ethylene (c) Ascorbic Acid (d) Oxalic Acid

1722. "Konkan Amrit" is an improved variety of
(a) Kokum (b) Vanilla (c) Juniper (d) Saffron

1723. 90% of the pungency in chili is present in which part of fruit
(a) Seed (b) Placenta (c) Pericarp (d) None of these

1724. Muskmelon variety suited long distant transport, good self life good flesh quality- (ARS, 2018)
(a) Arka Rajhans (b) Arka Jeet (c) Kashi madhu (d) Pusa Madhuars

1725. Vanilla is propagated through
(a) Stem cutting (b) Corm (c) Root cuttings (d) Tubers

1726. Which country is the leading producer of Vanilla
(a) Madagascar Republic (b) Sri Lanka
(c) India (d) Indonesia

1727. ----------------- are also known as marrow or musky gourd.
(a) Bitter gourd (b) Sponge gourd (c) Squashes (d) Snake gourd

1728. The other relatives of pumpkin is
(a) Cucurbita pepo (b) Momordica charanta
(c) Citurllus vulgaris (d) Benincasa hispida

1729. Which is a flowering hormone? (ARS, 2016)
(a) Cytokinins (b) Gibberellins (c) Auxins (d) Florigen

1730. About ______ tons of dry chillies are obtained from one hectare irrigated area.
(a) 1.0 (b) 1.5-2.5 (c) 4-5 (d) 7-8

1731. Short days, low temperature and humid climate are good for production of --------- flowers in ash gourd.
(a) Male (b) Female
(c) hermaphrodite (d) Both male and female

1732. Bhendi variety tolerant to salinity is --------------
(a) Kalyanapur green (b) Type -3
(c) Pusa Sawani (d) Pusa Dwarf

1733. Which is the proper stage of tapping asfoetida
(a) Foliage just appear (b) Foliage turns green to yellow
(c) Foliage lush green (d) Foliage senescence

1734. Ivy gourd is propagated by – (ARS, 2018)
(a) seed (b) cutting (c) layering (d) offshoot

1735. Fruit drop in chilli is prevented by application of
(a) NAA (b) Ethrel (c) CCC (d) GA

1736. The seed rat per hectare of okra for rainy season is -----------kg
(a) 8-10 (b) 10-12 (c) 12-14 (d) 14-16

1737. Which of the following is a dioecious spice
(a) Nutmeg (b) Juniper (c) Kokum (d) All of these

1738. Aroma of green leaf of tomato is due to
(a) Sirigrini (b) Glycoisocinate (c) Capsaicin (d) None of the above

1739. The most suitable time of transplanting sweet pepper is
(a) March – April (b) Mid February (c) April – May (d) None of the above

1740. Arka Abhay is a variety of
(a) Brinjal (b) Chilli (c) Tomato (d) Bhendi

1741. The chromosome number (2n) in bitter gourd is
(a) 24 (b) 22 (c) 14 (d) 32

1742. --------------------- is also known as anti sterility vitamin
(a) Vitamin C (b) Vitamin E (c) Vitamin K (d) Vitamin B

1743. Tomato fruit aroma is due to
(a) Sulfonium (b) Chlorophyll (c) Carotenoid (d) Polyphenols

1744. Bharat is the first hybrid of capsicum released by
(a) MAHYCO (b) IAHS (c) Sugrow (d) Namdhari

1745. MDV-1 is a chilli variety developed through
(a) Pure line selection (b) Mutation breeding
(c) Mass selection (d) Pedigree method

1746. Tomato produces ----flowers in cluster on the stem
(a) Proandry (b) Protogynous (c) Cleistogamous (d) Hercogammy

1747. The vines of Pointed gourd come in dormancy in
(a) spring season (b) Rainy season (c) Winter season (d) Summer season

1748. A hermaphrodite variety of ridge gourd is known as Satputia is widely cultivated in ----------state.
(a) Gujarat (b) Bihar
(c) Madhya Pradesh (d) Delhi

1749. A volatise oil a crystalline glycoside "koenigin" from the leaves and a glucoside" from the flowers are a few industrial products from the trees of
(a) Curry leaf (b) Drumstick (c) Chekkurmans (d) Winged bean

1750. Degree days is related to which crop: (ARS, 2016)
(a) Onion (b) Tomato (c) Chilli (d) Cauliflower

1751. The name of the crop derives from the shape of pod resembling the slender and curved stick used for beating the drum.
(a) Drumstick (b) Curry leaf (c) Wined bean (d) None of above

1752. Scientific name of summer squash is
(a) *Cucurbita moschata* (b) *Cucurbita maxima*
(c) *Cucurbita pepo* (d) *Cucurbita mixta*

1753. Seeds of Moringa contains ---------------- % of non drying oil.
(a) 10-15 (b) 15-20 (c) 20-30 (d) 35-40

1754. French bean is said to be a __________ resistant
(a) Salinity (b) Drought (c) Frost (d) Cold

1755. Pusa Nasdar is a variety of
(a) Ridge gourd (b) Sponge gourd (c) Bitter gourd (d) Bottle gourd

1756. Germination of water melon seed is
(a) Hypogeal (b) Epigeal (c) Semi hypogeal (d) both a & b

1757. Perennial drumstick types are propagated by.
(a) Seeds (b) Tubers (c) Root cuttings (d) Limb cuttings

1758. ------------- is also known as four angled gourd or angled loofah.
(a) Ridge gourd (b) Sponge gourd (c) Bitter gourd (d) Pointed gourd

1759. __________is also known as kidney bean or snap bean.
(a) Pea (b) Indian bean (c) French bean (d) Broad bean

1760. Non heading type Chinese cabbage is: (ARS, 2016)
(a) Pak-Choi (b) Yoko (c) Okido (d) Kasumi

1761. Tender crop and cascade are cultivars of ________suited for processing purpose.
(a) Indian bean (b) French bean (c) Cowpea (d) Winged bean

1762. Krishna is the variety of (JRF, 1997)
(a) Tomato (b) Brinjal (c) Chilli (d) Potato

1763. Pink colour of onion is due to (JRF, 1998)
(a) Qucertin (b) Anthocynin (c) Carotene (d) None

1764. In watermelon, ----------------------performs pollination
(a) wind (b) wasp (c) Butterfly (d) Honey bee

1765. Phomopsis blight is serious disease of (JRF, 2004)
(a) Tomato (b) Brinjal (c) Chilli (d) Onion

1766. Indian bean is _____________ pollinated crop
(a) Cross (b) self (c) Often cross (d) None of these

1767. Dhanaraj is variety of.
(a) Curry leaf (b) Moringa (c) Okra (d) French bean

1768. Cluster bean contains a mucilaginous substance known as
(a) Galactomanon (b) Manon (c) Safforin (d) None of the above

1769. ----------- is also known as guar
(a) Rice bean (b) Cluster bean (c) French bean (d) Broad bean

1770. The guar meal (Dry seeds) contains about ---------- per cent protein
(a) 20 (b) 10 (c) 15 (d) 33.3

1770. ----------------- is the most serious disease of cluster bean
(a) Fusarium wilt (b) Bacterial wilt (c) Leaf spot (d) Rust

1770. Cluster bean plant required ------------------conditions for induction of flowering
(a) Long day (b) Short day (c) Day neutral (d) Both a and b

1770. Cluster bean originated from
(a) America (b) Europe (c) Asia (d) Africa

1770. --------------- is botanically known as Cyamopsis tetragonolobus L.
(a) French bean (b) cluster bean (c) Pea (d) Broad bean

1770. In ---------------, cluster bean is grown throughout the year.
(a) North India (b) Western India (c) South India (d) Central India

1770. Pusa Mausami, Pusa Sadabahar, Pusa Navbahar and Sharad Bahar are varieties of
(a) French bean (b) Cow pea (c) Cluster bean (d) Winged bean.

1770. Snake gourd is ----------------- in nature
(a) Monoecious (b) Dioecious (c) Gynoecious (d) Androecious

ANSWERS

1. (a)	2. (b)	3. (c)	4. (c)	5. (d)	6. (b)	7. (b)	8. (b)	9. (a)	10. (d)
11. (c)	12. (d)	13. (c)	14. (d)	15. (b)	16. (b)	17. (d)	18. (a)	19. (d)	20. (a)
21. (c)	22. (a)	23. (b)	24. (a)	25. (b)	26. (a)	27. (b)	28. (d)	29. (c)	30. (b)
31. (a)	32. (c)	33. (c)	34. (c)	35. (a)	36. (a)	37. (b)	38. (a)	39. (b)	40. (a)
41. (c)	2. (a)	43. (a)	44. (b)	45. (c)	46. (c)	47. (a)	48. (a)	49. (d)	50. (b)
51. (b)	52. (c)	53. (b)	54. (c)	55. (a)	56. (b)	57. (c)	58. (c)	59. (b)	60. (a)
61. (c)	62. (c)	63. (b)	64. (a)	65. (b)	66. (c)	67. (a)	68. (a)	69. (c)	70. (b)
71. (a)	72. (c)	73. (b)	74. (b)	75. (c)	76. (a)	77. (c)	78. (d)	79. (b)	80. (b)
81. (a)	82. (d)	83. (a)	84. (c)	85. (a)	86. (b)	87. (c)	88. (b)	89. (b)	90. (a)
91. (b)	92. (c)	93. (a)	94. (d)	95. (b)	96. (c)	97. (a)	98. (b)	99. (d)	100. (d)
101. (c)	102. (b)	103. (c)	104. (b)	105. (c)	106. (d)	107. (d)	108. (b)	109. (a)	110. (d)
111. (c)	112. (a)	113. (c)	114. (b)	115. (a)	116. (a)	117. (d)	118. (b)	119. (d)	120. (a)
121. (b)	122. (d)	123. (d)	124. (a)	125. (c)	126. (d)	127. (a)	128. (c)	129. (a)	130. (b)
131. (a)	132. (b)	133. (b)	134. (b)	135. (c)	136. (a)	137. (b)	138. (a)	139. (b)	140. (c)
141. (a)	142. (b)	143. (c)	144. (a)	145. (c)	146. (d)	147. (a)	148. (a)	149. (c)	150. (b)
151. (b)	152. (b)	153. (d)	154. (c)	155. (b)	156. (b)	157. (d)	158. (c)	159. (b)	160. (d)
161. (c)	162. (b)	163. (a)	164. (a)	165. (b)	166. (c)	167. (c)	168. (d)	169. (b)	170. (c)
171. (b)	172. (d)	173. (b)	174. (d)	175. (d)	176. (b)	177. (c)	178. (b)	179. (a)	180. (b)
181. (b)	182. (a)	183. (c)	184. (a)	185. (d)	186. (a)	187. (d)	188. (a)	189. (b)	190. (a)
191. (b)	192. (b)	193. (b)	194. (c)	195. (d)	196. (b)	197. (d)	198. (a)	199. (c)	200. (b)
201. (d)	202. (b)	203. (d)	204. (b)	205. (c)	206. (a)	207. (d)	208. (d)	209. (a)	210. (a)
211. (a)	212. (d)	213. (b)	214. (a)	215. (a)	216. (b)	217. (c)	218. (a)	219. (a)	220. (b)
221. (d)	222. (a)	223. (b)	224. (d)	225. (b)	226. (c)	227. (b)	228. (b)	229. (d)	230. (a)
231. (a)	232. (a)	233. (d)	234. (a)	235. (d)	236. (d)	237. (a)	238. (a)	239. (a)	240. (c)
241. (a)	242. (d)	243. (b)	244. (a)	245. (b)	246. (b)	247. (d)	248. (a)	249. (b)	250. (b)
251. (a)	252. (c)	253. (b)	254. (a)	255. (a)	256. (d)	257. (b)	258. (a)	259. (b)	260. (b)
261. (a)	262. (a)	263. (a)	264. (c)	265. (b)	266. (d)	267. (a)	268. (c)	269. (c)	270. (d)
271. (d)	272. (d)	273. (d)	274. (d)	275. (b)	276. (d)	277. (a)	278. (c)	279. (a)	280. (b)
281. (c)	282. (a)	283. (d)	284. (b)	285. (b)	286. (a)	287. (c)	288. (d)	289. (b)	290. (b)
291. (c)	292. (a)	293. (a)	294. (a)	295. (a)	296. (b)	297. (d)	298. (b)	299. (d)	300. (d)
301. (b)	302. (d)	303. (c)	304. (d)	305. (c)	306. (d)	307. (d)	308. (c)	309. (a)	310. (b)
311. (c)	312. (a)	313. (a)	314. (b)	315. (a)	316. (d)	317. (b)	318. (a)	319. (b)	320. (c)
321. (b)	322. (b)	323. (a)	324. (a)	325. (b)	326. (a)	327. (a)	328. (a)	329. (a)	330. (b)
331. (b)	332. (d)	333. (d)	334. (a)	335. (c)	336. (b)	337. (b)	338. (a)	339. (d)	340. (b)
341. (a)	342. (d)	343. (a)	344. (c)	345. (c)	346. (a)	347. (d)	348. (b)	349. (a)	350. (c)
351. (c)	352. (c)	353. (d)	354. (c)	355. (c)	356. (c)	357. (b)	358. (d)	359. (a)	360. (a)
361. (c)	362. (b)	363. (b)	364. (b)	365. (a)	366. (b)	367. (a)	368. (a)	369. (a)	370. (b)
371. (b)	372. (c)	373. (c)	374. (c)	375. (d)	376. (c)	377. (c)	378. (b)	379. (d)	380. (a)
381. (a)	382. (b)	383. (b)	384. (a)	385. (c)	386. (a)	387. (d)	388. (c)	389. (d)	390. (b)

391. (b)	392. (a)	393. (a)	394. (a)	395. (d)	396. (b)	397. (b)	398. (a)	399. (b)	400. (b)
401. (a)	402. (c)	403. (d)	404. (b)	405. (a)	406. (a)	407. (b)	408. (d)	409. (c)	410. (d)
411. (c)	412. (d)	413. (c)	414. (b)	415. (a)	416. (b)	417. (d)	418. (b)	419. (d)	420. (b)
421. (b)	422. (a)	423. (b)	424. (c)	425. (d)	426. (a)	427. (b)	428. (c)	429. (c)	430. (c)
431. (c)	432. (d)	433. (b)	434. (b)	435. (d)	436. (c)	437. (d)	438. (a)	439. (c)	440. (b)
441. (c)	442. (a)	443. (a)	444. (a)	445. (d)	446. (c)	447. (a)	448. (b)	449. (a)	450. (c)
451. (c)	452. (d)	453. (d)	454. (d)	455. (b)	456. (b)	457. (b)	458. (b)	459. (b)	460. (b)
461. (a)	462. (a)	463. (d)	464. (b)	465. (b)	466. (d)	467. (b)	468. (b)	469. (d)	470. (a)
471. (a)	472. (b)	473. (a)	474. (c)	475. (b)	476. (d)	477. (a)	478. (a)	479. (a)	480. (c)
481. (c)	482. (c)	483. (d)	484. (d)	485. (d)	486. (b)	487. (a)	488. (c)	489. (a)	490. (a)
491. (c)	492. (b)	493. (c)	494. (b)	495. (d)	496. (c)	497. (b)	498. (d)	499. (a)	500. (b)
501. (c)	502. (d)	503. (b)	504. (c)	505. (a)	506. (a)	507. (d)	508. (c)	509. (c)	510. (d)
511. (c)	512. (d)	513. (a)	514. (d)	515. (a)	516. (a)	517. (b)	518. (a)	519. (c)	520. (c)
521. (b)	522. (c)	523. (a)	524. (b)	525. (b)	526. (a)	527. (c)	528. (a)	529. (a)	530. (c)
531. (b)	532. (c)	533. (d)	534. (c)	535. (a)	536. (b)	537. (b)	538. (c)	539. (a)	540. (a)
541. (b)	542. (a)	543. (b)	544. (d)	545. (a)	546. (d)	547. (d)	548. (d)	549. (c)	550. (b)
551. (c)	552. (a)	553. (b)	554. (b)	555. (a)	556. (b)	557. (b)	558. (b)	559. (c)	560. (a)
561. (b)	562. (b)	563. (c)	564. (a)	565. (a)	566. (a)	567. (a)	568. (c)	569. (c)	570. (a)
571. (a)	572. (a)	573. (b)	574. (b)	575. (b)	576. (b)	577. (a)	578. (a)	579. (d)	580. (a)
581. (d)	582. (d)	583. (b)	584. (b)	585. (c)	586. (c)	587. (a)	588. (d)	589. (c)	590. (b)
591. (c)	592. (a)	593. (c)	594. (b)	595. (a)	596. (a)	597. (b)	598. (c)	599. (d)	600. (c)
601. (b)	602. (b)	603. (d)	604. (a)	605. (a)	606. (d)	607. (b)	608. (b)	609. (a)	610. (a)
611. (a)	612. (c)	613. (a)	614. (a)	615. (a)	616. (c)	617. (b)	618. (b)	619. (d)	620. (c)
621. (b)	622. (d)	623. (c)	624. (a)	625. (a)	626. (d)	627. (a)	628. (c)	629. (c)	630. (d)
631. (d)	632. (a)	633. (c)	634. (c)	635. (b)	636. (c)	637. (d)	638. (a)	639. (b)	640. (b)
641. (d)	642. (b)	643. (d)	644. (c)	645. (c)	646. (d)	647. (a)	648. (d)	649. (c)	650. (c)
651. (c)	652. (c)	653. (c)	654. (b)	655. (a)	656. (d)	657. (b)	658. (d)	659. (d)	660. (b)
661. (b)	662. (c)	663. (c)	664. (b)	665. (a)	666. (d)	667. (c)	668. (c)	669. (b)	670. (b)
671. (d)	672. (b)	673. (a)	674. (a)	675. (b)	676. (b)	677. (b)	678. (c)	679. (b)	680. (b)
681. (c)	682. (a)	683. (a)	684. (c)	685. (b)	686. (d)	687. (a)	688. (b)	689. (d)	690. (b)
691. (d)	692. (a)	693. (c)	694. (d)	695. (b)	696. (b)	697. (a)	698. (b)	699. (c)	700. (c)
701. (c)	702. (a)	703. (c)	704. (b)	705. (b)	706. (a)	707. (d)	708. (b)	709. (b)	710. (c)
711. (a)	712. (b)	713. (a)	714. (d)	715. (a)	716. (d)	717. (d)	718. (b)	719. (c)	720. (b)
721. (b)	722. (b)	723. (a)	724. (d)	725. (b)	726. (c)	727. (b)	728. (d)	729. (b)	730. (b)
731. (a)	732. (b)	733. (c)	734. (c)	735. (c)	736. (b)	737. (a)	738. (a)	739. (b)	740. (a)
741. (b)	742. (b)	743. (c)	744. (b)	745. (c)	746. (c)	747. (b)	748. (c)	749. (d)	750. (a)
751. (c)	752. (b)	753. (c)	754. (b)	755. (b)	756. (c)	757. (d)	758. (c)	759. (c)	760. (c)
761. (d)	762. (b)	763. (d)	764. (c)	765. (d)	766. (c)	767. (b)	768. (c)	769. (c)	770. (a)
771. (a)	772. (c)	773. (b)	774. (a)	775. (b)	776. (b)	777. (b)	778. (b)	779. (a)	780. (a)

781. (b)	782. (a)	783. (b)	784. (b)	785. (a)	786. (b)	787. (b)	788. (c)	789. (a)	790. (b)
791. (c)	792. (d)	793. (b)	794. (d)	795. (c)	796. (c)	797. (b)	798. (c)	799. (b)	800. (a)
801. (b)	802. (b)	803. (d)	804. (a)	805. (b)	806. (c)	807. (a)	808. (a)	809. (d)	810. (a)
811. (c)	812. (d)	813. (b)	814. (b)	815. (c)	816. (b)	817. (b)	818. (d)	819. (a)	820. (c)
821. (d)	822. (b)	823. (b)	824. (c)	825. (d)	826. (d)	827. (c)	828. (a)	829. (b)	830. (d)
831. (d)	832. (b)	833. (b)	834. (a)	835. (a)	836. (b)	837. (a)	838. (a)	839. (a)	840. (b)
841. (b)	842. (b)	843. (b)	844. (a)	845. (c)	846. (d)	847. (b)	848. (d)	849. (a)	850. (b)
851. (a)	852. (a)	853. (a)	854. (a)	855. (c)	856. (b)	857. (c)	858. (a)	859. (c)	860. (b)
861. (d)	862. (b)	863. (b)	864. (c)	865. (b)	866. (c)	867. (c)	868. (b)	869. (c)	870. (b)
871. (b)	872. (b)	873. (a)	874. (a)	875. (b)	876. (d)	877. (a)	878. (d)	879. (d)	880. (a)
881. (d)	882. (b)	883. (a)	884. (b)	885. (b)	886. (a)	887. (b)	888. (a)	889. (c)	890. (b)
891. (c)	892. (d)	893. (c)	894. (b)	895. (b)	896. (d)	897. (a)	898. (d)	899. (c)	900. (c)
901. (b)	902. (d)	903. (b)	904. (c)	905. (a)	906. (c)	907. (a)	908. (c)	909. (b)	910. (c)
911. (d)	912. (a)	913. (d)	914. (d)	915. (a)	916. (a)	917. (a)	918. (b)	919. (a)	920. (a)
921. (b)	922. (c)	923. (c)	924. (a)	925. (c)	926. (d)	927. (c)	928. (b)	929. (a)	930. (c)
931. (b)	932. (b)	933. (b)	934. (a)	935. (b)	936. (c)	937. (d)	938. (a)	939. (a)	940. (a)
941. (a)	942. (b)	943. (a)	944. (d)	945. (b)	946. (a)	947. (d)	948. (c)	949. (c)	950. (c)
951. (c)	952. (b)	953. (a)	954. (d)	955. (a)	956. (d)	957. (b)	958. (a)	959. (b)	960. (a)
961. (b)	962. (b)	963. (a)	964. (d)	965. (a)	966. (d)	967. (b)	968. (d)	969. (d)	970. (a)
971. (b)	972. (c)	973. (a)	974. (a)	975. (d)	976. (b)	977. (d)	978. (a)	979. (b)	980. (c)
981. (c)	982. (c)	983. (a)	984. (b)	985. (b)	986. (b)	987. (a)	988. (b)	989. (d)	990. (d)
991. (d)	992. (a)	993. (c)	994. (d)	995. (d)	996. (b)	997. (d)	998. (a)	999. (b)	1000. (c)
1001. (d)	1002. (a)	1003. (d)	1004. (d)	1005. (d)	1006. (d)	1007. (d)	1008. (d)	1009. (b)	1010. (b)
1011. (a)	1012. (d)	1013. (d)	1014. (a)	1015. (b)	1016. (c)	1017. (b)	1018. (b)	1019. (c)	1020. (d)
1021. (b)	1022. (d)	1023. (c)	1024. (b)	1025. (c)	1026. (c)	1027. (b)	1028. (b)	1029. (b)	1030. (c)
1031. (d)	1032. (b)	1033. (b)	1034. (a)	1035. (a)	1036. (a)	1037. (a)	1038. (b)	1039. (a)	1040. (c)
1041. (b)	1042. (c)	1043. (a)	1044. (b)	1045. (c)	1046. (a)	1047. (c)	1048. (a)	1049. (b)	1050. (b)
1051. (b)	1052. (d)	1053. (b)	1054. (b)	1055. (c)	1056. (b)	1057. (d)	1058. (b)	1059. (a)	1060. (c)
1061. (a)	1062. (c)	1063. (d)	1064. (c)	1065. (d)	1066. (a)	1067. (c)	1068. (b)	1069. (a)	1070. (a)
1071. (d)	1072. (d)	1073. (a)	1074. (a)	1075. (a)	1076. (a)	1077. (b)	1078. (c)	1079. (a)	1080. (b)
1081. (a)	1082. (b)	1083. (c)	1084. (b)	1085. (c)	1086. (a)	1087. (c)	1088. (c)	1089. (a)	1090. (d)
1091. (c)	1092. (a)	1093. (a)	1094. (c)	1095. (a)	1096. (a)	1097. (d)	1098. (c)	1099. (b)	1100. (b)
1101. (a)	1102. (b)	1103. (b)	1104. ( )	1105. (c)	1106. (c)	1107. (b)	1108. (b)	1109. (d)	1110. (c)
1111. (c)	1112. (a)	1113. (b)	1114. (b)	1115. (b)	1116. (b)	1117. (b)	1118. (a)	1119. (a)	1120. (a)
1121. (b)	1122. (a)	1123. (c)	1124. (a)	1125. (a)	1126. (a)	1127. (a)	1128. (a)	1129. (b)	1130. (b)
1131. (b)	1132. (c)	1133. (b)	1134. (c)	1135. (a)	1136. (c)	1137. (d)	1138. (a)	1139. (c)	1140. (b)
1141. (a)	1142. (d)	1143. (d)	1144. (a)	1145. (d)	1146. (d)	1147. (b)	1148. (d)	1149. (d)	1150. (a)
1151. (a)	1152. (a)	1153. (a)	1154. (a)	1155. (b)	1156. (a)	1157. (a)	1158. (d)	1159. (a)	1160. (a)
1161. (a)	1162. (b)	1163. (a)	1164. (a)	1165. (b)	1166. (c)	1167. (a)	1168. (d)	1169. (b)	1170. (a)

1171. (c)	1172. (b)	1173. (a)	1174. (d)	1175. (d)	1176. (a)	1177. (a)	1178. (a)	1179. (d)	1180. (d)
1181. (b)	1182. (b)	1183. (c)	1184. (d)	1185. (a)	1186. (b)	1187. (b)	1188. (b)	1189. (d)	1190. (b)
1191. (a)	1192. (a)	1193. (c)	1194. (d)	1195. (a)	1196. (d)	1197. (c)	1198. (d)	1199. (a)	1200. ( )
1201. (a)	1202. (a)	1203. (a)	1204. (d)	1205. (a)	1206. (d)	1207. (c)	1208. (a)	1209. (b)	1210. (b)
1211. (c)	1212. (a)	1213. (b)	1214. (c)	1215. (b)	1216. (a)	1217. (b)	1218. (a)	1219. (a)	1220. (a)
1221. (a)	1222. (c)	1223. (b)	1224. (a)	1225. (a)	1226. (d)	1227. (c)	1228. (a)	1229. (a)	1230. (b)
1231. (b)	1232. (a)	1233. (a)	1234. (a)	1235. (a)	1236. (d)	1237. (a)	1238. (c)	1239. (b)	1240. (d)
1241. (c)	1242. (a)	1243. (a)	1244. (a)	1245. (a)	1246. (a)	1247. (a)	1248. (c)	1249. (a)	1250. (a)
1251. (a)	1252. (a)	1253. (d)	1254. (b)	1255. (d)	1256. (c)	1257. (c)	1258. (a)	1259. (a)	1260. (a)
1261. (b)	1262. (a)	1263. (a)	1264. (b)	1265. (a)	1266. (a)	1267. (a)	1268. (d)	1269. (b)	1270. (a)
1271. (c)	1272. (d)	1273. (c)	1274. (b)	1275. (a)	1276. (a)	1277. (a)	1278. (c)	1279. (d)	1280. (d)
1281. (c)	1282. (c)	1283. (a)	1284. (a)	1285. (d)	1286. (a)	1287. (a)	1288. (a)	1289. (a)	1290. (a)
1291. (b)	1292. (c)	1293. (a)	1294. (c)	1295. (a)	1296. (d)	1297. (c)	1298. (a)	1299. (b)	1300. (a)
1301. (a)	1302. (a)	1303. (b)	1304. (c)	1305. (b)	1306. (c)	1307. (c)	1308. (d)	1309. (c)	1310. (b)
1311. (b)	1312. (c)	1313. (d)	1314. (d)	1315. (a)	1316. (a)	1317. (d)	1318. (b)	1319. (b)	1320. (b)
1321. (c)	1322. (b)	1323. (b)	1324. (a)	1325. (a)	1326. (a)	1327. (a)	1328. (b)	1329. (d)	1330. (c)
1331. (b)	1332. (d)	1333. (b)	1334. (c)	1335. (b)	1336. (a)	1337. (d)	1338. (b)	1339. (c)	1340. (c)
1341. (c)	1342. (a)	1343. (b)	1344. (b)	1345. (c)	1346. (c)	1347. (d)	1348. (b)	1349. (c)	1350. (b)
1351. (b)	1352. (b)	1353. (c)	1354. (b)	1355. (a)	1356. (a)	1357. (c)	1358. (a)	1359. (a)	1360. (d)
1361. (d)	1362. (c)	1363. (c)	1364. (d)	1365. (b)	1366. (a)	1367. (c)	1368. (c)	1369. (c)	1370. (b)
1371. (b)	1372. (b)	1373. (c)	1374. (c)	1375. (a)	1376. (b)	1377. (b)	1378. (b)	1379. (b)	1380. (b)
1381. (c)	1382. (d)	1383. (b)	1384. (a)	1385. (c)	1386. (b)	7138. (b)	1388. (a)	1389. (b)	1390. (a)
1391. (b)	1392. (a)	1393. (d)	1394. (c)	1395. (c)	1396. (b)	1397. (c)	1398. (a)	1399. (b)	1400. (b)
1401. (d)	1402. (a)	1403. (a)	1404. (a)	1405. (c)	1406. (c)	1407. (a)	1408. (a)	1409. (c)	1410. (b)
1411. (d)	1412. (b)	1413. (d)	1414. (a)	1415. (c)	1416. (a)	1417. (a)	1418. (a)	1419. (a)	1420. (a)
1421. (b)	1422. (d)	1423. (a)	1424. (a)	1425. (a)	1426. (a)	1427. (a)	1428. (d)	1429. (a)	1430. (b)
1431. (a)	1432. (a)	1433. (a)	1434. (b)	1435. (a)	1436. (a)	1437. (a)	1438. (c)	1439. (b)	1440. (a)
1441. (c)	1442. (a)	1443. (d)	1444. (c)	1445. (d)	1446. (d)	1447. (b)	1448. (d)	1449. (a)	1450. (d)
1451. (a)	1452. (c)	1453. (c)	1454. (a)	1455. (b)	1456. (a)	1457. (b)	1458. (a)	1459. (b)	1460. (a)
1461. (d)	1462. (b)	1463. (c)	1464. (b)	1465. (b)	1466. (d)	1467. (a)	1468. (d)	1469. (b)	1470. (a)
1471. (d)	1472. (b)	1473. (b)	1474. (a)	1475. (a)	1476. (b)	1477. (d)	1478. (b)	1479. (d)	1480. (a)
1481. (b)	1482. (c)	1483. (a)	1484. (a)	1485. (a)	1486. (a)	1487. (b)	1488. (a)	1489. (c)	1490. (a)
1491. (d)	1492. (c)	1493. (a)	1494. (d)	1495. (a)	1496. (d)	1497. (c)	1498. (a)	1499. (a)	1500. (b)
1501. (b)	1502. (b)	1503. (c)	1504. (a)	1505. (a)	1506. (b)	1507. (c)	1508. (c)	1509. (b)	1510. (a)
1511. (a)	1512. (a)	1513. (b)	1514. (a)	1515. (d)	1516. (a)	1517. (b)	1518. (c)	1519. (b)	1520. (c)
1521. (a)	1522. (b)	1523. (a)	1524. (a)	1525. (a)	1526. (c)	1527. (b)	1528. (b)	1529. (c)	1530. (d)
1531. (c)	1532. (d)	1533. (a)	1534. (a)	1535. (c)	1536. (b)	1537. (c)	1538. (c)	1539. (a)	1540. (b)
1541. (c)	1542. (d)	1543. (b)	1544. (b)	1545. (a)	1546. (a)	1547. (c)	1548. (c)	1549. (d)	1550. (a)
1551. (a)	1552. (c)	1553. (a)	1554. (c)	1555. (b)	1556. (b)	1557. (b)	1558. (c)	1559. (c)	1560. (a)

1561. (b)	1562. (b)	1563. (b)	1564. (d)	1565. (c)	1566. (c)	1567. (b)	1568. (d)	1569. (b)	1570. (b)
1571. (b)	1572. (a)	1573. (b)	1574. (a)	1575. (d)	1576. (b)	1577. (c)	1578. (d)	1579. (c)	1580. (b)
1581. (a)	1582. (c)	1583. (a)	1584. (a)	1585. (d)	1586. (c)	1587. (c)	1588. (a)	1589. (d)	1590. (a)
1591. (b)	1592. (d)	1593. (b)	1594. (a)	1595. (b)	1596. (d)	1597. (c)	1598. (d)	1599. (b)	1600. (a)
1601. (c)	1602. (a)	1603. (c)	1604. (c)	1605. (b)	1606. (b)	1607. (b)	1608. (b)	1609. (c)	1610. (d)
1611. (a)	1612. (a)	1613. (d)	1614. (c)	1615. (b)	1616. (b)	1617. (a)	1618. (c)	1619. (b)	1620. (c)
1621. (a)	1622. (d)	1623. (b)	1624. (b)	1625. (b)	1626. (a)	1627. (c)	1628. (b)	1629. (d)	1630. (a)
1631. (b)	1632. (c)	1633. (c)	1634. (b)	1635. (b)	1636. (c)	1637. (c)	1638. (b)	1639. (b)	1640. (c)
1641. (d)	1642. (d)	1643. (a)	1644. (a)	1645. (b)	1646. (b)	1647. (c)	1648. (c)	1649. (b)	1650. (c)
1651. (b)	1652. (c)	1653. (c)	1654. (a)	1655. (c)	1656. (c)	1657. (a)	1658. (c)	1659. (b)	1660. (b)
1661. (b)	1662. (a)	1663. (c)	1664. (c)	1665. (b)	1666. (a)	1667. (b)	1668. (a)	1669. (a)	1670. (c)
1671. (a)	1672. (c)	1673. (c)	1674. (c)	1675. (c)	1676. (a)	1677. (b)	1678. (a)	1679. (a)	1680. (a)
1681. (b)	1682. (b)	1683. (b)	1684. (c)	1685. (c)	1686. (c)	1687. (a)	1688. (c)	1689. (c)	1690. (d)
1691. (d)	1692. (c)	1693. (c)	1694. (d)	1695. (b)	1696. (d)	1697. (a)	1698. (a)	1699. (a)	1700. (b)
1701. (c)	1702. (a)	1703. (c)	1704. (b)	1705. (b)	1706. (a)	1707. (a)	1708. (c)	1709. (c)	1710. (a)
1711. (b)	1712. (a)	1713. (a)	1714. (a)	1715. (b)	1716. (c)	1717. (a)	1718. (a)	1719. (d)	1720. (a)
1721. (b)	1722. (a)	1723. (b)	1724. (a)	1725. (a)	1726. (a)	1727. (c)	1728. (a)	1729. (d)	1730. (b)
1731. (b)	1732. (c)	1733. (b)	1734. (b)	1735. (a)	1736. (a)	1737. (d)	1738. (b)	1739. (b)	1740. (d)
1741. (b)	1742. (b)	1743. (a)	1744. (b)	1745. (b)	1746. (c)	1747. (c)	1748. (b)	1749. (a)	1750. (a)
1751. (a)	1752. (c)	1753. (d)	1754. (b)	1755. (a)	1756. (b)	1757. (d)	1758. (a)	1759. (c)	1760. (a)
1761. (b)	1762. (b)	1763. (a)	1764. (d)	1765. (b)	1766. (b)	1767. (b)	1768. (a)	1769. (b)	1770. (d)
1771. (b)	1772. (b)	1773. (d)	1774. (b)	1775. (c)	1776. (b)	1777. (a)			

ANNEXURE – II

Nutritive Value of Different Vegetables (per 100g edible portion on fresh weight basis)

S. No.	Name of Vegetable	Moisture (g)	Carbohydrates (g)	Protein (g)	Fat (g)	Calorie (energy)	Vit. A (IU)	Thiamine (mg)	Riboflavin (mg)	Ascorbic acid (mg)	Calcium (mg)	Iron (mg)	Phosphorus (mg)
1.	Potato	74.7	22.6	1.6	0.1	97	40	0.40	0.04	17.0	10.0	0.7	35.0
2.	Tomato (green)	93.1	3.6	1.9	0.1	23	307	0.07	0.01	31.0	20.0	1.8	-
3.	Tomato (ripe)	94.0	3.6	1.2	0.1	20	302	0.12	0.06	27.0	48.0	0.4	26.0
4.	Chilli	85.7	3.0	2.9	0.6	29	292	0.19	0.39	111.0	30.0	1.2	80.0
5.	Sweet pepper	92.4	4.3	1.3	0.3	24	683	0.06	0.03	175.0	10.0	1.2	-
6.	Brinjal	92.7	4.0	1.4	0.3	24	118	0.04	0.11	12.0	18.0	0.9	47.0
7.	Cabbage	92.4	5.3	1.4	0.2	29	80	0.06	0.05	100.0	46.0	0.8	38.0
8.	Cauliflower	91.7	4.9	2.4	0.2	31	70	0.04	0.03	75.0	30.0	17.0	76.0
9.	Knolkhol	90.1	6.7	2.1	0.1	36	20	0.05	0.10	50.0	20.0	0.4	60.0
10.	Broccoli	89.9	5.5	3.3	0.2	37	3500	0.05	0.12	137.0	80.0	0.8	79.0
11.	Brussels sprouts	84.9	8.9	4.4	0.5	58	400	0.05	0.13	105.0	40.0	1.6	80.0
12.	Kale	86.6	7.2	3.9	0.6	50	7540	0.05	0.16	115.0	85.0	1.6	62.0
13.	Wax gourd	96.5	1.9	0.4	0.1	10	0	0.06	0.01	1.0	30.0	0.08	20.0
14.	Bitter gourd (large)	92.4	4.2	1.6	0.2	25	210	0.07	0.09	88.0	20.0	1.8	70.0
15.	Bitter gourd (small)	83.2	10.6	2.1	1.0	60	210	0.07	0.06	96.0	23.0	2.0	38.0
16.	Bottle gourd	96.1	2.5	0.2	0.1	12	0	0.03	0.01	6.0	20.0	0.7	10.0
17.	Cucumber	96.1	2.5	0.4	0.1	13	0	0.03	0.01	7.0	10.0	1.5	25.0
18.	Sponge gourd	93.2	2.9	0.5	0.1	18	120	0.02	0.06	0.0	36.0	1.1	19.0
19.	Ridge gourd	95.2	3.4	0.5	0.1	17	56	0.01	0.01	5.0	18.0	0.5	26.0
20.	Snake gourd	94.6	3.3	0.5	0.3	18	160	0.04	0.06	0.0	26.0	0.3	20.0
21.	Pointed gourd	92.0	2.2	2.0	0.3	20	255	0.05	0.06	29.0	30.0	1.7	40.0
22.	Giant spine gourd	84.1	7.7	3.1	1.0	52	2592	0.05	0.18	0.0	33.0	4.6	42.0

S. No.	Name of Vegetable	Moisture (g)	Carbohydrates (g)	Protein (g)	Fat (g)	Calorie (energy)	Vit. A (IU)	Thiamine (mg)	Riboflavin (mg)	Ascorbic acid (mg)	Calcium (mg)	Iron (mg)	Phosphorus (mg)
23.	Ivy gourd	93.5	3.1	1.2	0.1	18	249	0.07	0.08	15.0	40.0	1.4	30.0
24.	Pumpkin (ripe)	86.0	4.6	1.4	0.1	25	2180	0.06	0.04	2.0	10.0	0.7	30.0
25.	Summer squash	94.8	3.5	1.0	0.1	19	260	0.05	0.03	18.0	10.0	0.6	30.0
26.	Winter squash (ripe)	86.0	6.6	1.1	0.2	32	3300	0.03	0.05	6.0	14.2	0.4	20.9
27.	Muskmelon	95.2	3.5	0.3	0.2	17	3420	0.11	0.08	26.0	32.0	1.4	14.0
28.	Watermelon	95.8	3.3	0.2	0.2	16	590	0.02	0.04	1.0	11.0	7.9	12.0
29.	Roundmelon	93.5	3.4	1.4	0.2	21	23	0.04	0.08	18.0	25.0	0.9	24.0
30.	Snapmelon	95.7	3.0	0.3	0.1	14	265	-	-	10.0	-	-	-
31.	Chowchow	92.5	5.7	0.7	0.1	27	8	0.01	0.04	4.0	140.0	0.6	30.0
32.	Radish (white)	94.4	3.4	0.7	0.1	17	50	0.06	0.02	15.0	50.0	0.04	22.0
33.	Radish (pink)	90.8	6.8	0.6	0.3	32	50	0.06	0.02	17.0	50.0	0.5	20.0
34.	Carrot	82.2	10.6	0.9	0.2	48	12000	0.04	0.02	3.0	48.0	0.6	30.0
35.	Beet	87.7	8.8	1.7	0.1	43	18	0.04	0.09	10.0	28.0	1.0	55.0
36.	Parsnip	78.6	18.2	1.5	0.5	78	0	0.08	0.12	18.0	57.0	0.6	80.0
37.	Turnip	91.6	6.2	0.5	0.2	28	4	0.04	0.04	43.0	30.0	0.4	40.0
38.	Rutabaga	89.1	8.9	1.1	0.1	38	330	0.07	0.08	36.0	55.0	1.0	41.0
39.	Onion	86.8	11.0	1.2	0.2	50	35	0.08	0.01	11.0	180.0	0.7	50.0
40.	Garlic	62.8	29.0	6.3	0.1	142	10	0.16	0.23	13.0	30.0	1.3	310.0
41.	Leek	78.9	5.0	1.8	0.1	28	30	0.23	-	11.0	-	2.3	70.0
42.	Shallot	-	-	2.6	-	-	-	0.06	0.02	1.0	37.0	1.3	60.0
43.	Lettuce	93.4	2.5	2.1	0.3	21	540	0.09	0.13	10.0	50.0	2.4	28.0
44.	Celery	88.0	1.6	6.3	0.6	38	0	0.01	0.11	7.0	50.0	0.6	38.0
45.	Sorrel	95.2	1.4	1.6	0.3	14	-	0.03	0.06	12.0	63.0	8.7	-
46.	Sweet corn	89.0	7.9	1.4	0.3	41	150	0.06	0.05	4.4	3.5	0.2	46.0

S. No.	Name of Vegetable	Moisture (g)	Carbohydrates (g)	Protein (g)	Fat (g)	Calorie (energy)	Vit. A (IU)	Thiamine (mg)	Riboflavin (mg)	Ascorbic acid (mg)	Calcium (mg)	Iron (mg)	Phosphorus (mg)
47.	Okra	89.6	6.4	1.9	0.2	35	88	0.07	0.10	13.0	66.0	1.5	56.0
48.	Pea	72.0	15.8	7.2	0.1	93	300	0.25	0.01	19.0	20.0	1.5	139.0
49.	French bean	91.4	4.5	1.7	0.1	25	321	0.08	0.06	16.0	50.0	1.7	28.0
50.	Cowpea	84.6	8.0	4.3	0.2	51	941	0.07	0.09	13.0	80.0	2.5	74.0
51.	Hycinth bean	86.1	6.7	3.8	0.7	48	312	0.1	0.06	9.0	210.0	1.7	68.0
52.	Cluster bean	81.0	10.8	3.2	0.4	59	316	0.09	0.09	47.0	130.0	5.0	50.0
53.	Winged bean	91.0	3.8	2.9	0.1	27	595	0.06	0.12	37.0	300.0	1.7	69.0
54.	Broad bean	85.4	7.2	4.5	0.1	48	14	0.08	-	12.0	50.0	1.4	-
55.	Lima bean	66.5	23.5	7.5	0.8	128	280	0.21	0.11	32.0	63.0	2.3	158.0
56.	Sweet potato	70.0	27.0	1.5	0.2	115	2800	0.08	0.04	24.0	46.0	0.8	49.0
57.	Taro corm	93.1	21.1	3.0	0.1	97	166	0.09	0.03	0.0	40.0	1.7	140.0
58.	Giant taro corm	81.2	16.9	0.6	0.1	71	-	0.10	0.02	0.0	50.0	0.5	46.0
59.	Tannia corm	75.4	21.0	2.2	0.1	93	-	0.12	0.04	8.0	34.0	1.2	62.0
60.	Elephant-foot yam	78.5	18.4	2.0	0.1	82	434	0.06	0.02	6.0	38.0	2.4	38.0
61.	Yam root	73.0	22.0	2.7	0.3	109	-	0.09	0.03	5.0	40.0	1.7	-
62.	Cassava	65.5	32.4	1.2	0.2	135	80	0.05	0.04	34.0	26.0	0.9	32.0
63.	Chinese potato	77.6	19.7	1.3	0.1	85	-	-	-	-	-	-	-
64.	Asparagus	91.7	2.9	1.7	0.2	20	762	0.12	0.13	25.0	15.8	0.7	46.9
65.	Globe artichoke	83.7	11.9	2.9	0.4	63	390	0.18	0.05	10.0	47.0	1.9	94.0
66.	Drumstick pod	86.9	3.7	2.5	0.1	25	176	0.05	0.07	120.0	30.0	3.3	110.0
67.	Jackfruit (unripe)	84.0	9.4	2.6	0.3	50	0	0.05	0.04	14.0	30.0	1.7	-
68.	Green banana	83.2	14.0	1.4	0.2	63	48	0.05	0.02	24.0	10.0	0.6	29.0

S. No.	Name of Vegetable	Moisture (g)	Carbohydrates (g)	Protein (g)	Fat (g)	Calorie (energy)	Vit. A (IU)	Thiamine (mg)	Riboflavin (mg)	Ascorbic acid (mg)	Calcium (mg)	Iron (mg)	Phosphorus (mg)
69.	Green papaya	92.0	5.7	0.7	0.2	27	0	0.01	0.01	12.0	28.0	0.9	-
70.	Green mango	87.5	10.1	0.7	0.1	44	144	0.04	0.01	3.0	10.0	-	-
71.	Amaranth	85.7	6.3	4.0	0.5	45	9200	0.03	0.10	99.0	397.0	25.5	83.0
72.	Palak	86.4	6.5	3.4	0.8	46	9770	0.26	0.56	70.0	380.0	16.2	30.0
73.	Spinach	92.1	2.9	2.0	0.7	27	9300	0.03	0.07	28.0	73.0	10.9	21.0
74.	Indian spinach	90.8	4.2	2.8	0.4	31	3250	0.03	0.16	87.0	200.0	1.0	35.0
75.	Water spinach	92.4	-	1.9	-	-	4800	-	-	58.0	90.0	4.8	-
76.	Fenugreek leaves	86.1	6.0	4.4	0.9	49	3744	0.05	-	54.0	360.0	17.2	51.0
77.	Mustard leaves	89.9	3.2	4.0	0.6	34	4195	0.03	-	33.0	155.0	16.3	26.0
78.	Bengal gram leaves	73.4	14.1	7.0	1.4	97	1564	0.09	0.10	61.0	340.0	23.8	-
79.	Bottle gourd leaves	87.9	6.1	2.3	0.7	40	-	-	-	-	80.0	-	59.0
80.	Pumpkin leaves	81.9	7.9	4.6	0.8	57	5760	-	-	-	392.0	-	112.0
81.	Pointed gourd leaves	80.5	5.8	5.4	1.1	54	-	-	-	-	531.0	-	73.0
82.	Coriander leaves	86.3	6.3	3.3	0.6	44	11168	0.50	0.06	135.0	184.0	18.5	-
83.	Pigweed	89.6	2.9	3.7	0.4	30	2784	0.01	0.14	35.0	150.0	4.2	-
84.	Portulaca	90.5	2.7	2.4	0.6	26	3664	0.10	0.22	29.0	111.0	14.8	-
85.	Curry leaves	63.8	18.7	6.1	1.0	36	12096	0.08	0.01	4.0	830.0	7.0	57.0
86.	Radish top	90.3	3.4	2.7	0.6	30	18	0.03	0.16	103.0	310.0	16.1	60.0
87.	Carrot top	83.9	8.3	5.1	0.5	58	-	-	-	-	340.0	8.3	110.0
88.	Beet top	90.4	5.6	2.0	0.3	27	6700	0.08	0.18	35.0	118.0	3.1	45.0
89.	Agathi leaves	73.1	11.8	8.4	1.4	93	8640	0.21	0.09	58.0	404.0	5.0	-

S. No.	Name of Vegetable	Moisture (g)	Carbohydrates (g)	Protein (g)	Fat (g)	Calorie (energy)	Vit. A (IU)	Thiamine (mg)	Riboflavin (mg)	Ascorbic acid (mg)	Calcium (mg)	Iron (mg)	Phosphorus (mg)
90.	Mint	84.9	5.8	4.8	0.6	48	2592	0.05	0.20	27.0	200.0	15.6	-
91.	Ceylon spinach	93.6	4.0	1.6	-	22	5600	-	-	86.0	106.0	1.6	43.2
92.	Swiss chard	91.8	4.4	1.6	0.2	21	3110	0.03	0.09	34.0	110.0	3.6	29.0
93.	Cowpea leaves	89.0	4.1	3.4	0.7	20	9715	0.05	0.18	4.0	290.0	20.1	-
94.	Drumstick leaves	75.9	12.5	6.7	1.7	92	10848	0.06	0.05	220.0	440.0	7.0	70.0
95.	Chekkurmanis	73.6	11.6	6.8	3.2	102	9510	0.48	0.32	247.0	570.0	28.0	200.0
96.	Onion scape	87.6	8.9	0.9	0.2	41	952	0.00	0.03	17.0	50.0	7.5	-
97.	Green onion	87.6	10.6	1.0	0.2	45	50	0.03	0.04	24.0	135.0	0.9	24.0
98.	Taro petioles	93.8	4.6	0.2	0.2	20	335	0.01	0.02	8.0	57.0	1.4	23.0
99.	Chicory	94.2	2.9	1.6	0.3	21	10000	0.05	0.20	15.0	18.0	0.7	21.0
100.	Chives	86.0	7.8	3.8	0.6	52	500	0.12	-	70.0	48.0	8.4	57.0
101.	Water cress	93.6	3.3	1.7	0.3	18	4720	0.08	0.16	77.0	195.0	2.0	46.0
102.	Endive	93.3	4.0	1.6	0.2	20	3000	0.07	0.12	11.0	79.0	1.7	56.0
103.	Parsley	83.9	9.0	3.7	1.0	50	8230	0.11	0.28	193.0	193.0	4.3	84.0
104.	Banana flower	89.9	5.1	1.7	0.7	33	43	0.05	0.02	16.0	32.0	1.6	-
105.	Pumpkin flower	89.1	5.8	2.2	0.8	39	-	-	-	-	120.0	-	60.0
106.	Agathi flower	87.4	-	1.8	0.6	-	-	0.13	-	41.0	-	-	-
107.	Lotus root	85.9	11.3	1.7	0.1	53	-	0.10	-	22.0	21.0	0.4	-
108.	Mushrooms	91.1	4.0	2.4	0.3	16	0	0.10	0.44	5.0	9.0	1.0	115.0

Note: 0.6 ug B carotene = 1 IU vitamin A

- = data not available

1 μg = 10^{-6} gram

Eight independent Vavilovian Centre of origin

S.No.	Centre of origin	Region
1.	The Chinese Centre	Mountainous regions of central and western China together with adjacent low lands.
2.	The Indian (Hindustani) centre	Indian subcontinent (dominated by India, the area also includes Bangladesh, Nepal, Sri Lanka and Bhutan) including Burma (now Myanmar).
2a.	The Indo-Malay centre	All of Malay archipelago (part of Malaysia which hangs off the corner of mainland southeast Asia stretching from the foot hills of the Himalayas in the north to the equator in the south), the Philippines and Indo-China. It is supplementary to Indian centre of origin.
3.	The Central Asiatic centre	Western Punjab (Now in Pakistan), northern part of Pakistan, all of Afghanistan, the Tadjik and Uzbk of former Soviet Republics (Now separate country Tajikistan, Uzbekistan) and Western Tien shan (China).
4.	The Near East Centre	Interior of Asia Minor (Syria, Lebanon, Turkey), high land of Turkmenistan, Iran and the whole Transcaucasia (a region to the south of the Caucasus mountains between the black sea and the Caspian, and comprising the present-day republics of Georgia, Armenia and Azerbaijan).
5.	The Mediterranean Centre	Israel, Jordon, Syria, Cyprus, Turkey, Lybia, Lebanon
6.	The Abyssinian centre	Ethiopia of Africa
7.	South Meican and central American centre	Southern part of Mexico, small countries of central America which lie within a neck of land that joins north and south America (Guatemala, El Salvador, Costa Rica, Panama, Nicaragua, Houdurus, Haiti, Trinidad and Tobago, Cuba, etc.)
8.	The South American Centre	Mountainous areas and plateaus of Columbia, Ecuador, Peru, Paraguay, Bolivia and Chilli

Centres of diversity of major vegetable crops

Gene centre	Primary Centre of origin and domestication	Secondary Centre
Chinese-Japanese	Eggplant, wax gourd, Chinese cabbage, welsh onion	Watermelon, amaranth
Indo-Chinese	Sponge ground, ridge gourd, bottle gourd, sword bean, winged bean, taro, chayote, cucumber, yams, bitter gourd	Chinese cabbage, bottle gourd, yam bean, amaranth, cucumber, yard long bean, shallot

Hindustani Centre	Eggplant, wax gourd, cucumber, ridge gourd, sponge gourd, bitter gourd, hyacinth bean, drumstick	Watermelon, melon, bottle gourd, amaranth, okra, vegetable cowpea
Central Asia	Onion, garlic, carot, spinach, *Brassica* spp.	Eggplant, watermelon, melon, cauliflower
Near East	Onion, garlic, leek, *Brassica* spp.	Okra
Mediterranean	Cabbage, cauliflower, broccoli, radish, beans	Sweet pepper, garlic, okra, pepper, onion, lima bean
African	Eggplant, watermelon, melon, bottle gourd, cowpea, okra	Amaranth, shallot,
European	Lettuce	Onion, cauliflower, cabbage, carrot, common bean, spinach
Central-American and Mexican region	Tomato, hot pepper, pumpkin, squashes, sweet potato, amaranth	-
South American	Tomato, hot pepper, cassava, Xanthosoma, pumpkin and squashes, lima bean chayote.	Common bean Tomato, eggplant, melon, watermelon, squashes, onion, lettuce, lima bean

Variability of major vegetable crops in different agro-ecological regions of India

S. No.	Agro-ecological region	Geographical ranges	Variability in major crops
1	Humid Western Himalayan Region	Jammu and Kashmir, Himachal Pradesh and hills of Uttar Pradesh	Cucurbits, radish, carrot, turnip, peas, cowpea, brinjal, okra, spinach, fenugreek, amaranth, *Solanum khasianum, S. hirsutum, Sechium edule, Basella rubra.*
2	Humid Bengal/ Assam Basin	West Bengal and Assam	Cucurbits, radish, cowpea, brinjal, okra, spinach beet, amaranth, *Abelmoschus manihot, S. indicum, S. indicum, S. khasianum, S. surattense, Cucumis sativus var. sikkimensis, Edgeria darjelingensis, Melothria assamica, Momordica cochinchinensis, Sechium edule, Tuladiantha cardifolia, Basella rubra.*
3	Humid Eastern Himalayan Region	Arunachal Pradesh, Nagaland, Manipur, Mizoram, Tripura, Meghalaya and Andaman & Nicobar Islands	Cucurbits, radish, peas, cowpea, brinjal, okra, spinach, beet, amaranth, *A. manihot spp. Tetraphyllus, S. khasianum, S. torvum, S. sisymbriifolium, S. forex, S. vervascifolium, Cucumis histrix, Luffa echinata, Sechium edule.*

4	Sub-humid Sutlej Ganga Alluvial Plants	Punjab, Plains of Uttar Pradesh and Bihar	Cucurbits, radish, carrot, peas brinjal, okra, spinach, beet, fenugreek, onion, garlic, *A. manihot spp. Tetraphyllus var. pungens, A. tuberculatus, S. indicum, S. torvum, S. khasianum, S. surrattense, S. hispidum, Cucumis hardwickii, C. trigonus.*
5	Humid Eastern and Southern Eastern Uplands	East Madhya Pradesh, Orissa and Andhra Pradesh	Cucurbits, radish, carrot, cowpea, chillies, brinjal, okra spinach, amaranth, garlic, *A. manihot* var. *manihot, Solanum surattense, S. torvum.*
6	Arid Western Plains	Haryana, Rajasthan and Gujarat	Cucurbits, cauliflower, radish, carrot, peas, cowpea, chillies, brinjal, okra, spinach beet, fenugreek, onion, garlic, *A. tuberculatus, A. ficulneus, A. manihot ssp. Tetraphyllus, A. tuberculatus, Solanum surattense, S. nigrum, Citrullus colocynthis*
7	Semi-Arid Lava Plateau and Central Highlands	Maharashtra and West Madhya Pradesh	Cucurbits, cauliflower, radish, carrot, cowpea, chillies, brinjal, okra, spinach beet, fenugreek, amaranth, onion, *S. surattense, S. torvum, S. nigrum, S. khasianum, Cucumis setosus, Luffa acutangula* var. *acutangula*
8	Humid to Semi-Arid Western Ghats and Karnataka Plateau	Karnataka, Tamil Nadu, Kerala and Lakshadweep Islands	Cucurbits, chillies, brinjal, okra, amaranth, *A. crunitus, A. angulosus, A. ficulneus, A. moschatus, A. manihot* var. *tetraphyllus, Solanum trilobatum, S. indicum, S. insamum, S. pubescens, S. surattense, S. torvum, Luffa acutangula var. acutangula, Melothria angulata, Basella rubra.*

Classification of vegetables based on nutrient content

Nutrients	Sources
Calories	Immature seeds of lima bean, broad bean and peas, tapioca, sweet potato, yams, colocasia (corms), potato, garlic, brussel's sprouts and onion
Protein	Peas, double bean, winged bean, garlic, brussel's sprouts, cowpea, lima bean, lima bean seeds, amaranth leaves, drumstick leaves, and mathi.
Precursor of Vitamin-A (β-Carotene)	Carrot, spinach, turnip green, palak, mustard green, amaranth, coriander, Colocasia leaves, sweet potato (yellow), pumpkin (yellow) and tomato
Vitamin-B complex	Peas, broad bean, lima bean, garlic, asparagus, Colocasia corms and tomato

Vitamin-C (Ascorbic acid)	Turnip green, green chillies, brussel's sprouts, mustard green and amaranths, coriander, drumstick leaves, cauliflower, knoll khol, spinach, cabbage, bitter gourd and radish leaves.
Calcium	Curry leaves, amaranth leaves, drumstick leaves, methi, turnip green, mustard green, coriander and palak.
Iron	Drumstick leaves and fruits, amaranth, methi, mint, coriander, spinach, palak and mustard green leaves.
Roughages	Spinach, lettuce, cabbage, amaranth and root vegetables.
Vegetable milk	Pea pods and cabbage leaves.

Classification of vegetables based on tenderness

Tender Crops	Very tender crops	
Cowpea	Cucumber	Bitter gourd
New Zealand spinach	Brinjal	Watermelon
French bean	Limba bean	Ridge gourd
Soybean	Cluster bean	Muskmelon
Sweet corn	Winged bean	Cassava
Tomato	Amaranth	Pointed gourd
Sweet potato	Hyacinth bean	Indian spinach
Sweet pepper	Okra	Elephant foot yam
Summer squash	Chilli	
Pumpkin	Bottle gourd	

Classification of vegetables based on Respiratory activity

	Respiratory activity of the produce	Crop
(i)	Very low	Potato, onion
(ii)	Low	Cabbage, sweet potato, turnip, cucumber
(iii)	Moderate	Chilli, carrot, tomato, beet, sweet pepper
(iv)	High	Hyacinth bean, French bean, peas, lettuce, lima bean, radish
(v)	Very high	Green onion, muskmelon, watermelon, cauliflower, okra, leafy greens, broccoli, asparagus, parsley

Classification of vegetable crops based on their Temperature Requirement

Monthly mean temperature (^{0}C) (day and night)			Vegetable crops
Minimum	**Maximum**	**Optimum**	
Group: Hot			
18.0	35.0	25.0-27.0	Okra, watermelon, muskmelon, chilli, sweet potato, martynia, yam, cassava, amaranth, winged bean, cluster bean

Group: Warm			
10.0	35.0	20.0-25.0	Tomato, sweet pepper, brinjal, lima bean, French bean, hyacinth bean, cowpea, sweet corn, palak, Newzealand spinach, cucumber, pumpkin, pointed gourd, bottle gourd, bitter gourd, ridge gourd, snake gourd, summer squash, drumstick, Indian spinach, portulaca, elephant foot yam
Group: Cool-hot			
5.0	30.0	20.0-25.0	Onion, leek, garlic, shallot, chicory, chives, salsify, scorzonera, pak-choi, globe artichoke, taro, Indian cauliflower.
Group: Cool-warm			
5.0	25.0	18.0-25.0	Pea, broad bean, cauliflower, Chinese cabbage, cabbage, broccoli, Brussels sprouts, Kale, Knolkhol, turnip, radish, carrot, beet, rutabaga, horse radish, parsnip, potato, parsley, fennel, dill, Swiss chard, spinach, lettuce, celery, celeriac, asparagus, garden cress, mustard, cardoon.

Classification of Cool Season Vegetable Crops

Hardy Crops		**Half hardy crops**	
Broad bean	Brussels sprouts	Salsify	Asparagus
Broccoli	Chives	Carrot	Rutabaga
Cabbage	Garlic	Cauliflower	Beet
Collards	Kale	Celery	
Horse radish	Leek	Chard	
Knolkhol	Onion	Chicory	
Mustard	Peas	Chinese cabbage	
Parsley	Rhubarb	Globe artichoke	
Radish	Turnip	Endive	
Spinach	Pak-choi	Parsnip	
Sorrel		Potato	

Classification of vegetable crops based on their Relative Tolerance to Soil Acidity

Less tolerant (pH 6.8-6.0)	Moderately tolerant (pH 6.8-5.5)	Very tolerant (pH 6.8-5.0)
Asparagus	French bean	Chicory
Beet	Lima bean	Dendelion
Broccoli	Brussels sprouts	Endive
Cabbage	Carrot	Fennel
Cauliflower	Collard	Potato
Celery	Cucumber	Rhubarb
Palak	Brinjal	Shallot
Swiss chard	Garlic	Sorrel
Cress	Horse radish	Sweet potato
Chinese cabbage	Kale	Watermelon
Leek	Knolkhol	
Onion	Parsley	
Lettuce	Pea	
Muskmelon	Chilli	
New Zealand spinach	Pumpkin	
Okra	Radish	
Onion	Rutabaga	
Orach	Summer squash	
Parsnip	Winter squash	
Salsify	Tomato	
Spinach	Turnip	
Water cress		

Classification of vegetable crops based on their water requirement

High	Moderate	Low	Very Low
Palak	Onion	Pea	Watermelon
Amaranth	Cucumber	French bean	Muskmelon
Lettuce	Chilli	Cowpea	Pumpkin
Celery	Brinjal	Broad bean	Wax gourd
Other leafy greens	Tomato	Cluster bean	
Sweet pepper	Carrot	Winged bean	
Cabbage	Potato	Hyacinth bean	
Cauliflower			
Asparagus			

Broccoli			
Rhubarb			
Brussels sprouts			
Horse radish			
Radish			
Beet			
Turnip			
Ridge gourd			
Rutabaga			
Green onion			

***Note:** The water requirement is expressed as the units of absorbed water required for the production of one unit of dry matter.

Classification of vegetable crops based on seed longevity in normal storage condition

Longevity	Vegetable Seeds
Up to 1 year	Onion, French bean
Up to 2 years	Okra, carrot, chilli, sweet pepper, cauliflower, cabbage, broccoli
Up to 3 years	Radish, beet, peas, ridge gourd, turnip, palak
Up to 4 years	Tomato, brinjal, pumpkin, summer squash, bottle gourd, bitter gourd, watermelon, muskmelon, cucumber, spinach

Critical Stages of Irrigation in Some Vegetable Crops

Crop	Stages of growth
Tomato	Flower development, fruitset and after each harvest
Brinjal	Same
Chilli	Tenth leaf to flowering, to fruiting and after periodical harvests
Potato	Stolon formation, tuberization and tuber enlargement
Pumpkin, Cucumber, Watermelon, Gourds, Muskmelon, Summer Squash, etc.	Flower bud development and early fruit development
Cabbage	Head formation and enlargement
Cauliflower	Throughout the whole vegetation period
Broccoli	Early stages of development
Onion	Bulb formation and enlargement
Garlic	Bulbing
Radish, Carrot, Beet, Turnip	Constant water supply during the whole vegetation period

Pea	Beginning of flower bud development and pod filling
French bean, Hyacinth bean	Flowering and pod development
Cowpea	Prior to flowering and after podset
Lettuce (head type)	Head formation
Celery	Throughout the whole vegetation period
Sweet potato	40-50 days after planting at tuber formation
Okra	After fruitset
Leafy vegetables	During the whole vegetation period
Elephant-foot yam	Shoot emergence

*Growth stage(s) when water stress reduces the yield most.

Classification of vegetable crops based on their Root Depth

Very shallow rooted (15-30 cm)	Shallow-rooted(30-60 cm)	Moderately deep-rooted (60-90 cm)	Deep-rooted (90-120 cm)	Very deep-rooted (120-180 cm)
Onion	Broccoli	French bean	Pea	Artichoke
Lettuce	Brussels sprouts	Beet	Chilli	Asparagus
Small radish	Cabbage	Carrot	Rutabaga	Lima bean
	Cauliflower	Chard	Summer squash	Parsnip
	Celery	Cucumber	Turnip	Pumpkin
	Chinese cabbage	Brinjal		Winter squash
	Endive	Muskmelon		Sweet potato
	Garlic			Tomato
	Leek			
	Parsley			
	Potato			
	Palak			
	Spinach			
	Radish			
	Cowpea			

Vegetatively propagated crops

Crop	Plant parts for regeneration
Potato	Underground stem tuber
Sweet potato	Underground root tuber and above ground vines
Elephant-foot yam	Corm, corm pieces and cormel
Taro	Cormel
Garlic	Clove (bulblet), bulbil

Pointed gourd	Fleshy root and vine cuttings
Giant spine gourd	Fleshy root
Cassava	Stem cutting
Drumstick	Limb cutting
Jerusalem artichoke	Root tuber
Globe artichoke	Root suckers or offshoots and division of old crown
Yam	Pieces of stem tuber and vine cuttings
Rhubarb	Division of crowns

Types for variation in chromosome number of vegetables

Type	Number	Description
Euploids		
Monoploids	n	Genetically hemizygous
Diploids	2n	Two homologous sets
Polyploids		
Triploids	3n	Three homologous sets
Tetraploids	4n	Four homologous sets
Hexaploid	6n	Sex homologous sets
Aneuploids		
Hypoploids	2n – x	Loss of one or more chromosomes
Nullisomics	2n – 2	Loss of homologous pair
Monosomic	2n – 1	Loss one member of pair
Double monosomic	2n – 1 – 1	Loss of two non-homologous chromosomes
Hyperploids	2n + x	Addition of one or more chromosomes
Trisomic	2n + 1	Addition of one chromosome of a pair
Tetrasomic	2n + 2	Addition of two chromosomes of a pair
Double tetrasomic	2n + 2 + 2	Addition of two chromosomes of two separate pair

Soil Temperature Requirement for Seed Germination

Crop	**Soil Temperature**			
	Minimum	**Optimum range**	**Optimum**	**Maximum**
Tomato	10.0	15.5-25.0	25.0	30.0
Brinjal	15.5	24.0-32.0	28.0	35.0
Chilli	15.5	18.0-35.0	29.0	35.0
Potato	10.0	15.0-25.0	25.0	30.0
Cabbage	4.0	10.0-30.0	20.0	37.8
Cauliflower	4.0	10.0-30.0	20.0	37.8

Radish	4.0	7.0-32.0	28.0	35.0
Carrot	4.0	7.0-26. 0	24.0	35.0
Turnip	4.0	15.0-35.0	29.0	40.0
Beet	4.0	10.0-29.0	26.0	35.0
Parsnip	1.6	10.0-21.0	18.0	29.0
Asparagus	10.0	25.0-30.0	30.0	35.0
Muskmelon	15.5	24.0-35.0	30.0	37.8
Cucumber	15.5	15.5-35.0	30.0	40.5
Watermelon	15.5	21.0-35.0	30.0	40.5
Summer squash	15.5	21.0-35.0	30.0	37.8
Pumpkin	15.5	21.0-32.0	30.0	37.8
Bitter gourd	15.5	20.0-35.5	32.0	37.8
Ridge gourd	15.5	21.0-35.5	30.0	40.5
Sweet corn	10.0	15.5-35.0	35.0	40.5
Okra	17.0	21.0-35.0	29.0	40.5
Pea	4.4	4.4-24.0	22.0	29.4
French bean	15.5	15.5-29.0	26.0	35.0
Lima bean	15.5	18.0-29.0	26.0	29.0
Cowpea	15.5	15.5-29.0	28.5	40.5
Hyacinth bean	15.5	15.5-29.0	28.0	35.0
Cluster bean	15.5	15.5-35.0	29.0	35.0
Palak	10.0	15.0-30.0	25.0	35.0
Amaranth	10.0	15.0-35.0	32.0	35.0
Spinach	1.6	7.2-23.9	21.1	29.4
Celery	4.4	15.5-21.1	21.1	29.4
Swiss chard	4.4	10.0-29.4	29.4	35.0
Lettuce	1.6	4.4-26.6	23.9	29.4
Parsley	4.4	10.0-29.4	23.9	32.2
Onion	1.6	10.0-35.0	23.9	35.0
Drumstick	15.0	20.0-35.0	29.0	40.0

Names and Structure of Some Plant Growth Regulators Used in Vegetable Production

Abbreviation or Common Name	Chemical Name or Structure	Major uses
ABA	Abscisic acid	Defoliant
Chlormequat (CCC)	2-chloroethyl-trimethyl ammonium chloride	LP, SR
3-CPA	2-(3-chlorophenoxy) propionic acid	FT
4-CPA	4-chlorophenoxy-acetic acid	FT
2,4-D	(2,4-dichlorophenoxy) acetic acid	Herbs.
Daminozide (B 9, Alar)	Succinic acid-2, 2-demethyl hydrazine	GR
Ethephon	2-chloroethyl phosphonic acid	SR
GA_3	Gibberellic acid	GE
Kinetin	6-furfuryl aminopurine	DB
IAA	Indole-3-acetic acid	Enlarge.
IBA	Indole-3-butyric acid	Enlarge.
Maleic hydrazide	1,2-dihydro-3, 6-pyridazinedione	Herb., GR
Mepiquat chloride	N,N-dinethyl piperidinium chloride	GR
NAA	1-naphthaleneacetic acid	FT, GR
NOXA BNOA	2-naphthalenyloxyacetic acid	GS
Paclobutrazol	(2RS, 3RS)-1-(4-chlorophenyl)-4, 4-dimethyl-2-(1 ,2,4-triazol-1 -yl)-pentan-3-ol	GR
2,4,5-T	(2,4,5-trichlorophenoxy) acetic acid	Herb.
TIBA	2,3,5-triiodo-benzoic acid	GR
Uniconazol	(E)-(4-chlorophenyl)-4, 4-dimethyl-2-(1,2,4-triazol-1 -yl)-1-penten-3-ol	GR, Fung.

Abbreviations: DB-Dormancy breaker; Enlarg-Plant cell enlarger; FT-Fruit thinner; Fung-Fungicide; GR-Growth retardant; GS-Growth stimulant; Herb -Herbicide; LP-Lodging preventor; GE-Grape enlarger; SR-Sugarcane ripener.

Mode of application and concentration of plant growth regulation in vegetable production.

S. No.	Growth Regulators	Conc- (mg/1)	Method of applications	Crops	Attributes affected
1	Cycocel (CCC)	250-500	Foliar Sprays	Cucurbits Tomato, Okra	Flowering sex-expression fruit yield.
		250	Seed treatment	Okra	Resistant to salt. Fruiting and yield.
		250	Seedling Treatment	Tomato	Resistant to virus, fruiting and yield.

2.	Isopropl-N	5000	Dust or tuber dip	Potato	Storage at room temperature.
3	p-Chlophenoxy	50	Foliar spray	Tomato	Fruit set, fruiting and yield
4	2.4-Dichloro-phenoxy acetic acid	0.5	Foliar spray	Tomato	Fruit set, fruiting and yield
5	Ethephon (CEPA)	100-500	Foliar spray	Cucurbit, Okra, Tomato	Flowering ,Fruiting sex-expression and yield.
		1000	Pre-harvest	Tomato Chillies	Fruit ripening earliness and yield
		2000	Post-harvest	Tomato Chillies	Fruit ripening
6	Ethylene chlorohydrin	(11/20g)	Vapour treatment	Potato	Breaking dormancy
7	Gibberellic acid (GA)	1	Tuber dip.	Potato	Breaking dormancy
		10	Foliar sprays	Watermelon, tomato	Sex expression, fruiting and yields
		40-100	Seed or foliar	Okra ,Tomato Brinjal	Seed germination fruit set and yield.
8	Indole acetic acid (IAA)	10-15	Seed or Foliar	Spinach, Fenugreek,Okra Tomato ,Brinjal Cowpea, Onion	Seed germination fruit set and yield
9	Indolebutyric acid (IBA)	0.2	Seedling roots dip	Cabbage Cauliflower	Seedlings growth and yield
		25-100	Foliar sprays	Lettuce, Chinese Cabbage	Growth and yield
		250	Foliar Sprays	Cow Pea	Fruit set and yield
10	Maleic hydrazide (MH)	25	Foliar sprays	Peas	Growth and yield
		50-150	Foliar sprays	Cucurbits	Flowering sex-expression and yield
		2500-3000	Foliar Sprays	Onion, Garlic	Reducing storage losses
11	Methyl ester of NAA	5000	Foliar Sprays	Tomato Chilli, Brinjal Potato	Flowering, fruit set and yield.
12	Naphthalene acetic acid (NAA)	0.2	Seedling, root dip	Cauliflower Cabbage Tomato Brinjal Onion Chillies	Seedling set, growth and yield
		10-20	Foliar sprays	Tomato Chillies	Flower drop,fruit set and yield
		25-30	Seed/Foliar	Okra, Tomato Brinjal, Onion Cucurbits	Seed germination growth and yield
		100	Foliar sprays	Cabbage Cauliflower	Plant growth and yield

14	Napthoxyacetic acid	25-100	Seed/Foliar	Tomato Okra	Germination ,growth and yield
15	2,4,5-Trichlorophenoxyacetic acid (2,4,5-T)	75-125	Foliar sprays	Potato	Improving shelf life
16	Silver nitrate	500	Foliar sprays	Cucumber	Induction of male flowers in gynoeceous lines
17	Silver thio Sulphate	400	Foliar sprays	Muskmelon	Induction of male flowers in gynoeceous lines
18	2,3,5-Triiodobenzoic acid (TIBA)	25-50	Foliar sprays	Cucurbits	Flowering, sex-expression and yield
		700	Foliar sprays	Soybean	Fruit set and yield
19	Triacontant	2	Foliar sprays	Chillies, Peas	Fruit set and yield
20	Thiourea	100	Tuber dip	Potato	Breaking dormancy

ɪant parts to be sampled and Growth stage of the crop for Plant Tissue Nuient Analysis

Crop	Plant parts to be sampled	Stage of growth
Potato	Fourth to sixth leaf from growing tip	Early growth (35-40 days after planting)
Tomato	Fourth to sixth leaf from growing tip	Early bloom
Chilli and Sweet pepper	Young mature leaves	Early fruit set
Cauliflower	Young mature outside leaves	Button stage of curd
Cabbage	First mature leaf from central whorl	Prior to heading
Broccoli	Young mature leaves	First bud formation stage
Brussels sprouts	Young mature leaves	Midgrowth
Watermelon	Mature leaves near the base	Early growth
Muskmelon, Cucumber, Pumpkin, etc.	Portion of plant on main stem	Prior to fruit set
Beans (French bean, Cowpea, Lima bean, Hyacinth bean, etc.)	Two to three fully develop leaves at top of the plant	Initial flowering
Pea	Leaves from third node down from top of the plant	Initial flowering
Root crops (Carrot, Radish, Beet, Turnip)	Young mature leaves from central rings	Prior to root enlargement
Bulb crops (Onion, Garlic etc.)	Young mature leaves from central	Prior to bulbing

Celery	Petiole of youngest fully elongated leaf	Midgrowth (30-35 cm tall)
Lettuce (head type)	Middrib of wrapper leaf	Heading
Lettuce (leaf type)	Youngest mature leaf	Midgrowth
Leafy greens (Palak, Spinach, etc.)	Youngest mature leaf	Midgrowth
Sweet potato	Fourth to sixth leaf from the growing tip	Prior to root enlargement
Asparagus	Top 10 cm of the new fern branch	Midgrowth of the fern
Sweet corn	Entire leaf at the ear node	Tasseling

Potential Vegetables for export and quality specifications:

Item	Specifications
Onion	
Big	Light to dark red, globose, 4-6 cm in size and strongly pungent (for Banglades South-East Asia and Gulf countries): yellow or brown, globose and 4-6 cm i size (for European and Japan markets).
Multiplier	Bigger cloved bulb of bright red colour
Small	Dark red, round, 2-3 cm in size
Okra	Green, tender, 6-8 cm, 4-5 ridges, slender and roundish with pointed tips.
Potato	White-skinned long and extra-long tubers with yellow to light yellow fles (for Iraq and Iran); both white and red types of oval shape and 4-5 cm i diameter size (for Bangladesh).
Bitter gourd	Complete green, lustrous, more prickly, medium long and spindle shaped.
Chillies	Slender, dark green, lustrous, 7-8 cm, smooth without wrinkles and bends.
Garlic	White, round, 3.5-4 cm having 15-20 big-sized cloves.
Gherkin	Green, small sized (7.5 cm x 2.5 cm), 160-300 fruits/kg is considered a premium grade.
Brinjal	Round, thornless, shiny purple with dark green calyx.
Baby corn	Yellow seeded, thick walled, predominantly lobed, firm and shining green.
Lettuce	Round, medium to large head (750g to 1.0 kg) with wrinkled and green no wrapper leaves, brittle textured and tight folded.
Cabbage	Round to flat-round head with dark green, smooth outer leaves, solid, sho cone.
Bottle gourd	Green, tender, cylindrical, 20-25 cm with soft hairs.
French bean	Tender, green, 15 cm in length
Tomato	Round, medium-sized and red in colour
Cluster bean	Green, tender, 7-10 cm in length

'arieties Suitable for export

Crop	Variety
Onion: Big	N-53, Nasik Red, Agrifound Dark Red, Agrifound Light Red, Arka Niketan (for Gulf and other countries); Arka Pitambar, Granex-33 and Spanish Brown (for Japan and Europe); Arka Bindu (for dried onion).
Okra	Parbhani Kranti, Arka Anamika, Arka Abhay, Varsha Uphar, Pusa Sawani, Kufri Sinduri, Kufri Jyoti (for Bangladesh)
Bitter Gourd	Arka Harit, Pusa Vishesh, MBTH-101, Hirkani, Konkan Tara, Prethi
Chilli	Bhagyalakshmi (G-4), Andhra Jyoti (G-5), G-3, Pant C-1, Arka Lohit, X-235, Panvel, Ansari, Musalwadi
Garlic	G-41, Jamnagar Ladwa, Malek. Agrifound Parvati, Yamuna Safed.
Bottle gourd	Arka Bahar, Pusa Naveen, Varad, Samrat (Hybrid)
Tomato	Round-fruited hybrids
Brinjal	Pant Rituraj, Pusa Uttam, Pusa Upkar, Pusa Kranti, Arka Navneet and Pusa Anmol (hybrid)
Gherkin	Ajax, Calypsoplus, BSS-168, BSS-169 (Hybrid)
Baby corn	VL42 (OP), MEH 114 (Hybrid), COBC-1 (Synthetic).
Sweet pepper	Bharath, Green Gold, Indira, Hira, Lario (Hybrid).
Watermelon	Sugar Baby, Arka Manik
Cluster bean	Pusa Sadabahar, Pusa Musami, Pusa Naubahar
French bean	Contender, Masterpiece, Premier, Arka Komal
Elephant-foot yam	Kovour, Gajendra

Seed related facts

Physical purity test: Physical purity denotes the percentage (by weight) of seeds belonging to the variety under certification. The working sample from the seed lot is classified into the components like (a) pure seed i.e., seeds of the variety under certification, (b) seeds of other varieties of the same crop, (c) seeds of other crops, (d) seeds of weeds and (e) inert matters like sand, straw, stone, pebbles, soil particles, etc. So, by purity analysis, the composition of the sample is ascertained and the results are expressed as weight percentage.

$$\textbf{Purity (\%)} = \frac{\text{Weight of pure seed}}{\text{Total weight of working sample}} \times 100$$

Similarly, weight percentage of the impurities can be determined separately.

Determination of real value of seed: Real value of the seed ascertains the percentage of seed of the working sample of the seed lot that would produce normal seedlings of the variety under certification.

$$\textbf{Real value of seed (\%)} = \frac{\text{Purity (\%)} \times \text{Germination (\%)}}{100}$$

Storage of certified seeds: The storage period ranges from 8-9 months. If the see are stored in moisture vapour permeable containers like, cloth bag, paper packet, multiw paper bags, etc. the safe seed moisture content is around 9 per cent at ambient temperatur If the seeds are stored in moisture vapour-proof containers like, tin cans, low densi polyethylene bags, rigid plastic containers, aluminium foil pouches, etc. the seed moistu content must be within 4-6 per cent. Short-lived seeds like onion, French bean are al stored in cold storage condition. So, adequate facilities should be available for seed dryi so that such low seed moisture content is kept in the seeds. Generally, hybrid seeds a packed in moisture vapour-proof containers.

Storage of germplasm seeds: Germplasm seeds are required to be kept for many yea Basic requirements of such long-term storage are to provide cold temperature and lo relative humidity in the storage atmosphere. Properly dried seeds to perfect moisture lev of below 8 per cent is preferably stored at 5^0 to 10^0C and 30% relative humidity conditio

Storage of breeder or foundation stock seeds: Such stock seeds need to be stored f 4-5 years to minimize genetic drift in the varietal population upon reproducing every ye Properly dried seeds having around 9 per cent moisture under moisture vapour permeab packaging should be stored under a storage atmosphere with 25% relative humidity 30^0C or less or 45% relative humidity at 20^0C or less temperature. The required relati humidity in the storage room can be achieved by making the room moisture-proof and l using dehumidifier.

Classification of seed and tag colour

S. No.	Category of seed	Description	Colour of tag
	Nucleus seed	The seed or the research material produced the breeder by various selection procedures in a pure line variety or clone is known as nucleus	-
	Breeder seed	Breeder seed is produced form nucleus seed under the supervision of a qualified plant breeder in a research institute of Agricultural University	Yellow
	Foundation seed	Foundation seed is the progeny of breeder seed and is produced by State Farm Corporation of India. National Seed Corporation, State Seed Corporation under technical control of qualified plant breeders or technical officers.	White
	Certified seed	Certified seed is the progeny of foundation seed and its production is supervised and approved by certification agency.	Azar blue
	Labelled seed	There is another category of seed known as Labelled Seed in which case there is no certification but labelling is compulsory.	Opel green

Effect of seed moisture content on seeds in storage

Seed moisture content	Effect on seed
33-60%	Germination occurs on imbibition, quick loss of viability
18-20%	High respiratory rate, high heating with inadequate ventilation, high activity of moulds and storage pests, seeds susceptible to mechanical damage in handling
13-17%	Still high respiratory rate, moulds and insects can be damaging, fumigation can affect seed viability, resistant to mechanical damage
10-12%	Seeds can be stored reasonably well for 6-18 months in open storage in temperate climate, insects can be a problem, fumigation can safely be done.
8-10%	Seed can safely be stored for 1-3 years in open storage, very little insect activity, seeds susceptible to mechanical damage.
4-8%	Safe moisture content for sealed storage
0-4%	Excessive desiccation is damaging to seeds, seeds may lose viability.

Approximate moisture content of vegetable seeds in equilibrium in air having different relative humidities at room temperature (20^{0}C)

Seeds of the vegetable				Per cent relative humidity in air		
	10	20	30	45	60	75
Pea	5.4	7.3	8.6	10.1	11.9	15.0
Broad bean	4.2	5.8	7.2	9.3	11.1	14.5
Lima bean	4.6	6.6	7.7	9.2	11.0	13.8
French bean	3.0	4.8	6.8	9.4	12.0	15.0
Cabbage	3.2	4.6	5.4	6.4	7.6	9.6
Chinese cabbage	2.4	3.4	4.6	6.3	7.8	9.4
Radish	2.6	3.8	5.1	6.8	8.3	10.2
Carrot	4.5	5.9	6.8	7.9	9.2	11.6
Beet	2.1	4.0	5.8	7.6	9.4	11.2
Turnip	2.6	4.0	5.1	6.3	7.4	9.0
Tomato	3.2	5.0	6.3	7.8	9.2	11.1
Brinjal	3.1	4.9	6.3	8.0	9.8	11.9
Chillies	2.8	4.5	6.0	7.8	9.2	11.0
Cucumber	2.6	4.3	5.6	7.1	8.4	10.1
Winter squash	3.0	4.3	5.6	7.4	9.0	10.8
Watermelon	3.0	4.8	6.1	7.6	8.8	10.4
Okra	3.8	7.2	8.3	10.0	11.2	13.1

Onion	4.6	6.8	8.0	9.5	11.2	13.4
Welsh onion	3.4	5.1	6.9	9.4	11.8	14.0
Parsnip	5.0	6.1	7.0	8.2	9.5	11.2
Celery	5.8	7.0	7.8	9.0	10.4	12.4
Lettuce	2.8	4.2	5.1	5.9	7.1	9.6
Spinach	4.6	6.5	7.8	9.5	11.1	13.2
Mustard	1.8	3.2	4.6	6.3	7.8	9.4
Sweet corn	3.8	5.8	7.0	9.0	10.6	12.8

Temperature range for multiplication of the organisms associated with seed storage

Organism	Temperature range (^{0}C)		Relative humidity %
	For multiplication	Optimum	
Insects	21-42	27-37	30-95
Mites	8-31	19-31	60-100
Fungi	8-80	20-40	60-100
Bacteria	8-80	26-28	91-100

Methods of germination test of the seeds of some vegetable crops

Crop	Substrate	Temp(^{0}C)	First count (days)	Final count (days)	Additional recommendation to break dormancy
Okra	TP; BP	20-30	4	21	-
Onion	BP; TP	20	6	12	Pre-chill at 5^{0}C
Beet	TP; BP	20-30	4	14	Pre-wash 2-4 hours i running water
Cabbage	TP	20-30	5	10	Pre-chill, soaking the pape in 0.2% KNO_3 solution
Cauliflower	TP	20-30	5	10	Do
Chilli	TP, BP	20-30	7	14	Soaking paper in 0.2% KNO_3 solution
Muskmelon	BP	20-30	4	8	-
Cucumber	TP;BP	20-30	4	8	-
Winter squash	BP	20-30	4	8	-
Pumpkin	BP	20-30	4	8	-
Summer squash	BP	20-30	4	8	-
Carrot	TP; BP	20-30	7	14	-
Lettuce	TP; BP	20	4	7	Pre-chill at 5^{0}C

Tomato	TP; BP	20-30	5	14	Soaking the paper in 0.2% KNO_3 solution
Pea	BP	20	5	8	-
Brinjal	TP; BP	20-30	7	14	-
Broad bean	BP	20	4	14	-
Cowpea	BP	20-30	5	8	-

BP: Between two or more layers of germination or filter paper

TP: Top of the one or more layer of filter or germination paper

Note: Potassium nitrate (KNO_3) solution at 0.2% concentration is to be applied to the papers (substrate) at the point of saturation only at the beginning of the test. Subsequent moistening is to be done normally with water.

Seed standards of genetic purity

Class	Standards for Minimum Genetic Purity (%)
Foundation	99.00
Certified	
Varieties, composites, synthetics & multilines	98.00
Hybrids	95.00
Hybrids of cotton, TPS, muskmelon, brinjal & tomato	90.00
Hybrid castor	90.00

Important rouging stages and permitted off types for production of different vegetable seeds

S. No.	Vegetable crops	Minimum number of field inspections and rouging states	Off type (%) Maximum permitted	
			C	F
	Cowpea	2: (JRF, 2014) From flowering to fruiting	0.10*	0.20
	Garden pea	3: (JRF, 2014) first before flowering, second at flowering and third at edible pod stage	0.10*	0.50*
	Chilli and Capsicum	3: (JRF, 2015) First before flowering, second 8 flowering and third at maturity fruit stage	0.10*	0.20*
	Cauliflower, Cabbage and Knol khol	4: (JRF, 2012) First before the marketable stage, second start of curd head formation, third when most plants have formed curd and forth at flowering stage	0.10*	0.20*

	Lettuce	3: (JRF, 2014) first before full grown stages in non-heading types, second full-grown stage inn non-heading type and third at flowering	0.10*	0.20*
	Carrot	4: (JRF, 2017) First 20 - 30 days after sowing, second at lifting and replanting, third at flowering and fourth at maturity to verify the true nature of umbels	0.10*	0.20**
	Onion	4: (JRF, 2016) (When seed crop is raised by the transplanting method) First early (20 - 30 days after sowing), second when bulbs are lifted, third when bulbs are replanted and fourth at flowering 3: (JRF, 2015) (When seed crop is raised by seed-to-seed method) First 20 – 30 days after sowing, second when bulbs are formed and third at flowering	0.10**	0.20**
	Radish and Turnip	3: (JRF, 2014) First 20 -30 days after sowing, second when lifted and replanted and third at flowering	0.10**	0.20**
	Okra	3: (JRF, 2010) First before flowering, second at full flowering and fruiting and third at mature fruit stage	0.10**	0.20**
	Tomato	3: (JRF, 2011) First before flowering, second during flowering and the immature fruit stage and third at mature fruit stage	0.10**	0.20**
	Brinjal	3: (JRF, 2007) First before flowering, second during flowering and the immature fruit stage and third at mature fruit stage	0.10**	0.20**
	Cucurbits	3: (JRF, 2013) First before flowering, second during flowering and the immature fruit stage and third at mature fruit stage	0.10**	0.20**
	Spinach beet (Palak)	3: (JRF, 2018) First at cotyledon leaf stage, second during vegetable growth, third at bolding and flowering	0.10**	0.20**

*Maximum permitted at the final inspection; **Maximum permitted at and after flowering F= Foundation seed; C+ Certified seed.

Seed standards for different vegetables

Crop	Class of seed	Germination (%)	Pure seed (%)	Inert matter (%)	Other crop seed Max.	Weed seed Max.	Moisture (%)	
		(Min.)	(Min.)	(Max.)	(no/kg)	(no/kg)	Ordinary pack	Vapour proof pack
Tomato	FS	70	98	2	5	None	8	6
	CS	70	98	2	10	None	8	6
Brinjal	FS	70	98	2	None	None	8	6
	CS	70	98	2	None	None	8	6
Chilli	FS	60	98	2	5	5	8	6
Capsicum	CS	60	98	2	10	10	8	6
	FS	65	99	1	None	None	10	8
	CS	65	99	1	5	5	10	8
Cabbage	FS	70	98	2	5	5	7	5
	CS	70	98	2	10	10	7	5
Cauliflower	FS	65	98	2	5	5	7	5
Knol khol	CS	65	98	2	10	10	7	5
Radish	FS	98	98	2	5	5	6	5
	CS	98	98	2	10	10	6	5
Carrot	FS	60	95	5	5	5	8	7
	CS	60	95	5	10	10	8	7
Beet	FS	60	96	4	5	5	9	8
	CS	60	96	4	10	10	9	8
Spinach	FS	60	96	4	5	5	9	8
	CS	60	96	4	10	10	9	8
Amaranthus	FS	70	95	5	5	10	8	6
	CS	70	95	5	10	20	8	6
Lettuce	FS	70	98	2	None	5	8	7
	CS	70	98	2	None	10	8	7
Onion	FS	70	98	2	5	5	8	6
	CS	70	98	2	10	10	8	6
Garden pea	FS	75	98	2	None	None	9	8
	CS	75	98	2	5	None	9	8
Cowpea	FS	75	98	2	5	5	9	8
	CS	75	98	2	10	10	9	8
French bean	FS	75	98	2	None	None	9	7
	CS	75	98	2	None	10	9	7

Cucurbits	FS	75	98	2	-	-	7	6
	CS	75	98	2	-	-	7	6

Recommended Storage life for Different Vegetables

Vegetables	Temperature (^{0}C)	Relative humidity (%)	Storage life (weeks)
Brinjal	10.0-11.1	92	2-3
Tomato (unripe)	8.9-10.0	85-90	4-5
Tomato (ripe)	7.2	90	1.0
Chilli (green)	7.2	85-90	3-5
Chilli (ripe)	5.6-7.2	90-95	2.0
Potato	3.0-4.4	85	34.0
Cabbage (early)	0.0-1.7	92-95	4-6
Cabbage (late)	1.0-1.7	92-95	12.0
Cauliflower (late)	0.0-1.7	85-95	7.0
Bitter gourd	0.6-1.7	85-90	4.0
Cucumber	10.0-11.7	92	2.0
Muskmelon	1.7-3.3	85-90	1.5
Pumpkin	10.0-12.0	70-75	24-36
Winter squash	12.8-15.6	70-75	24-36
Watermelon	7.2-15.6	80-90	2.0
Beet (topped)	0.0-1.7	90-95	8-14
Beet (bunched)	0.0	90	1.5
Carrot (topped)	0.0	95	20-24
Radish (topped)	0.0	88-92	3-5
Turnip	0.0	90-95	8-16
Hyacinth bean	0.6-1.7	90	3.0
Lima bean	4.4-7.2	90-95	1.5-2.0
Pea	0.0	88-92	2-3
Garlic	0.0	65	28-36
Onion (bulb)	0.0	70-65	20-24
Onion (leaves)	0.0	90-95	2.0
Taro	11.1-12.8	85-90	21
Cassava	0.0-1.7	85	23
Sweet potato	10.0-12.8	80-90	13-20
Yam	26.7	66-70	3-5
Ginger	7.2-10.0	75	16-24

Okra	8.9	90	2.0
Sweetcorn	0.6-1.7	90-95	1.0
Celery	0.0-0.6	92-95	8.0
Lettuce (head type)	0.0	90-95	3.0
Lettuce (leaf type)	0.0	95	1.0
Coriander leaves	0.0-1.7	90	5.0
Asparagus	0.0	95	3-4

Prescribed Field Standards (Isolation Requirements, Minimum Field Inspections and Maximum Permitted off-Types) for Vegetables Seed Production

Crops	Minimum isolation distance (m)			Minimum field inspections		Maximum permitted off-types (%)	
	F	C	To be isolated from	Number	Stages of crop growth	F	C
Potato	5	5	Other varieties	3	45 days after planting, just before haulm cutting, after haulm cutting	0.2	0.5
Tomato	50	100	Other varieties	3	Before flowering, flowering to immature fruit stage, mature fruit stage	0.1	0.1
Brinjal	200	100	Other varieties	3	Before flowering, flowering to immature fruiting, mature fruit stage	0.1	0.2
Chilli	400	200	Other varieties and chillies from sweet pepper	3	Before flowering, flowering to fruiting, mature fruit stage	0.1	0.2
Sweet pepper	400	200	Other varieties and sweet pepper from chillies	3	Before flowering, flowering to fruiting, mature fruit stage	0.1	0.2
Okra	400	200	Other varieties and wild *Abelmoschus spp.*	3	Before flowering, flowering to fruiting, mature fruit stage	0.1	0.2
Pea	20	10	Other varieties	3	Before flowering, flowering, edible pod stage	0.1	0.2
Other beans like cowpea, French bean, cluster bean, hyacinth bean, etc.	50	25	Other varieties of the crop	2	Before flowering, flowering to fruiting	0.1	0.2
Cabbage	1600	1000	Other varieties of the crop and other crops like, cauliflower, Brussels sprouts, knolkhol, kale, collards	3	Head formation stage, fully developed head stage, flowering stage	0.1	0.2

Cauliflower	1600	1000	Other varieties of the crop and other crops like, cabbage, sprouting broccoli, Brussels sprouts, knolkhol, collard, kale	4	Before initiation of curds, curd initiative on stage, fully developed curd stage, flowering stage	0.1	0.2
Knolkhol	1600	1000	Other varieties of the crop and other crops like cabbage, cauliflower, Brussels sprouts, broccoli, collards, kale	3	Before initiation of stem knob, fully developed knob stage, flowering stage	0.1	0.2
Chinese cabbage	1600	1000	Same	3	Head formation stage, fully developed head stage, flowering stage	0.1	0.2
Onion	1000	500	Other varieties	3*	20-30 days after sowing, fully developed bulb stage, flowering stage. (*seed-to-seed method)	0.1	0.2
				4**	20-30 days after sowing, after lifting of bulbs, at replanting of bulbs, flowering stage (*bulb-to-seed method)		
Radish	1600	1000	Other varieties of the crop	3	20-30 days after sowing, at lifting and replanting, flowering, (root-to-seed method)	0.1	0.2
Turnip	1600	1000	Other varieties of the crop and other crops, like Chinese cabbage, repeseed, mustard and rutabaga	3	20-30 days after sowing, at lifting and replanting, flowering, fully developed umbel stage (root-to-seed method)	0.1	0.2
Carrot	1000	800	Other varieties	4	20-30 days after sowing, at lifting and replanting, flowering, fully developed umbel stage (root-to-seed method)	0.1	0.2
Beet	1600	1000	Other varieties of the crop and other crops, like palak, sugar beet, swiss chard	3	20-30 days after sowing, at lifting and replanting, flowering	0.1	0.2

Amaranth	400	200	Other varieties and wild *Amaranthus* spp.	2	Fully developed stage before flowering, flowering stage	0.1	0.2
Palak	1600	1000	Other varieties of the crops and other crops like beet, sugar beet, Swiss chard	2	Fully developed stage before flowering, flowering stage	0.1	0.2
Fenugreek	50	25	Other varieties	2	Fully developed stage before flowering, flowering stage	0.1	0.2
Lettuce	50	25	Other varieties	3	Before head formation, fully developed head stage, flowering stage (head type lettuce) and before fully grown stage, at full grown stage, flowering stage (non-head type lettuce)	0.1	0.2
Cucurbits	800	400	Other varieties and other crops Like long melon, snap melon	3	Berore flowering, flowering to immature fruiting, mature fruit stage	0.1	0.2
Muskmelon	-do-	-do-	Other varieties and other crops Like long melon, snap melon	-do-	-do-	-do-	-do-
Pumpkin	-do-	-do-	Other varieties and other crops like Summer squash and winter squash	-do-	-do-	-do-	-do-
Summer squash	-do-	-do-	Other varieties and other crops like pumpkin and winter squash	-do-	-do-	-do-	-do-
Winter squash	-do-	-do-	Other varieties and other crops like pumpkin and summer squash	-do-	-do-	-do-	-do-
Other cucurbits, like bitter gourd, bottle gourd, ridge gourd, sponge gourd, snake	-do-	-do-	Other varieties of the crop	-do-	-do-	-do-	-do-

Specified Seed Standards for different Vegetable Crops

Crop	Minimum physical purity (%)		Minimum germination (%)		Maximum moisture (%) in ordinary container		Maximum moisture (%) in vapour-proof container	
	F	C	F	C	F	C	F	C
Tomato	98	98	70	70	8	8	6	6
Brinjal	98	98	70	70	8	8	6	6
Chilli	98	98	70	70	8	8	6	6
Sweet pepper	98	98	70	70	8	8	6	6
Okra	99	99	65	65	10	10	8	8
Pea	98	98	75	75	9	9	8	8
Cowpea	98	98	75	75	9	3	-	-
Hyacinth bean	98	98	75	75	9	9	7	7
French bean	98	98	75	75	9	9	7	7
Cluster bean	98	98	70	70	9	9	-	-
Cabbage	98	98	70	70	7	7	5	5
Cauliflower	98	98	65	65	7	7	5	5
Knolkhol	98	98	70	70	7	7	5	5
Chinese cabbage	98	98	70	70	7	7	5	5
Onion	98	98	70	70	8	8	5	5
Radish	98	98	70	70	6	6	5	5
Turnip	98	98	70	70	6	6	5	5
Carrot	95	95	60	60	8	8	7	7
Beet	96	96	60	60	9	9	8	8
Amaranth	95	95	70	70	8	8	6	6
Palak	96	96	60	60	9	9	8	8
Fenugreek	98	98	70	70	8	8	-	-
Lettuce	98	98	70	70	8	8	6	6
Cucumber	99	99	60	60	7	7	6	6
Muskmelon	99	99	60	60	7	7	6	6
Longmelon	99	99	60	60	7	7	6	6
Watermelon	99	99	60	60	7	7	6	6
Pumpkin	99	99	60	60	7	7	6	6
Summer squash	99	99	60	60	7	7	6	6
Winter squash	99	99	60	60	7	7	6	6
Bitter gourd	99	99	60	60	7	7	6	6
Bottle gourd	99	99	60	60	7	7	6	6
Ridge gourd	99	99	60	60	7	7	6	6

Sponge gourd	99	99	60	60	7	7	6	6
Snake gourd	99	99	60	60	7	7	6	6
Round melon	99	99	60	60	7	7	6	6

F = Foundation and C = Certified seed.

Breeding methods of vegetable improvement

Self-pollinated crops: In these crops, pollination involves the stigma of the sam flower that produces the pollens or other flowers of the same plant, which effect natura self-fertilization. In these crops, natural cross-pollination is normally less than 10 per cent Sexform of these crops are hermaphrodite e.g., potato, tomato, brinjal, chilli, sweet pepper peas, cowpea, hyacinth bean, French bean, cluster bean, runner bean, fenugreek, lettuce etc.

Cross-pollinated crops: In these crops, pollens from the flowers of one plant i transferred by some agents (insects, wind, etc.) to the stigma of the flowers of anothe plant to effect pollination. Cross-pollination is indicative of high efficiency of reproductio systems based on cross-fertilization. In cross-pollinated vegetable crops, natural cross pollination generally exceeds 60 per cent to the extent of even 100 per cent. Sexform different adaptations and some inherent mechanisms favour cross-pollination. Differen cross-pollinated vegetable crops are cabbage, cauliflower, knolkhol, Brussels sprouts sprouting broccoli, radish, beet, turnip, carrot, onion, leek, amaranth, palak, spinach asparagus, pumpkin, bottle gourd, watermelon, muskmelon, ridge gourd, snake gourd wax gourd, summer squash, bitter gourd, cucumber, sweet corn, pointed gourd, giant spin gourd, cassava, sweet potato, coriander, parsnip, etc.

Often-cross-pollinated crops: These crops are predominantly self-pollinated bu natural cross-pollination normally exceeds 25 %. Okra, lima bean, chilli, sweet pepper an potato are sometimes regarded as often-cross-pollinated crops due to high natural cross pollination.

Floral organization favouring self-pollination: Bisexual flower structure an homogamy (stigmas and stamens maturing at the same time) are the main aspects favourin self-pollination. In vegetable crops normally homogamy occurs after flowers have opene (Chasmogamy). Homogamy in closed flowers (Cleistogamy) is not generally met with i vegetable crops. In peas and bean, stamens and pistil in opened flower remain envelope inside a tube-like structure of jointed petals called keel.

Floral organism favouring cross-pollination:

(a) Sexform favoring cross-pollination:

Monoecy: Male and female flowers are borne separately in the same plant. Thi unisexualism favours cross-pollination e.g., most of the cucurbitaceous vegetable crops amaranth, sweet corn, cassava, etc. In muskmelon, most of the varieties are andromonoeciou (separate male and hermaphrodite flowers in the same plant) while some are monoecious

Dioecy: Male and female flowers are borne on separate plants meaning that the mal and female plants are separate e.g., pointed gourd, asparagus, spinach, giant spine gourd yam, etc.

(b) Inherent mechanisms favouring cross-pollination: Despite hermaphrodite sexform, self-fertilization is hindered favouring cross-pollination due to some genetically controlled mechanisms.

Self-incompatibility: It is genetically controlled physiological hindrance to self-fertilization, which marks inability of the functional pollens to fertilize the vaule after self-pollination. It is found in cabbage, cauliflower, knolkhol, Brussels sprouts, radish, turnip and beet.

Male sterility: In this phenomenon, male parts of the flower are non-functional without affecting the female fertility. It is controlled by nuclear genes or sterile cytoplasm or by the interaction between nuclear gene and sterile cytoplasm. It is found in tomato, brinjal, chilli, muskmelon, onion, carrot, sprouting broccoli, Brussels sprouts, cabbage, cauliflower, sweet potato, etc.

Dichogamy

Protandry: In protandrous flowers, pollens are shed before stigma becomes receptive e.g., onion, carrot, beet, parsley, leek, etc.

Protogyny: In protogynous flowers, stigma becomes receptive before pollens are shed from the anthers of the same flower e.g., elephant-foot yam.

Breeding methods of vegetable improvement

Conventional Methods	Advance Breeding techniques	Non conventional methods
•Introduction • Pure line selection • Mass selection • Pedigree method • Single seed descent method • Back cross method • Bulk method • Recurrent selection • Heterosis breeding • Synthetic breeding • Clonal Selection etc.	• Mutation breeding • Polyploidy breeding	• Genetic Engineering • Molecular breeding (MAS) • Tissue culture • Somatic hybridization etc.

Breeding of self-pollinated vegetable crops introduction of elite varieties

Crop	Variety	Introduced from
Tomato	Roma	USA
	Labonita	USA
	Sioux	USA
	Marvel	USA
	Best of All	USA
	Money Maker	USA

Sweet pepper	California Wonder	USA
	Yolo Wonder	USA
	World Beater	USA
	Chinese Giant	USA
	Golden Cal Wonder	USA
	Bullnose	USA
Okra	Perkins Long Green	Exotic
Pea	Early Superb	UK
	Meteor	UK
	Arkel	UK
	Little Marvel	UK
	Alaska	UK
	Early Badger	USA
	Bonneville	USA
	Lincon	USA
	Alderman	USA
	Perfection New Line	USA
	Sylvia	Sweden
French bean	Contender	USA
	Gaint Stringless	USA
	Ketucky Wonder	USA
	Bountiful	USA
	Master piece	USA
	Premier	USA
	Jampa	Mexico
Cowpea	Philippines Early	Philippines

Other Varieties through introduction in India

Crop	**Introduction**
Cucumber	Poinsette and Japnese Long Green
Onion	Early Grano, Red Grano, Bermuda Yellow
Cauliflower	Snowball-16, Improved Japanese, Early Snowball
Cabbage	Golden Acre, Drum Head
Radish	Japanese White, Rapid Red White Tipped, China Red, Red Tail Radish
Carrot	Nantes, Chanteny
Turnip	Snowball, Golden Ball, Purple Top White Globe

Varieties developed through selection

Crop	Selection
Tomato	Solan Vajar, Solan Lalima, Solan Red Round, Pusa 120, HS-110
Egg Plant	Pusa Purple Long, Pusa Purple Round, Pusa Purple Cluster, Pant samrat, Arka Sheel
Chilli	NP 46 A, Sindhur, Patna Red
Cauliflower	Pusa Katki and Pusa Dipali C
Muskmelon	Hara Madhu, Arka Jeet, Arka Rajhans, MH-1
Onion	Pusa Red, Pusa Ratnar, Arka Niketan, Arka Pragati
Bottle Gourd	Pusa Summer Prolific Long, Pusa Summer Prolific Round

Radiation Dose for different vegetables

Crop	Dose (Krd)	Crop	Dose (Krd)	Crop	Dose (Krd)
Potato	5-6	Peas	15-25	Carrot	30-50
Tomato	15-25	French bean	10-25	*Brassica* sp.	60-80
Brinjal	20-30	Broad bean	10-25	Radish	70-80
Chilli	10-20	Onion	10-15	Okra	30-60
				Sweet potato	3-15

Estimated seed requirement in India (hybrid and open pollinated (0P) varieties

Crops	Area under cultivation (ha)			Seed requirement (tones)		
	OP	Hybrid	Total	OP	Hybrid	Total
Tomato	199780	299670	499450	119.87	89.91	209.77
Brinjal	384390	108420	492810	307.52	43.37	350.88
Chilli	378886	42100	420985	378.89	21.05	399.94
Cabbage	156431	156430	312860	125.15	78.22	203.36
Cauliflower	265935	46930	312860	212.75	23.47	236.21
Gourds	273640	147345	420985	1368.20	736.73	2104.92
Melons	130650	43550	174197	522.59	130.65	653.24
Okra	286730	95575	282302	2867.27	955.76	3823.03
Total	2076435	940015	3016449	5902.22	2079.13	7981.34

The commonly utilized mechanisms/methods for developing commercial hybrids

Mechanism	Commercially exploited crops
Hand emasculation + MP	Tomato, eggplant, okra, sweet pepper
Bagging/protection of staminate and pistillate flowers + MP	Cucurbits (bottle gourd, pumpkin, watermelon)
Pinching of staminate flowers + MP/NP	Cucurbits (bitter gourd, bottle gourd, pumpkin watermelon)
Male sterility + MP	Tomato, sweet pepper, chilli
Male sterility + NP	Onion, cabbage, carrot, radish, muskmelon, chilli
Self-incompatibility + NP	Cabbage, cauliflower, broccoli
Gynoecism + NP	Cucumber, muskmelon
PGR and pinching of staminate flowers + NP	Summer squash, winter squash, etc.

MP –Manual pollination NP – Natural pollination

Premium quality attributes of the hybrids

Crop/hybrid	**Targeted quality attributes**
Tomato	High temperature tolerance, better self-life, high TSS, acidity and high lycopene content.
Chilli	High oleoresin, specific fruit size, shape and pungency for various markets.
Sweet pepper	Adaptability to higher temperature condition
Cabbage	High temperature tolerance, longer staying capacity in field, narrow, short and soft core.
Cauliflower	High temperature tolerance, non ricy, compact, bract free and self-blanching curd.
Okra	Attractive light green fruit with spineless, smooth skin.
Muskmelon	High TSS content, thick flesh.
Watermelon	High TSS content, less seed, pink flesh.
Bitter gourd	Spindle shaped green fruit.
Bottle gourd	Uniform shape, non-ridged fruit surface.
Cucumber	Bunching fruit habit, finger size fruit (for gherkin), less seeds, spineless smooth skin.
Pumpkin	Fruits with high flesh recovery and beta carotene content.

Research priorities for developing multiple resistant hybrids against biotic stresses

Crops	Target for biotic stresses
Tomato	Tomato leaf curl virus + Early blight + Bacterial wilt + Root knot nematode
Brinjal	Phomopsis + Bacterial wilt + Fruit and shoot borer.
Chilli	Leaf curl + Thrips + Mites + Anthracnose.
Capsicum	Phytophthora + Thrips + Mites.
Okra	Yellow vein mosaic virus + Fruit borer.
Onion	Stenophyllium +Purple blotch + Thrips.
Cucumber	Downey mildew + Mosaic.
Muskmelon	Powdery mildew + Downey mildew + Anthracnose + Fusarium. + fruit fly.
Watermelon	Powdery mildew + Downey mildew + Anthracnose.
Cabbage	Black rot + Diamond back moth.
Cauliflower	Black rot + Diamond back moth.
Bittergourd	Mosaic + Fruit fly.
Bottlegourd	DM + PM
Cucumber	CMV + PM
Pumpkin	CMV.

Research Priorities for Hybrid Seed Production System and Specific Trait

Crops	F_1 seed production system*	Heterosis breeding aims
Tomato	Use of genetic cytoplasmic male sterile (CMS) and genie male sterile lines High oleoresin	Tolerant to high temperature (off-season), high TSS and high lycopene.
Chilli	Use of cytoplasmic male sterile or genetic male lines.	High oleoresin.
Capsicum	Standardization of techniques for plains (use of protected infrastructure).	Adapted to the plains of North India.
Cauliflower	Use of strong self-incompatible or CMS lines.	Heat tolerance (off-season).
Cabbage	Use of strong self-incompatible or CMS lines.	Tolerant to high temperature (tropical cabbage).
Onion	Development and use of short day CMS lines.	Photoperiod insensitive (off-season).

Carrot	Use of cytoplasmic male sterile lines.	High carotene per cent.
Watermelon	Use of male sterile lines.	High TSS and seedlessness.
Muskmelon	Use of male sterile lines.	High TSS.
Cucumber	Use. of gynoecious line.	Pickling type (for export).

*Initiative may be taken to identify, characterize and utilize conditional mutants of male sterility in all the vegetables.

F1 Hybrids of vegetables

Crop	F1 Hybrid	Genetic Mechanism
Cabbage	KGMR-1(Pusa cabbage Hybrid 1), KTCBH 51, KTCBH 81	SI
Cauliflower	Pusa Hybrid-2, Pusa Kartik Sankar	SI
Cabbage	KCH-5, Hybrid 991-5, Hybrid 854-6	CMS
Cauliflower	Hybrid 8401 ×31022	CMS
Chilli	Arka Sweta, Arka Meghna, Arka Harita, Arka Khyati, Kashi Surkh, CH-1, CH-3	CGMS and GMS
Onion	Arka Kirthiman, Arka Lalima	CMS
Carrot	Pusa Nayanjyoti, Pusa Vasudha	CMS
Cucumber	Solan Khira Hybrid-1, Solan Khira Hybrid-2	Gynoecious based F1 hybrids

Hybrids recommended for release and cultivation

Crop	Name of identified hybrids
Tomato	ARTH-4, MTH-6 ARTH-3, Pusa Hybrid-2, Arka Vardhan, NA-501, DTH-4, KT-4, NA-601, FMH-2, BSS-20, Avinash-2, HOE-303, Sun-496, DTH-8, Swarn Baibhav (CHTH-1), ARTH-128, KTH-2.
Brinjal	Arka Navneet, Pusa Hybrid-6, Pusa Hybrid-5, HRBH-201, NDBH-1, ABH-1, MHB-10, MHB-39, NDBH-6, ABH-2, Phule Ḥybrid-2, Pusa Hybrid-9, ARBH-541, PBH-6, JBH-1, BH-1, BH-2, Kashi Sandesh.
Chilli	HOE-888, ARCH-236, Sungro-86-235, ARCH-228.
Capsicum	KT-1, SEL-11, Lario, DARL-202.
Okra	DVR-1, DVR-2, DVR-3, DVR-4.
Cauliflower	Pusa Hybrid-1, DCH-541; **Synthetics**: Synthetic-1, Early Synthetic. Shri
Cabbage	Ganesh Gol, Nath-401, BSS-32, Nath-501, Quisto; **Synthetics**: Pusa Synthetic
Carrot	Hybrid-1.
Muskmelon	Hybrid M-3, MHY-5
Watermelon	Arka Jyoti

Bitter gourd	Pusa Hybrid-2
Cucumber	PCUCH-1
Bottle gourd	PBOG-2, PBOG-1, NDBH-4

Some popular vegetable hybrids in India

Crops	**Name of Popular hybrids**
Tomato	Rupali, Naveen, Rashmi, Tolstoi, Minakshi, NS-815, NS-2530, TH-2312, TH-802
Brinjal	Pusa Hybrid-5, Pusa Hybrid-6, Manju, Kalpataru
Chilli	CH-1, CH-3, NS-1101, ARCH-236, Tejashwani
Sweet pepper	Bharat, Kt-1, Indira, Lario
Cauliflower	NS-60, Punam, Priya, Summer King, Agati Himlata, Sweta, Pawas, NS-66
Cabbage	Sumit, Nav Kranti, Manisha, Sri Ganesh Gol, Bajrang, NS 25
Cucumber	Pant Sanker, Khira-1, Malini
Bitter gourd	Pusa hybrid-1
Bottle gourd	Pusa Hybrid-3, Varad (MGH-4)
Muskmelon	Punjab Hybrid-1
Watermelon	NS-295

Public sector hybrids in solanaceous vegetable crops

Name of hybrid/crop	**Source**
Tomato	
Pusa Hybrid-1, Pusa Hybrid-2, DTH-4, DTH-8, Pusa Hybrid-4, KT-4, Arka Vishal, Arka Vardan, FMH-1, Arka Shrestha, Arka Abhijit Pant Hybrid-1, Pant Hybrid-2, Pant Hybrid-10, Pant Hybrid-11, NDTH-1, NDTH-2, NDTH-6, NDTH-4, Rajashree, Phule Hybrid-1, Hybrid-37, TH-2312	IARI IIHR Pantnagar Faizabad Rahuri PAU, Ludhiana
Brinjal	
Arka Navneet, Pusa Anmol, PusaHybrid-6, Pusa Hybrid-5, Pusa Hybrid-9, Vijay Hybrid, Azad Hybrid NDBH—1, NDBH-6, NDTH-11, NDBH-7 Hybrid-2 Punjab Hybrid, BH-1 ABH-1, ABH-2 Pant Hybrid-2	IIHR IARI Kanpur Faizabad Rahuri Ludhiana Anand Pantnagar
Chilli	
CH-1	Ludhiana
CH-3	Ludhiana

Sweet pepper	
Solan Hybrid-1	Solan
KT-1 (Pusa Deepti), KT-2	Katrain
Sel-2	Srinagar
Okra	
DOH-3, DOH-4	IARI
JOH-5	Parvani
DVR-1, DVR-2	Varanasi

Public sector hybrids/synthetics of cole and bulb crops

Name of hybrid	Source
Cauliflower	
Pusa Hybrid-2 (F_1 hybrid)	IARI
Pusa Early Synthetic, Pusa Synthetic (synthetic)	IARI
Pant Gobi-3 (synthetic)	Pantnagar
Onion	
Arka Kirtiman, Arka Lalima	IIHR
Hybrid 63, Hybrid 35	IARI
Cabbage	
Pusa Synthetic (synthetic)	IARI
H-64 (Hybrid), BRH-5	Katrain

Classification of male sterility in vegetable crops

S. No.	Type of male sterility		Inheritance	Crops in which Identified
1.	**Pollen sterility**	(a)	Cytoplasmic	Carrot, sweet pepper, turnip, Chinese cabbage, cucumber (in cauliflower, cabbage and broccoli this male sterility has been introduced from sterile cytoplasm of radish)
		(b)	Genic Single recessive gene	Tomato, brinjal, pea, muskmelon, watermelon, chilli, lima bean, pumpkin, cucumber, cabbage, cauliflower, Brussels sprouts, broccoli, Chinese cabbage
			Duplicate dominant genes	Cauliflower
		(c)	Cytoplasmic-genic	
			Single recessive gene	Onion, radish, sweet pepper
			Two recessive gene	Beet
			Single dominant gene	Carrot

2.	**Staminal Sterility**			
	Male sterility arises because stamens are either altogether absent or present in rudimentary form. So, pollens are not formed. The flowers are having normal corolla and style.		Genic (Single recessive gene)	Tomato
3.	**Functional male sterility**			
(a)	Positional sterility			
	Male sterility arises because another fails to dehisce to liberate the normal and fertile pollens so that the pollens cannot reach the stigma. This male sterility is maintained easily by manual self-pollination. The male sterile plants can easily be identified by examining the anthers.		Gene (single recessive)	Tomato, brinjal, sweet pepper
(b)	**Exserted stigma sterility**			
	Male sterility is caused due to long style where stigma protrudes through the anther cone and becomes out of the reach of the pollens of the same flower. However, long styly is very much influenced by environment		Gene (single recessive)	Tomato

Use of CMS for hybrid seed production

Crop	F1 Hybrid	Genetic Mechanism
Cabbage	KGMR-1(Pusa cabbage Hybrid 1), KTCBH 51, KTCBH 81	SI
Cauliflower	Pusa Hybrid-2 (Nov maturing, Group-II), Pusa Kartik Sankar (group-I)	SI
Cabbage	KCH-5, Hybrid 991-5, Hybrid 854-6	CMS
Cauliflower	Hybrid 8401 ×31022	CMS

Vegetable varieties developed through various breeding methods

SN	Vegetable	Introduction	Selection			(Hybridization) Pedigree	Hybrids/ Synthetics
1	Tomato	Variety	Country	Pure line/ SPS/ Clonal selection	Mass selection		Pusa Hybrid – 1
		Sioux	USA	Meeruti		Pusa Ruby (Sioux x Improved Meeuti)	Pusa Hybrid – 2
		Lebonita	USA	Pusa 120		Pusa Early Dwarf (Improved Meeruti x Red Cloud)	Arka Vardan
		Marglobe	USA			Punjab Chhuhara (EC 55005 x Punjab Tropic)	Arka Vishal
		Roma	USA			Arka Meghali (Arka Vikash x IHR 584)	Arka Shreshta
		Dwarf Money Maker	Israel			Hisar Lalima (Pusa Early Dwarf x HS 101)	Arka Abhijit
		Balkan	Bulgaria				DTH – 4
		Fireball	Canada				KT – 4
2	Pea	Arkel	England			Jawahar Matar 3 (T 9 x Early Badger)	
		Meteor	England			P 88 (Pusa 2 x Morasis 55)	
		Bonneville	USA			VL Ageti Matar 7 (Pant Uphar x Arkel)	
		Sylvia	Sweden			VL 3 (Old Sugar x Early Wrinkled Dwarf)	
		Little Marvel	England			Pant Sabji Matar 3 (Arkel x GC 141)	
3	French bean	Contender	USA	Pusa Himlata		Pant Bean 2 (Turkish Brown x Contender)	
		Bountiful	USA	Arka Komal			

4	Dolichos bean			CO 1		CO – 2 (CO 8 x CO 1)	
5	Cluster bean			Sharad Bahar		Pusa Navbahar (Pusa Mausami x Pusa Sadabhar)	
				IC 11388			
6	Chilli			NP 46 A		Pusa Jwala (NP 46 A x Puri Red)	Punjab Hybrid
				Patna Red		Pusa Sadabahar (Pusa Jwala x IC 31339)	KT – 1
7	Brinjal			Pant Samrat	Pusa Purple Long	Pusa Kranti [(PPL x Hyderpur) x Wynad Giant]	Pusa Hybrid – 5
					Pusa Purple Round	Arka Navneet (IIHR 21 x Supreme)	Pusa Hybrid – 6
					Arka Sheel	Pant Rituraj (T 3 x PPL)	Pusa Hybrid – 9
8	Okra					Pusa Sawani (IC 1542 x Pusa Makhmali)	
						Parbhani Kranti (Pusa Sawani x *A. manihot*) x Pusa Sawani	
9	Onion	Early Grano	USA	Pusa Red	Arka Pragati		Arka Kirtiman
		Red Grano	USA	Hisar - 2	Arka Bindu		Arka Lalima
		Bermuda Yellow	Philippines	Punjab Selection	Arka Niketan		
10	Cauliflower	Snowball - 16	Holland	Dania		Pusa Snowball - 1 (EC 12013 x EC 12012)	Pusa Synthetic
		Improved Japanese	Israel	Hisar – 1		Pusa Subhra (MGS 23 x 15-1) x (D 96)	Pusa Hybrid – 2
11	Cabbage	September	Germany	Pusa Drum Head		Pusa Mukta (EC. 24855 x EC. 10109)	Pusa Synthetic
12	Watermelon	New Hampshire Midget	USA			Arka Manik (IIHR 21 x Crimson Sweet)	
		Asahi Yamato	Japan				
		Sugar baby	USA				
		Shipper	USA				
13	Cucumber	Poinsette	USA				Pusa Sanyog
14	Summer Squash	Patty Pen	USA				Pusa Alankar

15	Carrot	Jeno	Germany			Pusa Kesar (Local Red x Nantes)	Hybrid – 1
						Pusa Meghali (Pusa Kesar x Nantes)	
						Pusa Yamdagni (EC 9981 x Nantes)	
16	Muskmelon			Arka Jeet		Pusa Sharbati (Kutana x PMR 6)	Punjab Hybrid
				Arka Rajhans		Punjab Sunehri (Hara Madhu x Edisto)	Pusa Rasraj
17	Bottle Gourd			Pusa Summer Prolific Long			
				Pusa Summer Prolific Round			
18	Turnip					Pusa Chandrima (Japanese White x Snowball) (Asiatic x European)	
						Pusa Swarnima (Japanese Whitev x Golden Ball) (Asiatic x European)	
						Pusa Kanchan (local Red Round x Golden Ball)	
19	Radish					Punjab Safed (White 5 x Japanese White)	
						Pusa Himani (Radish Black x Japanese White)	
20	Sweet Potato			Pusa Safed			
				Pusa Lal			
				Pusa Sinduri			
21	Potato			Kufri Red round			
				Kufri Safed			

Mode of inheritance of resistance:

Oligogenic resistant: Single or few major genes govern this disease resistance of the host plant, and both dominant and recessive genes may condition the resistance reaction. Such oligogenic resistance is specific to some pathotype of the pathogen.

Polygenic resistance: This disease resistance is governed by many genes (polygenes) with small effect expressed in cumulative manner thus showing continuous variation for disease reaction. In this case several genes contribute to a quantitative trait loci (QTL) that show additive effect against different races of the pathogen.

Oligogenic Resistance

Crops	Genetic control for resistance to disease
Tomato	Single dominant gene for Fusarium wilt (I), Verticillium wilt (Ve), early blight (Ad), late blight (Ph-1), leaf mould (Cf), leaf spot (Sm), mosaic virus (TM), leaf curl virus (Tlc); Two independent genes for Fusarium wilt.
Potato	Single dominant gene for late blight (R), potato virus X, potato virus Y and few dominant major genes for bacterial wilt.
Brinjal	Single dominant gene for bacterial wilt (Rps); A pair of dominant genes for phomopsis blight
Chilli	Single dominant gene for Phytophthora fruit rot, bacterial leaf spot, tobacco mosaic, single recessive gene for leaf curl virus (lc), two independent dominant genes for phytophthora root rot.
Okra	Two complementary dominant genes for yellow vein mosaic virus in A. manihot ssp. Manihot.
Pea	Single dominant gene for Fusarium wilt (Fnw) and ascohyta blight and single recessive gene for powdery mildew (er-1), pea-seed-born mosaic virus (sbm)
Cowpea	Single dominant gene for cercospora leaf spot (Cls-1) and bean yellow mosaic, single recessive gene for powdery mildew, bacterial blight and cercospora leaf spot.
Cucumber	Single dominant (P) for downey mildew; two pairs of duplicate genes and three complementary recessive genes for powdery mildew.
Muskmelon	Two incompletely dominant genes for downey mildew, single dominant gene for powdery mildew and watermelon mosaic, three recessive genes for cucumber mosaic.
Watermelon	Single dominant gene for fusarium wilt and anthracnose.

Polygenic Resistance

Crop	Polygenic control for resistance to disease
Tomato	Bacterial wilt, bacterial spot, bacterial canker, leaf curl in *L. pimpinellifolium,* *L.peruvianum* (3-5 recessive genes); Early blight (two or more recessive genes).
Potato	Potato leaf roll virus
Brinjal	Phomopsis blight
Cauliflower	Black rot (dominant polygenes)

Wild and allied species of the vegetable crops carrying resistance attribute

Crops	Disease	Pathogen	Species
Potato	Late blight	*Phytophthora infestans*	*Solanum tuberosum ssp. andigena, S.demissum S.curtilobum, S.stoloniferum*
Tomato	Fusarium wilt	*Fusarium oxysporum f.sp.lycopersici*	*L. pimpinelifolium, L. hirsutum*
	Early blight	*Alternaria solani*	*L. pimpinelifolium,*
			L. hirsutum f.glabratum
	Bacterial wilt	*Pseudomonas solanacearum*	*L. pimpinellifolium*
	Leaf curl	Tomato leaf curl virus	*L. peruvianum, L. pimpinellifolium, L. hirsutum f. glabratum*
Brinjal	Bacterial wilt	*Pseudomonas solanacearum*	*S. sisymbrifolium, S. integrifolium*
	Phomopsis blight	*Phomopsis vexans*	*S.xanthocarpum, S.indicum, S. gilo, S.khasianum, S.nigrum, S.sisymbrifolium*
	Little leaf	Phytoplasma	*S.viarum, S.incanum*
Chilli	Tobacco mosaic	Tobacco mosaic virus	*C. chacoense*
	Phytophthora rot	*Phytophthora capsici*	*C. frutescens*
	Leaf curl	Leaf curl virus	*C. frutescens*
	Anthracnose	*Colletotrichum capsici*	*C. baccatum, C.chinense*
Okra	Yellow vein mosaic	Yellow vein mosaic virus	*A.manihot ssp. manihot A.manihot ssp. Tetraphylus*
Onion	Purple blotch	*Alternaria porri*	*Allium fistulosum*
Cabbage and cauliflower	Club root White rust	*Plasmodiophora brassicae* *Albugo candida*	*B. napus, B. campestris B. carinata*

SOURCES OF DROUGHT TOLERANCE

Vegetable Crops	Resistance Source
Tomato	*S. habrochaites, S. pennelli ,S. pimpinellifoloium, S. esculentum var. cerasiforme, S. hirsutum, S. cheesmanii, S. chilense, S. habrochaites, S. sitiens* *L. pennellii, L. chilense* Arka Vikas, RF- 4 A, Solan Vajr, Arka Meghali
Brinjal	*S. microcarpon, S. gilo, S. macrosperma, S. integrifolium, S. sodomaeum (syn. S. linneanum)* Bundelkhand Deshi, Violette Round, Supreme
Chilli	*C. chinense, C. baccatum var. pendulum, C. eximium* Arka Lohit, IIHR - Sel.-132
Potato	*S. acaule, S. demissum and S. stenotonum,* *S.ajanhuiri, S.curtilobum, S.xjuzepczukii* *Solanum chacoense,* Kufri Sindhuri ,Kufri Sheetman ,Alpha, Bintje
Okra	*A. caillei, A. rugosus, A. tuberosus*
Onion	*Allium fistulosum, A. munzii,* Arka Kalyan
French bean	*P. acutifolius*
Water melon	*Citrullus colocynthis (L.) Schrad*
Cucumber	AHC-13
Winter Squash	*Cucurbita maxima*
Cucumis Spp.	*Cucumis melo var. momordica* AHS-10, AHS-82 *Cucumis pubescens*, AHK-119 *Cucumis melo var. callosus,* AHK- 200 *Cucumis melo var. chat*, Arya
Cassava	CE-54, CE-534, CI-260, CI-308, CI-848, 129, 7, 16, TP White, Narukku-3, Ci-4, Ci-60, Ci-17, Ci-80
Sweet potato	VLS6, IGSP 10, IGSP 14, Sree Bhadra

SALT STRESS TOLERANT SPECIES AND GENOTYPES OF VEGETABLES

Crop	Resistance Source
Tomato	*S. cheesmanii, S. pimpinellifolium, S. peruvianum, S. hirsutum* and *S. pennellii, Sabour Suphala*
Okra	Pusa Sawani
Onion	Hissar 2, Punjab Selection
Bean	Gevas Sirik 57 (GS57)
Brinjal	Pusa Bindu, CO-1, *S. macrocarpon, S. gilo*

SOURCES OF HEAT TOLERANCE

Crop	Resistance Source
Tomato	*S. cheesmanii, S. cerasiformae* (Cherry tomato), Philippines, Punjab Tropics, HS-101, HS-102 and Pusa Hybrid 1
Okra	*A. manihot var manihot*
Potato	*S. chaconense*, Kufri Surya
Brinjal	*S. macrocarpon, S. gilo*

SOURCES OF COLD TOLERANCE

Crop	Resistance Source
Tomato	*S. habrochites, S. pimpinellifolium*, Cold Set, Pusa Sheetal
French bean	*P. filliformis, P. angustissimus*
Potato	*S. acaule*
Okra	*A. angulosus*

Disease Resistant Varieties Developed/Identified (ICAR Institutes and State Agricultural Universities)

Crops	Diseases	Name of the resistant Varieties or breeding lines	Name of the centre
Tomato	Bacterial wilt (*Pseudomonas solanacearum*)	LB-79 (Shakti)	KAU
		BWR-5 (Arka Alok) BWR-1 (Arka Abha)	IIHR
		BT- 1 (Utkal Palilavi) BT-2 (Utkal Deepti)	OUAT
	Late blight (*Phytophora infestans*)	TRB-l, TRB-2	PAU
	Leaf curl virus	H-24 (Hisar Anmol), H-36,H-86, H-88	CCSHAU
	Verticillium wilt and Fusarium wilt	Pant Bahar	GBPUAT
	Root Knot Nematodes	Sel-120, PNR-7	

Brinjal	Bacterial wilt (*Pseudomonas solanacearum*)	BWR-12 (Arka Keshav) Arka Nidhi Arka Nellkanth	IIHR
		BB-7 (Utkal Tarini)	OUAT
		Pant Rituraj,	
		Pusa Purple Cluster	
	Phomopsis rot (*Phomopsis vexans*)	SM-6-6 (Soorya)	KAU
		JC- I, JC-2	AAU
		Pusa Bhairav	IARI
		Pant Samrat	
Okra	Yellow vein mosaic virus	Pusa Sawani ,91-4	IARI
		Sel-2	NBPGR
		Punjab Padmini P-7	PAU
		PB-57, Parbhani Kranti	MAU
		Sel-4 (Arka Abhay) Sel- 10 (Arka Anamika)	IIHR
		HRB-9-2 (Varsha Uphar) HRB55 (Hisar Barsati)	CCSHAU
		BO-2, Utkal Gaurav	OUAT
		KS-404	VRS
Chilli	Leaf curl virus	Pusa Jwala Pusa Sadabahar	IARI
		Pant C-4	GBPUAT
	CMV, TMV & Leaf curl	Punjab Lal	PAU
	Mosaic, Wilt & Dieback	Punjab Surkh	
	Virus Complex (Bacterial wilt) (*Pseudomonas solanacearum*)	LCA 235,LCA 305 ,BC-21-2(Utkal Rashmi)	OUAT
		AAUM-1 ,AAUM-2	AAU
Capsicum	Bacterial wilt (*Pseudomonas solanacearum*)	Arka Gaurav	IIHR
Muskmelon	Downey mildew (*Pseudoperonospora cubensis*)	MR-12 (Punjab Rasila)	PAU
	Cucumber green mottle mosaic virus	DVRM-I DVRM-2	IARI
	Powdery mildew (*Erysiphae cichoracearuin*)	Arka Rajhans	IIHR

Watermelon	Powdery mildew (*Erysiphae cichoracearuin*) and Anthracnose (*Colletotrichum lagenarium*)	Arka Manik	IIHR
Garden Pea	Powdery mildew (*Erysiphe polygoni*)	PRS-4, Rachna, KS-225, KS-245	VRS
		JP-4, JP-83, JP-7l and JP-885	JLNKVV
		FC-1 (Arka Ajit)	IIHR
		IP-3, Pant P-5, PMR-21	GBPUAT
		DPP-62	HPKVV
		VP-9003, VP-8902	VPKAS
		DMR-7 (Alankar)	IARI
		HFP-4 & HFP-12 (Jayanti & Ultra)	CCSHAU
		HUP-l	BHU
Cauliflower	Black rot (*Xanthomonas compestris*) & Curd and Inflorescence blight (brassicicola)	Pusa Shubhra Pusa Snowball K-1	IARI
Cabbage	Black rot (*Xanthomonas compestris*)	Pusa Mukta	IARI RS
Cowpea	Bacterial blight (*Xanthomonas compestris cv. vignicola*)	Pusa Komal	IARI RS
French bean	Common mosaic virus and rust (Uromyces)	UPF 191 (Pant Anupama)	GBPUAT
	Angular leaf spot	SVM-1	

KAU = Kerala Agricultural University, Vellanikkara.

IIHR = Indian Institute of Horticultural Research, Hessarghatta.

GBPUAT = GB Pant University of Agriculture and Technology, Panthagar.

IARIRS = IARI, Regional Station, Katrain

IARI = Indian Agricultural Research Institute, New Delhi.

BHU = Banaras Hindu University, Varanasi.

CCSHAU = CCS, Haryana Agricultural University, Hissar.

HPKVV = Himachal Pradesh Krishi Vishwa Vidyalaya, Palampur.

VPKAS = Vivekananda Parvatiya Krishi Anusaudhan Shala, Almora.

VRS = Vegetable Research Station, Kalyanpur,

JLNKVV = Jawahar Lal Nehru Krishi Vishwa Vidyalaya, Jabalpur.

PAU = Punjab Agricultural University, Ludhiana.

AAU = Assam Agricultural University, Jorhat.

OUAT = Orissa University of Agriculture & Technology, Bhubaneshwar.

MAU = Marathwada Agricultural University, Parbhani

NBPGR = NBPGR, New Delhi.

IIVR = Indian Institute of Vegetable Research Varanasi.

Pest Resistant Varieties Hybrids/Lines of Major Vegetable Crops

Crop	Pest	Varieties/hybrids
Brinjal	Shoot and fruit borer (*Leucenodes orbonalis*)	SM 17-4, PBr 129-5, Doli 5, ARU 2-C, Punjab Jamuni Gola, Punjab Sadbahar, Punjab Barsati Pusa Purple Round, Kalyanpur-2 Punjab Neelam, Punjab Chamkila
	Jassids (*Amrasa undalis*)	Black Beauty, Ram Nagar Giant Vijay Hybrid.
	Nematode	Pusa Purple Cluster, Pant Rituraj Arka Nidhi, Arka Kesav, Arka Neelkanth.
Chilli	Thrips (*Thrips tabaci*)	Bhaskar
Cabbage	Aphid (*Brevícorync brassicae*)	All Season, Red Drum Read, Sure Head, Express Mail.
Cauliflower	Stem borer (*Hellula undcjljs*)	EMS-3, KW-5, KW-8 Early Patna, Kathmandu Local.
Okra	Jassid (*Amrasa biguttula*)	IC-7194, IC-139999, New Selection, Punjab Padmini.
Onion	Thrips (*Thrips tabaci*)	PBR-2, PBR-6, PBR-4, PBR-5, PBR-6. Arka Niketan, Pusa Ratnar
Round gourd	Fruitfly (*Dacus cucurbitae*)	Arka Tinda.
Pumpkin	Fruitfly (*Dacus cucurbitac*)	Arka Suryamukhi.
Bitter gourd	Fruitfly (*Dacus cucurbitac*)	Hissar-II.

Biotic Stresses and their Source of Resistance in Brinjal

Biotic Stress	Resistance Sources
Bacterial wilt	West coast Green Round 112-8 (WCGR 128-8) S. melongena var. incanum, S. integrifolium, S. torvum
Little leaf	Solanum viarum, S. incanum, S. sisymbrifolium, Pusa Purple Cluster, Katrai
Phomopsis blight	S. xanthocarpum, S. sisymbrifolium, S. indicum, S. khasianum, S. nigram, S. gilo, Florida Market, Florida Beauty, BPL-1, Ornamental brinjal, Pusa Uttam, IC-316237 RKN Solanum sisymbrifolium, Co-1, Solanum torvum

Shoot and fruit borer	Solanum sisymbrifolium, S. integrifolium, S. xanthoranpum, S. nigrum, S. khasianum, Pusa Purple Long, H-128, H-129, Azcabey, Thorn Pendy, Black Pendy, Banaras Long Purple.

Biotic stresses and their source of resistance in Tomato

Biotic Stress	Resistance Sources
Tomato Leaf Curl Virus (ToLCV)	*L. hirsutum f. glabratum, L. peruvianum, L pimpinellifolium* and *L. hirsutum,* HS 101, *L. hirsutum (*LA386, LA 1777, PI 390513), *L. glandulosum (EC 68003) and L. peruvianum* (PI 127830 and PI 127831), H-88-78-1, H-88-78-2, H-88-87
Fusarium Wilt	*L. hirsutum f. glabratum (*Wir 4172), *L. hirsutum* (PI 13448) and L. peruvianum (EC 148898), Columbia, Roma, HS 110, Fla. 7547, Fla. 7481
Early Blight	*L. hirsutum* (PI 134437), P-1, EC 529061, WIR-3928, H-88-28-1, H- 86-11 and H86-7
Bacterial wilt	*Lycopersicon pimpinellifolium,* Acc 99, Sweet 72, Acc 151, Hyb 54, IIHR 663-12-3, BWR 1, BWR 5, LE 79 BT 1, BT-10, H 24, BRH-2, LE-415, H-86, Capitan, Caravel, Ga. 1565, Ga 219, CRA 66
Fruit Borer	*L. hirsutum f. glabratum*
White fly	*L. hirsutum and L. hirsutum f. glabratum*

Resistance Breeding

Crop	Disease	Resistance source
Musk melon	Powdery mildew	PMR 45, PMR 450, PMR 5, PMR 6, PI 124111
	Downey mildew	MR-1, PI 414723, DMDR-1, DMDR-2 CGMMV DVRM-1, 2, *C. africanus, C. ficifolius, C. anguria*
	Fruit fly	*C. callosus*
	Nematodes	*C. metuliferus*
Watermelon	Fusarium wilt	Summit, Conqueror, Charleston Gray, Dixilee, Crimson Sweet
	Anthracnose	Fair, Charleston gray, Congo, PI 189225
Cucumber	Anthracnose	PI 175111, PI 175120, PI 179676, PI 182445, wise 2757 (USA)
	Downey mildew	B-184, B159, wise 2757 (USA)
	Powdery mildew	PI 200815, PI 200818, C.hardwikkii, Wise 2757 (USA)
	CMV	Wise ,SMR-12, SMR-15, SMR-18, wise 2757 (USA)
	PM and Viruses	*C. lundelliana, C. martenezii* ,ZYMV, WMV, *C. ecuadorensis, C. faetidistima, C. martenezii*

Symptoms of Nutrient Deficiency

Nutrient	Plant symptoms	Occurrence
Nitrogen	The deficiency symptoms are mostly generalized over the whole plant; the deficiency results in decrease in carbohydrate synthesis and loss in chlorophyll; slow and stunted growth; stem thin, erect and hard; leaves smaller than the normal; premature yellowing of lower leaves and new leaves are light green in colour; in acute deficiency flowering is greatly reduced. Some specific nitrogen deficiency symptoms are yellow lower leaves with purple veins and light green young leaves, (tomato); yellow lower leaves, light green young leaves and slender upright stem (potato); older leaves orange to purple pigmented and young leaves pale green (cole crops); leaves pale green and turning reddish to purple (beet); leaves light green followed by yellowing (carrot and radish); stem hard and leaves yellowish green to yellow (cucurbits); stiff, upright, short and light green leaves (onion); small, sickly and pale green leaves (okra); yellowish green leaves with reddish tinged areas (sweet potato); yellow older leaves and pale green young leaves (celery and lettuce).	Occur in light soil where leaching is more due to heavy rainfall.
Phosphorus	The deficiency symptoms are mostly generalized over the whole plant; the deficiency causes delay in starch development, accumulation of sugars and consequent development of anthocyanin pigment; early growth gets restricted with poor root development; stem thin and shortened; mature leaves have characteristic dark to blue-green colouration and often purple colouration is developed; the purple colouration first appears on underside of leaves and later throughout; lower leaves sometimes turn yellow or greenish brown in colour; delayed flowering and maturity. Some specific phosphorus deficiency symptoms are olive-green leaves with purple colouration in the undersurface (tomato); forward rolling of leaves with marginal scorch (potato); dull green leaves with purple colouration in the undersurface (cole crops); small dull purple leaves (beet); dull green leaves with purple colouration in lower surface (carrot, radish); dark green leaves with dropping of older leaves (pea, bean); tip die back with wilting of older leaves (onion); dull green leaves with purple tinge in the lower surface (lettuce).	Occur in acid soil (pH below 6.0) due to fixation of soluble phosphates. It also occurs in cold and moist soils.

Nutrient	Plant symptoms	Occurrence
Potassium	Symptoms of potassium deficiency progressively appear on younger leaves; slow and stunted growth with weak stem; in older leaves interveinal chlorosis near the margins occur followed by the development of grey or tan areas near the leaf margin and eventually a scorch appears around the entire leaf margin; leaf edges may become twisted or curled; chlorotic areas may develop throughout the leaf; necrotic areas may develop at the tip of the plant; potassium deficiency interferes the absorption of water and nutrient from soil, so the plant looks withered. Some specific potassium deficiency symptoms are bronze or grayish green lower leaves with tip necrosis starting from tip (potato); yellowing of older leaves with browning at margin (cabbage, cauliflower); leaves bluish green with surface crinkling (beet); leaves bluish green with pale yellow to brown margin (radish); dull green leaves followed by browning (carrot); leaves bluish green near veins followed by browning and necrosis at margins (cucurbits); chlorotic leaflet with necrotic brown areas at margins (beans); mottled leaflet followed by necrosis (cowpea); young leaves at margins (pea); young leaves brown-yellow followed by scorching (okra); yellowing and drying of older leaves from tip (onion); chlorosis followed by necrosis at margin in older leaves (sweet potato); dark green leaves followed by curling and browning (celery).	Occur on light soil where leaching is high due to heavy rainfall.
Calcium	The young leaves are affected first which often become small and get distorted and twisted; tip of the innermost unfolding leaves get twisted, gelatinized and finally die back at tips and margin; hooking appearance of the dry leaves of terminal bud is the characteristic symptom; leaf blades may be whitish green and sometimes brown to light brown spots appear on the leaves; stem becomes weak and brittle; root tip dies and root growth is restricted; specific deficiency disorder includes blossom end rot of tomato, sweet pepper and watermelon; cavity in root phloem immediately below the epidermis in carrot and parsnip; black heart of celery; brown heart of endive; tip burn of lettuce and internal breakdown of cabbage. Other symptoms manifested in the foliage are: young leaves turn yellow, brown or purple, roll upwards at margins and become necrotic (tomato); small, pale green leaves which roll upwards (potato); leaves rolled up at margins and become discoloured (cabbage, cauliflower); leaves turn pale green around margins which curl upward (beet); leaves chlorotic and scorched (carrot); narrow white band at leaf margin (radish); chlorotic older leaves and curled young leaves (pea); young leaves light green and older leaves show some reddish areas (sweet potato).	Occur on acidic and highly leached sandy soil. It also occurs on very dry soil and soil with very high potassium and nitrogen level.

Nutrient	Plant symptoms	Occurrence
Magnesium	The symptom begins with a loss of green colour at the tips of the lower leaves between the veins; as the deficiency develop, larger portions of the lower leaves are affected, and the leaves further up in the plant become bleached out and turn whitish in colour; leaf becomes lifeless in texture, small in size and tints of orange, red and purple colour may develop unevenly; leaves may curve upwards at margin; older leaves may fall; stem become thin and brittle; maturity gets delayed. Some specific magnesium deficiency symptoms are interveinal chlorosis and necrosis in older leaves (tomato, brinjal, chilli); interveinal chlorosis and puckering of older leaves (cabbage); chlorotic leaves with reddish tint between veins (beet); pale green leaves with light yellow or brown spots at tips (carrot); interveinal chlorosis in older leaves (radish); interveinal yellow spots followed by chlorosis in leaves (okra); leaves die back at tips (onion); browning of leaf tips (pea).	Occur on acidic soils, on highly leached light soils and on soils having high potassium levels.
Sulphur	Characteristic uniform yellowish green appearance of the younger leaves that resembles nitrogen starvation; in severe cases, the older leaves may turn pale green; the stem and petiole become brittle and collapse; shoot growth gets restricted; stem becomes stiff, woody and smaller in diameter; the terminal bud commonly remains alive and necrotic spots may or may not be present over the leaf.	Occur in the soil having low organic matter; also occur due to excessive leaching on light soils.
Iron	Typical symptoms are distinctive green vein colour and yellow interveinal areas on the young leaves; points and margins of the leaves keep their green colour for long time; in severe cases, the entire leaf, veins and interveinal area turn yellow which sometimes become bleached.	Occur on soil with pH above 6.8; on soil which contains considerable sodium and calcium; commonly occur on calcareous soil; due to excess phosphorus application.

Nutrient	Plant symptoms	Occurrence
Copper	Yellowing of young leaves; leaves may become elongated and tips may curl; leaf edges may become ragged; plant top may wither; in onion, bulbs become soft with thin pale yellow scales.	Occur on soils with 5 per cent or higher organic matter content and on peat soil; it may also occur on heavily cropped sandy soil with acidic reaction and on alkaline soil.
Zinc	Deficiency symptoms mostly appear on the 2nd or 3rd fully mature leaves from the top of the plants; leaves become mottled or chlorotic and thick: rosette growth due to shortened internodes. Characteristic symptoms include small reddish brown spots on cotyledonary leaves of bean; green and yellow broad striping at the base of the leaves of corn; interveinal yellowing with marginal burning in beet; 'fernleaf' disorder in potato; tomato leaves become smaller, chlorotic and inwardly curved; onion leaves turn yellow; whole plant turns yellow in garlic.	Occur in neutral to alkaline sandy soils; also occur due to heavy phosphorus fertilization; occur in calcareous soils after leaching and erosion and acid-leached soils.
Manganese	Chlorosis between the veins and appearance of chlorotic and necrotic spots in the interveinal areas; grayish areas may appear near the base of the younger leaves; leave become smaller; typical symptoms include densely red foliage in beet; narrow, yellow striping in the leaves of onion and corn; chlorosis followed by necrosis and yellowing in the leaves of bean: small, yellowish leaves with interveinal yellow mottling in cabbage; interveinal chlorosis in pea and chlorotic and forwardly rolled leaves in tomato.	Occur on soil with pH above 6.7; it may also occur in sandy, peat and muck soil; often occur in calcareous soils.

Nutrient	Plant symptoms	Occurrence
Boron	Breakdown of the growing tip resulting in death of shoot tip; rosette appearance of the plant due to shortening of terminal growth; young leaves of terminal buds are light green, thick, curled and brittle; root growth gets restricted; flowers become barren. Some typical symptoms include browning of cauliflower; hollow stem in cole crops, particularly in cabbage; black heart of beet and internal browning of turnip due to breakdown of internal tissues; cracked stem of celery; stem resetting and open fruit locule in tomato.	Occur on soil with pH above 6.8; it may occur due to high soil phosphate levels; occur in crops having high boron requirement; occur on acidic soil where native boron has been leached; also occur on coarse textured soil, peats and mucks; water stressed conditions and over limed acid soil.
Molybdenum	Impairment of growth; deformation of shoots, older leaves show interveinal yellowing with necrotic margins; leaves may crinkle as found in cole crops; in legume crops, the symptoms are identical with those of nitrogen deficiency. Typical symptoms include whiptail leaf and small, open, loose curd in cauliflower; bleached, scorched and mottled older leaves in cabbage; pale, upwardly rolled leaves with diffused marginal and interveinal in tomato; decomposition and granulation in veins of chilli; leaf chlorosis with prominent red vein in beet; marginal yellowing followed by inward red colouration in carrot; leaf cupping; chlorosis, mottling and whiptail in turnip; discoloured leaf and death of growing tip in radish; yellow or white veins in leaves of pea; pale green leaves with interveinal mottling and scorching in bean; deep blue leaves with conspicuous yellow and green mottling in onion.	Occur on very acid and sandy soil having pH below 5.2 and on highly leached sandy soil.

ASRB-NET-2021

- Vitamins are lost during - Heating
- CFTRI is located at - Mysore
- Headquarter of National Biodiversity – Chennai
- Mulching Sheet thickness for vegetable cultivation – 15-30 microns
- Vegetable belonging to monocot family – Asparagus
- Family of Parsnip/Parsley – Apiaceae
- Bush type French bean are – Short day
- Toxin content in beans – Trypsin inhibitor
- Phytoalexins (Antifungal compounds) present in cowpea – Kievitone
- Colour of "Palam vichitra" (a variety of Broccoli) - Purple
- Isolation distance of foundation seed production in okra – 400m
- Arka Nikita is F_1 okra hybrid developed through – GMS
- Both GMS and CGMS exploited for hybrid seed production in – Chilli F_1 hybrid
- Chilli male sterility gene – msms
- Brown seed marker in seed production of – Onion
- *Amaranthus hypochondriacus* is an ornamental plant which is– Day neutral
- *Amaranthus dubius* is – Tetraploid
- Major flavonoid in onion – Quercetin
- Shallow root is used as crop – Onion
- TSS of white onion variety for processing - >18%
- Source of male sterility in onion – *Allium galanthum*
- Palam Paushtik is a variety of – Leek
- Pusa Asita is a variety of – Black Carrot
- Fufu is prepared by – Yam or Cassava
- Sauerkrat is prepared from – White Cabbage
- Edible part of Rhubarb– Fleshy petioles (stalks)
- Edible part of Jerusalem artichoke– Tuber
- Indian Spinach – Basella
- Perianth flowers are seen in - Onion

- Sago is prepared from – Tapioca
- Hardiest crop – Kale (*B. oleracea acephala*)
- Thermo sensitive crop – Cauliflower (in option)
- Temperature need for Snowball type cauliflower – (10-16^{0}C)
- Whip Tail in cauliflower have – Undeveloped leaf blades
- Napiform root is present in which crop – Turnip
- 'B' deficiency in cole crop leads to – Browning or Brown rot
- Hypobaric (Low Pressure Storage) storage means – Reduce oxygen supply and slow down the respiration
- No of ovule in Chow-Chow – one
- Butternut squash – Pumpkin
- Pinnately lobed leaf found in cucurbit – Watermelon
- The genome of the cucumber *cucumis sativus* reported by - Huaug *et al.*
- Gynoecious line in bitter gourd is maintained by – $AgNo_3$
- Cucumber variety grown in other than protected condition (open field) – Non-parthenocarpic type
- Multiple disease resistant cucumber variety – Poinsett
- Ice box is in section of – watermelon
- Half slip stage is maturity stage indices in – Muskmelon
- Oriental pickling melon variety – (Mudicode, Arunima and Saubhagya)
- Fluted pumpkin – *Telfaria occidentalis*
- Bitter gourd seed multiplication ratio – 1:41
- 1gm cauliflower seed (seed weight) – 250-350 seeds
- Fruit weight of spine gourd - (5-20g)
- Wet method of seed extraction in – Sweet pepper
- The female and male ratio in Ivy gourd field should be – 10:1
- Arka Vikram is variety of – Ridge gourd
- Kashi Kushi is variety of – Ridge gourd
- The anthesis time of Spine gourd is – Early Morning
- Multi vitamin green is nickname of – Chekurmanis
- The Germination type of Garden Pea is – Hypogeal
- Pea maturity is measure by – Tenderometer
- Firmness of any material is measure by – Penetrometer
- Which instrument is used for knowing for ploidy level - Cytometer
- Pigweed is a serious weed in seed production of - Amaranthus
- *Canavalia gladiata* is scientific name of – Sword bean
- Asparagus propagation is done by – seed & offshoots
- Ben oil is extracted by– Moringa (Drumstick)

- Low ethylene producing vegetable – Moringa
- Recently in India which tomato chromosome sequence has completely studied – chromosome 5
- Compounds that confer insect resistance to tomato - Sesquiterpenes
- Red colour (Lycopene) development stops at - >30^0C
- Truss hooking in polyhouse is done in -Tomato
- For long distant market Tomato is harvest at– Mature green stage
- Tomato gene for long shelf life - Ripening inhibitor (rin) gene
- Seed rate of hybrid tomato– 125-175g/ha
- Which species is source of Ty2 resistant gene for Tomato yellow leaf curl virus (TY-LCV) – *S. habrochaites*
- B-carotene rich (ten times than normal) Caro-Red is variety of which crop –Tomato
- Chromosome number in Trisomic Tomato is – 26
- Days of fruit maturity from pollination to hybrid seed production in Brinjal – 70 days
- The Colour of Brinjal variety KKM - 1 is – White
- Which part of chilli has content maximum amount of capsaicin - Placenta
- Jalapeno pepper belongs to species of – *Capsicum annum*
- Number of domesticated species in capsicum genus is - 5
- Single seed decent method for selection is used in - Pea
- Photo insensitive variety of tomato is - Pusa Sadabahar
- Characteristic of processing potato - High dry matter and low sugar content
- Greening of potato due to – Exposure of sun light
- Cultivated tetraploid crop – Potato
- Protato (protein rich potato) gene – AmA1
- Sprout inhibiter in potato - CIPC (isopropyl N-(3-chlorophenyl) carbamate)
- Heat tolerant variety of potato – Kufri Surya
- Potato variety for French fry - Kufri Frysona
- Legumes are rich nutrient of – protein
- Isogenic line develop by- Anther culture
- Composite and synthetic varieties are – Heterogeneous
- Riverbed cultivation is done in – muskmelon
- Marsh spot of pea due to – Manganese (Mg) deficiency
- Number of sepals or petals in cole crops – 4
- Tetraploid (2n = 32) - leek
- Tropical carrot hybrid – Pusa Vasuda
- Heteromorphic sex chromosomes present in – Ivy gourd (*Coccinia Indica*)
- Maturity index of Elephant Foot Yam (EFY) – yellowing and drooping down of leaves
- Powdery mildew resistant onion species – *Allium roylei*

- Arka Nikitha (F_1 hybrid) okra variety developed by- GMS
- Pest of potato in storage – Potato Tuber moth
- Bulblets planting in onion done in - Summer season
- Abiotic stress – Salinity
- Non PCR based marker – RFLP
- Which growth hormone are retardant or inhibitor – ABA
- Apomixis is - asexual reproduction without fertilization
- Hormone responsible for bolting is – Gibberellins
- Main objective of pre- cooling is - Remove of field heat
- Antitranspirant chemical widely used is- PMA (Phenyl Mercuric Acetate)
- Selection of extreme by not selecting intermediates – Disruptive selection
- The edible portion of the globe artichoke is – flower bud
- Gajendra is – Elephant foot yam (EFY) Variety
- In India post harvest losses in onion is – about 35-40%
- Nutrient direct role in nodulation in legumes – Mo
- Flower stimulant hormone is – Florigen
- Vine turning operation is done in - Sweet potato
- Pinching for hybrid seed production in – Pumpkin
- Selfing and massing technique done in - Onion
- Bud necrosis in watermelon spread by the vector – Thrips
- Rainfed tomato variety – Arka Meghali
- Acid soil encourage – Club root of cabbage
- Base collection seeds ware stored - under long term (50 to 100 years), at about -20^{0}C with 5% moisture content
- Kashi Shishir is variety of – Pumpkin
- In India seed certification - Compulsory
- Detection of viral disease is done through – ELISA Test
- Maximum Tuberization in potato occurs at – 20^{0}C
- Agency which is Involved in export of fruit and vegetables in India – APEDA (1985)